Stress Response and Immunity: Links and Trade Offs

Authored by

Nadia Danilova

Department of Molecular, Cell & Developmental Biology, University of California, Los Angeles CA, USA

STRESS RESPONSE AND IMMUNITY: LINKS AND TRADE OFFS

Author: Nadia Danilova

ISBN (Online): 978-981-14-3717-5

ISBN (Print): 978-981-14-3716-8

Bentham Science Publishers Pte. Ltd.
80 Robinson Road #02-00
Singapore 068898
Singapore
Email: subscriptions@benthamscience.net

BENTHAM SCIENCE

CONTENTS

PREFACE i

CHAPTER 1 LIFE IS STRESSFUL 1
INTRODUCTION 1
1.1. Conditions that Support Life 1
1.2. Life is Fueled by ATP 3
1.3. Sensing Environment 7
1.4. Homeostasis 8
1.5. Proteostasis 10
1.6. Stress and Stress Response 11
1.6.1. Shortage of Energy and Nutrients 12
1.6.2. Oxidative Stress 12
1.6.3. Radiation 15
1.6.4. Heat and Cold Stress 16
1.6.5. Osmotic Stress 18
1.6.6. pH Stress 19
1.6.7. Toxic Compounds 20
1.6.8. DNA Damage 22
CONCLUDING REMARKS 23
CONSENT FOR PUBLICATION 23
CONFLICT OF INTEREST 23
ACKNOWLEDGEMENTS 23
REFERENCES 23

CHAPTER 2 PATHOGENS 36
INTRODUCTION 36
2.1. Viruses 38
2.2. Viroids and Prions 43
2.3. Mobile Genetic Elements 43
2.3.1. Insertion Sequences 44
2.3.2. Transposons 44
2.3.3. Plasmids 46
2.3.4. Group I and Group II Introns 46
2.3.5. Homing Endonucleases 46
2.3.6. Chromosomal Islands 47
2.3.7. Gene Cassettes and Integrons 47
2.3.8. The Role of MGEs 47
2.4. Bacterial Pathogens 48
2.5. Archaea 53
2.6. Fungal Pathogens 53
2.7. Protozoan Pathogens 54
2.8. Helminths 55
2.9. Life is a Communal Enterprise 56
CONCLUDING REMARKS 58
CONSENT FOR PUBLICATION 58
CONFLICT OF INTEREST 59
ACKNOWLEDGEMENTS 59
REFERENCES 59

CHAPTER 3 HOW BACTERIA COPE WITH STRESS 71
INTRODUCTION 71

3.1. Sensing Environment ... 71
3.2. Heat and Cold Shock Responses ... 73
3.3. General Stress Response ... 74
3.4. Envelope Stress Response ... 75
3.5. Starvation ... 78
3.6. Oxidative Stress ... 79
3.7. Osmotic Stress ... 80
3.8. pH Stress ... 80
3.9. Metals ... 82
3.10. DNA Damage Response ... 84
3.11. Multiple Stresses ... 85
3.12. Programmed Cell Death in Bacteria ... 86
3.13. Phenotypic Diversity and Persisters ... 87
3.14. Stress of Living Together ... 88
3.15. Archaea Stress Responses ... 89
CONCLUDING REMARKS ... 89
CONSENT FOR PUBLICATION ... 89
CONFLICT OF INTEREST ... 89
ACKNOWLEDGEMENTS ... 89
REFERENCES ... 90

CHAPTER 4 EXTREMOPHILES ... 101
INTRODUCTION ... 101
4.1. Thermophiles ... 103
4.2. Psychrophiles ... 104
4.3. Halophiles ... 107
4.4. Alkaliphiles ... 109
4.5. Acidophiles ... 111
4.6. Piezophiles ... 113
4.7. Polyextremophiles ... 115
CONCLUDING REMARKS ... 116
CONSENT FOR PUBLICATION ... 116
CONFLICT OF INTEREST ... 116
ACKNOWLEDGEMENTS ... 116
REFERENCES ... 117

CHAPTER 5 IMMUNODEFENSES OF PROKARYOTES ... 125
INTRODUCTION ... 125
5.1. Secreted Extracellular Molecules ... 126
5.2. Preventing Cell Entry ... 127
5.3. Non-specific Silencing of Foreign Genes ... 130
5.4. Restriction Modification Systems ... 130
5.5. Bacteriophage Exclusion (BREX) Systems ... 132
5.6. CRISPR-Cas Immunity ... 133
5.7. Argonaute-based Nucleic Acid Interference ... 138
5.8. Abortive Infection ... 138
5.9. Superinfection Exclusion ... 139
5.10. New Immune Systems ... 140
CONCLUDING REMARKS ... 141
CONSENT FOR PUBLICATION ... 141
CONFLICT OF INTEREST ... 141
ACKNOWLEDGEMENTS ... 141

REFERENCES ... 142

CHAPTER 6 INTERACTION OF STRESS AND IMMUNE RESPONSES IN PROKARYOTES 155
INTRODUCTION ... 155
6.1. Mobile Genetic Elements are Induced by Stress ... 156
6.2. Mobile Genetic Elements can Improve the Host's Fitness ... 158
6.3. Infection Induces Envelope Stress Response ... 160
6.4. CRISPR-Cas Activation in Response to Stress ... 161
6.5. CRISPR-Cas Responds to Metabolic Changes ... 162
6.6. Effect of Stress on R/M Systems ... 163
6.7. MGEs and Pathogenic Bacteria ... 164
6.8. The Role of Stress Response in Bacterial Pathogenesis ... 166
6.9. Role of Stress in Evolution of Prokaryotic Immunity ... 168
CONCLUDING REMARKS ... 170
CONSENT FOR PUBLICATION ... 170
CONFLICT OF INTEREST ... 170
ACKNOWLEDGEMENTS ... 170
REFERENCES ... 170

CHAPTER 7 RESPONSE OF EUKARYOTES TO STRESS ... 182
INTRODUCTION ... 182
7.1. Integrated Stress Response ... 183
7.2. Transcriptional Stress Response ... 185
7.3. Endoplasmic Reticulum Stress and Unfolded Protein Response ... 187
7.4. Autophagy ... 189
7.5. Mitochondrial Stress ... 190
7.6. Starvation ... 191
7.7. Oxidative Stress ... 193
7.8. Heat Stress ... 196
7.9. Osmotic Stress ... 197
7.10. DNA Damage Response ... 197
7.11. Regulated Cell Death ... 199
7.12. Regulation of Stress Response at the Level of Organism ... 202
7.12.1. Non-autonomous Regulation of Stress Response in Animals ... 202
7.12.2. Stress Response in Plants ... 205
7.13. Cross-Stress Resistance ... 205
CONCLUDING REMARKS ... 206
CONSENT FOR PUBLICATION ... 206
CONFLICT OF INTEREST ... 206
ACKNOWLEDGEMENTS ... 206
REFERENCES ... 206

CHAPTER 8 P53, GUARDIAN OF THE GENOME ... 224
INTRODUCTION ... 224
8.1. Genes of p53 Family ... 225
8.2. p53 Activity ... 226
8.3. p53 Network Regulation ... 228
8.4. p53 Dictates Cell Fate ... 232
8.5. Genome Protection by p53 ... 234
8.6. p53 as a Tumor Suppressor ... 236
8.7. p53 Interaction with NFkB ... 240
CONCLUDING REMARKS ... 241

CONSENT FOR PUBLICATION 242
CONFLICT OF INTEREST 242
ACKNOWLEDGEMENTS 242
REFERENCES 242

CHAPTER 9 AUTONOMOUS IMMUNITY 257
INTRODUCTION 257
9.1. RNA Interference 258
9.2. Pathogen-associated Molecular Patterns and Pattern Recognition Receptors 260
9.3. Danger Theory 262
9.4. Interferons 263
9.5. Inflammasome 265
9.6. Major Sensors of Autonomous Immune System 269
9.6.1. Sensing Foreign RNA 269
9.6.2. Sensing Foreign DNA 271
9.6.3. Nucleotide-binding Domain (NBD) and Leucine-rich Repeat (LRR)-containing Receptors (NLRs) 273
9.6.4. Uncommon PRRs 274
9.7. Activation of Immune Mechanisms by Genetic Defects 275
9.8. Immune Effectors 275
9.9. Toxic Compounds as Immune Effectors 280
9.10. Regulated Cell Death as Immunodefense 280
CONCLUDING REMARKS 282
CONSENT FOR PUBLICATION 283
CONFLICT OF INTEREST 283
ACKNOWLEDGEMENTS 283
REFERENCES 283

CHAPTER 10 STRESS RESPONSE MEETS AUTONOMOUS IMMUNITY 307
INTRODUCTION 307
10.1. Homeostatic Perturbations and Immune Response 308
10.2. Integrated Stress Response and Immune Response 309
10.3. Unfolded Protein Response and Autonomous Immunity 310
10.4. DDR and Autonomous Immunity 312
10.5. p53 and Autonomous Immunity 314
10.6. Oxidative Stress and Autonomous Immunity 316
10.7. Autophagy and Immunodefense 319
10.8. Balancing Stress and Immune Responses 322
CONCLUDING REMARKS 323
CONSENT FOR PUBLICATION 323
CONFLICT OF INTEREST 324
ACKNOWLEDGEMENTS 324
REFERENCES 324

CHAPTER 11 SPECIALIZED IMMUNE SYSTEMS OF ANIMALS 339
INTRODUCTION 339
11.1. Inflammation 341
11.1.1. Inflammation Triggers and Mediators 341
11.1.2. Fever 343
11.1.3. Inflammation Control and Resolution 344
11.1.4. Memory of Inflammation 346
11.1.5. Inflammation is a Necessary Step in Tissue Repair 346

11.1.6. Inflammation in Immunodefense 347
11.1.7. Inflammatory Diseases 347
11.2. Phagocytosis 349
11.3. Natural Killer Cells and Missing-Self Recognition 352
11.4. Complement System 354
11.5. Adaptive Immunity 358
11.5.1. Adaptive Antigen Receptors 358
11.5.2. DNA Damage Response in T and B Cells 359
11.5.3. Regulation of T and B Cells 360
CONCLUDING REMARKS 362
CONSENT FOR PUBLICATION 362
CONFLICT OF INTEREST 362
ACKNOWLEDGEMENTS 362
REFERENCES 362

CHAPTER 12 BRAIN, STRESS, AND IMMUNITY CONNECTIONS 380
INTRODUCTION 380
12.1. Immune System of the Brain 382
12.2. Hypothalamic-pituitary-adrenal (HPA) Axis 384
12.3. How Brain Controls Immunity 386
12.4. Sickness Behavior 389
12.5. Psychological Stress and Immunity 391
12.6. Effects of Stress and Inflammation on Brain Development and Health 393
12.7. Inflammation, Microglia, and Mental Illness 397
12.7.1. Depression 397
12.7.2. Anxiety 400
12.7.3. Neurodegenerative Disorders 401
12.7.4. Schizophrenia 405
12.8. Physical Illnesses, the Brain, and Immunity 407
12.9. Stress and Lymphoid Cells 409
CONCLUDING REMARKS 410
CONSENT FOR PUBLICATION 410
CONFLICT OF INTEREST 410
ACKNOWLEDGEMENTS 410
REFERENCES 410

SUBJECT INDEX 434

PREFACE

All living organisms face two major challenges: to adjust to constantly changing environment and to protect themselves from pathogens. How organisms integrate responses to these challenges is the subject of this book. Cellular machinery can function properly only in a narrow range of condition. The same is true for multicellular organisms. When conditions deviate from the acceptable range, that creates stress and requires change. Physical stress can be caused by starvation, heat, cold, irradiation, and other factors. In addition, higher animals can experience mental stress caused by fear, neglect, isolation *etc*. Stress response is a set of measures that preserve homeostasis in the face of environmental changes. Pathogens are another challenge for most life forms. Viruses and mobile genetic elements infect all organisms. Multicellular organisms can also be infected by bacterial and eukaryotic pathogens. These subjects are presented in the first two chapters of the book.

The next section presents the elaborate mechanisms of stress and immune responses in bacteria and archaea. A common response to stress in prokaryotes includes, among other means, switching to an alternative transcriptional mode. Prokaryotic immunodefense mechanisms are built on two strategies that are also conserved in eukaryotes. One is innate immunity based on genetically encoded molecules/receptors. The other — adaptive immunity is based on unique molecules/receptors that are created *de novo* in response to infection.

Eukaryotic stress response is discussed next. Global inhibition of translation, called integrated stress response, is a common reaction to many stresses in eukaryotic cells. In multicellular organisms, most individual cells have autonomous immunodefense mechanisms which function in collaboration with stress response. Some stress responses can participate in immunodefense. A notable example is unfolded protein response. It cleanses the cell of misfolded proteins plus also targets viral proteins because of their difference from cellular proteins. In animals, cellular stress response can trigger cytokine production and systemic response, which includes inflammation and engagement of specialised immune systems. Even subtle changes in homeostasis can activate such a response. The incredible sensitivity of cellular machinery to changes has a dark side; stress and ensuing immune mechanisms such as inflammation and complement can be induced without infection or substantial injury and lead to pathology.

In complex organisms with specialised immune systems, discussed next, the relationship between stress and immunity becomes more complex and sometimes antagonistic. Mental stress can cause activation of immune mechanisms, which, in turn, can affect the brain's functioning, and behavior. In the recent decade, science has discovered the paramount importance of interaction of all levels of stress response with immunity in the etiology of many human diseases from atherosclerosis to Alzheimer's.

Nadia Danilova
Department of Molecular
Cell & Developmental Biology,
University of California, Los Angeles CA,
USA

CHAPTER 1

Life is Stressful

Abstract: Life is ubiquitous on Earth wherever liquid water is present. The three domains of life Eukaryota, Bacteria, and Archaea, although very different, have a common origin. They share the basic enzymatic machinery, and they use ATP as an energy source. Every organism is adapted to a specific set of environmental conditions. Divergence from such conditions disturbs the normal functioning of the organism and generates stress - a pressure to adapt to the new conditions. Stress can be caused by many factors, from lack of nutrients to heat, cold, radiation, and toxins. Organism's reaction to stress is called stress response. It involves modification of cellular functioning, which may include activation of specific transcriptional programs, modification of membranes and proteins, production of protective compounds, and metabolic adaptations. Every type of stress causes similar problems in all organisms; for example, heat causes protein denaturation. A universal response to this problem is the production of chaperones that help proteins to refold and of other compounds that stabilize protein structure. Accordingly, diverse organisms share many features of their stress responses.

Keywords: Archaea, ATP, Bacteria, DNA damage, Eukaryotes, Homeostasis, Heat stress, LUCA, Nutrient stress, Osmotic stress, Oxidative stress, pH stress, Radiation, Stress response.

INTRODUCTION

Earth is booming with life. Living organisms populate water, soil, and air. They are found in cold permafrost, in hot deserts, deep in the ocean under high pressure, on mountaintops under low pressure, and even in radioactive mines. Living organisms created many current features of our planet from the oxygenic atmosphere to mineral deposits and soil. A single living organism is fragile and easy to destroy but life, as a whole, seems unstoppable. Modeling of astrophysical events shows that even extreme conditions like the boiling of oceans are unlikely to lead to global sterilization of our planet [1]. What makes life so resilient? We will discuss this in the following chapters.

1.1. Conditions that Support Life

Currently, three domains of life are recognized, Eukaryota, Bacteria, and Archaea

Nadia Danilova

although some data suggest Eukaryota belong to the Archaea domain [2 - 4] (Fig. **1A**). The origin of eukaryotes is one of the central events in evolution since it led to complex multicellular organisms. Eukaryotes differ from bacteria and archaea by the presence of the endomembrane system, the nucleus, and other organelles as well as by mitosis, and meiotic sex in their cells. They are thought to be chimeric organisms derived from an archaeal ancestor that incorporated the bacterial ancestors of the mitochondria and chloroplast. Animal mitochondria and plant chloroplast arose from proteobacteria and cyanobacteria, respectively [5 - 8]. They preserved part of their DNAs although most of their proteins are encoded in the nucleus. These organelles are used by eukaryotes to produce energy. Archaeal host cell could belong to the superphylum "Asgard", which has the closest association with eukaryotes in phylogenomic analyses [9, 10]. It is still debated at what evolutionary stage mitochondria were acquired and if it was captured through phagocytosis or through symbiotic interaction [11, 12].

It is believed that all living organisms are descendants from the last universal common ancestor (LUCA), which lived >3.5 billion years (Ga) ago [13]. The earliest signs of life are found in 3.95 Ga sedimentary rocks in Labrador [3, 14]. The Earth is 4.5 Ga old but was lifeless until it cooled down by 4.2-4.3 Ga and oceans formed. It means that life originated in 200 million years window. The earliest signs of life on land are found in the 3.5 Ga old hot spring deposits in New Zealand [15]. At approximately 3.4 Ga, Bacteria and Archaea lineages formed [3]. This time is also marked by the earliest signs of oxygen in Earth atmosphere, which continued to increase for the next billion years leading to the emergence of oxygen-tolerant organisms and eukaryotes at ~ 2. 2 Ga, and multicellular life at ~1.5 Ga. All modern organisms still share many basic cellular processes. A number of human genes have recognizable homologs in bacteria and archaea [9, 16]. Amazingly, there is not much difference in the number of genes between human and bacteria. The Human genome has ~19,000 protein-coding genes while the genome of a free-living bacteria *Sorangium cellulosum* has 11,599 genes [17, 18].

The universal prerequisite for life as we know it is the presence of liquid water. At the normal atmospheric pressure, water is liquid between 0°C and 100°C. At high pressure such as at the seafloor hydrothermal vents, water stay liquid up to 150°C. Increased salinity may extend freezing temperature to –20°C. Based on water properties, –20°C and 150°C could be the theoretical cut-off points for life.

In addition to water, also required is a source of the major building elements of life known as CHNOPS (carbon, hydrogen, nitrogen, oxygen, phosphorus, and sulfur), several metals, and a source of energy to support cellular functions [19]. Living cells generate energy by oxidation of mineral or organic compounds or by absorbing light. Some organisms are autotrophs ("self-feeding", from the Greek *autos* "self" and *trophe* "nourishing"). They make organic compounds from simple inorganic substances such as CO_2, using the energy of the sun or chemical reactions. The universal common ancestor LUCA might have been an autotroph organism that fixed CO_2 using H_2 as an electron donor and lived in an alkaline hydrothermal vent [13]. Photoautotrophs such as cyanobacteria and plants use photosynthesis both for energy and CO_2 acquisition. Chemoautotrophs get nutrients and energy directly from thermal sources and compounds such as methane, hydrogen sulfide, ammonia, and metal ions. In contrast, heterotrophs use organic compounds made by other organisms to get nutrients and energy. Animals are typical heterotrophs. Some heterotrophs need organic compounds as a source of carbon but are able to use light or inorganic compounds as a source of energy. For example, photoheterotrophs use light for energy but organic compounds as a carbon source. Some organisms require oxygen for survival while for anaerobic organisms oxygen is toxic.

In addition to nutrients and energy, life is possible only inside a certain range of environmental conditions such as temperature, pressure, salinity, pH, *etc.* [19]. The acceptable range of conditions is different between living forms. Most organisms are adapted to a stable environment with moderate conditions. They are mesophiles. However, there are a lot of extreme environments on Earth including Polar regions, deserts, seafloors, volcano springs, and high mountains. Organisms called extremophiles populate these places [20, 21]. Archaea are especially common among extremophiles and versatile in their energy sources from sugars to hydrogen gas. Not only microorganisms can tolerate extreme environment. Antarctic ocean, for example, has rich biodiversity with organisms of various complexities from bacteria to fish [22]. Therefore organisms differ in ways they obtain energy and carbon and which specific environmental conditions they adapted to.

1.2. Life is Fueled by ATP

The living cell needs energy to fuel cellular processes and growth. Adenosine triphosphate (ATP) is a universal energy carrier for the majority of cellular functions including synthesis of macromolecules, transport, maintenance of cell structure, motility, and cell division. Energy is stored in ATP as phosphoanhydride bonds between adjacent phosphates, especially between the second and third phosphate (Fig. **1B**). When cells need energy for some processes, they use ATP hydrolysis or phosphorylation by ATP.

ATP is produced by phosphorylation of ADP or AMP by various enzymes in energy-releasing processes. Metabolic processes that use ATP convert it back to its precursors so that ATP is constantly recycled in cells. It is calculated that a marathon runner needs to re-synthesize approximately 60 kg of ATP during a 2-hour run [23].

ATP production and use depend on several molecules that can be easily oxidized and reduced. Coenzyme Q can cycle between oxidized ubiquinone (UQ) state and fully reduced ubiquinol (UQH2) state [24] (Fig. **1C**). The intermediate semiquinone (QH) form can result from partial reduction. Another important molecule involved in energy production is nicotinamide adenine dinucleotide (NAD) [25]. NAD also can accept or donate electrons in reduction-oxidation (redox) reactions and continuously cycles between two forms, NAD+ and NADH. NAD+ is an oxidant; when it oxidizes some molecule, it is converted into NADH; when NADH reduces another molecule it is re-oxidized back to NAD+. Flavin adenine dinucleotide (FAD) also exists in different redox states by accepting or donating electrons [26].

The major pathways of ATP biosynthesis include substrate phosphorylation and oxidative phosphorylation. In the substrate level phosphorylation, the phosphate group from a phosphorylated reactive substrate is transferred to ADP to form ATP. Phosphorylated reactive substrates are generated in processes such as glycolysis, stepwise oxidation of glucose into pyruvate. Two molecules of ATP and two molecules of NADH are generated during the oxidation of one molecule of glucose. Pyruvate can be further oxidized in mitochondria to generate more ATP and NADH.

Mitochondria, bacteria, and archaea use their membranes to generate ATP by oxidative phosphorylation. To explain this process, Peter D. Mitchell proposed the chemiosmotic hypothesis in 1961, which was confirmed later and awarded the Nobel prize in 1978 [27]. He suggested that mitochondria built an electrochemical gradient across their inner membranes by pumping protons from the mitochondria interior called matrix into intermembrane space using the energy of NADH and FADH2. Backflow of protons to the matrix through ATP synthase drives ATP synthesis.

Protons are pumped across the membrane by the electron transport chain [28]. The chain starts with NADH: ubiquinone oxidoreductase (complex I) that catalyzes the transfer of two electrons from NADH to ubiquinone (UQ). As the electrons pass through this complex, four protons are pumped from the mitochondrial matrix into the intermembrane space. In addition, the reduction of ubiquinone to ubiquinol UQH2 takes two more protons from the matrix.

Complex II, succinate dehydrogenase, also supplies electrons to the electron chain by oxidation of succinate to fumarate and reduction of ubiquinone. Electrons that originate in beta-oxidation of fatty acids and catabolism of amino acids and choline also enter the electron chain and reduce ubiquinone to ubiquinol. Therefore ubiquinol (UQH_2) is generated by several mechanisms.

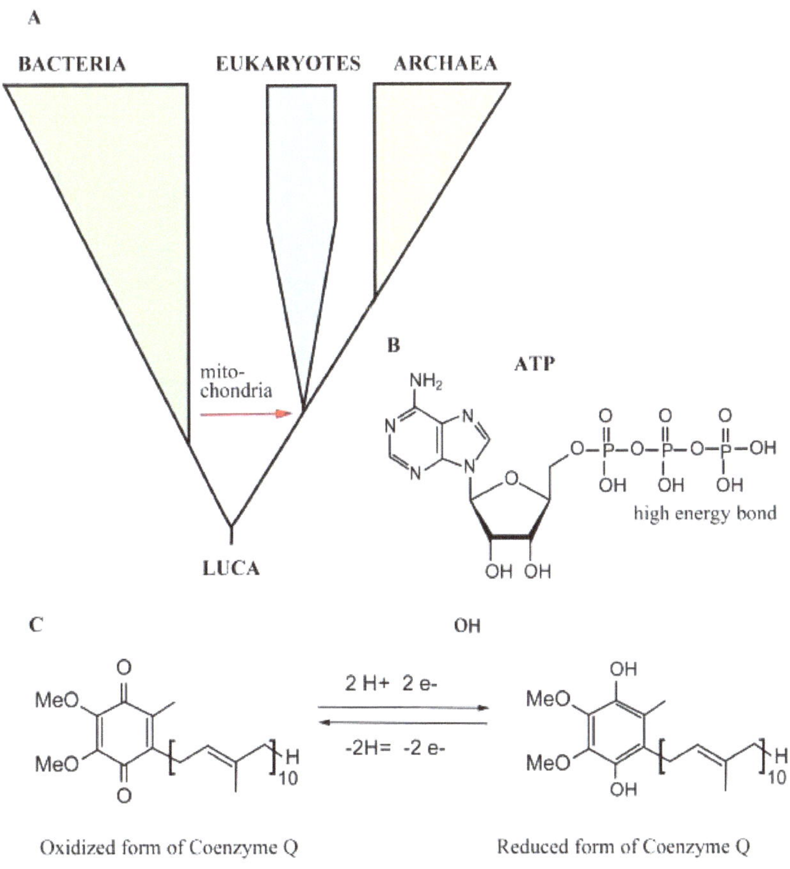

Fig. (1). A. The three domains of life. Eukaryotes evolved inside Archaea domain and incorporated bacteria as their cellular power stations. LUCA – the last universal common ancestor. **B.** ATP is a universal energy carrier for all life forms. The high-energy bond is highlighted. **C.** Coenzyme Q can exist in oxidized or reduced states by accepting or donating electrons.

Next, ubiquinol-cytochrome c reductase complex III catalyzes the transfer of electrons from UQH2 to cytochrome c; this reaction results in ubiquinone UQ, reduced cytochrome c, and transfer of four H+ to the intermembrane space.

The final protein in the electron chain is cytochrome c oxidase (complex IV) that transfers electrons from cytochrome c to oxygen while pumping protons across the membrane. Different eukaryotes have some unique features in their mitochondrial electron chains.

Moving protons across the mitochondrial membrane creates a negative charge on the inside of the membrane $\Delta\psi$ and a difference in H^+ concentration across the membrane (ΔpH). Together they form electrochemical gradient across the membrane, which is called proton-motive force (PMF) [29].

Oxidative phosphorylation has a drawback; it can produce reactive oxygen species (ROS) as side products when complex III leaks electrons to O_2 and generates superoxide [30]. Many species possess alternative oxidases that can transfer electrons directly from UQH2 to oxygen, which lowers the production of ROS. Such oxidases are induced in response to stress and infection [31].

An alternative way to generate PMF is by using light as a source of energy. In plants, after absorption of light, activated chlorophyll strips an electron from the water molecule oxidizing it into O_2 and H^+ through several intermediates. Then chlorophyll donates an electron to the electron transport chain. This flow of electrons is used to transport H^+ across the membrane to create PMF and synthesize ATP and NADPH [32]. ATP and NADPH are used in the next stage of photosynthesis to capture and reduce carbon dioxide from the air to form sugars. Photosynthesis emerged early in life history. The first photoautotroph oxidized not water but chemical substances such as hydrogen sulfide. Water-oxidizing organisms evolved by 3.5–3.8 Ga ago [33].

Some microorganisms use a simpler method to exploit the energy of light for building the chemiosmotic potential. They use rhodopsins, photochemically reactive membrane-embedded proteins bound to the chromophore retinal [34]. Light absorption by retinal triggers photoisomerization of rhodopsin and a proton pumping action. This produces a proton gradient more directly.

Instead of oxygen, some organisms use other electron acceptors such as nitrate, iron (III), carbon dioxide, sulfate and even arsenate [35]. Ecosystems at deep seafloor often rely on oxidizing organic compounds or molecular hydrogen (H_2) while reducing sulfate (SO_4^{2-}) to hydrogen sulfide (H_2S) [36]. Cooperation between bacteria and archaea has been observed in marine sediments where anaerobic oxidation of methane is performed by an association of methanotrophic archaea and sulfate-reducing bacteria with direct transfer of electrons between species [37 - 39].

Next, electrochemical PMF created by various means is used by ATP synthases to generate ATP from ADP and phosphate. Paul Boyer and John E Walker shared the 1997 Nobel prize for the discovery of rotary catalysis employed by ATP synthases for energy conversion [40]. ATP synthases are found in membranes of all living organisms. They consist of multiple proteins that form catalytic heads and membrane ring rotors [41]. Two aqueous channels, each spanning one half of the membrane, pass protons to and from glutamates in the ring rotor [42, 43]. Intermembrane space has pH higher that matrix. A proton enters the intermembrane channel at pH 7.2 and protonates glutamate of the ring causing its rotation. Rotation brings the protonated glutamate to the matrix channel that has pH 8.0, which causes glutamate deprotonation. The rotation is propagated through the central stalk to the catalytic head causing a series of conformational changes leading to ATP synthesis. Ring motor can rotate at a speed up to 400 rps [44]. ATP synthases can also work in reverse and generate PMF *via* ATP hydrolysis.

Prokaryotes often have several diverse electron chains adapted to specific conditions and energy sources. *Escherichia coli* lives both in the anaerobic environment of the gut and aerobic environments of the soil. Its electron-transport chain has many alternative dehydrogenases, terminal reductases, oxidases, and quinones [45]. It also has isoenzymes for various electron acceptors such as O_2 and nitrate and electron acceptors such as formate, H_2, NADH, and glycerol-3-phosphate. In the aerobic environment, O_2 is the preferred electron acceptor and it represses the terminal reductases of anaerobic respiration. In anaerobic respiration, aerobic metabolism is repressed and among terminal reductases, those that support the best growth rate are expressed while others are repressed. For example, if nitrate is present in the environment, it represses other terminal reductases, such as fumarate or DMSO reductases. Electron acceptors also regulate the expression of dehydrogenases. This elaborate system let *E. coli* survive in a diverse environment with different oxygen levels and different food sources.

1.3. Sensing Environment

Any cell needs information about its environment, including the levels of nutrients and toxins, temperature, salinity, pH, and presence of other cells and pathogens. Surface or cytoplasmic molecules sense conditions important for cell survival and signals are transmitted inside the cell up to DNA. Most signals recognized by cells are chemical in nature. Some cells can also sense light and mechanical stimuli.

Prokaryotes use various sensing systems [46, 47]. The simplest signaling system is one-component that combines in one molecule a sensory domain and a regulatory domain. The typical two-component system consists of the sensor histidine kinase that responds to intra- and extracellular signals by catalyzing the phosphorylation of a corresponding response regulator that can bind DNA, RNA, another protein, or have enzymatic activity [48 - 50]. These regulators adjust gene expression or cellular physiology to environmental changes. Two-component systems in bacteria are the major instrument to sense environmental changes and adjust to them. Some signaling systems induce secondary messengers such as cAMP that have global effects on cell physiology. Two-component systems exist in some eukaryotes such as plants and yeasts [51].

In eukaryotic cells, most sensors are G-protein coupled receptors, ion channel receptors, and enzyme-linked receptors. Signals are usually amplified in signal transduction cascades and involve second messengers [52 - 55].

1.4. Homeostasis

Living organisms populate extremely diverse environments with different temperature, pressure, salinity, pH, radiation levels, *etc*. Moreover, their external conditions constantly change with the day/night cycle, seasons, and other factors. On the other hand, cellular enzymatic machinery can function optimally only within a narrow range of conditions, which creates a necessity for organisms to maintain the stability of their internal environment despite external changes. Claude Bernard first recognized this requirement in 1865 and suggested that "milieu interieur" of the body needs to be actively stabilized against external disturbances [56]. Walter Bradford Cannon expanded Bernard's ideas and coined the term 'homeostasis' in 1926. This term is composed of Greek *homeo*, meaning "similar", and *stasis* meaning "staying still".

Organisms have various systems that maintain homeostasis at the cellular, tissue, and organismal levels. Most current models postulate that these mechanisms maintain physiological variables within an acceptable range by comparing the actual value of the variable with an optimal value or "set point" or rather a range of values [57, 58]. Integral feedback steers the physiological variable to the desired set point despite stochastic fluctuations or "noise" in genetic circuits [59, 60].

Two types of variables are distinguished in homeostatic systems; regulated variables such as blood glucose, core temperature, or blood pressure and controlled variables such as the rate of gluconeogenesis. Multiple controlled variables usually contribute to the stability of a particular regulated variable. For example, the blood glucose level is affected by the rate of gluconeogenesis and insulin activity among other factors. Blood pressure also depends on many factors including blood volume, vascular tone, and cardiac output [61]. In turn, blood volume depends on the concentration of salt in the blood, which is regulated through sodium ions reabsorption in the kidney (controlled variable).

Homeostatic regulation of living organisms is often described in terms of system dynamics [62]. In these terms, regulated variables are described as stocks of the living system while controlled variables are system's flows. In order to keep a regulated variable/stock within the optimal range, it is monitored by so-called Controller that compares its value with a set point and generates a signal proportional to the difference. The signal acts on so-called Plants that adjust flows into and out of the system to bring a regulated variable closer to the set point. Controllers are usually endocrine cells or sensory neurons. Pancreatic beta cells are a Controller for blood glucose; they detect elevated glucose level and releases insulin, which acts on various tissues (Plants) that increase glucose uptake. Blood pressure is sensed through baroreceptors, the mechanosensitive afferent nerve endings in the wall of blood vessels. They detect deviation of blood pressure from the optimal range and induce changes in sympathetic signaling to adjust it.

Variables that are maintained at the systemic level are referred to as "System stocks", while those maintained at the plant level are referred to as "Plant stocks". Blood glucose system stock is maintained by insulin, glucagon, and catecholamines while glucose Plant stock in skeletal muscle is monitored by intracellular sensors and maintained by metabolic activity. System stocks are often monitored by specialized receptors such as the transient receptor potential (TRP) superfamily of channels that sense various variables from temperature to membrane stretch [63]. Some Plant stocks represent storage substances such as glycogen or fat that buffer regulated variables such as glucose or fatty acids from variations caused by their intake and expenditure. Energy expenditure is also maintained within a narrow physiological range [64].

Some regulated variables can have adjustable set points. For example, the temperature of a human body increases during infection to a higher set point [65]. During winter, many mammals can enter a state of hibernation with low metabolic activity and a drop of temperature [66]. Homeostatic systems with adjustable set points are more flexible than systems with fixed set points but are more susceptible to dysregulation. Body weight and adiposity have adjustable set points that let an organism to accumulate nutrients during food abundance, which has an adaptive value when food is not regularly available but in modern society may lead to an obesity epidemic [67].

Organism integrates numerous signals from stocks and flows to maintain homeostasis [62]. They include signals from system Controllers such as insulin and glucagon reporting blood glucose level, signals reporting Plants stocks, those of Storage stocks, and values of flows such as a signal from gut hormone GLP1 that reports dietary glucose inflow. Monitoring flows allows the organism to anticipate changes in a regulated variable and elicit a response that prevents dramatic changes. Production, inflow, and expenditure of the stock can be adjusted at multiple points. Homeostatic signals can directly affect the flows of the system such as insulin stimulation of glucose conversion into glycogen. Homeostatic signals can change the sensitivity of the flows such as when catecholamines change the sensitivity of tissues to insulin. Finally, homeostatic signals can change the gain of Controllers so that the same input such as a change in blood glucose would lead to a variable amount of insulin produced by beta cells.

The organism can prioritize some tissues over others. For example, the brain does not tolerate large variations in glucose levels while other tissues are more permissive. Correspondingly, the brain has tighter regulation of glucose levels than have other tissues.

The homeostatic state can change in response to the external environment or internal cellular program. Bacteria or mammalian cells that migrate along a chemotactic gradient would adjust their homeostasis along the way. Differentiation of stem cells into specialized tissue cells would require change in their homeostasis.

Sometimes, internal balance in the organism can only be reached at conditions outside the homeostatic range, which is known under multiple names "allostasis", "heterostasis", and "adaptive homeostasis" [68]. These balanced states are not optimal for the organism and usually do not last long. When the buffering capacity of a homeostatic system is overwhelmed, this may lead to a cascading failure of multiple circuits, disease, and death.

1.5. Proteostasis

An important part of homeostasis is proteostasis: protein production, turnover, and recycling. An average living cell contains billions of protein molecules. Processes associated with the production of proteins consume a large proportion of cellular resources, up to 50% in a rapidly growing bacterial cell and 30% in a differentiating mammalian cell [69].

Misfolded or aggregated protein can cause cell death, therefore all organisms have molecular chaperones such as heat shock proteins (HSPs) that assist protein folding. Some chaperones guide protein folding in an ATP-dependent manner, others maintain the proper folding of their client proteins thereby increasing their stability [70]. They also assist in removing defective or damaged proteins [71].

In eukaryotes, proteins removal is mostly done by the ubiquitin-proteasomal system (UPS) and microautophagy [72]. The ubiquitin system includes a small 76 aa protein ubiquitin, ubiquitin-activating enzymes (E1), ubiquitin-conjugating enzymes (E2), and ubiquitin-protein ligases (E3). In human, there are two genes for E1 enzyme, 41 genes for E2, and more than 600 genes for E3. The cascade starts with ATP-dependent formation of a thioester bond between E1 and the C end of the ubiquitin. Activated ubiquitin then is transferred to E2 from which E3 transfers it to the amino group of lysine on the protein substrate [73, 74]. Ubiquitin itself can be polyubiquitinated at several lysines, most often at K48, 11, 29, and 63. The fate of the ubiquitinated protein depends on the site and the level of ubiquitination among other factors. Typically proteins polyubiquitinated at the ubiquitin K48 residue undergo proteasomal degradation. A chain of at least four ubiquitins is required for proteasomal targeting. Other types of ubiquitination are involved in other processes such as regulation of protein activity and in signal transduction. Archaea have analogs of the ubiquitin system [16] and some bacteria have ubiquitin-like proteins [75].

1.6. Stress and Stress Response

Hans Selye defined stress as "response of the body to any demand for change". The definition suggests that life is stressful by design, which is emphasized by the title of Selye's book "The Stress of Life" [76]. Indeed, life is far from static. Cells and organisms are always in a dynamic interaction with the environment.

Organisms adapt to stress through stress response mechanisms that serve to preserve homeostasis in face of changes, adjust to these changes, and survive periods of extreme conditions. Stress response includes the activation of genes that stop growth and proliferation and activate repair of macromolecules. This response is conserved in species from all domains of life. In addition, stress response includes species and stressor-specific adaptations. Stress tolerance at the cellular and molecular levels differs between species since, during evolution, every species has adapted to a specific set of environmental conditions.

Adaptation to stress can also take place at the populational level. The bacterial population can increase its fitness by bet hedging when stress leads to a regulated increase in stochasticity in a bacterial signaling network, which permits the emergence of subpopulations that have phenotypes not optimally adapted to the current environmental conditions but to conditions that likely to be encountered [49]. Such subpopulations can survive well at conditions that are stressful and maybe even deadly for the rest of the population. Several typical stresses that organisms often encounter are discussed below.

1.6.1. Shortage of Energy and Nutrients

It is essential for life that production and consumption of the energy carrier ATP are kept in balance. The deficit of energy is a major stress for all organisms. Different organisms developed various ways to sense the energy level. Eukaryotic cells use AMP-activated protein kinase (AMPK) that is activated by increased levels of AMP or ADP relative to ATP [77, 78]. AMPK can also sense the level of glucose. In response to low energy, cells adjust their metabolic pathways that produce and use ATP. They shut down processes that consume ATP and boost ones that generate ATP. These processes include induction of alternative pathways of ATP production, an increase of usage of stored nutrients, upregulation of glucose transporters, autophagy, and induction of mitochondrial fission [77 - 79].

Low levels of essential nutrients such as carbon, nitrogen, phosphorus, and iron also cause stress. In application to plants, Justus von Liebig in 1840 suggested "law of the minimum" that states that growth of a plant is limited by the one essential mineral that is in the shortest supply. For example, iron is an essential element in plants and its shortage would cause nutritional stress even if all other elements are abundant [80]. This notion is applicable to all other phyla.

Low level of amino acids is a common stress signal. During starvation, the level of amino acids is not sufficient to load all tRNAs and many organisms can sense the extent of aminoacylation of tRNAs [81]. In eukaryotes, a serine/ threonine kinase target of rapamycin (TOR) protein complex integrates the information about nutrient availability [82]. Arrest of growth and activation of catabolic processes is a common response to starvation among all species. Other responses range from activation of autophagy to switch to an alternative source of food, and in bacteria, to forming biofilms and sporulation [83, 84].

In animals with a brain, it controls food consumption, it also directs appetite to foods rich in the specific nutrients that organism needs, and restricts consumption of nutrients that are in excess [85, 86].

1.6.2. Oxidative Stress

Oxygen is a strong oxidant. It is believed that the early Earth atmosphere contained little or no oxygen; it started to accumulate after the emergence of oxygenic photosynthesis [33]. An expansion of photosynthetic organisms led to a sharp increase in the oxygen level around 2.3 billion years ago, which is called the great oxygenation event. It is also called the oxygen catastrophe since oxygen is toxic to species not adapted to it and its accumulation likely caused the mass extinction of most anaerobic species populating the Earth at that time. Obligate anaerobic organisms became restricted to niches free from oxygen such as deep seafloor. Oxygenic atmosphere led to the evolution of aerobes that use oxygen as a terminal electron acceptor in oxidative phosphorylation and ultimately to multicellular life [27].

Oxidative phosphorylation generates a low level of reactive oxygen species (ROS) during the 4-electron reduction of O_2 to water [87]. Two-electron reduction produces hydrogen peroxide H_2O_2. One electron reduction produces superoxide O_2^-. A highly reactive hydroxyl radical OH* is generated by decomposition of organic peroxides, by H_2O_2 reaction with reduced metal ions such as Fe^{2+}, and by other mechanisms. These are the most common ROS. H_2O_2 is a major ROS product under aerobic condition [88]. It is produced by NADPH oxidases and other enzymes. Contrary to many short-lived ROS, H_2O_2 is relatively stable and has selective reactivity. It can diffuse through cellular membranes and induce signaling cascades even in tissues remote from the site of its production.

Nitric oxide (NO) and peroxynitrite ($ONOO^-$), a product of NO reaction with a superoxide anion are representatives of reactive nitrogen species (RNS). Peroxynitrite can react with other molecules and produce additional RNS. Reactive substances can also be based on other elements such as chlorine, bromine, and sulfur.

Reactive molecules can also be generated during the metabolism of environmental toxins and drugs by the cytochrome P450 enzymes [89]. Antibiotics often induce ROS production by various mechanisms [90 - 92]. Reactive species are also produced as effectors in immune defense [93, 94]. Moreover, reactive species are produced by many organisms as signaling molecules [88, 95, 96].

Balance of oxidation-reduction (redox) reactions is important for cellular health. Many cellular processes are regulated by the redox state of the cell [97, 98]. Disruption of redox homeostasis in favor of oxidants can lead to oxidative stress, a term introduced by Helmut Sies in 1985 [99]. For example, the physiological intracellular level of H_2O_2 in human cells is in the range 1–10 nM when its production and deactivation are balanced. But when it increases to ~ 1 μM or higher the balance is broken and the cell experiences oxidative stress [88]. Physiological extracellular H_2O_2 level is much higher than intracellular and is usually in the 1–5 μM range. Oxidative stress may lead to damage of macromolecules, induction of inflammation, cell cycle arrest, and cell death. It can also interfere with ROS signaling [100].

H_2O_2 and other reactive substances can interact with biological macromolecules including DNA, carbohydrates, proteins, and lipids. Proteins that contain cysteines and metals are especially sensitive to oxidative conditions. Cysteine SH group is a common target for modifications. It can be oxidized to an unstable sulfenic acid (SOH), which can form disulfide bond or can be further oxidized to sulfinic (SO_2H) or sulfonic (SO_3H) acid. Cysteine can also undergo S-nitrosylation, persulfidation, metalation, and other modifications. The versatility of cysteine modifications gave rise to the concept of cysteine proteome [101, 102]. Specific cysteine modifications with regulatory function have been reported for a number of proteins [102, 103].

Excessive oxidation of cysteines can cause protein damage. Disruption of their disulfide bonds can lead to protein unfolding and loss of activity. Oxidized proteins can refold in an abnormal configuration or form bonds with other proteins, DNA, or lipids. Non-enzymatic glycation of amino groups with sugars and aldehydes can lead to insoluble products known as the advanced glycation end products (AGEs) [104]. Glycation of histones affects the stability of nucleosomes and chromatin architecture [105].

Carbonylation of side chains of cysteine and several other amino acids such as lysine is another modification that can be irreversible [106]. Hydroxyl radical reaction often leads to carbonyl products [107].

RNS also can damage proteins by cysteine oxidation, tyrosine nitration, and in proteins containing metals also by oxidation of metal ions [95]. Oxidized proteins need to be degraded by cells to preserve homeostasis but highly oxidized cross-linked protein aggregates are resistant to proteolysis, which makes them highly cytotoxic.

Reactive substances cause various DNA damage from oxidized bases to abasic sites and DNA breaks [108]. Guanine is the most sensitive base with 8-oxo-7-8-dihydroguanine (8-oxoGua) as the most abundant modification. It promotes mutagenesis through the incorporation of dATP instead of dCTP opposite the lesion during replication. Oxidative base lesions are corrected mostly by the base excision repair mechanism with the use of specific glycosylases [109].

Unsaturated lipids are oxidized by ROS into lipid hydroperoxides leading to diverse products that change membrane properties. It also generates reactive substances such as isolevuglandins that can form adducts with nucleotides and proteins [110]. Some products such as 4-hydroxy-2-nonenal can participate in the regulation of redox homeostasis [111].

Sensing of oxidative stress often employs compounds that can quickly alternate between reduced and oxidized states. Cysteine thiol groups have unique properties that let them easily bind nucleophiles, metals, and form reversible disulfide bonds.

Reversible oxidation of cysteine residues is used to sense oxidative stress by bacterial OxyR, yeast Yap1, plant NPR1/TGA, and animal Keap1/Nrf2 protein systems [112].

Many mechanisms protecting organisms from oxidative stress have evolved [113, 114]. They can be roughly subdivided into three classes: compounds that scavenge reactive species and prevent damage, enzymes that detoxify them into less harmful molecules, and enzymes that repair damage to macromolecules. The first line of defense includes low molecular weight compounds such as uric acid, glutathione (GSH), vitamins C and E, lipoic acid, carotenes, and ubiquinol. These compounds can be easily oxidized and reduced and they directly neutralize reactive substances. Hydroxyl radicals are inactivated mostly by such compounds. Recent data suggest that Coenzyme A is used by many species along with glutathione for protection of thiol groups [115].

Several enzymatic systems that detoxify reactive substances are present in nearly all organisms. They include enzymes thioredoxins, which act as antioxidants by reducing oxidized proteins through thiol-disulfide exchange [116]. Reduced state of thioredoxins is maintained by thioredoxin reductases. Peroxiredoxins are a family of peroxidases with a conserved cysteine residue called "peroxidatic" Cys, which is oxidized by peroxides into sulfenic acid, which then reacts with "resolving" Cys to form a disulfide bond. The bond is then reduced by electron donors such as thioredoxins, which completes the cycle [88]. Glutaredoxins is another family of enzymes important in redox homeostasis. They contain intramolecular disulfide bond; it can change to the reduced form by oxidation of glutathione [117]. The oxidized glutathione is then regenerated by glutathione reductase. Another conserved way to detoxify H_2O_2 is through its conversion to water and oxygen by catalases [118]. Superoxide anions are deactivated through conversion into H_2O_2 by superoxide dismutases [119].

The oxidative stress response is characterized by upregulation of production and activity of these factors. It also involves upregulation of mechanisms that repair macromolecules and destroy those that are not reparable. For example, in human cells, multiple enzymes repair oxidative DNA damage including several DNA glycosylases that specifically recognize and remove 8-oxoGua [109].

1.6.3. Radiation

Radiation is a common name for various high-energy particles or waves (X-rays, gamma rays, beta particles, neutrons, protons, UV light, *etc*). The atmospheric ozone layer reduces the biological hazard of cosmic and UV radiation by 3 orders of magnitude, still, radiation remains a deleterious factor for living organisms, especially in the equatorial regions and at high altitudes.

High-energy radiation such as X-rays and gamma rays can break chemical bonds and cause the formation of free radical or ionization. The UV light is the most common radiation. It can directly excite molecules such as thymidine in DNA, which then enters in a chemical reaction with other pyrimidines forming dimers [120]. Thymidine dimers are difficult to repair; they can interfere with replication and lead to mutations and double-strand breaks. UV can also be absorbed by endogenous or exogenous photosensitizers, which then produce free radicals, singlet oxygen, superoxide and other ROS [121]. ROS generated by UV can oxidize DNA and proteins. Recent studies have shown that radiation perturbs DNA methylation, which alters epigenetic regulation and may be one of the pathways from radiation to cancer [122].

Organisms protect themselves against radiation in two major ways. First is the production of photoprotective pigments such as melanins that are able to absorb UV radiation and dissipate it as heat [123]. Such pigments have been found in species from microorganisms to humans.

The most important protection against radiation is efficient DNA repair. Species differ in their ability to repair DNA and survive radiation [124]. In humans, 5 gray (the standard measure of absorbed dose of radiation) causes death, whereas *Deinococcus radiodurans*, one of the hardiest of the radioresistant microorganisms can survive radiation doses of up to 15,000 gray. DNA repair is extremely efficient in this microorganism [124]. *D. radiodurans* is not alone, many other microorganisms such as cyanobacteria *Chroococcidiopsis* are radiation resistant because of efficient DNA repair [125]. Life has been found even in containers with radioactive waste.

Fast proliferating cells are more sensitive to radiation. For this reason, rapidly renewing mammalian blood cells and intestinal cells are most vulnerable [126]. Dormancy, on the other hand, increases resistance. *Bacillus subtilis* endospores can survive up to 6 years in space especially if coated in dust particles protecting them from radiation [127].

1.6.4. Heat and Cold Stress

The temperature has a profound effect on physiology of living cells. All organisms have mechanisms for sensing variation in temperature. Many bacteria use sensors that are based on changes in the secondary structure of DNA, RNA, and proteins induced by temperature [128]. Temperature can also be sensed by changes in membrane properties. Animals mostly use transmembrane ion channels for thermal sensing. They include members of the transient receptor potential (TRP) cation channel, the anoctamin family of chloride channels, and the TREK-1 potassium channels. They are expressed on neural and non-neural cells. Heat-responsive TRPV1 channel can also be activated by capsaicin from chili pepper, whereas menthol activates the cold-responsive TRPM8 channel, which explains the heat and cold sensation produced by these compounds.

The accepted range of temperatures is different for each species. Organisms adapted to mild conditions often can live only in a narrow temperature range. For example, many plants cannot tolerate temperatures below 0°C or above 45°C. Mammals and birds are endothermic homeotherms, they produce heat and can maintain their temperature higher than the environment. They usually do not tolerate well even small deviations from their normal temperature. In contrast, the internal temperature of poikilotherms such as fish or reptiles can vary significantly. Many microorganisms tolerate wide variation of temperature associated with day-night and seasonal changes or with infection of the warm-blooded host [129].

Temperature increase above the normal range can cause disruption of protein folding, unwinding of DNA double-strand helixes leading to dissociation of DNA into separate chains, melting of RNA secondary structure and spontaneous hydrolysis of RNA phosphodiester bonds and of the N-glycosidic bonds between the sugar and a base. Temperature increase also changes membrane permeability to protons. Disruption of protein folding is especially dangerous; it can lead to the formation of protein aggregates, disruption of cellular functions, and cell death. All organisms respond to high temperatures by heat shock response mediated in eukaryotes by heat shock factors [51, 130]. In bacteria, mRNAs of some regulatory factors that induce expression of heat shock proteins are thermosensors [131]. Heat shock response normally includes increased production of heat shock proteins (HSPs). HSPs prevent and repair protein misfolding and aggregation [132 - 134].

HSPs were initially discovered in cells exposed to heat shock (42–45°C) but they are also expressed constitutively and can be induced not only by heat but by many proteotoxic conditions such as hypoxia, oxidative stress, toxins, *etc.* [135]. Heat shock response also involves upregulation of proteases since some proteins damaged by heat cannot be repaired and need to be destroyed. Heat shock response protects other cellular components including DNA, cytoplasmic membranes, ribosomes, and rRNA.

Low temperatures are also stress for cells; they decrease water viscosity, diffusion, and dissociation to H^+ and OH^- ions, which slows down enzymatic reactions. Diffusion of other molecules onto the cell also slows down. DNA becomes more negatively supercoiled, and secondary structures of RNA become stabilized, which negatively affects transcription, translation, and mRNA decay [136]. Ribosome assembly and other cellular processes are also affected. Initiation of translation is especially compromised at low temperatures.

Both prokaryotes and eukaryotes respond to a drop in temperature by cold shock response. In mammalian cells, temperature decrease leads to transcriptional upregulation of cold-inducible proteins (CIPs) such as cold-inducible RNA-binding protein (CIRP) [137]. They adjust the cellular response to cold such as inhibition of cell proliferation [138] and stimulation of translation [139].

1.6.5. Osmotic Stress

Osmotic pressure depends on the total concentration of salts and other solutes in a solution. Cells experience osmotic stress when osmotic pressure inside the cell significantly differs from that on the outside. There can be hypo- and hyperosmotic shock. Water comes out of cells if the outside pressure is higher leading to shrinkage of the cytoplasm, wrinkling, and detachment of membranes, and crowding of cytoplasmic biopolymers. Desiccation has a similar effect. When the outside pressure is significantly lower, water comes into the cell, which leads to an increase of the cytoplasmic volume, increase of the turgor pressure onto membranes, and may lead to cell lysis. Hypo-osmotic shock is used in research to break cells in order to release cellular components.

Maintenance of osmotic homeostasis requires mechanisms that sense deviations of osmotic pressure from the optimal set point as well as mechanisms that restore the pressure. Various ion and solute channels and transporters serve this role. They can sense a change in external and internal osmotic pressure, change in membrane tension and in turgor, or they can react to specific signaling molecules such as cyclic di-AMP in bacteria [140]. Mechanosensitive channels sense membrane tension through interaction with phospholipid acyl chains inside pockets formed by transmembrane helices [141]. In animals that have brains, dehydration is felt as thirst. The brain fine-tunes the feeling of thirst after incorporation of signals from the body such as blood solute concentration, blood volume, and hormone levels with the anticipatory response to the expected eating and drinking [142, 143].

In order to survive an osmotic upshift, the cell needs to increase the internal concentration of osmolytes [144]. Two major strategies are widely used in all domains of life: the first is to import solutes from the environment including salts such as KCl and to adjust cellular processes to high intracellular salt; the second is to increase the synthesis of osmoprotectants [145]. Accumulation of K^+ inside cells is a fast response, which immediately prevents dehydration, but it is not compatible with cellular functions in all organisms. Osmoprotectants or compatible solutes are low molecular weight uncharged compounds that do not disrupt the structure and function of macromolecules [146]. They include sugars (mannitol, trehalose, myo-inositol), amino acids and their derivatives (proline, glycine betaine), urea, amines (trimethylamine-N-oxide), dimethyl-sulfoniopropionate, sarcosine, ectoine, glycerophosphorylcholine, and others.

Usage of specific osmoprotectants varies depending on species and the type of osmotic stress they experience. Cellular solutes differ in their properties and need to be compatible with each other and with the structures they protect. For example, urea can replace the water in membranes during hyperosmolarity, which would protect the lipid assembly and prevent phase transition [147]. Urea, however, can damage proteins. Sharks have a high blood urea concentration and use a compatible osmoprotectant trimethylamine N-oxide; it stabilizes proteins against osmotic pressure and neutralizes the negative effect of urea on protein structure [148].

In order to survive osmotic downshift, cells need to decrease the concentration of solutes on the inside, which is achieved by the passive response of mechanosensitive channels. They open under increased membrane tension and let the release of cytoplasmic solutes [149].

Organic compatible solutes protect protein structure and thereby increase the survival not only during osmotic stress but during other stresses as well [150]. For example, increased production of trehalose is a common part of heat and cold-shock responses [151]. Ectoine also protects against heat and cold stress [152]. Accumulation of solutes in plant cells is a prime adaptive mechanism that supports plant survival during drought [153].

1.6.6. pH Stress

Highly acidic or alkaline environments characterized respectively by low or high pH can directly damage macromolecules of cellular membranes. In addition, low or high pH affects the proton gradient across the cellular membrane and its conductivity for H^+. Extreme pH can inactivate the enzymatic machinery that generates ATP [154].

In humans, the normal cytoplasmic pH is 7.3 whereas outside cells it is strictly maintained between pH 7.35 and 7.45. Acidosis develops below this range and alkalosis above it. pH is especially important for the brain since it affects neuronal functions. pH in humans is sensed primarily by acid-sensing ion channels [155]. pH homeostasis is achieved at the organismal level by respiration (CO_2 removal), renal excretion, and bone buffering by Ca^{2+} reabsorption. At the cellular level, ion exchange (Na^+/H^+, Cl^-/HCO_3^-) is used for this purpose [156].

In contrast to strict pH requirements for human cells, many prokaryotes can grow in environments with fluctuating pH. For example, some neutrophilic bacteria can grow at ~pH 5.5–9, while maintaining cytoplasmic pH within the range ~7.5–7.7 [154]. Pathogenic bacteria can survive the extremely low pH of the human stomach or phagocytic vacuoles.

Various species developed diverse strategies to adjust their homeostasis to acidic or alkaline stress. The acidic stress response can include a decrease in conductivity of membranes for H^+, active extrusion of proton ions from the cell, alkalization of the cytoplasm, and other homeostatic adjustments [157 - 159]. A response to alkaline stress can include acidification of cytoplasm by extrusion of K^+ and Na^+ from cells and other adjustments [160]. Plants are also sensitive to pH. Soil pH affects the solubility of iron salts. Most soil iron is in the Fe^{3+} form inaccessible to plants. In response to iron deficiency, some plants can release protons into the rhizosphere to reduce Fe^{3+} to accessible Fe^{2+}. Soils in arid regions are alkaline, which keeps iron in the Fe^{3+} form and creates combined stress of alkaline condition and iron deficiency. Plants there adjust to such environment by secreting coumarins from their roots. The coumarins reduce and chelate the soil iron, making it plant accessible [161].

1.6.7. Toxic Compounds

Many substances are poisonous to cells. These include inorganic compounds, such as heavy metals, and toxins that are poisons produced by living organisms. Poisons interfere with intracellular processes and can be deadly for one type of organisms but harmless for others. For example, carbon monoxide (CO) is highly toxic to humans and other animals that have blood cells with hemoglobin as an oxygen carrier. CO binds hemoglobin preventing its binding to oxygen. At the same time, CO is harmless to plants. In another example, sodium cyanide is highly toxic for organisms that use oxidative phosphorylation, because it inhibits respiration by binding cytochrome oxidase. But some plants produce sodium cyanide to deter herbivores. Apricot and apple seeds, for instance, have high amounts of it. Mercury is an irreversible inhibitor of selenoenzymes such as thioredoxin reductase; so it is highly toxic to all organisms that have selenoenzymes [162]. On the other hand, microorganisms are very versatile in their sources of nutrients and energy and some of them depend on carbon monoxide, mercury or cyanide for their growth and survival.

Many organisms produce toxins designed to kill or weakened competitors, kill parasites, or paralyze prey. Some toxins are peptides or proteins; others are low molecular weight compounds with diverse chemical structure. Most antibiotics belong to the latter category. Antibiotics inhibit crucial cellular functions such as cell-wall assembly, protein synthesis, and DNA replication [91, 163]. For example, fluoroquinolones such as ciprofloxacin inhibit DNA topoisomerases blocking DNA replication and causing DNA breaks. Penicillin and related beta-lactam antibiotics target enzymes that build the bacterial cell wall. Beta-lactams do more than inhibition of enzymes, their action results in a pathological cycle of cell wall synthesis and degradation, thereby depleting the cellular resources, which enhances the killing capability of antibiotics [164]. As a countermeasure, many bacteria produce enzymes that can catabolize antibiotics [165].

Some studies show that many antibiotics disrupt cellular redox homeostasis leading to an increase in ROS, which also contributes to cell death [166 - 168]. Maladaptive DNA repair contributes to cell death induced by the antibiotic trimethoprim [90]. Trimethoprim inhibits thymidine biosynthesis. At aerobic condition, the DNA is damaged by ROS production in response to thymidine deficiency and the entailing replication stress, whereas under anaerobic conditions the DNA is damaged by byproducts of nitrate respiration.

A single species usually produce multiple diverse antibiotics, for example, bacteria *Rheinheimera japonica* produces eight low weight antibacterial compounds [169]. Multicellular organisms also produce antibacterial compounds. For example, human skin cells produce sapienic acid that causes membrane depolarization in staphylococci species [170].

Proteinaceous toxins are also produced by all life domains [171]. They include nucleases cleaving DNA, tRNAs and rRNA, glycosylases that cleave sugar bases of nucleic acids, enzymes modifying nucleic acids, peptidases, and protein-modifying enzymes. Antibacterial peptides (AMPs) are an ancient and ubiquitous weapon that is used mostly by multicellular organisms to kill microbes [172, 173]. AMPs typically target bacterial cell wall, which is rich in negatively charged acidic phospholipids such as phosphatidylglycerol and cardiolipin. Most AMPs are cationic or amphipathic, so they can bind the negatively charged cell wall and disrupt its integrity. Subsets of AMPs have alternative or additional mechanisms of action [174].

The diverse toxins that target animals have been suggested to be classified into 13 categories based on their targets [175]. For example, one target is the plasma membrane; toxins from this category degrade lipids, disrupt membrane organization, or form pores, such as antibacterial peptides do. Another target comprises ion pumps; an example is cardiac glycosides such as digoxin that inhibit Na^+K^+ATPase. G-protein coupled receptors are also frequent targets for toxins such as cholera toxin and anthrax toxin. There is a group of toxins targeting small GTP-binding proteins; it includes botulinum C3 exoenzyme toxin, which ADP-ribosylates its targets. Another group of toxins target integrins and C-type lectins that function in cell adhesion. Mycotoxins and histone deacetylase toxins target DNA, rotenone toxin targets mitochondria electron transport chain, diphtheria toxin interferes with translation. Okadaic acid from marine microalgae responsible for the red tides inhibits protein phosphatases, which may lead to diarrhea in humans. The most toxic to humans are botulinum and tetanus neurotoxins; they act on exocytic vesicles, interfering with neurotransmitter release at the presynaptic sites.

To survive poisons, organisms use various strategies including the block of cell entry, pumping the poisons out or metabolizing them into less toxic derivatives. Each organism typically uses multiple means of protection; some are common, others are species-specific. Some bacteria, for example, use compounds such as H_2S for detoxification [176, 177]. Secretion of proteases is commonly used for protection against protein toxins.

Low doses of toxins can have a beneficial effect, a phenomenon known as hormesis [178]. Low toxin doses may activate general stress response mechanisms such as autophagy and induce epigenetic modification of the genome that contributes to the increased fitness.

Widespread use of antibiotics in modern society frequently exposes bacteria to their non-lethal concentrations, which may lead to the selection of antibiotic-resistant bacteria [179]. Genes responsible for detoxification of poisons can spread in the microbial world by horizontal gene transfer, which is the basis of multi-drug resistance [180].

Organisms may encounter other toxins such as organic solvents, xenobiotic (synthetic) compounds, and metals [181]. Exposure to heavy metals induces a specific response, induction of metallothioneins that bind and sequester metal ions. It also induces heat shock response and NFkB, which increases general the tolerance to metal stress.

1.6.8. DNA Damage

All macromolecules can be damaged in stressful conditions, but DNA damage is the most dangerous. Life depends on the preservation of genetic information encoded in DNA and its accurate transmission through generations. Therefore all organisms have elaborate systems that sense DNA damage and repair it [182 - 184]. Some DNA damage happens during normal cellular processes such as DNA replication, other has external cause such as nucleotide dimerization from UV irradiation. In a human cell, up to 100,000 various DNA damage events caused by environmental and endogenous factors happen every day [185]. Most ongoing DNA damages are repaired without induction of general stress response pathways but extensive, prolonged, or unrepairable DNA damage induces a strong stress response. Eukaryotic DNA damage response (DDR) includes mechanisms for DNA damage detection, induction of repair, and induction of cell death in cells that cannot be repaired [182]. Different mechanisms are used to repair specific DNA lesions. Mismatch repair (MMR) replaces nucleotides erroneously inserted during replication; base excision repair (BER) replaces modified nucleotides, nucleotide excision repair (NER) fixes helix-distorting lesions. Double strand breaks are the most dangerous damage. It is repaired by homologous recombination (HR) or by non-homologous end joining (NHEJ) [186, 187]. DDR response also increases resistance against other stresses [188].

CONCLUDING REMARKS

All living organisms have a common origin, inheriting cellular enzymatic machinery that functions optimally within a limited range of internal conditions. External conditions, on the other hand, including temperature, salinity, radiation, pressure, and pH, among others, can vary in a much wider range. Such variations produce stress and elicit stress response, whose function is to adjust to the environmental variations, survive extreme conditions, and restore homeostasis.

CONSENT FOR PUBLICATION

Not applicable.

CONFLICT OF INTEREST

The author confirms that this chapter contents have no conflict of interest.

ACKNOWLEDGEMENTS

Declare none.

REFERENCES

[1] Sloan D, Alves Batista R, Loeb A. The Resilience of Life to Astrophysical Events. Sci Rep 2017; 7(1): 5419.
[http://dx.doi.org/10.1038/s41598-017-05796-x] [PMID: 28710420]

[2] Woese CR, Kandler O, Wheelis ML. Towards a natural system of organisms: proposal for the domains Archaea, Bacteria, and Eucarya. Proc Natl Acad Sci USA 1990; 87(12): 4576-9.

[http://dx.doi.org/10.1073/pnas.87.12.4576] [PMID: 2112744]

[3] Betts HC, Puttick MN, Clark JW, Williams TA, Donoghue PCJ, Pisani D. Integrated genomic and fossil evidence illuminates life's early evolution and eukaryote origin. Nature ecology & evolution 2018; 10: 1556-62.
[http://dx.doi.org/10.1038/s41559-018-0644-x]

[4] Castelle CJ, Banfield JF. Major New Microbial Groups Expand Diversity and Alter our Understanding of the Tree of Life. Cell 2018; 172(6): 1181-97.
[http://dx.doi.org/10.1016/j.cell.2018.02.016] [PMID: 29522741]

[5] Ku C, Nelson-Sathi S, Roettger M, *et al.* Endosymbiotic origin and differential loss of eukaryotic genes. Nature 2015; 524(7566): 427-32.
[http://dx.doi.org/10.1038/nature14963] [PMID: 26287458]

[6] Sánchez-Baracaldo P, Raven JA, Pisani D, Knoll AH. Early photosynthetic eukaryotes inhabited low-salinity habitats. Proc Natl Acad Sci USA 2017; 114(37): E7737-45.
[http://dx.doi.org/10.1073/pnas.1620089114] [PMID: 28808007]

[7] Martijn J, Vosseberg J, Guy L, Offre P, Ettema TJG. Deep mitochondrial origin outside the sampled alphaproteobacteria. Nature 2018; 557(7703): 101-5.
[http://dx.doi.org/10.1038/s41586-018-0059-5] [PMID: 29695865]

[8] Nowack ECM, Weber APM. Genomics-Informed Insights into Endosymbiotic Organelle Evolution in Photosynthetic Eukaryotes. Annu Rev Plant Biol 2018; 69: 51-84.
[http://dx.doi.org/10.1146/annurev-arplant-042817-040209] [PMID: 29489396]

[9] Zaremba-Niedzwiedzka K, Caceres EF, Saw JH, *et al.* Asgard archaea illuminate the origin of eukaryotic cellular complexity. Nature 2017; 541(7637): 353-8.
[http://dx.doi.org/10.1038/nature21031] [PMID: 28077874]

[10] Spang A, Eme L, Saw JH, *et al.* Asgard archaea are the closest prokaryotic relatives of eukaryotes. PLoS Genet 2018; 14(3)e1007080
[http://dx.doi.org/10.1371/journal.pgen.1007080] [PMID: 29596421]

[11] Pittis AA, Gabaldón T. Late acquisition of mitochondria by a host with chimaeric prokaryotic ancestry. Nature 2016; 531(7592): 101-4.
[http://dx.doi.org/10.1038/nature16941] [PMID: 26840490]

[12] Zachar I, Szilágyi A, Számadó S, Szathmáry E. Farming the mitochondrial ancestor as a model of endosymbiotic establishment by natural selection. Proc Natl Acad Sci USA 2018; 115(7): E1504-10.
[http://dx.doi.org/10.1073/pnas.1718707115] [PMID: 29382768]

[13] Weiss MC, Preiner M, Xavier JC, Zimorski V, Martin WF. The last universal common ancestor between ancient Earth chemistry and the onset of genetics. PLoS Genet 2018; 14(8)e1007518
[http://dx.doi.org/10.1371/journal.pgen.1007518] [PMID: 30114187]

[14] Tashiro T, Ishida A, Hori M, *et al.* Early trace of life from 3.95 Ga sedimentary rocks in Labrador, Canada. Nature 2017; 549(7673): 516-8.
[http://dx.doi.org/10.1038/nature24019] [PMID: 28959955]

[15] Djokic T, Van Kranendonk MJ, Campbell KA, Walter MR, Ward CR. Earliest signs of life on land preserved in ca. 3.5 Ga hot spring deposits. Nat Commun 2017; 8: 15263.
[http://dx.doi.org/10.1038/ncomms15263] [PMID: 28486437]

[16] Hennell James R, Caceres EF, Escasinas A, *et al.* Functional reconstruction of a eukaryotic-like E1/E2/(RING) E3 ubiquitylation cascade from an uncultured archaeon. Nat Commun 2017; 8(1): 1120.
[http://dx.doi.org/10.1038/s41467-017-01162-7] [PMID: 29066714]

[17] Han K, Li ZF, Peng R, *et al.* Extraordinary expansion of a Sorangium cellulosum genome from an alkaline milieu. Sci Rep 2013; 3: 2101.
[http://dx.doi.org/10.1038/srep02101] [PMID: 23812535]

[18] Southan C. Last rolls of the yoyo: Assessing the human canonical protein count. F1000 Res 2017; 6: 448.
[http://dx.doi.org/10.12688/f1000research.11119.1] [PMID: 28529709]

[19] Schulze-Makuch D, Airo A, Schirmack J. The Adaptability of Life on Earth and the Diversity of Planetary Habitats. Front Microbiol 2017; 8: 2011.
[http://dx.doi.org/10.3389/fmicb.2017.02011] [PMID: 29085352]

[20] Rothschild LJ, Mancinelli RL. Life in extreme environments. Nature 2001; 409(6823): 1092-101.
[http://dx.doi.org/10.1038/35059215] [PMID: 11234023]

[21] Rampelotto PH. Extremophiles and extreme environments. Life (Basel) 2013; 3(3): 482-5.
[http://dx.doi.org/10.3390/life3030482] [PMID: 25369817]

[22] Chown SL, Clarke A, Fraser CI, Cary SC, Moon KL, McGeoch MA. The changing form of Antarctic biodiversity. Nature 2015; 522(7557): 431-8.
[http://dx.doi.org/10.1038/nature14505] [PMID: 26108852]

[23] Buono MJ, Kolkhorst FW. Estimating ATP resynthesis during a marathon run: a method to introduce metabolism. Adv Physiol Educ 2001; 25(2): 70-1.

[24] James AM, Smith RA, Murphy MP. Antioxidant and prooxidant properties of mitochondrial Coenzyme Q. Arch Biochem Biophys 2004; 423(1): 47-56.
[http://dx.doi.org/10.1016/j.abb.2003.12.025] [PMID: 14989264]

[25] Xiao W, Wang RS, Handy DE, Loscalzo J. NAD(H) and NADP(H) Redox Couples and Cellular Energy Metabolism. Antioxid Redox Signal 2018; 28(3): 251-72.
[http://dx.doi.org/10.1089/ars.2017.7216] [PMID: 28648096]

[26] Saleem-Batcha R, Stull F, Sanders JN, *et al.* Enzymatic control of dioxygen binding and functionalization of the flavin cofactor. Proc Natl Acad Sci USA 2018; 115(19): 4909-14.
[http://dx.doi.org/10.1073/pnas.1801189115] [PMID: 29686059]

[27] Mitchell P, Moyle J. Chemiosmotic hypothesis of oxidative phosphorylation. Nature 1967; 213(5072): 137-9.
[http://dx.doi.org/10.1038/213137a0] [PMID: 4291593]

[28] Davies KM, Blum TB, Kühlbrandt W. Conserved in situ arrangement of complex I and III_2 in mitochondrial respiratory chain supercomplexes of mammals, yeast, and plants. Proc Natl Acad Sci USA 2018; 115(12): 3024-9.
[http://dx.doi.org/10.1073/pnas.1720702115] [PMID: 29519876]

[29] Zorova LD, Popkov VA, Plotnikov EY, *et al.* Mitochondrial membrane potential. Anal Biochem 2018; 552: 50-9.
[http://dx.doi.org/10.1016/j.ab.2017.07.009] [PMID: 28711444]

[30] Usselman RJ, Chavarriaga C, Castello PR, *et al.* The Quantum Biology of Reactive Oxygen Species Partitioning Impacts Cellular Bioenergetics. Sci Rep 2016; 6: 38543.
[http://dx.doi.org/10.1038/srep38543] [PMID: 27995996]

[31] Rogov AG, Zvyagilskaya RA. Physiological role of alternative oxidase (from yeasts to plants). Biochemistry (Mosc) 2015; 80(4): 400-7.
[http://dx.doi.org/10.1134/S0006297915040021] [PMID: 25869356]

[32] Bailey-Serres J, Pierik R, Ruban A, Wingler A. The Dynamic Plant: Capture, Transformation, and Management of Energy. Plant Physiol 2018; 176(2): 961-6.
[http://dx.doi.org/10.1104/pp.18.00041] [PMID: 29438068]

[33] Cardona T. Early Archean origin of heterodimeric Photosystem I. Heliyon 2018; 4(3)e00548
[http://dx.doi.org/10.1016/j.heliyon.2018.e00548] [PMID: 29560463]

[34] Govorunova EG, Sineshchekov OA, Li H, Spudich JL. Microbial Rhodopsins: Diversity, Mechanisms, and Optogenetic Applications. Annu Rev Biochem 2017; 86: 845-72.

[http://dx.doi.org/10.1146/annurev-biochem-101910-144233] [PMID: 28301742]

[35] Rascovan N, Maldonado J, Vazquez MP, Eugenia Farías M. Metagenomic study of red biofilms from Diamante Lake reveals ancient arsenic bioenergetics in haloarchaea. ISME J 2016; 10(2): 299-309. [http://dx.doi.org/10.1038/ismej.2015.109] [PMID: 26140530]

[36] Kato S, Ikehata K, Shibuya T, Urabe T, Ohkuma M, Yamagishi A. Potential for biogeochemical cycling of sulfur, iron and carbon within massive sulfide deposits below the seafloor. Environ Microbiol 2015; 17(5): 1817-35. [http://dx.doi.org/10.1111/1462-2920.12648] [PMID: 25330135]

[37] Krukenberg V, Riedel D, Gruber-Vodicka HR, *et al.* Gene expression and ultrastructure of meso- and thermophilic methanotrophic consortia. Environ Microbiol 2018; 20(5): 1651-66. [http://dx.doi.org/10.1111/1462-2920.14077] [PMID: 29468803]

[38] McGlynn SE, Chadwick GL, Kempes CP, Orphan VJ. Single cell activity reveals direct electron transfer in methanotrophic consortia. Nature 2015; 526(7574): 531-5. [http://dx.doi.org/10.1038/nature15512] [PMID: 26375009]

[39] Wegener G, Krukenberg V, Riedel D, Tegetmeyer HE, Boetius A. Intercellular wiring enables electron transfer between methanotrophic archaea and bacteria. Nature 2015; 526(7574): 587-90. [http://dx.doi.org/10.1038/nature15733] [PMID: 26490622]

[40] Boyer PD. A research journey with ATP synthase. J Biol Chem 2002; 277(42): 39045-61. [http://dx.doi.org/10.1074/jbc.X200001200] [PMID: 12181328]

[41] He J, Ford HC, Carroll J, *et al.* Assembly of the membrane domain of ATP synthase in human mitochondria. Proc Natl Acad Sci USA 2018; 115(12): 2988-93. [http://dx.doi.org/10.1073/pnas.1722086115] [PMID: 29440398]

[42] Allegretti M, Klusch N, Mills DJ, Vonck J, Kühlbrandt W, Davies KM. Horizontal membrane-intrinsic α-helices in the stator a-subunit of an F-type ATP synthase. Nature 2015; 521(7551): 237-40. [http://dx.doi.org/10.1038/nature14185] [PMID: 25707805]

[43] Klusch N, Murphy BJ, Mills DJ, Yildiz Ö, Kühlbrandt W. Structural basis of proton translocation and force generation in mitochondrial ATP synthase. eLife 2017; 6e33274 [http://dx.doi.org/10.7554/eLife.33274] [PMID: 29210357]

[44] Nakanishi-Matsui M, Sekiya M, Nakamoto RK, Futai M. The mechanism of rotating proton pumping ATPases. Biochim Biophys Acta 2010; 1797(8): 1343-52. [http://dx.doi.org/10.1016/j.bbabio.2010.02.014] [PMID: 20170625]

[45] Unden G, Bongaerts J. Alternative respiratory pathways of *Escherichia coli*: energetics and transcriptional regulation in response to electron acceptors. Biochim Biophys Acta 1997; 1320(3): 217-34. [http://dx.doi.org/10.1016/S0005-2728(97)00034-0] [PMID: 9230919]

[46] Jung K, Fabiani F, Hoyer E, Lassak J. Bacterial transmembrane signalling systems and their engineering for biosensing. Open Biol 2018; 8(4)180023 [http://dx.doi.org/10.1098/rsob.180023] [PMID: 29695618]

[47] Kellogg SL, Kristich CJ. Convergence of PASTA kinase and two-component signaling in response to cell wall stress in Enterococcus faecalis. Journal of bacteriology 2018; JB.00086-18. [http://dx.doi.org/10.1128/JB.00086-18]

[48] Bi S, Jin F, Sourjik V. Inverted signaling by bacterial chemotaxis receptors. Nat Commun 2018; 9(1): 2927. [http://dx.doi.org/10.1038/s41467-018-05335-w] [PMID: 30050034]

[49] Carey JN, Mettert EL, Roggiani M, Myers KS, Kiley PJ, Goulian M. Regulated Stochasticity in a Bacterial Signaling Network Permits Tolerance to a Rapid Environmental Change. Cell 2018; 173(1): 196-207.e14. [http://dx.doi.org/10.1016/j.cell.2018.02.005] [PMID: 29502970]

[50] Galperin MY, Makarova KS, Wolf YI, Koonin EV. Phyletic Distribution and Lineage-Specific Domain Architectures of Archaeal Two-Component Signal Transduction Systems. J Bacteriol 2018; 200(7): e00681-17.
[PMID: 29263101]

[51] Krakowiak J, Zheng X, Patel N, *et al.* Hsf1 and Hsp70 constitute a two-component feedback loop that regulates the yeast heat shock response. eLife 2018; 7e31668
[http://dx.doi.org/10.7554/eLife.31668] [PMID: 29393852]

[52] Gent L, Forde BG. How do plants sense their nitrogen status? J Exp Bot 2017; 68(10): 2531-9.
[http://dx.doi.org/10.1093/jxb/erx013] [PMID: 28201547]

[53] Roper SD, Chaudhari N. Taste buds: cells, signals and synapses. Nat Rev Neurosci 2017; 18(8): 485-97.
[http://dx.doi.org/10.1038/nrn.2017.68] [PMID: 28655883]

[54] Iannacone MJ, Beets I, Lopes LE, *et al.* The RFamide receptor DMSR-1 regulates stress-induced sleep in *C. elegans.* eLife 2017; 6e19837
[http://dx.doi.org/10.7554/eLife.19837] [PMID: 28094002]

[55] Ishchenko A, Gati C, Cherezov V. Structural biology of G protein-coupled receptors: new opportunities from XFELs and cryoEM. Curr Opin Struct Biol 2018; 51: 44-52.
[http://dx.doi.org/10.1016/j.sbi.2018.03.009] [PMID: 29554543]

[56] Ayala RA. Welcome to the New Age. Claude Bernard's "Introduction to the Study of Experimental Medicine" and the Shift of Medical Thought Towards Science: 150 Years Later. Arch Med Res 2017; 48(4): 393-6.
[http://dx.doi.org/10.1016/j.arcmed.2017.08.006] [PMID: 28886874]

[57] Speakman JR, Levitsky DA, Allison DB, *et al.* Set points, settling points and some alternative models: theoretical options to understand how genes and environments combine to regulate body adiposity. Dis Model Mech 2011; 4(6): 733-45.
[http://dx.doi.org/10.1242/dmm.008698] [PMID: 22065844]

[58] Mairet F. A biomolecular proportional integral controller based on feedback regulations of protein level and activity. R Soc Open Sci 2018; 5(2)171966
[http://dx.doi.org/10.1098/rsos.171966] [PMID: 29515895]

[59] Eldar A, Elowitz MB. Functional roles for noise in genetic circuits. Nature 2010; 467(7312): 167-73.
[http://dx.doi.org/10.1038/nature09326] [PMID: 20829787]

[60] Briat C, Gupta A, Khammash M. Antithetic Integral Feedback Ensures Robust Perfect Adaptation in Noisy Biomolecular Networks. Cell Syst 2016; 2(1): 15-26.
[http://dx.doi.org/10.1016/j.cels.2016.01.004] [PMID: 27136686]

[61] Oparil S, Acelajado MC, Bakris GL, *et al.* Hypertension. Nat Rev Dis Primers 2018; 4: 18014.
[http://dx.doi.org/10.1038/nrdp.2018.14] [PMID: 29565029]

[62] Kotas ME, Medzhitov R. Homeostasis, inflammation, and disease susceptibility. Cell 2015; 160(5): 816-27.
[http://dx.doi.org/10.1016/j.cell.2015.02.010] [PMID: 25723161]

[63] Moran MM. TRP Channels as Potential Drug Targets. Annu Rev Pharmacol Toxicol 2018; 58: 309-30.
[http://dx.doi.org/10.1146/annurev-pharmtox-010617-052832] [PMID: 28945977]

[64] Müller MJ, Geisler C. From the past to future: from energy expenditure to energy intake to energy expenditure. Eur J Clin Nutr 2017; 71(3): 358-64.
[http://dx.doi.org/10.1038/ejcn.2016.231] [PMID: 27901032]

[65] Evans SS, Repasky EA, Fisher DT. Fever and the thermal regulation of immunity: the immune system feels the heat. Nat Rev Immunol 2015; 15(6): 335-49.

[http://dx.doi.org/10.1038/nri3843] [PMID: 25976513]

[66] Cerri M. The Central Control of Energy Expenditure: Exploiting Torpor for Medical Applications. Annu Rev Physiol 2017; 79: 167-86. [http://dx.doi.org/10.1146/annurev-physiol-022516-034133] [PMID: 27813827]

[67] Klip A. Homeostasis - the Walter B. Cannon's Legacy - Applied to the Metabolic Syndrome and the Scientific Enterprise. Physiology (Bethesda) 2016; 31(4): 246-7. [http://dx.doi.org/10.1152/physiol.00011.2016] [PMID: 27252158]

[68] Davies KJ. Adaptive homeostasis. Mol Aspects Med 2016; 49: 1-7. [http://dx.doi.org/10.1016/j.mam.2016.04.007] [PMID: 27112802]

[69] Li GW, Burkhardt D, Gross C, Weissman JS. Quantifying absolute protein synthesis rates reveals principles underlying allocation of cellular resources. Cell 2014; 157(3): 624-35. [http://dx.doi.org/10.1016/j.cell.2014.02.033] [PMID: 24766808]

[70] Zuehlke AD, Moses MA, Neckers L. Heat shock protein 90: its inhibition and function. Philos Trans R Soc Lond B Biol Sci 2018; 373(1738): 373. [http://dx.doi.org/10.1098/rstb.2016.0527] [PMID: 29203712]

[71] Fernández-Fernández MR, Gragera M, Ochoa-Ibarrola L, Quintana-Gallardo L, Valpuesta JM. Hsp70 - a master regulator in protein degradation. FEBS Lett 2017; 591(17): 2648-60. [http://dx.doi.org/10.1002/1873-3468.12751] [PMID: 28696498]

[72] Kwon YT, Ciechanover A. The Ubiquitin Code in the Ubiquitin-Proteasome System and Autophagy. Trends Biochem Sci 2017; 42(11): 873-86. [http://dx.doi.org/10.1016/j.tibs.2017.09.002] [PMID: 28947091]

[73] Mészáros B, Kumar M, Gibson TJ, Uyar B, Dosztányi Z. Degrons in cancer. Sci Signal 2017; 10(470)eaak9982 [http://dx.doi.org/10.1126/scisignal.aak9982] [PMID: 28292960]

[74] Koren I, Timms RT, Kula T, Xu Q, Li MZ, Elledge SJ. The Eukaryotic Proteome Is Shaped by E3 Ubiquitin Ligases Targeting C-Terminal Degrons. Cell 2018; 173(7): 1622-1635.e14. [http://dx.doi.org/10.1016/j.cell.2018.04.028] [PMID: 29779948]

[75] Delley CL, Müller AU, Ziemski M, Weber-Ban E. Prokaryotic Ubiquitin-Like Protein and Its Ligase/Deligase Enyzmes. J Mol Biol 2017; 429(22): 3486-99. [http://dx.doi.org/10.1016/j.jmb.2017.04.020] [PMID: 28478282]

[76] Selye H. The Stress of Life. New York: McGraw-Hill Book 1978.

[77] Garcia D, Shaw RJ. AMPK: Mechanisms of Cellular Energy Sensing and Restoration of Metabolic Balance. Mol Cell 2017; 66(6): 789-800. [http://dx.doi.org/10.1016/j.molcel.2017.05.032] [PMID: 28622524]

[78] Hardie DG. Keeping the home fires burning: AMP-activated protein kinase. J R Soc Interface 2018; 15(138)20170774 [http://dx.doi.org/10.1098/rsif.2017.0774] [PMID: 29343628]

[79] Toyama EQ, Herzig S, Courchet J, *et al.* Metabolism. AMP-activated protein kinase mediates mitochondrial fission in response to energy stress. Science 2016; 351(6270): 275-81. [http://dx.doi.org/10.1126/science.aab4138] [PMID: 26816379]

[80] Curie C, Mari S. New routes for plant iron mining. New Phytol 2017; 214(2): 521-5. [http://dx.doi.org/10.1111/nph.14364] [PMID: 27918629]

[81] Winther KS, Roghanian M, Gerdes K. Activation of the Stringent Response by Loading of RelA-tRNA Complexes at the Ribosomal A-Site. Mol Cell 2018; 70(1): 95-105.e4. [http://dx.doi.org/10.1016/j.molcel.2018.02.033] [PMID: 29625042]

[82] Maegawa K, Takii R, Ushimaru T, Kozaki A. Evolutionary conservation of TORC1 components, TOR, Raptor, and LST8, between rice and yeast. Mol Genet Genomics 2015; 290(5): 2019-30.

[http://dx.doi.org/10.1007/s00438-015-1056-0] [PMID: 25956502]

[83] Dunlop EA, Tee AR. mTOR and autophagy: a dynamic relationship governed by nutrients and energy. Semin Cell Dev Biol 2014; 36: 121-9.
[http://dx.doi.org/10.1016/j.semcdb.2014.08.006] [PMID: 25158238]

[84] Hüttener M, Prieto A, Espelt J, Bernabeu M, Juárez A. Stringent Response and AggR-Dependent Virulence Regulation in the Enteroaggregative *Escherichia coli* Strain 042. Front Microbiol 2018; 9: 717.
[http://dx.doi.org/10.3389/fmicb.2018.00717] [PMID: 29692772]

[85] Joseph RM, Sun JS, Tam E, Carlson JR. A receptor and neuron that activate a circuit limiting sucrose consumption. eLife 2017; 6e24992
[http://dx.doi.org/10.7554/eLife.24992] [PMID: 28332980]

[86] Steck K, Walker SJ, Itskov PM, Baltazar C, Moreira JM, Ribeiro C. Internal amino acid state modulates yeast taste neurons to support protein homeostasis in *Drosophila.* eLife 2018; 7e31625
[http://dx.doi.org/10.7554/eLife.31625] [PMID: 29393045]

[87] Lushchak VI. Free radicals, reactive oxygen species, oxidative stress and its classification. Chem Biol Interact 2014; 224: 164-75.
[http://dx.doi.org/10.1016/j.cbi.2014.10.016] [PMID: 25452175]

[88] Sies H. Hydrogen peroxide as a central redox signaling molecule in physiological oxidative stress: Oxidative eustress. Redox Biol 2017; 11: 613-9.
[http://dx.doi.org/10.1016/j.redox.2016.12.035] [PMID: 28110218]

[89] Veith A, Moorthy B. ROLE OF CYTOCHROME P450S IN THE GENERATION AND METABOLISM OF REACTIVE OXYGEN SPECIES. Curr Opin Toxicol 2018; 7: 44-51.
[http://dx.doi.org/10.1016/j.cotox.2017.10.003] [PMID: 29527583]

[90] Giroux X, Su WL, Bredeche MF, Matic I. Maladaptive DNA repair is the ultimate contributor to the death of trimethoprim-treated cells under aerobic and anaerobic conditions. Proc Natl Acad Sci USA 2017; 114(43): 11512-7.
[http://dx.doi.org/10.1073/pnas.1706236114] [PMID: 29073080]

[91] Kohanski MA, Dwyer DJ, Collins JJ. How antibiotics kill bacteria: from targets to networks. Nat Rev Microbiol 2010; 8(6): 423-35.
[http://dx.doi.org/10.1038/nrmicro2333] [PMID: 20440275]

[92] Takahashi N, Gruber CC, Yang JH, *et al.* Lethality of MalE-LacZ hybrid protein shares mechanistic attributes with oxidative component of antibiotic lethality. Proc Natl Acad Sci USA 2017.201707466
[http://dx.doi.org/10.1073/pnas.1707466114] [PMID: 28794281]

[93] Nathan C, Cunningham-Bussel A. Beyond oxidative stress: an immunologist's guide to reactive oxygen species. Nat Rev Immunol 2013; 13(5): 349-61.
[http://dx.doi.org/10.1038/nri3423] [PMID: 23618831]

[94] Kadota Y, Shirasu K, Zipfel C. Regulation of the NADPH Oxidase RBOHD During Plant Immunity. Plant & cell physiology 2015; 56(8): 1472-80.
[http://dx.doi.org/10.1093/pcp/pcv063]

[95] Astuti RI, Nasuno R, Takagi H. Nitric oxide signaling in yeast. Appl Microbiol Biotechnol 2016; 100(22): 9483-97.
[http://dx.doi.org/10.1007/s00253-016-7827-7] [PMID: 27722918]

[96] Roy J, Galano JM, Durand T, Le Guennec JY, Lee JC. Physiological role of reactive oxygen species as promoters of natural defenses. FASEB J 2017; 31(9): 3729-45.
[http://dx.doi.org/10.1096/fj.201700170R] [PMID: 28592639]

[97] Cortese-Krott MM, Koning A, Kuhnle GGC, *et al.* The Reactive Species Interactome: Evolutionary Emergence, Biological Significance, and Opportunities for Redox Metabolomics and Personalized Medicine. Antioxid Redox Signal 2017; 27(10): 684-712.

[http://dx.doi.org/10.1089/ars.2017.7083] [PMID: 28398072]

[98] Wible RS, Ramanathan C, Sutter CH, *et al.* NRF2 regulates core and stabilizing circadian clock loops, coupling redox and timekeeping in *Mus musculus*. eLife 2018; 7e31656 [http://dx.doi.org/10.7554/eLife.31656] [PMID: 29481323]

[99] Sies H. Oxidative stress: a concept in redox biology and medicine. Redox Biol 2015; 4: 180-3. [http://dx.doi.org/10.1016/j.redox.2015.01.002] [PMID: 25588755]

[100] Martinez Guimera A, Welsh CM, Proctor CJ, McArdle A, Shanley DP. 'Molecular habituation' as a potential mechanism of gradual homeostatic loss with age. Mech Ageing Dev 2018; 169: 53-62. [http://dx.doi.org/10.1016/j.mad.2017.11.010] [PMID: 29146308]

[101] Go YM, Chandler JD, Jones DP. The cysteine proteome. Free radical biology & medicine 2015; 84: 227-45. [http://dx.doi.org/10.1016/j.freeradbiomed.2015.03.022]

[102] van der Reest J, Lilla S, Zheng L, Zanivan S, Gottlieb E. Proteome-wide analysis of cysteine oxidation reveals metabolic sensitivity to redox stress. Nat Commun 2018; 9(1): 1581. [http://dx.doi.org/10.1038/s41467-018-04003-3] [PMID: 29679077]

[103] Keyes JD, Parsonage D, Yammani RD, Rogers LC, Kesty C, Furdui CM, *et al.* Endogenous, regulatory cysteine sulfenylation of ERK kinases in response to proliferative signals. Free radical biology & medicine 2017; 112: 534-43. [http://dx.doi.org/10.1016/j.freeradbiomed.2017.08.018]

[104] Fournet M, Bonté F, Desmoulière A. Glycation Damage: A Possible Hub for Major Pathophysiological Disorders and Aging. Aging Dis 2018; 9(5): 880-900. [http://dx.doi.org/10.14336/AD.2017.1121] [PMID: 30271665]

[105] Zheng Q, Omans ND, Leicher R, *et al.* Reversible histone glycation is associated with disease-related changes in chromatin architecture. Nat Commun 2019; 10(1): 1289. [http://dx.doi.org/10.1038/s41467-019-09192-z] [PMID: 30894531]

[106] Fedorova M, Bollineni RC, Hoffmann R. Protein carbonylation as a major hallmark of oxidative damage: update of analytical strategies. Mass Spectrom Rev 2014; 33(2): 79-97. [http://dx.doi.org/10.1002/mas.21381] [PMID: 23832618]

[107] Uranga J, Mujika JI, Grande-Aztatzi R, Matxain JM. Oxidation of Acid, Base, and Amide Side-Chain Amino Acid Derivatives *via* Hydroxyl Radical. J Phys Chem B 2018; 122(19): 4956-71. [http://dx.doi.org/10.1021/acs.jpcb.7b12450] [PMID: 29676577]

[108] Cadet J, Davies KJA, Medeiros MH, Di Mascio P, Wagner JR. Formation and repair of oxidatively generated damage in cellular DNA. Free radical biology & medicine 2017; 13-34. [http://dx.doi.org/10.1016/j.freeradbiomed.2016.12.049]

[109] Van Houten B, Santa-Gonzalez GA, Camargo M. DNA repair after oxidative stress: current challenges. Curr Opin Toxicol 2018; 7: 9-16. [http://dx.doi.org/10.1016/j.cotox.2017.10.009] [PMID: 29159324]

[110] Dixon KB, Davies SS, Kirabo A. Dendritic cells and isolevuglandins in immunity, inflammation, and hypertension. Am J Physiol Heart Circ Physiol 2017; 312(3): H368-74. [http://dx.doi.org/10.1152/ajpheart.00603.2016] [PMID: 27986660]

[111] Luczaj W, Gegotek A, Skrzydlewska E. Antioxidants and HNE in redox homeostasis. Free radical biology & medicine 2017; 111: 87-101.

[112] Poole LB. The basics of thiols and cysteines in redox biology and chemistry. Free radical biology & medicine 2015; 80: 148-57. [http://dx.doi.org/10.1016/j.freeradbiomed.2014.11.013]

[113] Imlay JA. The molecular mechanisms and physiological consequences of oxidative stress: lessons from a model bacterium. Nat Rev Microbiol 2013; 11(7): 443-54.

[http://dx.doi.org/10.1038/nrmicro3032] [PMID: 23712352]

[114] Ezraty B, Gennaris A, Barras F, Collet JF. Oxidative stress, protein damage and repair in bacteria. Nat Rev Microbiol 2017; 15(7): 385-96.
[http://dx.doi.org/10.1038/nrmicro.2017.26] [PMID: 28420885]

[115] Tsuchiya Y, Zhyvoloup A, Baković J, *et al.* Protein CoAlation and antioxidant function of coenzyme A in prokaryotic cells. Biochem J 2018; 475(11): 1909-37.
[http://dx.doi.org/10.1042/BCJ20180043] [PMID: 29626155]

[116] Lu J, Holmgren A. The thioredoxin antioxidant system. Free radical biology & medicine 2014; 66: 75-87.
[http://dx.doi.org/10.1016/j.freeradbiomed.2013.07.036]

[117] Ukuwela AA, Bush AI, Wedd AG, Xiao Z. Glutaredoxins employ parallel monothiol-dithiol mechanisms to catalyze thiol-disulfide exchanges with protein disulfides. Chem Sci (Camb) 2017; 9(5): 1173-83.
[http://dx.doi.org/10.1039/C7SC04416J] [PMID: 29675162]

[118] Zámocký M, Gasselhuber B, Furtmüller PG, Obinger C. Molecular evolution of hydrogen peroxide degrading enzymes. Arch Biochem Biophys 2012; 525(2): 131-44.
[http://dx.doi.org/10.1016/j.abb.2012.01.017] [PMID: 22330759]

[119] Schatzman SS, Culotta VC. Chemical Warfare at the Microorganismal Level: A Closer Look at the Superoxide Dismutase Enzymes of Pathogens. ACS Infect Dis 2018; 4(6): 893-903.
[http://dx.doi.org/10.1021/acsinfecdis.8b00026] [PMID: 29517910]

[120] Pattison DI, Davies MJ. Actions of ultraviolet light on cellular structures. EXS 2006; (96): 131-57.
[http://dx.doi.org/10.1007/3-7643-7378-4_6] [PMID: 16383017]

[121] Baptista MS, Cadet J, Di Mascio P, *et al.* Type I and Type II Photosensitized Oxidation Reactions: Guidelines and Mechanistic Pathways. Photochem Photobiol 2017; 93(4): 912-9.
[http://dx.doi.org/10.1111/php.12716] [PMID: 28084040]

[122] Koturbash I. 2017 Michael Fry Award Lecture When DNA is Actually Not a Target: Radiation Epigenetics as a Tool to Understand and Control Cellular Response to Ionizing Radiation. Radiat Res 2018; 190(1): 5-11.
[http://dx.doi.org/10.1667/RR15027.1] [PMID: 29697303]

[123] Marchetti B, Karsili TN. Theoretical insights into the photo-protective mechanisms of natural biological sunscreens: building blocks of eumelanin and pheomelanin. Phys Chem Chem Phys 2016; 18(5): 3644-58.
[http://dx.doi.org/10.1039/C5CP06767G] [PMID: 26753793]

[124] Pobegalov G, Cherevatenko G, Alekseev A, *et al.* Deinococcus radiodurans RecA nucleoprotein filaments characterized at the single-molecule level with optical tweezers. Biochem Biophys Res Commun 2015; 466(3): 426-30.
[http://dx.doi.org/10.1016/j.bbrc.2015.09.042] [PMID: 26365346]

[125] Singh OV, Gabani P. Extremophiles: radiation resistance microbial reserves and therapeutic implications. J Appl Microbiol 2011; 110(4): 851-61.
[http://dx.doi.org/10.1111/j.1365-2672.2011.04971.x] [PMID: 21332616]

[126] Komarova EA, Chernov MV, Franks R, *et al.* Transgenic mice with p53-responsive lacZ: p53 activity varies dramatically during normal development and determines radiation and drug sensitivity *in vivo.* EMBO J 1997; 16(6): 1391-400.
[http://dx.doi.org/10.1093/emboj/16.6.1391] [PMID: 9135154]

[127] Horneck G, Klaus DM, Mancinelli RL. Space microbiology. Microbiol Mol Biol Rev 2010; 74(1): 121-56.
[http://dx.doi.org/10.1128/MMBR.00016-09] [PMID: 20197502]

[128] Sengupta P, Garrity P. Sensing temperature. Curr Biol 2013; 23(8): R304-7.

[http://dx.doi.org/10.1016/j.cub.2013.03.009] [PMID: 23618661]

[129] Schumann W. Thermosensor systems in eubacteria. Adv Exp Med Biol 2012; 739: 1-16. [http://dx.doi.org/10.1007/978-1-4614-1704-0_1] [PMID: 22399392]

[130] Gomez-Pastor R, Burchfiel ET, Thiele DJ. Regulation of heat shock transcription factors and their roles in physiology and disease. Nat Rev Mol Cell Biol 2018; 19(1): 4-19. [http://dx.doi.org/10.1038/nrm.2017.73] [PMID: 28852220]

[131] Meyer S, Carlson PD, Lucks JB. Characterizing the Structure-Function Relationship of a Naturally Occurring RNA Thermometer. Biochemistry 2017; 56(51): 6629-38. [http://dx.doi.org/10.1021/acs.biochem.7b01170] [PMID: 29172455]

[132] Richter K, Haslbeck M, Buchner J. The heat shock response: life on the verge of death. Mol Cell 2010; 40(2): 253-66. [http://dx.doi.org/10.1016/j.molcel.2010.10.006] [PMID: 20965420]

[133] Park CJ, Seo YS. Heat Shock Proteins: A Review of the Molecular Chaperones for Plant Immunity. Plant Pathol J 2015; 31(4): 323-33. [http://dx.doi.org/10.5423/PPJ.RW.08.2015.0150] [PMID: 26676169]

[134] Radons J. The human HSP70 family of chaperones: where do we stand? Cell stress & chaperones 2016; 21(3): 379-404. [http://dx.doi.org/10.1007/s12192-016-0676-6]

[135] Jacob P, Hirt H, Bendahmane A. The heat-shock protein/chaperone network and multiple stress resistance. Plant Biotechnol J 2017; 15(4): 405-14. [http://dx.doi.org/10.1111/pbi.12659] [PMID: 27860233]

[136] Rodrigues DF, Tiedje JM. Coping with our cold planet. Appl Environ Microbiol 2008; 74(6): 1677-86. [http://dx.doi.org/10.1128/AEM.02000-07] [PMID: 18203855]

[137] Fujita T, Liu Y, Higashitsuji H, *et al.* Involvement of TRPV3 and TRPM8 ion channel proteins in induction of mammalian cold-inducible proteins. Biochem Biophys Res Commun 2018; 495(1): 935-40. [http://dx.doi.org/10.1016/j.bbrc.2017.11.136] [PMID: 29175331]

[138] Roilo M, Kullmann MK, Hengst L. Cold-inducible RNA-binding protein (CIRP) induces translation of the cell-cycle inhibitor p27Kip1. Nucleic Acids Res 2018; 46(6): 3198-210. [http://dx.doi.org/10.1093/nar/gkx1317] [PMID: 29361038]

[139] Zhang Y, Burkhardt DH, Rouskin S, Li GW, Weissman JS, Gross CA. A Stress Response that Monitors and Regulates mRNA Structure Is Central to Cold Shock Adaptation. Mol Cell 2018; 70(2): 274-286.e7. [http://dx.doi.org/10.1016/j.molcel.2018.02.035] [PMID: 29628307]

[140] Devaux L, Sleiman D, Mazzuoli MV, *et al.* Cyclic di-AMP regulation of osmotic homeostasis is essential in Group B Streptococcus. PLoS Genet 2018; 14(4)e1007342 [http://dx.doi.org/10.1371/journal.pgen.1007342] [PMID: 29659565]

[141] Pliotas C, Dahl AC, Rasmussen T, *et al.* The role of lipids in mechanosensation. Nat Struct Mol Biol 2015; 22(12): 991-8. [http://dx.doi.org/10.1038/nsmb.3120] [PMID: 26551077]

[142] Zimmerman CA, Lin YC, Leib DE, *et al.* Thirst neurons anticipate the homeostatic consequences of eating and drinking. Nature 2016; 537(7622): 680-4. [http://dx.doi.org/10.1038/nature18950] [PMID: 27487211]

[143] Gizowski C, Bourque CW. The neural basis of homeostatic and anticipatory thirst. Nat Rev Nephrol 2018; 14(1): 11-25. [http://dx.doi.org/10.1038/nrneph.2017.149] [PMID: 29129925]

[144] Hoffmann EK, Lambert IH, Pedersen SF. Physiology of cell volume regulation in vertebrates. Physiol

Rev 2009; 89(1): 193-277.
[http://dx.doi.org/10.1152/physrev.00037.2007] [PMID: 19126758]

[145] Gunde-Cimerman N, Plemenitaš A, Oren A. Strategies of adaptation of microorganisms of the three domains of life to high salt concentrations. FEMS Microbiol Rev 2018; 42(3): 353-75. [http://dx.doi.org/10.1093/femsre/fuy009] [PMID: 29529204]

[146] Yancey PH, Clark ME, Hand SC, Bowlus RD, Somero GN. Living with water stress: evolution of osmolyte systems. Science 1982; 217(4566): 1214-22. [http://dx.doi.org/10.1126/science.7112124] [PMID: 7112124]

[147] Pham QD, Wolde-Kidan A, Gupta A, *et al.* Effects of Urea and TMAO on Lipid Self-Assembly under Osmotic Stress Conditions. J Phys Chem B 2018; 122(25): 6471-82. [http://dx.doi.org/10.1021/acs.jpcb.8b02159] [PMID: 29693387]

[148] Yancey PH, Gerringer ME, Drazen JC, Rowden AA, Jamieson A. Marine fish may be biochemically constrained from inhabiting the deepest ocean depths. Proc Natl Acad Sci USA 2014; 111(12): 4461-5. [http://dx.doi.org/10.1073/pnas.1322003111] [PMID: 24591588]

[149] Buda R, Liu Y, Yang J, *et al.* Dynamics of *Escherichia coli's* passive response to a sudden decrease in external osmolarity. Proc Natl Acad Sci USA 2016; 113(40): E5838-46. [http://dx.doi.org/10.1073/pnas.1522185113] [PMID: 27647888]

[150] Yancey PH. Organic osmolytes as compatible, metabolic and counteracting cytoprotectants in high osmolarity and other stresses. J Exp Biol 2005; 208(Pt 15): 2819-30. [http://dx.doi.org/10.1242/jeb.01730] [PMID: 16043587]

[151] Magazù S, Migliardo F, Gonzalez MA, Mondelli C, Parker SF, Vertessy BG. Molecular mechanisms of survival strategies in extreme conditions. Life (Basel) 2012; 2(4): 364-76. [http://dx.doi.org/10.3390/life2040364] [PMID: 25371270]

[152] Czech L, Hermann L, Stöveken N, *et al.* Role of the Extremolytes Ectoine and Hydroxyectoine as Stress Protectants and Nutrients: Genetics, Phylogenomics, Biochemistry, and Structural Analysis. Genes (Basel) 2018; 9(4)E177
[http://dx.doi.org/10.3390/genes9040177] [PMID: 29565833]

[153] Blum A. Osmotic adjustment is a prime drought stress adaptive engine in support of plant production. Plant, cell & environment 2017; 40(1): 4-10.
[http://dx.doi.org/10.1111/pce.12800]

[154] Krulwich TA, Sachs G, Padan E. Molecular aspects of bacterial pH sensing and homeostasis. Nat Rev Microbiol 2011; 9(5): 330-43.
[http://dx.doi.org/10.1038/nrmicro2549] [PMID: 21464825]

[155] Rash LD. Acid-Sensing Ion Channel Pharmacology, Past, Present, and Future.... Adv Pharmacol 2017; 79: 35-66.
[http://dx.doi.org/10.1016/bs.apha.2017.02.001] [PMID: 28528673]

[156] Chesler M. Regulation and modulation of pH in the brain. Physiol Rev 2003; 83(4): 1183-221. [http://dx.doi.org/10.1152/physrev.00010.2003] [PMID: 14506304]

[157] Lund P, Tramonti A, De Biase D. Coping with low pH: molecular strategies in neutralophilic bacteria. FEMS Microbiol Rev 2014; 38(6): 1091-125.
[http://dx.doi.org/10.1111/1574-6976.12076] [PMID: 24898062]

[158] De Biase D, Lund PA. The *Escherichia coli* Acid Stress Response and Its Significance for Pathogenesis. Adv Appl Microbiol 2015; 92: 49-88.
[http://dx.doi.org/10.1016/bs.aambs.2015.03.002] [PMID: 26003933]

[159] Fingermann M, Hozbor D. Acid tolerance response of Bordetella bronchiseptica in avirulent phase. Microbiol Res 2015; 181: 52-60.
[http://dx.doi.org/10.1016/j.micres.2015.09.001] [PMID: 26640052]

[160] Nyanga-Koumou AP, Ouoba LI, Kobawila SC, Louembe D. Response mechanisms of lactic acid bacteria to alkaline environments: a review. Crit Rev Microbiol 2012; 38(3): 185-90. [http://dx.doi.org/10.3109/1040841X.2011.640978] [PMID: 22168378]

[161] Tsai HH, Schmidt W. Mobilization of Iron by Plant-Borne Coumarins. Trends Plant Sci 2017; 22(6): 538-48. [http://dx.doi.org/10.1016/j.tplants.2017.03.008] [PMID: 28385337]

[162] Carvalho CM, Chew EH, Hashemy SI, Lu J, Holmgren A. Inhibition of the human thioredoxin system. A molecular mechanism of mercury toxicity. J Biol Chem 2008; 283(18): 11913-23. [http://dx.doi.org/10.1074/jbc.M710133200] [PMID: 18321861]

[163] Liu Y, Imlay JA. Cell death from antibiotics without the involvement of reactive oxygen species. Science 2013; 339(6124): 1210-3. [http://dx.doi.org/10.1126/science.1232751] [PMID: 23471409]

[164] Cho H, Uehara T, Bernhardt TG. Beta-lactam antibiotics induce a lethal malfunctioning of the bacterial cell wall synthesis machinery. Cell 2014; 159(6): 1300-11. [http://dx.doi.org/10.1016/j.cell.2014.11.017] [PMID: 25480295]

[165] Crofts TS, Wang B, Spivak A, *et al.* Shared strategies for β-lactam catabolism in the soil microbiome. Nat Chem Biol 2018; 14(6): 556-64. [http://dx.doi.org/10.1038/s41589-018-0052-1] [PMID: 29713061]

[166] Kohanski MA, Dwyer DJ, Hayete B, Lawrence CA, Collins JJ. A common mechanism of cellular death induced by bactericidal antibiotics. Cell 2007; 130(5): 797-810. [http://dx.doi.org/10.1016/j.cell.2007.06.049] [PMID: 17803904]

[167] Dong TG, Dong S, Catalano C, Moore R, Liang X, Mekalanos JJ. Generation of reactive oxygen species by lethal attacks from competing microbes. Proc Natl Acad Sci USA 2015; 112(7): 2181-6. [http://dx.doi.org/10.1073/pnas.1425007112] [PMID: 25646446]

[168] Dwyer DJ, Collins JJ, Walker GC. Unraveling the physiological complexities of antibiotic lethality. Annu Rev Pharmacol Toxicol 2015; 55: 313-32. [http://dx.doi.org/10.1146/annurev-pharmtox-010814-124712] [PMID: 25251995]

[169] Kalinovskaya NI, Romanenko LA, Kalinovsky AI. Antibacterial low-molecular-weight compounds produced by the marine bacterium Rheinheimera japonica KMM 9513^{T}. Antonie van Leeuwenhoek 2017; 110(5): 719-26. [http://dx.doi.org/10.1007/s10482-017-0839-1] [PMID: 28176144]

[170] Moran JC, Alorabi JA, Horsburgh MJ. Comparative Transcriptomics Reveals Discrete Survival Responses of *S. aureus* and *S. epidermidis* to Sapienic Acid. Front Microbiol 2017; 8: 33. [http://dx.doi.org/10.3389/fmicb.2017.00033] [PMID: 28179897]

[171] Zhang D, de Souza RF, Anantharaman V, Iyer LM, Aravind L. Polymorphic toxin systems: Comprehensive characterization of trafficking modes, processing, mechanisms of action, immunity and ecology using comparative genomics. Biol Direct 2012; 7: 18. [http://dx.doi.org/10.1186/1745-6150-7-18] [PMID: 22731697]

[172] Zasloff M. Antimicrobial peptides of multicellular organisms. Nature 2002; 415(6870): 389-95. [http://dx.doi.org/10.1038/415389a] [PMID: 11807545]

[173] Rodríguez-Rojas A, Makarova O, Rolff J. Antimicrobials, stress and mutagenesis. PLoS Pathog 2014; 10(10)e1004445 [http://dx.doi.org/10.1371/journal.ppat.1004445] [PMID: 25299705]

[174] Guilhelmelli F, Vilela N, Albuquerque P, Derengowski LdaS, Silva-Pereira I, Kyaw CM. Antibiotic development challenges: the various mechanisms of action of antimicrobial peptides and of bacterial resistance. Front Microbiol 2013; 4: 353. [http://dx.doi.org/10.3389/fmicb.2013.00353] [PMID: 24367355]

[175] Lahiani A, Yavin E, Lazarovici P. The Molecular Basis of Toxins' Interactions with Intracellular Signaling *via* Discrete Portals. Toxins (Basel) 2017; 9(3)E107 [http://dx.doi.org/10.3390/toxins9030107] [PMID: 28300784]

[176] Shatalin K, Shatalina E, Mironov A, Nudler E. H2S: a universal defense against antibiotics in bacteria. Science 2011; 334(6058): 986-90. [http://dx.doi.org/10.1126/science.1209855] [PMID: 22096201]

[177] Luhachack L, Nudler E. Bacterial gasotransmitters: an innate defense against antibiotics. Curr Opin Microbiol 2014; 21: 13-7. [http://dx.doi.org/10.1016/j.mib.2014.06.017] [PMID: 25078319]

[178] Agathokleous E, Kitao M, Calabrese EJ. Environmental hormesis and its fundamental biological basis: Rewriting the history of toxicology. Environ Res 2018; 165: 274-8. [http://dx.doi.org/10.1016/j.envres.2018.04.034] [PMID: 29734028]

[179] Andersson DI, Hughes D. Microbiological effects of sublethal levels of antibiotics. Nat Rev Microbiol 2014; 12(7): 465-78. [http://dx.doi.org/10.1038/nrmicro3270] [PMID: 24861036]

[180] Metcalf JA, Funkhouser-Jones LJ, Brileya K, Reysenbach AL, Bordenstein SR. Antibacterial gene transfer across the tree of life. eLife 2014; 3 [http://dx.doi.org/10.7554/eLife.04266] [PMID: 25422936]

[181] Park C, Jeong J. Synergistic cellular responses to heavy metal exposure: A minireview. Biochim Biophys Acta, Gen Subj 2018; 1862(7): 1584-91. [http://dx.doi.org/10.1016/j.bbagen.2018.04.003] [PMID: 29631058]

[182] Ciccia A, Elledge SJ. The DNA damage response: making it safe to play with knives. Mol Cell 2010; 40(2): 179-204. [http://dx.doi.org/10.1016/j.molcel.2010.09.019] [PMID: 20965415]

[183] Baharoglu Z, Mazel D. SOS, the formidable strategy of bacteria against aggressions. FEMS Microbiol Rev 2014; 38(6): 1126-45. [http://dx.doi.org/10.1111/1574-6976.12077] [PMID: 24923554]

[184] Shin DS, Pratt AJ, Tainer JA. Archaeal genome guardians give insights into eukaryotic DNA replication and damage response proteins Archaea 2014; in press(206735):206735. [http://dx.doi.org/10.1155/2014/206735]

[185] Friedberg EC, Aguilera A, Gellert M, *et al.* DNA repair: from molecular mechanism to human disease. DNA Repair (Amst) 2006; 5(8): 986-96. [http://dx.doi.org/10.1016/j.dnarep.2006.05.005] [PMID: 16955546]

[186] Marnef A, Legube G. Organizing DNA repair in the nucleus: DSBs hit the road. Curr Opin Cell Biol 2017; 46: 1-8. [http://dx.doi.org/10.1016/j.ceb.2016.12.003] [PMID: 28068556]

[187] Ranjha L, Howard SM, Cejka P. Main steps in DNA double-strand break repair: an introduction to homologous recombination and related processes. Chromosoma 2018; 127(2): 187-214. [http://dx.doi.org/10.1007/s00412-017-0658-1] [PMID: 29327130]

[188] Williams AB, Schumacher B. DNA damage responses and stress resistance: Concepts from bacterial SOS to metazoan immunity. Mech Ageing Dev 2017; 165(Pt A): 27-32.

CHAPTER 2

Pathogens

Abstract: Pathogens live inside other organisms called hosts, exploiting them and causing disease. Viruses and other mobile genetic elements such as transposons are pathogens that parasite on all life forms. They are obligate pathogens since they replicate only inside their hosts. Many bacterial pathogens are facultative since they can live outside their host. Viruses recognize host cells by specific molecules that serve as viral receptors. Only cells expressing them can be infected. Multicellular organisms host pathogenic bacteria and eukaryotic pathogens such as fungi, protozoa, and helminths in addition to viruses. Infection depends on pathogen's ability to survive inside the host. Some pathogens quickly kill their host with large numbers of new pathogens released, whereas others establish long-lasting infection. Lytic viruses replicate and then lyse their host to get released. Other viruses do not lyse their host; instead, they establish chronic infection by replicating slowly and exiting the host cells by extrusion. Temperate viruses integrate into host DNA and stay silent for many cell divisions until the cell experiences stress. Then, the lytic program is activated, and the virus replicates and kills the cell. Similar to viruses, some intracellular bacterial pathogens are obligate and can only replicate inside their host. Pathogens differ in their host range; some can infect only a single species, while others are generalists. Hosts constantly evolve mechanisms preventing infection and destroying pathogens. In turn, pathogens evolve to overcome host defenses.

Keywords: Bacteria, Fungi, Helminths, Infection, Iytic virus, Mobile genetic elements, Pathogen, Protozoan pathogens, Symbiosis, Transposon, Temperate virus, Virulence, Virus.

INTRODUCTION

Every organism is a source of food for other organisms, so parasitism and predation likely have deep roots in life history. Pathogens are organisms that live inside other organisms called hosts and cause disease to them [1]. Viruses are pathogens that infect all life forms. A great variety of other pathogens from bacteria to fungi and worms infect multicellular organisms.

In order to successfully colonize other organisms, pathogens need to have means to avoid recognition by the host immune system. In addition, life inside other organisms is associated with various stresses. The human body, for example, has a higher temperature than its usual environment, so pathogens that come from such

Nadia Danilova

an environment need to overcome the stress of heat. The human immune system uses ROS and acidic pH to kill pathogens, so in order to survive, they need to handle well oxidative and pH stress.

The severity of disease caused by a pathogen depends on its virulence. Virulence factors are proteins and other molecules that ensure successful colonization of the host, suppression or manipulation of the host immune system, manipulation of the host metabolism (for example, to help to scavenge nutrients within the host), and organization of the pathogen's exit from the host. The pathogenic potential of an organism is revealed only in its interaction with its host. Moreover, the same infection can have a different outcome in different individuals.

Pathogens are divided into obligate and facultative. Obligate pathogens can complete their life cycle only inside the host. All viruses are obligate pathogens since they are able to replicate only inside their host cells. Certain bacteria are also obligate pathogens. They include intracellular bacteria that, like viruses, can reproduce only inside host cells, *e.g. Rickettsia* that causes typhus and bacteria of phylum *Chlamydia.* Bacteria that cause tuberculosis and syphilis are also obligate pathogens. Some human obligate pathogens such as Zika virus have intermediate invertebrate hosts like mosquitoes. The natural environment also can serve as a reservoir for pathogens. For instance, Anthrax spores remain infectious in soils for decades [2].

Facultative pathogens live outside the host and cause infection only occasionally in a susceptible host. So-called opportunistic pathogens cause disease only in an injured or immunocompromised host. Among them, colonizing opportunistic pathogens such as *Candida albicans, Staphylococcus aureus,* and *Escherichia coli* can asymptomatically colonize the human body and cause disease only when the body is weakened [3].

Pathogens cause disease by damaging host cells during replication and by releasing toxins. For example, a toxin produced by *Vibrio cholerae* causes exhausting diarrhea. Overactivation of the immune system can also cause considerable damage to the organism [4]. 1918 influenza pandemic preferentially killed young adults with a peak of mortality at age 28, which was explained by the immune system overactivation [5].

Some pathogens are species-specific; others are generalists and can infect a broad range of hosts. For example, *Shigella flexneri* infects a small group of related primate species, *Yersinia pestis*, the agent of plague, infects all mammals, whereas *Pseudomonas aeruginosa* can cause disease in plants as well as animals [4].

About 1400 pathogens can infect humans. It is necessary to note that most microbial species have no negative effect on human health and many are beneficial. Recent estimates suggest that the total number of microorganisms in a typical human body is in the same order ($3.8x10^{13}$) as the number of human cells ($3.0x10^{13}$) [6]. Only a tiny fraction of microorganisms that reside in the human body are pathogens. Despite this fact, during human history, death from infectious diseases far exceeded all other causes of death [4]. In this way, pathogens strongly influenced human evolution and history. Infectious diseases played a major role in human migrations, outcomes of wars, and rise and fall of civilizations. Smallpox, which is caused by the variola virus is one example. Originated > 4000 years ago in Africa, it killed pharaoh Ramses V in 1157 B. C., killed Europeans at the rate of 500 million/century between 1400 and 1800, and even in the 20th century killed between 300 and 500 million people worldwide [7]. Discovery of vaccination by Jenner in 1798 saved millions and eventually culminated in the eradication of smallpox declared by the World Health Organization in 1980.

Pathogens constantly evolve to overcome host defenses while hosts evolve to keep pathogens away. Leigh Van Valen called this request for constant change just to stay alive in pathogens-hosts co-evolution a Red Queen's race [8]. Red Queen in Lewis Carroll's book "Through the looking glass" said to Alice that in her land "it takes all the running you can do to keep in the same place".

Pathogens have shaped the evolution of the human genome. Viruses are major drivers of human protein evolution. In human proteome, conserved with other mammals, up to 30% of all amino acid changes may have been driven by viruses [9]. Genes involved in immune response and pathogen genes often co-evolve [10].

Some human pathogens such as *Helicobacter pylori* cospeciated with their human host for a long time, while others, such as measles and HIV jumped on from animals or from the wider environment relatively recently [4]. New pathogens continue to evolve from commensal species and from pathogens of other species through mutations and transfer of virulence genes [11 - 13]. Ebola virus is one of the deadliest emerging infections, which caused a massive outbreak in 2014 when more than 11000 died [14].

2.1. Viruses

Viruses are the most abundant biological entities on earth. In many marine and soil environments, the number of virus particles is 10–100 times greater than the number of cells [15]. Viruses are also very diverse and carry many novel, uncharacterized genes.

Three domains of life: Bacteria, Eukarya, and Archaea, have different viruses.

Those infecting Bacteria are called “bacteriophages,” or “phages”. They comprise the majority of viruses with new species that are being discovered at a high rate [16]. More than 200 viruses cause diseases in humans from a relatively rare AIDS to a common cold, which affects millions.

Viral particles or virions are composed of genetic material, which can be either single-stranded (ss) or double-stranded (ds)RNA or DNA, a protein coat called capsid, and in some cases an envelope of lipids. Viral genomes range in size from less than 2 kb to over 2.5 Mb, and virion size ranges from 1/100 of an average bacterium to a size comparable to it [17].

Viruses are currently classified according to the International Committee on Taxonomy of Viruses (ICTV) (http://www.ictvdb.org) scheme or by Baltimore scheme. The 2018 ICTV report classifies viruses into orders based on their evolutionary relationship [18]. ICTV orders are *Caudovirales* (tailed dsDNA, bacteria), *Herpesvirales* (dsDNA, eukaryotes), *Ligamenvirales* (dsDNA, archaea), *Mononegavirales* (-ssRNA, plant and animal), *Nidovirales* (+ssRNA), *Ortervirales* (ssRNA and ssDNA), *Picomavirales* (+ssRNA), *Tymovirales* (+ ssRNA, plants), *Bunyavirales* (-ssRNA). Baltimore classification places viruses into seven groups based on the nature of their genome and their replication strategy: I–dsDNA, II–ssDNA, III–dsRNA, IV– (+)ssRNA, V– (-)ssRNA, VI– (+)ssRNA that replicate through DNA intermediate, VII – dsDNA that replicate through ssRNA intermediate.

DNA viruses are replicated in the nucleus; some of them use host enzymatic machinery, some encode all necessary enzymes themselves. Giant viruses of amoebae encode a significant part of translational machinery except for ribosomes [19]. RNA viruses usually replicate in the cytoplasm. They encode their own replicases. Retroviruses that have ssRNA genome use reverse transcriptase to generate DNA intermediate that integrates into the host DNA as a provirus.

A special group comprises satellite viruses that depend on the presence of other viruses for propagation. Hepatitis delta virus has an RNA genome that does not code for the coat proteins. This virus can, however, use the coat proteins of hepatitis B virus; therefore it propagates only when the hepatitis B virus is also present in the cell. Virophages need the presence of the translational machinery of giant viruses to propagate in protozoan cells [20].

The relative abundance of various classes of viruses differs between prokaryotes and eukaryotes. In prokaryotes, most viruses have dsDNA genomes, with less abundant ssDNA viruses and only a few RNA viruses. On the contrary, in eukaryotes, including humans, a majority of viruses have RNA genomes [21].

Morphology of viral particles is also used to differentiate viruses. Viral capsids are assembled around the viral genome from several identical protein subunits and, therefore, typically have geometrical shapes. Helical and icosahedral viruses are the most frequent morphology. In helical viruses, ssRNA or ssDNA in some cases is coiled in a helix of a single type of protein capsomeres forming a rod that can be short and rigid or long and flexible. Filamentous virus particles such as Ebola virus can look like knots. Icosahedral viruses have a spherical symmetry; which requires at least twelve capsomeres composed of five protein subunits. Additional capsomeres are positioned at the apices and on the triangular faces. Many bacteriophages have prolate heads in the shape of an elongated icosahedron. Bacteriophage T4 also has a helical tail. Other complex shapes are reported; viruses infecting archaea have an especially large diversity of viral morphotypes, including linear, spindle-shaped, spherical, drop, and bottle-shaped [22].

After infection, the virus uses host cellular machinery to generate replicas of its genome and to produce capsid proteins that then self-assemble to form new viruses [23]. The first step in virus infection is an attachment to the cell surface, which can take place only if capsid proteins can bind strongly enough to cell surface proteins that serve as viral receptors. Receptors determine the range of hosts. It can be restricted to one cell type in one species or include many cell types in multiple species. Binding to a receptor induces receptor-mediated endocytosis or membrane fusion in viruses that have membranes. This is how a virus enters the cell. Bacterial cells have protective cell walls; therefore, many bacteriophages developed mechanisms for injecting their genome through the wall leaving the capsid outside.

Viruses differ in their interaction with the host cell. Some viruses replicate and lyse their host during release; others do not lyse their hosts but leave them by extrusion, thus establishing a chronic infection. T4 phage of *E. coli* is the most studied lytic virus [24]. Lytic cycle of T4 is very fast. T4 proteins injected together with the phage DNA block association of *E. coli* mRNAs with ribosomes and end synthesis of host proteins within one minute after infection. Next, T4 directs host metabolism to the production of phage progeny. It takes less than an hour from entry into the cell toward cell lysis and release of around 100 of new virions.

In contrast to lytic viruses, temperate viruses coexist with the host as a non-active provirus either integrated into the host DNA or as an independent cytoplasmic entity. Bacteria harboring virus are called lysogens. Ball and Bordet discovered them in 1925.

Phage lambda of *E. coli* is the most studied temperate virus [25 - 27]. Its infection can go along a lytic or lysogenic pathway (Fig. **2**). Most infections follow the lytic pathway. Alternatively, lambda phage can integrate its DNA into the host genome entering a lysogenic pathway.

The choice between lysogeny and lysis depends on several key genes. Transcription of viral genome starts with the 'immediate early' transcripts' that include *cro* gene responsible for the suppression of the lysogenic pathway. Then, the 'late early' viral transcripts follow, which include cII. It activates the promoters of the lysogenic pathway [28]. cII concentration determines the fate of the infection; high concentration of cII switches the lysogeny on.

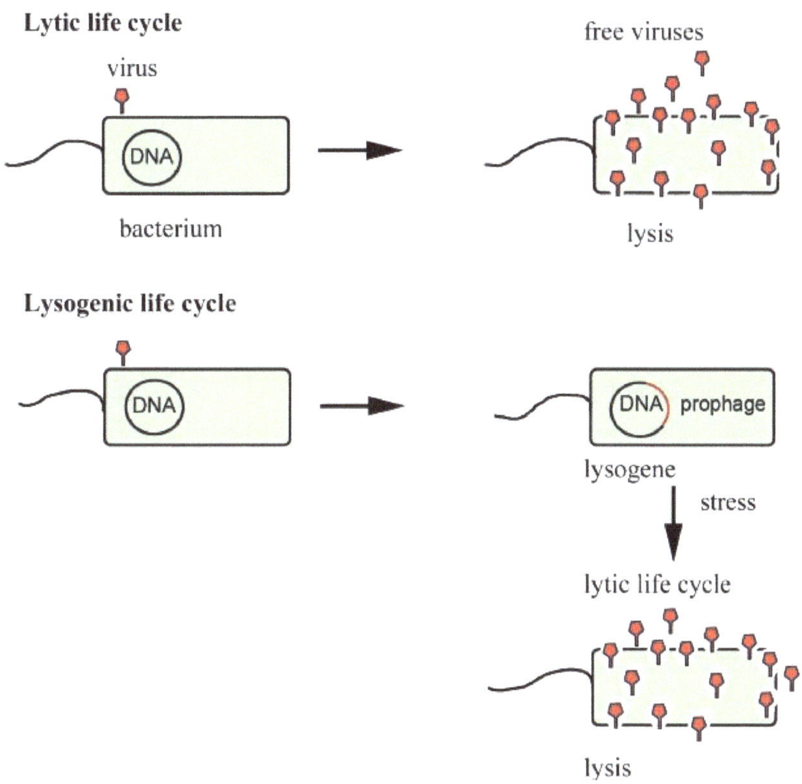

Fig. (2). Virus infection of bacteria can go by lytic or lysogenic pathway. Integrated prophages are replicated together with the cellular genome and remain quiescent. They can be reactivated by stress, which switches infection to a lytic pathway.

After integration of the virus into *E. coli* genome, the repressor protein Cl keeps it in a dormant state. Cl clams two distal operators OL and OR bending the intervening DNA into a loop [29, 30]. This lets CI suppress the lytic promoters and maintain its own expression [31]. Cro is an antagonist of Cl; it's binding to O(R)3 leads to prophage induction [32].

The lysogeny/lysis choice was initially explained by stochastic variations in concentrations of products of the above-mentioned genes in individual cells [33]. Later it was shown that the switch could be influenced by many factors including bacterial growth rate, starvation, and the number of viral particles that enter the cell [27, 34, 35]. The phage of the SPbeta group produces during infection of its Bacillus host a short communication peptide. In subsequent infection, phages recognize this peptide and lysogenize if its concentration is sufficiently high [36].

Several human viruses can establish latent infection through integration into chromosomal DNA or as an episome. Most people are infected with at least one type of herpes virus [37, 38]. Herpes viruses have tropism to neural cells. However, in ~1% of the human population, herpesvirus 6A and B are integrated into the germline, which means every cell has them [39]. Integrated viruses are replicated together with the cellular genome and can remain quiescent for a long time.

Virus integrated into a genome can influence the host in many ways. Often it promotes cell survival. For example, both productive and latent HIV-1 infection of human T cells activates cellular survival program by upregulation of an inhibitor of apoptosis BIRC5 [40]. Other effects on the host include enhancement of host metabolism, improvement of stress resistance, and prevention of infection by other viruses [41, 42]. Up to 20% of the bacterial genome might have originated from phages that lost the ability to propagate and even these cryptic prophages have an effect on the host's resistance to antibiotics and stress [43].

Proviruses can switch from latent to productive lytic infection in response to stress such as DNA damage, nucleolar stress, UV light, cold, *etc.* [44, 45]. Multiple latent viruses reactivate in astronauts during space shuttle missions [46]. The exact mechanisms that reactivate proviruses remain largely unknown. The general response to many stresses is cell cycle arrest, which may play a role in switching proviruses to the lytic program [47]. Dormant cells are protected from provirus induction [48].

Quiescent proviruses express a limited amount of viral products and therefore are 'invisible' to the immune system. This allows long-term persistence of human infections caused by herpes viruses and HIV [49]. Antiretroviral therapy also targets only active viruses. During HIV infection, the virus stays latent in a small number of resting T cells and in macrophages, which means the viremia can rebound as soon as therapy stops. This problem is a central obstacle on the way to completely cure HIV infected people. A new class of therapeutics called latency-reversing agents emerges that activate quiescent proviruses [50 - 52]. This is a part of the "shock and kill" strategy. The virus is activated and targeted by antiviral drugs, and then virus-producing cells are killed by the immune system.

Viruses create diversity in bacterial and archaeal genomes by moving adaptive genes between microbial communities through horizontal gene transfer [53]. Eukaryotic RNA viruses also evolve by incorporation of genes from diverse hosts. For example, new genes can be brought to vertebrate viruses through horizontal gene transfer from nematodes or arthropods [54]. Half of the genome of Vaccinia virus is devoted to antagonizing innate immune responses [55].

2.2. Viroids and Prions

Living forms have a fascinating ability for self-assembly and self-organization, which is manifested not only in viruses but even in living macromolecules as seen in the examples of viroids and prions. Viroids are infection agents of plants composed solely of small circular RNAs that interact with host cell and use the host RNA polymerase II for self-replication. The first viroid, a potato pathogen, was discovered in 1971 by Theodor Diener, who later suggested that viroids are the "living fossils" of the pre-cellular RNA world [56]. In accord with this hypothesis, many viroids are ribozimes.

Another example of a living infectious macromolecule is a prion. This term denotes a small proteinaceous infectious particle [57]. Prion is a protein that can adopt an alternative fold that induces conversion of the existing, properly folded proteins into the misfolded prion form. Prion form thus becomes self-propagating. Prions aggregate into amyloids. Amyloid fibrils grow at the ends and replicate when break. Prion aggregates are extremely stable. After infection, accumulation of prions in the altered conformation leads to polyubiquitination, activation of the complement system, cell death, and tissue damage. Prions are the cause of several diseases such as scrapie in sheep, bovine spongiform encephalopathy (mad cow disease) and Kuru and Creutzfeldt-Jakob disease in humans. Recent evidence suggests that synucleinopathies including Parkinson's disease, Lewy body dementia, multiple system atrophy, and others also have prion-like mechanisms [58].

2.3. Mobile Genetic Elements

Mobile genetic elements (MGEs) are pieces of genetic material ranging from one to tens of thousands of base pairs capable of moving within genomes [59]. In 1948, Barbara McClintock discovered increased moving of such elements in the maize genome in response to stress. Movement of MGEs in response to stress can modify genes and change gene regulation; these changes in some cases can be adaptive. McClintock was awarded the Nobel prize in 1983 for the discovery of MGEs and their role in species evolution [60]. MGEs are common in both prokaryotic and eukaryotic genomes including humans. More than half of the human genome derives from mobile genetic elements [61]. A variety of mobile genetic elements has been described including transposons, plasmids, insertion sequences, genomic islands, and introns. Some of MGEs encode all enzymes necessary for transposition and are called autonomous elements. Non-autonomous elements rely on the enzymes from the autonomous elements. Some MGEs use DNA-dependent DNA synthesis to replicate themselves while others use reverse transcriptase to transcribe their RNA into DNA.

2.3.1. Insertion Sequences

Insertion sequences or insertion-sequence (IS) elements are small segments of host DNA that can move around the genome by transposition. They are flanked by inverted terminal repeats recognized by a transposase during transposition. IS elements usually encode their transposases. They can cause large-scale rearrangements in their hosts [62]. IS can exist autonomously or as a part of complex transposons.

2.3.2. Transposons

Transposons carry genes necessary for moving within the genome and usually additional genes coding for functions other than transposition, *e.g.* antibiotic resistance genes. Some transposons are flanked by inverted terminal repeats necessary for transposition whereas composite transposons are flanked on both sides by IS elements that mediate their transposition. Some transposons rely on Cas1 protein for integration and excision [63].

Class I transposons or retrotransposons are first transcribed into RNA intermediate and then reverse transcribed into complementary DNA, which then is inserted into another position in the genome. Class I retrotransposons are essentially a form of endogenous viruses. Class II elements move directly from one position in the genome to another by cut-and-paste mechanism catalyzed by transposases. Most DNA of the human genome is derived from transposons whereas protein-coding genes comprise only 1.5% [61].

Class I transposons include transposable elements with long terminal repeats (LTRs), LINEs, and SINEs elements. LTR retrotransposons encode reverse transcriptases that mediate their integration into the genome. A subset of these elements originated from retroviruses. Elements with LTRs make up 8.3% of the human genome [61]. One of the recently acquired elements is HERVK(HML-2). It is expressed in preimplantation blastocysts [64]. Its products upregulate expression of an antiviral protein IFITM1 [64].

LINEs (long interspersed nuclear elements) have an internal promoter for RNA polymerase II and encode reverse transcriptase. During transposition, LINE RNA in complex with proteins moves to the nucleus and assembles on DNA. Then, an endonuclease makes a single-strand nick and the reverse transcriptase uses the nicked DNA to prime reverse transcription of the LINE RNA. Often, such reverse transcription fails to generate a full-length product and most LINE-derived repeats are truncated. LINEs make up 21% of the human genome [61]. From 1.5 million human LINE-1 elements about 100 are still fully functional. LINE-1 retrotransposons are expressed and mobile in germline and early embryo but are silenced in somatic tissues by epigenetic and post-transcriptional mechanisms [65]. Their activity in germline and embryo introduces individual genome variations that in some cases may lead to genetic diseases. LINE-1 can also be activated in some pathological condition such as tumorigenesis or HIV infection.

LINE-1 can retrotranspose not only its own RNA to new genomic locations but also other RNAs such as cellular RNAs, which may lead to mutations and contribute to cancer and other diseases [66].

SINEs (short interspersed nuclear elements) have a promoter for RNA polymerase III, which transcribes them into RNA intermediates. SINEs are non-autonomous elements since they do not encode reverse transcriptases but are retrotransposed by active elements like LINEs. SINEs are especially common in mammals. More than a million Alu-SINE copies populate the human genome making up to 15–17% of the genome with Alu family especially abundant [61].

Transposable elements often have specific patterns of integration and can change the expression of genes. For example, Tf1, a long-terminal repeat retrotransposon in *Schizosaccharomyces pombe*, integrates into promoters with a preference for the promoters of stress response genes. It increases their expression by inserting enhancer activity [67]. Another retrotransposon, Tf2, is activated in response to changes in the environmental level of oxygen and direct expression of adjacent genes [68].

Transposons are regulated by epigenetic mechanisms such as DNA methylation and histone modifications. Stress can lead to changes in DNA methylation and loss of epigenetic control, which activates transposition [69].

2.3.3. Plasmids

Plasmids are DNA molecules within cells, distinct from the chromosomal DNA that can replicate autonomously [70]. They are most often found in bacteria as small circular dsDNAs but they can also be integrated into chromosomal DNA. Plasmids vary from 1 to 200 Kbp in lengths and can be present in cells in various numbers. When cells divide, each daughter cell inherits plasmids as well. Some archaea and eukaryotes also have plasmids [71].

Conjugative plasmids such as fertility factor F in *E. coli* encode genes that mediate sexual conjugation between cells. During conjugation, plasmids are transferred from one cell to another. If a conjugative plasmid was previously integrated into the chromosomal DNA, it can carry pieces from that DNA. Cells can also pick up plasmids from the environment. Therefore, similar to viruses, plasmids can serve as vehicles for horizontal gene transfer.

Plasmids often carry information useful for the host such as genes for antibiotic resistance, genes that code for bacteriocins, genes necessary for utilization of unconventional nutrients, or genes needed for the degradation of toxins [72]. Toxin-antitoxin (TA) pairs are also often encoded by plasmids. A stable toxin inhibits one of the basic cellular processes. Antitoxins are unstable and need to be constantly produced to compensate for their degradation. If a daughter cell does not inherit a plasmid encoding antitoxin, the remaining toxin will kill it. This serves as a mechanism of plasmid maintenance.

2.3.4. Group I and Group II Introns

Group I and group II introns are intervening sequences that act as self-splicing ribozymes. Group I introns invade tRNA, rRNA, and protein-coding genes [73 - 75]. Some group I introns that encode homing endonucleases are considered to be mobile genetic elements. Group II introns are both catalytic RNAs capable of self-splicing and retrotransposons [76]. They are thought to be evolutionary ancestors of the eukaryotic spliceosome. A fully functional group II intron forms a ribonucleoprotein complex with the intron-encoded protein, which can reverse transcribe the intron RNA into the target DNA [77].

2.3.5. Homing Endonucleases

Homing endonucleases (HEs) are DNAses encoded within self-splicing elements [78]. HEs catalyze the hydrolysis of a degenerated recognition sequence of 12-40 bp, which are rare in the genome. The repair of the HE cut by recombination results in copying of the gene encoding the homing endonuclease from the intact allele into the cleavage site. In this way, HEs make the sequences that encode them mobile. If only one allele carries HE, the other has the intact recognition sequence and therefore is the major HE target.

2.3.6. Chromosomal Islands

Chromosomal islands (CI) are chromosomal regions that are originated by horizontal gene transfer and differ from the rest of the genome. They often carry a specific set of genes that let the host organism metabolize novel nutrients, degrade novel compounds, resist antibiotics and heavy metals (resistance islands), or infect other organisms (pathogenicity islands). Chromosomal islands range from 13 to 17 kB. They encode integrase and DNA replication genes and may use a helper phage for packaging and dissemination. CIs can excise from the bacterial chromosome, replicate as episomes, and re-integrate back [79]. SpyCIM1 (*Streptococcus pyogenes* chromosomal island M1) can excise and re-integrate into the DNA repair gene *mutL* [80]. During rapid growth, this CI replicates as episome allowing the expression of *mutL* gene, but when the cells reach stationary phase, it re-integrates and disrupts the expression of *mutL* and the downstream genes. This prevents DNA repair and leads to an increase in the spontaneous mutation rate.

2.3.7. Gene Cassettes and Integrons

Gene cassettes are MGEs that contain a gene and a recombination site. They can exist as free circular DNA or be a part of an integron. Integrons contain a gene for the site-specific tyrosine recombinase (integron integrase), a recombination site, and a promoter. They can capture gene cassettes by site-specific recombination [81]. Rearrangement of gene cassettes can lead to new functions. Integrons become mobile by association with transposable elements and conjugative plasmids. They are playing a role in the rise of the antibiotic multidrug resistance.

2.3.8. The Role of MGEs

All MGEs share the ability to move inside and between genomes. Proviruses, plasmids, and transposons of archaea and bacteria often use similar genes for their mobility and can exchange genes horizontally between the elements as well as with their hosts. Viruses and plasmids often interact inside the host cell; they may depend on each other or compete. For example, inoviruses infect *E. coli* by binding the conjugative pili encoded by the F plasmid [82]. On the other hand, *E. coli* plasmids with the genes abpA and abpB block the propagation of several families of bacteriophages [83]. Transposons can move from chromosomal DNA to plasmids and back contributing to gene transfer. A close evolutional relationship between MGEs has been noted: if a virus loses the ability to make a virion, it becomes a plasmid; a plasmid that acquires genes to build a virion becomes a virus, and a virus that acquires a transposase may become a transposon.

In eukaryotes, viruses serve as vectors in the horizontal transmission of transposable elements [54, 84]. MGEs are a source of variation and innovation in host organisms [85]. For example, horizontal gene transfer is responsible for the independent integration of bacterial lysozyme family into all domains of life [86]. One well-known example of host domestication of a transposon is Rag1/2 enzymes that are involved in the recombination of lymphoid receptors in jaw vertebrates [87]. They are thought to originate from a transposon from a Transib family that infected a vertebrate ancestor some 500 million years ago [88]. Data suggest that the adaptive immune system of archaea and bacteria CRISPR-Cas also originated from a transposon [89].

2.4. Bacterial Pathogens

Bacteria are widespread; human gut and skin harbor trillions of them [90]. Only about 100 bacterial species are pathogenic for humans. However bacterial diseases such as tuberculosis, leprosy, diphtheria, tetanus, cholera, pneumonia, syphilis and others have been a major cause of mortality worldwide until the invention of antibiotics [4]. Even in 21st century, one-fourth of the world's population is infected with *Mycobacterium tuberculosis* that kills 1.7 million people a year worldwide.

Bacteria are very diverse and are classified into several phyla based on their 16S rRNA or, more recently, on the complete genome sequences [91, 92]. They can be subdivided into photosynthetic Cyanobacteria and non-photosynthetic Spirochetes and Proteotic phyla. Often bacteria are classified as gram-negative or gram-positive based on their staining by the method developed by Hans Gram. Gram-positive bacteria such as *Bacillus subtiles* have a thick layer of peptidoglycan spanned by chains of teichoic and lipoteichoic acids that covers their cytoplasmic membrane. An example of gram-negative bacteria is *E. coli*, which have their inner membrane composed a phospholipid bilayer and their outer membrane composed of phospholipids and lipopolysaccharides (LPS). Between these layers is a periplasm containing peptidoglycan.

The morphology of bacteria ranges from spherical (cocci) to rod (bacilli), spiral (spirilla), comma (vibrios), and corkscrew (spirochaetes). Some bacterial species exist as single cells, others form associations; for example, *Staphylococcus* form clusters resembling a bunch of grapes. A number of bacteria species can form spores when exposed to starvation and other stresses. Spores can stay viable for millions of years [93, 94]. Some bacteria can turn into a dormant state in which they can respire and uptake substrates but cannot replicate, *e.g. Actinobacteri*a in the permafrost samples from Antarctica, Canada and Siberia [95].

Whole genome analysis shows that bacterial evolution has often proceeded by horizontal gene transfer (HGT) between different species and genera [96]. HGT is especially often used to transfer such useful traits as antibiotic resistance or ability to fix nitrogen. In pathogenic bacteria, virulence genes that are especially important for the infection are also frequently moved by HGT. They include genes encoding toxins, secretion systems, adhesins, and invasins. Virulence factors are usually concentrated in pathogenicity islands that are genetically mobile and are transferred between bacteria by HGT. New pathogens can be generated in this way [97]. The ability of pathogenic bacteria to easily acquire anti-microbial resistance is a fundamental threat to human health, which has revived interest in the use of bacteriophages as antibacterial agents [98].

Bacteria can acquire new DNA by transduction through bacteriophage infection, conjugation (transfer of plasmids during direct cell contact), and transformation, direct uptake of nucleic acids from the environment. Transformation takes place when the cell reaches a state of competence. Development of competence is usually associated with stressful environmental conditions and seems to be an adaptation that facilitates repair of DNA damage in the recipient cells and generates genetic diversity [99]. The high rate of gene exchange means that bacterial genome composition varies between individual cells [100].

Many bacterial cellular circuits are associated with sensing and responding to the environment. Bacteria are remarkably equipped for survival. They can sense nutrients, toxins, oxygen, pH, pressure, light, temperature, presence of other bacteria, phages, *etc*. They respond by chemotaxis to attractants such as nutrients and to repellents such as toxins and can achieve an astonishing speed of up to 100 body lengths per second using their flagella. They can also move as flagella-independent [101]. An example of diverse bacterial adjustments is the use of Earth's magnctic ficld by magnctotactic bacteria to navigate towards an environment with the optimal oxygen level [102]. This is based on intracellular membrane-coated magnetic nanoparticles, which let bacteria align along the magnetic field lines. Bacteria integrate various external signals with signals from their interior such as ATP levels to optimally adapt to the environment [103].

They can secrete various compounds to communicate with conspecifics as well as toxins to kill enemies and competitors, and products that increase nutrients availability. The latter include iron chelating compounds siderophores to scavenge iron, and various enzymes such as amylase, pectinase, and others to break macromolecules into smaller more bio-available units.

Pathogenic bacteria evolved a variety of tactics aimed to colonize their hosts and avoid host defenses. They can be illustrated by the example of *Staphylococcus aureus*, a typical extracellular pathogen. *S. aureus* is a common commensal but also causes diverse illnesses ranging from skin infections such as carbuncles to pneumonia, endocarditis, osteomyelitis, and even sepsis. *S. aureus* often infects wounds after surgery and forms biofilms on implanted medical devices [104]. The spread of multi-drug resistant strains of *S. aureus* creates a serious public health threat [105].

The versatility of *S. aureus* as a pathogen is attributed to the variety of factors it can produce to support infection [106, 107]. They include such enzymes as proteases that digest host proteins, hyaluronidase, sphingomyelinase and lipase that damage host membranes, deoxyribonuclease that digests host DNA, staphylokinase that digests fibrin, and coagulase that places protective clots on the pathogen's surface. Production of surface-bound proteins for attachment to the extracellular matrix or to host cells is another important part of *S. aureus* pathogenicity [108].

In addition, *S. aureus* strains produce an assortment of toxins. Some of them, such as Panton-Valentine leukocidin are encoded by phages that infect *S. aureus*. Toxins can be transferred between strains by HGT. Several *S. aureus* toxins damage cellular membranes. Those that form pores in erythroid cells are called hemolysins and those that target white blood cells are called leukotoxins [106]. Alpha, gamma and several bi-component toxins form beta-barrel pores, which leads to lysis of host cells [109]. These toxins use receptors for their cytolytic activity. Delta toxin belongs to a group of toxins called phenol-soluble modulins (PSMs). They are small amphipathic peptides with detergent-like properties and a non-specific cytolytic activity. They contribute to the neutrophil lysis after *S. aureus* phagocytosis. Pain induced during *S. aureus* infection depends on the binding of the membrane-targeting toxins to neurons [110].

A distinctive group of *S. aureus* toxins is superantigens. They are secreted proteins of 20-30 kD that cross-link the major histocompatibility complex (MHC) class II with T cell receptor and stimulate a pathological uncontrollable T cell activation, which diverts the immune system from attacking *S. aureus* and causes a severe illness with vomiting, intestine dysfunction and diarrhea [111]. The most known *S. aureus* superantigen is the 22 kD toxic shock syndrome toxin (TSST); it stimulates the release of multiple cytokines causing severe shock. Several other toxins block immune receptors that recognize *S. aureus* and prevent activation of the complement system by binding to the receptor for C5a [106].

D-amino acids produced by *S. aureus* suppress sinonasal innate immunity through sweet taste receptors [112]. They also bind TRPV1 nociceptors and suppress the recruitment of neutrophils through the release of calcitonin gene-related peptide (CGRP).

Neutrophils are the first line of defense in human blood. *S. aureus* has developed multiple mechanisms that block neutrophils' chemotaxis, activation, phagocytosis, and pathogen killing [113]. To block the antibody response, *S. aureus* produces several antibody-binding proteins. Protein A binds the constant region of IgG, preventing it from binding to *S. aureus* antigens. Superantigen-like SSL7 binds IgA. *S. aureus* membranes are negatively charged as all bacterial membranes. To avoid recognition by host cationic antimicrobial peptides, *S. aureus* incorporates positively charged D-alanine into membrane's teichoic acid, which decreases the membrane's negative charge. *S. aureus* also produces factors that degrade antimicrobial peptides. The oxidative killing of *S. aureus* by the host is prevented by the production of enzymes detoxifying ROS. Although *S. aureus* is mostly extracellular pathogen, it can survive phagocytosis and replicate inside macrophages [114]. Thus many adaptations make *S. aureus* a successful pathogen.

Intracellular bacterial pathogens such as *Mycobacterium tuberculosis, Salmonella enterica,* or *Listeria monocytogenes* can live and replicate inside host cells. After infection, some intracellular pathogens remain in the vacuolar compartment while others escape into the cytosol [115]. Specialization of bacteria for parasitism in eukaryotic cells is associated with a massive gene loss, which irreversibly associates them with the host and compromises their survival outside the host cell [116]. A common set of about 100 genes is lost in all obligate intracellular bacteria, which suggests a convergent evolution.

Pathogenic bacteria cause damage to the host by multiple mechanisms including direct cell destruction, toxins effects, and sequestration of iron by siderophores among other means. For example, human infection by *Vibrio cholerae* is associated with massive life-threatening diarrhea caused by a phage-encoded cholera toxin. Toxin's B subunit binds to epithelial cells in the gut and transfers the enzymatic subunit into the cytoplasm, which then induces cyclic AMP overproduction, ion imbalance, and water loss [117].

Some bacteria predate on prokaryotes as solitary or group predators. BALO

(*Bdellovibrio* and like organisms) bacteria are solitary obligatory predators of other gram-negative bacteria [118]. These pathogens are strain specific and have two-phase life cycles. During the free-living phase, they use chemotaxis to search for the prey and die if cannot find it because they cannot replicate their DNA or divide. In the growth phase, they attach to a prey cell. Periplasmic predators enter the prey cell periplasm while epibiotic predators stay attached to the cell from the outside. Periplasmic predators secrete enzymes that kill the prey and modify the prey's cell wall. They grow as a polynucleoid filament, which then Separates into individual cells. They grow a flagellum and burst away to start a new life cycle. Epibiotic predators consume the prey cell from the outside and divide. Predatory bacteria are highly resistant to antibiotics due to multiple antibiotic-resistance genes that they carry [118].

δ-proteobacteria *Myxococcus xanthus* is a social predator. When food is present, these bacteria spread forming a swarm and hunt in coordination [119]. They use a mixture of antibiotics and hydrolytic enzymes to kill and lyse the prey cells [120]. The cargo of their outer membrane vesicles includes at least 46 proteins. Many other antibiotic-producing bacteria kill neighboring cells and use them as food although their predation is not obligatory [121].

In bacteria, cooperative behavior is regulated in a population density-dependent manner by small signaling molecules such as peptides and N-acyl homoserine lactones. This phenomenon is called quorum sensing (QS) [122, 123]. For example, a single bacterium usually does not secrete antibiotics or other products because the secretion products would be quickly diluted. The secretion is induced only when the bacterial density reaches a certain threshold that let secreted compounds to reach an effective concentration. This can be achieved only when most neighboring cells are genetically related. Therefore QS ensures that costly metabolic traits benefit only conspecifics [124]. Many bacterial activities such as cell aggregation, group motility, the formation of biofilms, exopolysaccharide secretion, and virulence are regulated in dependence on cell density.

Formation of biofilms is one of the cooperative bacterial behaviors regulated by QS. In response to environmental stresses, many bacteria adhere to various surfaces forming complex communities encased in the self-produced extracellular matrix. Various QS signal molecules coordinate biofilm initiation and maturation [125]. The extracellular matrix contains exopolysaccharides, secreted proteins, and cell surface adhesins and can have an elaborate secondary structure with networks of channels that enable better diffusion of nutrients. Biofilm contains multiple layers of cells, where different layers are exposed to different conditions and have a different metabolism. Peripheral biofilm cells can potentially starve interior cells of nutrients. This conflict is resolved by an oscillation with periodical halts in biofilm growth to increase nutrient availability for the sheltered interior cells [126]. Cells inside the biofilm are coordinated by spatially propagated waves of potassium ions [127, 128]. A coupling of two *Bacillus subtilis* biofilm communities through electrical signaling has recently been described. They undergo metabolic oscillations in antiphase where each community takes a turn in consuming nutrients [129]. Distant cells are attracted to biofilm [130]. The attraction is species-independent and is mediated by potassium ion channel-mediated signaling that alters the membrane potential of distant cells and affect their motility.

Some data suggest that subpopulations of bacteria can perform different tasks to achieve a common goal. During starvation, about 100,000 *Myxobacteria* aggregate to form fruiting bodies; some cells then migrate to the top and differentiate into myxospores. In this process, cells perform different functions, which is a feature typical of multicellular organisms [131].

Many cyanobacteria form complex three-dimensional structures such as weed-like thalli or bundles of filaments and spirals. In these structures, individual filaments continually move back and forth, a mechanism that may be mediated by secretion of mucilage/slime from junctional pores aligned on both sides of each septum [132]. Thus, in some instances, prokaryotes behave as multicellular organisms.

2.5. Archaea

Archaea are present in human and animal microbiota [133]. However, no definitive archaeal pathogens have been so far identified. Archaea may have some beneficial effects on human health by consuming methanol and other toxic compounds in the intestine, but they may also contribute to pathological conditions such as constipation [134].

2.6. Fungal Pathogens

Fungi are eukaryotic heterotrophs living on dead organisms or as parasites especially on plants, where they cause diseases such as mildew. Fungi are extremely diverse organisms and can live in different habitats including extreme environments such as dry mountaintops with a high level of radiation. There are between 2.2 to 3.8 million of fungi species divided into 8 phyla [135, 136].

Some fungi are single cell microscopic yeasts, whereas others are multicellular filamentous molds or macroscopic filamentous organisms. Yeasts typically multiply by budding a daughter cell from the parent cell. Molds grow as hyphae, long and branching strands that form a network of threads called a mycelium. Hyphal tips secrete digestive enzymes that break down organic material into small molecules, which are consumed by the fungus. Some of the hyphal branches grow into the air and produce spores that are disseminated by wind and other means. Macroscopic filamentous fungi grow by producing a mycelium underground but they produce visible fruiting bodies that are known as mushrooms. Dimorphism is a common feature of fungi. Fungi produce many biologically active compounds including antibiotics such as penicillins and mycotoxins such as aflatoxins that cause necrosis in liver cells.

About 300 fungi cause disease in humans [137]. Most fungal pathogens are opportunistic. Fungal diseases are typically called after pathogen. Aspergillosis is caused by a common mold *Aspergillus*; it causes allergic reaction and infection of lungs and other organs. Candidiasis is caused by any type of yeasts from the genus *Candida*; its most frequent manifestation is white patches on the tongue. Coccidioidomycosis is caused by fungi from *Coccidioides* family [138]. *Coccidioides immitis* is a dimorphic saprophytic fungus. It is common in the southwestern United States, most notably in Arizona and in Central Valley in California. It grows as mycelium in the soil during the wet season and forms spores during dry periods. Inhaled by humans, it produces a spherule form, which causes pneumonia and also can cause malfunction of other organs when spread to other parts of the body. Histoplasmosis is a lung infection by the fungus *Histoplasma*. Onychomycosis is a nail infection caused by several types of fungi. Chromoblastomycosis is a subcutaneous infection caused by black fungi [139]. Recently, the fast spread of pathogenic fungi resistant to antifungal drugs has been noted [140].

Horizontal gene transfer (HGT) common in bacteria, is rare in fungi [141]. The structure of the eukaryotic cell with a tightly packed genome and the difference in genome architecture restricts HGT. Nevertheless, HGT is observed for large genomic fragments called metabolic gene clusters (MGCs), gene clusters participating in the same metabolic pathway. HGT of MGCs may be advantageous at specific environments such as in the presence of uncommon nutrients. HGT is frequent for MGCs involved in fungal virulence.

2.7. Protozoan Pathogens

Protozoa or protists are one-celled eukaryotic organisms. Most species are free-living but some are animal pathogens. Around 100 protozoan species can infect humans, including the causative agents of malaria, amoebiasis, leishmaniasis, and sleeping sickness. Protozoan parasites often have elaborate life cycles involving more than one host.

Malaria is the most common protozoan disease. Several hundred thousand people die from it every year. It is caused by several species of *Plasmodium* and is transmitted by mosquitoes [142]. Infected mosquito, which is the definitive host, introduces motile infectious sporozoites into human blood during a bite. They move from the blood into the liver and multiply in the hepatocytes. Infected cells eventually burst to release non-motile infectious merozoites. Merozoites infect red blood cells and reproduce asexually, destroying the host cells in the process. Their progeny can infect new red blood cells or differentiate into immature gametocytes that have male and female gametes. When a mosquito bites an infected person, gametocytes mature in the mosquito gut; male and female gametocytes fuse and form an ookinete, which develops into a sporozoite. New sporozoites move into the mosquito salivary gland completing the cycle. Several genetic defects provide partial resistance against malaria, including sickle cell trait, thalassemia trait, glucose-6-phosphate dehydrogenase deficiency, and absence of Duffy antigens on red blood cells [143]. Immune cells promptly eliminate infected red blood cells with these defects, which slows infection spread.

Amoebiasis is an intestinal disease caused by *Entamoeba histolytica*. The disease is caused by motile trophozoite form, which undergoes encystation under stress (*e.g.* starvation) to form a non-motile cyst form [144]. Cysts can survive in the environment for several months. Infection of a new host occurs through ingestion of food or water contaminated with cysts or through dirty hands.

Several free-living amoebae can also cause serious human diseases. For example "brain-eating" amoeba *Naegleria fowleri* causes primary amoebic meningoencephalitis [145].

2.8. Helminths

Helminths are parasitic worms from various taxa that live in the animal intestine, blood, and other organs. Most helminths belong to nematodes (roundworms), cestodes (tapeworms), platyhelminths (flatworms), and trematodes (flukes). Helminths that live in a digestive tract are called intestinal parasites. From those, *Ascaris lumbricoides* infect about a billion people. Helminths produce a large number of eggs, which are extremely resistant to adverse conditions due to their multilayered shell and can remain viable and infectious for several years in feces or sewage sludge. A heavy burden of parasitic worms can cause serious health problems and even death, but low numbers typically cause only limited or no pathology and can even be beneficial to the host.

Long co-evolution of helminths with their human host resulted in the adaptation of the human immune system and metabolism to the presence of helminths.

Helminths developed an ability to modulate the human immune system. They induce regulatory T cells that suppress inflammation and reduce allergic reactions [146, 147]. Some studies suggest that eradication of helminths in the developed world resulted in an increase of allergies and inflammatory diseases. Increase in metabolic diseases, such as diabetes, could also be ascribed to the eradication of helminths. Helminths induce eosinophils, which are necessary for the maintenance of glucose homeostasis. In the absence of helminths, eosinophils number is decreased, resulting in impaired glucose tolerance and insulin resistance [148]. Currently, products secreted by helminths are studied as potential anti-inflammatory drugs [149]. Moreover, several clinical trials are using live helminths for treatment of inflammatory bowel disease and show efficacy.

2.9. Life is a Communal Enterprise

Animals, including humans, live in symbiosis with a complex gastrointestinal microbiota that includes organisms from all three domains of life [150]. Interactions between symbionts can be mutualistic with benefits for both species, commensalistic with benefit for one species and no harm to the other, or parasitic. Interaction of most gut microbiota with humans is mutualistic [151]. Mutualistic microbiota benefits from human food and secure habitat, and in turn, improve human digestion by breaking down indigestible carbohydrates, converting sugars to lactic acid, producing vitamins, and inhibiting the growth of potentially pathogenic bacteria. Other gut microbiota are commensalists or pathogens.

Colonization of human intestine starts at birth during contact with vaginal microbiota. Microbiota composition gradually changes during development [152]. Human milk shapes microbiota composition. Milk oligosaccharides support the growth of bifidobacteria while milk antibodies prevent the proliferation of pathogens [153]. Delayed development of microbiota was noted in children fed with formula milk. Altered microbiota is associated with risk of asthma and can be compensated for by supplementation with *Lactobacillus rhamnosus* [154].

Along its length, intestine forms several distinctive habitats with different ecosystems/biocenosis [155]. Typically all species in the biocenosis are interdependent. The size of each species is regulated by environmental conditions and by competition and cooperation between species and the host.

Organisms compete through direct interference by killing or weakening competitors, or exploitatively by consuming common resources [156]. In human microbiota, *E. coli* depletes *C. albicans* from magnesium killing it [157]. *Vibrio cholerae* use a type VI secretion system to produce toxins that kill members of the host commensal microbiota [158]. If competing species have exactly the same requirements for survival and reproduction, one species would eventually drive the other to extinction. It is called the competitive exclusion principle. Consequently, most species in the human gut have different niches.

Gut species also cooperate in defense and nutrient acquisition. A majority of human gut microorganisms are bacteria such as *Bacteroides thetaiotaomicron* that degrade various polysaccharides such as pectin and cellulose that humans cannot digest. *B. thetaiotaomicron* encodes multiple paralogs of genes involved in carbohydrate metabolism including 226 glycoside hydrolases. This bacterium degrades polysaccharides into short-chain fatty acids, CO_2, and hydrogen gas. Methanogenic archaea such as *Methanobrevibacter smithii* then join the metabolic chain of polysaccharide processing. They consume hydrogen and produce methane [159]. On the other hand, *B. thetaiotaomicron* benefits from the efficient removal of its waste products. Hence, *B. thetaiotaomicron* and *M. smithii* syntrophy is beneficial for both species.

Even more complex cooperative strategy is observed in *Bacteroides ovatus*. It expresses enzymes that extracellularly digest inulin polysaccharide, even though it prefers to feed on inulin without digestion. Moreover, digestion products interfere with its inulin consumption. It turns out that extracellular digestion of inulin feeds other gut species such as *Bacteroides vulgaris* [160]. In return, these species produce factors that increase *B. ovatus* fitness and promote its growth.

A balanced representation of various species is necessary for a healthy ecosystem including gut microbiota. It was shown that biodiverse communities maintained multiple functions at higher levels than depauperate ones [161]. The components of human microbiota change during sickness or antibiotic use [162]. Its species composition may shift substantially, as one species or another is affected. Eventually, the microbiota returns to a state of equilibrium, even if different from the previous one.

Additional members of the gut community are bacteriophages and other viruses [163]. The bacteriophage crAssphage and related phages are the most abundant in the human gut [164]. crAssphage infects phylum *Bacteroides*. Most gut bacteria carry prophages (are lysogens). Correspondently, many stressors including drugs, pollutants, and some nutrients can activate these prophages [98]. Bacteriophages can regulate the relative abundance of bacteria. In the intestine, bacteriophages concentrate in mucus, which may restrict bacteria presence on mucosal surfaces. Interestingly, bacteriophages are found not only in the gut and skin but also in regions that previously were considered sterile [165].

Microorganisms live in the animal intestinal tract, skin, and other surfaces for millions of years and the animals' metabolism and immune system have evolved in interactions with these organisms. Metabolites produced by microbiota enter the host body and can affect its metabolism, immune responses, organ function, neural signaling, and behavior. The composition of microorganisms affects human health from immunity to mental health. Phageomes of people with inflammatory diseases differ from phageomes of healthy people, which suggests that changes in the viral composition may contribute to intestinal inflammation and dysbiosis [166]. Modern lifestyle may have caused a partial loss of the ancestral microbiota, which may have altered immunological, metabolic, and cognitive development and contributed to modern diseases [167].

Recent data suggest a connection between multiple sclerosis and the composition of microbes in the human gut. Microbiota from patients with multiple sclerosis (MS) causes an MS-like autoimmune disease in mice [168]. A detailed analysis of the microbiota composition revealed that *Akkermansia muciniphila* and *Acinetobacter calcoaceticus* (increased in MS patients) induced a proinflammatory response in human mononuclear cells and in mice. In contrast, *Parabacteroides distasonis* that is reduced in MS patients stimulates the production of anti-inflammatory IL10 by human and mice T cells [169].

Similar to animals, plants live in association with microorganisms. The composition of species living on roots is regulated by the interaction between plants and microorganisms in accordance with their traits, pathogenicity, and with nutrients availability [170]. Modern sequencing techniques reveal that human-made super niches such as big cities, also have their specific microbiota as illustrated by a snapshot of metagenomics of New York City [171].

CONCLUDING REMARKS

Parasitism and predation are an indispensable part of life. A great variety of organisms from viruses and bacteria to fungi and worms are pathogens. Life inside other organisms requires the ability to avoid recognition by the host immune system and to resist immune mechanisms, which are designed to kill invaders. Pathogens have evolved mechanisms facilitating colonization of the host, suppression or manipulation of its immune system, and exploitation of the host metabolism to their advantage. Pathogens constantly evolve to overcome host defenses, and so need the hosts.

CONSENT FOR PUBLICATION

Not applicable.

CONFLICT OF INTEREST

The author confirms that this chapter contents have no conflict of interest.

ACKNOWLEDGEMENTS

Declared none.

REFERENCES

[1] Casadevall A, Pirofski LA. WHAT IS A HOST II: THE ATTRIBUTES OF INDIVIDUAL SUSCEPTIBILITY. Infect Immun 2018; 86(2): e00636-17. [PMID: 29084893]

[2] Carlson CJ, Getz WM, Kausrud KL, *et al.* Spores and soil from six sides: Interdisciplinarity and the environmental biology of anthrax (Bacillus anthracis). Biol Rev Camb Philos Soc 2018; 93(4): 1813-31. [http://dx.doi.org/10.1111/brv.12420] [PMID: 29732670]

[3] Price LB, Hungate BA, Koch BJ, Davis GS, Liu CM. Colonizing opportunistic pathogens (COPs): The beasts in all of us. PLoS Pathog 2017; 13(8)e1006369 [http://dx.doi.org/10.1371/journal.ppat.1006369] [PMID: 28796836]

[4] Balloux F, van Dorp L. Q&A: What are pathogens, and what have they done to and for us? BMC Biol 2017; 15(1): 91. [http://dx.doi.org/10.1186/s12915-017-0433-z] [PMID: 29052511]

[5] Morens DM, Fauci AS. The 1918 influenza pandemic: insights for the 21st century. J Infect Dis 2007; 195(7): 1018-28. [http://dx.doi.org/10.1086/511989] [PMID: 17330793]

[6] Sender R, Fuchs S, Milo R. Revised Estimates for the Number of Human and Bacteria Cells in the Body. PLoS Biol 2016; 14(8)e1002533 [http://dx.doi.org/10.1371/journal.pbio.1002533] [PMID: 27541692]

[7] McFadden G. Killing a killer: what next for smallpox? PLoS Pathog 2010; 6(1)e1000727 [http://dx.doi.org/10.1371/journal.ppat.1000727] [PMID: 20126444]

[8] Van Valen L. A new evolutionary law. Evol Theory 1973; 1: 1-30.

[9] Enard D, Cai L, Gwennap C, Petrov DA. Viruses are a dominant driver of protein adaptation in mammals. eLife 2016; 5e12469 [http://dx.doi.org/10.7554/eLife.12469] [PMID: 27187613]

[10] Stucki D, Brites D, Jeljeli L, *et al.* Mycobacterium tuberculosis lineage 4 comprises globally distributed and geographically restricted sublineages. Nat Genet 2016; 48(12): 1535-43. [http://dx.doi.org/10.1038/ng.3704] [PMID: 27798628]

[11] Chauché C, Nogales A, Zhu H, *et al.* Mammalian Adaptation of an Avian Influenza A Virus Involves Stepwise Changes in NS1. J Virol 2018; 92(5)e01875-17 [http://dx.doi.org/10.1128/JVI.01875-17] [PMID: 29237841]

[12] Fouché S, Plissonneau C, Croll D. The birth and death of effectors in rapidly evolving filamentous pathogen genomes. Curr Opin Microbiol 2018; 46: 34-42. [http://dx.doi.org/10.1016/j.mib.2018.01.020] [PMID: 29455143]

[13] Petersen H, Mostafa A, Tantawy MA, *et al.* NS Segment of a 1918 Influenza A Virus-Descendent Enhances Replication of H1N1pdm09 and Virus-Induced Cellular Immune Response in Mammalian and Avian Systems. Front Microbiol 2018; 9: 526. [http://dx.doi.org/10.3389/fmicb.2018.00526] [PMID: 29623073]

[14] Sweileh WM. Global research trends of World Health Organization's top eight emerging pathogens. Global Health 2017; 13(1): 9.
[http://dx.doi.org/10.1186/s12992-017-0233-9] [PMID: 28179007]

[15] Parikka KJ, Le Romancer M, Wauters N, Jacquet S. Deciphering the virus-to-prokaryote ratio (VPR): insights into virus-host relationships in a variety of ecosystems. Biol Rev Camb Philos Soc 2017; 92(2): 1081-100.
[http://dx.doi.org/10.1111/brv.12271] [PMID: 27113012]

[16] Kauffman KM, Hussain FA, Yang J, *et al.* A major lineage of non-tailed dsDNA viruses as unrecognized killers of marine bacteria. Nature 2018; 554(7690): 118-22.
[http://dx.doi.org/10.1038/nature25474] [PMID: 29364876]

[17] La Scola B, Audic S, Robert C, *et al.* A giant virus in amoebae. Science 2003; 299(5615): 2033.
[http://dx.doi.org/10.1126/science.1081867] [PMID: 12663918]

[18] King AMQ, Lefkowitz EJ, Mushegian AR, *et al.* Changes to taxonomy and the International Code of Virus Classification and Nomenclature ratified by the International Committee on Taxonomy of Viruses (2018). Arch Virol 2018; 163(9): 2601-31.
[http://dx.doi.org/10.1007/s00705-018-3847-1] [PMID: 29754305]

[19] Colson P, La Scola B, Levasseur A, Caetano-Anollés G, Raoult D. Mimivirus: leading the way in the discovery of giant viruses of amoebae. Nat Rev Microbiol 2017; 15(4): 243-54.
[http://dx.doi.org/10.1038/nrmicro.2016.197] [PMID: 28239153]

[20] Koonin EV, Krupovic M. Polintons, virophages and transpovirons: a tangled web linking viruses, transposons and immunity. Curr Opin Virol 2017; 25: 7-15.
[http://dx.doi.org/10.1016/j.coviro.2017.06.008] [PMID: 28672161]

[21] Woolhouse MEJ, Brierley L. Epidemiological characteristics of human-infective RNA viruses. Sci Data 2018; 5180017
[http://dx.doi.org/10.1038/sdata.2018.17] [PMID: 29461515]

[22] Krupovic M, Cvirkaite-Krupovic V, Iranzo J, Prangishvili D, Koonin EV. Viruses of archaea: Structural, functional, environmental and evolutionary genomics. Virus Res 2018; 244: 181-93.
[http://dx.doi.org/10.1016/j.virusres.2017.11.025] [PMID: 29175107]

[23] Hoang HD, Graber TE, Alain T. Battling for Ribosomes: Translational Control at the Forefront of the Antiviral Response. J Mol Biol 2018; 430(14): 1965-92.
[http://dx.doi.org/10.1016/j.jmb.2018.04.040] [PMID: 29746850]

[24] Kutter E, Bryan D, Ray G, Brewster E, Blasdel B, Guttman B. From Host to Phage Metabolism: Hot Tales of Phage T4's Takeover of *E. coli*. Viruses 2018; 10(7)E387
[http://dx.doi.org/10.3390/v10070387] [PMID: 30037085]

[25] Blasche S, Wuchty S, Rajagopala SV, Uetz P. The protein interaction network of bacteriophage lambda with its host, *Escherichia coli*. J Virol 2013; 87(23): 12745-55.
[http://dx.doi.org/10.1128/JVI.02495-13] [PMID: 24049175]

[26] Casjens SR, Hendrix RW. Bacteriophage lambda: Early pioneer and still relevant. Virology 2015; 479-480: 310-30.
[http://dx.doi.org/10.1016/j.virol.2015.02.010] [PMID: 25742714]

[27] Gandon S. Why Be Temperate: Lessons from Bacteriophage λ. Trends Microbiol 2016; 24(5): 356-65.
[http://dx.doi.org/10.1016/j.tim.2016.02.008] [PMID: 26946976]

[28] Kobiler O, Rokney A, Oppenheim AB. Phage lambda CIII: a protease inhibitor regulating the lysis-lysogeny decision. PLoS One 2007; 2(4)e363
[http://dx.doi.org/10.1371/journal.pone.0000363] [PMID: 17426811]

[29] Lewis D, Le P, Zurla C, Finzi L, Adhya S. Multilevel autoregulation of λ repressor protein CI by DNA looping *in vitro*. Proc Natl Acad Sci USA 2011; 108(36): 14807-12.

[http://dx.doi.org/10.1073/pnas.1111221108] [PMID: 21873207]

[30] Michalowski CB, Little JW. Role of cis-acting sites in stimulation of the phage λ P(RM) promoter by CI-mediated looping. J Bacteriol 2013; 195(15): 3401-11. [http://dx.doi.org/10.1128/JB.02148-12] [PMID: 23708136]

[31] Sepúlveda LA, Xu H, Zhang J, Wang M, Golding I. Measurement of gene regulation in individual cells reveals rapid switching between promoter states. Science 2016; 351(6278): 1218-22. [http://dx.doi.org/10.1126/science.aad0635] [PMID: 26965629]

[32] Lee S, Lewis DEA, Adhya S. The Developmental Switch in Bacteriophage λ: A Critical Role of the Cro Protein. J Mol Biol 2018; 430(1): 58-68. [http://dx.doi.org/10.1016/j.jmb.2017.11.005] [PMID: 29158090]

[33] Arkin A, Ross J, McAdams HH. Stochastic kinetic analysis of developmental pathway bifurcation in phage lambda-infected *Escherichia coli* cells. Genetics 1998; 149(4): 1633-48. [PMID: 9691025]

[34] Joh RI, Weitz JS. To lyse or not to lyse: transient-mediated stochastic fate determination in cells infected by bacteriophages. PLOS Comput Biol 2011; 7(3)e1002006 [http://dx.doi.org/10.1371/journal.pcbi.1002006] [PMID: 21423715]

[35] Knowles B, Bailey B, Boling L, *et al.* Variability and host density independence in inductions-based estimates of environmental lysogeny. Nat Microbiol 2017; 2: 17064. [http://dx.doi.org/10.1038/nmicrobiol.2017.64] [PMID: 28452987]

[36] Erez Z, Steinberger-Levy I, Shamir M, *et al.* Communication between viruses guides lysis-lysogeny decisions. Nature 2017; 541(7638): 488-93. [http://dx.doi.org/10.1038/nature21049] [PMID: 28099413]

[37] Kinchington PR, Leger AJ, Guedon JM, Hendricks RL. Herpes simplex virus and varicella zoster virus, the house guests who never leave. Herpesviridae 2012; 3(1): 5. [http://dx.doi.org/10.1186/2042-4280-3-5] [PMID: 22691604]

[38] Gershon AA, Breuer J, Cohen JI, *et al.* Varicella zoster virus infection. Nat Rev Dis Primers 2015; 1: 15016. [http://dx.doi.org/10.1038/nrdp.2015.16] [PMID: 27188665]

[39] Tweedy J, Spyrou MA, Pearson M, Lassner D, Kuhl U, Gompels UA. Complete Genome Sequence of Germline Chromosomally Integrated Human Herpesvirus 6A and Analyses Integration Sites Define a New Human Endogenous Virus with Potential to Reactivate as an Emerging Infection. Viruses 2016; 8(1)E19 [http://dx.doi.org/10.3390/v8010019] [PMID: 26784220]

[40] Kuo HH, Ahmad R, Lee GQ, *et al.* Anti-apoptotic Protein BIRC5 Maintains Survival of HIV---Infected $CD4^+$ T Cells. Immunity 2018; 48(6): 1183-1194.e5. [http://dx.doi.org/10.1016/j.immuni.2018.04.004] [PMID: 29802019]

[41] Chaturongakul S, Ounjai P. Phage-host interplay: examples from tailed phages and Gram-negative bacterial pathogens. Front Microbiol 2014; 5: 442. [http://dx.doi.org/10.3389/fmicb.2014.00442] [PMID: 25191318]

[42] Breitbart M, Bonnain C, Malki K, Sawaya NA. Phage puppet masters of the marine microbial realm. Nat Microbiol 2018; 3(7): 754-66. [http://dx.doi.org/10.1038/s41564-018-0166-y] [PMID: 29867096]

[43] Wang X, Kim Y, Ma Q, *et al.* Cryptic prophages help bacteria cope with adverse environments. Nat Commun 2010; 1: 147. [http://dx.doi.org/10.1038/ncomms1146] [PMID: 21266997]

[44] Gelgor A, Gam Ze Letova C, Yegorov Y, Kalt I, Sarid R. Nucleolar stress enhances lytic reactivation of the Kaposi's sarcoma-associated herpesvirus. Oncotarget 2018; 9(17): 13822-33. [http://dx.doi.org/10.18632/oncotarget.24497] [PMID: 29568397]

[45] Washington SD, Edenfield SI, Lieux C, Watson ZL, Taasan SM, Dhummakupt A, *et al.* Depletion of the insulator protein CTCF results in HSV-1 reactivation *in vivo*. J Virol 2018. pii: JVI.00173-18

[46] Mehta SK, Laudenslager ML, Stowe RP, *et al.* Latent virus reactivation in astronauts on the international space station. NPJ Microgravity 2017; 3: 11. [http://dx.doi.org/10.1038/s41526-017-0015-y] [PMID: 28649633]

[47] Balistreri G, Viiliäinen J, Turunen M, *et al.* Oncogenic Herpesvirus Utilizes Stress-Induced Cell Cycle Checkpoints for Efficient Lytic Replication. PLoS Pathog 2016; 12(2)e1005424 [http://dx.doi.org/10.1371/journal.ppat.1005424] [PMID: 26891221]

[48] Pearl S, Gabay C, Kishony R, Oppenheim A, Balaban NQ. Nongenetic individuality in the host-phage interaction. PLoS Biol 2008; 6(5)e120 [http://dx.doi.org/10.1371/journal.pbio.0060120] [PMID: 18494559]

[49] Kumar S, Morrison JH, Dingli D, Poeschla E. HIV-1 activation of innate immunity depends strongly on the intracellular level of TREX1 and sensing of incomplete reverse transcription products. J Virol 2018; 92(16): e00001-18. [http://dx.doi.org/10.1128/JVI.00001-18] [PMID: 29769349]

[50] Cliffe AR, Arbuckle JH, Vogel JL, *et al.* Neuronal Stress Pathway Mediating a Histone Methyl/Phospho Switch Is Required for Herpes Simplex Virus Reactivation. Cell Host Microbe 2015; 18(6): 649-58. [http://dx.doi.org/10.1016/j.chom.2015.11.007] [PMID: 26651941]

[51] Margolis DM, Garcia JV, Hazuda DJ, Haynes BF. Latency reversal and viral clearance to cure HIV-1. Science 2016; 353(6297)aaf6517 [http://dx.doi.org/10.1126/science.aaf6517] [PMID: 27463679]

[52] Aneja KK, Yuan Y. Reactivation and Lytic Replication of Kaposi's Sarcoma-Associated Herpesvirus: An Update. Front Microbiol 2017; 8: 613. [http://dx.doi.org/10.3389/fmicb.2017.00613] [PMID: 28473805]

[53] Touchon M, Moura de Sousa JA, Rocha EP. Embracing the enemy: the diversification of microbial gene repertoires by phage-mediated horizontal gene transfer. Curr Opin Microbiol 2017; 38: 66-73. [http://dx.doi.org/10.1016/j.mib.2017.04.010] [PMID: 28527384]

[54] Dolja VV, Koonin EV. Metagenomics reshapes the concepts of RNA virus evolution by revealing extensive horizontal virus transfer. Virus Res 2018; 244: 36-52. [http://dx.doi.org/10.1016/j.virusres.2017.10.020] [PMID: 29103997]

[55] Smith GL. Vaccinia Virus Protein C6: A Multifunctional Interferon Antagonist. Adv Exp Med Biol 2018; 1052: 1-7. [http://dx.doi.org/10.1007/978-981-10-7572-8_1] [PMID: 29785476]

[56] Diener TO. Circular RNAs: relics of precellular evolution? Proc Natl Acad Sci USA 1989; 86(23): 9370-4. [http://dx.doi.org/10.1073/pnas.86.23.9370] [PMID: 2480600]

[57] Prusiner SB. Novel proteinaceous infectious particles cause scrapie. Science 1982; 216(4542): 136-44. [http://dx.doi.org/10.1126/science.6801762] [PMID: 6801762]

[58] Ayers JI, Giasson BI, Borchelt DR. Prion-like Spreading in Tauopathies. Biol Psychiatry 2018; 83(4): 337-46. [http://dx.doi.org/10.1016/j.biopsych.2017.04.003] [PMID: 28506438]

[59] McLaughlin RN Jr, Malik HS. Genetic conflicts: the usual suspects and beyond. J Exp Biol 2017; 220(Pt 1): 6-17. [http://dx.doi.org/10.1242/jeb.148148] [PMID: 28057823]

[60] McClintock B. The significance of responses of the genome to challenge. Science 1984; 226(4676): 792-801.

[http://dx.doi.org/10.1126/science.15739260] [PMID: 15739260]

[61] Lander ES, Linton LM, Birren B, *et al.* International Human Genome Sequencing Consortium. Initial sequencing and analysis of the human genome. Nature 2001; 409(6822): 860-921. [http://dx.doi.org/10.1038/35057062] [PMID: 11237011]

[62] Lee H, Doak TG, Popodi E, Foster PL, Tang H. Insertion sequence-caused large-scale rearrangements in the genome of *Escherichia coli*. Nucleic Acids Res 2016; 44(15): 7109-19. [PMID: 27431326]

[63] Krupovic M, Makarova KS, Forterre P, Prangishvili D, Koonin EV. Casposons: a new superfamily of self-synthesizing DNA transposons at the origin of prokaryotic CRISPR-Cas immunity. BMC Biol 2014; 12: 36. [http://dx.doi.org/10.1186/1741-7007-12-36] [PMID: 24884953]

[64] Grow EJ, Flynn RA, Chavez SL, *et al.* Intrinsic retroviral reactivation in human preimplantation embryos and pluripotent cells. Nature 2015; 522(7555): 221-5. [http://dx.doi.org/10.1038/nature14308] [PMID: 25896322]

[65] Pizarro JG, Cristofari G. Post-Transcriptional Control of LINE-1 Retrotransposition by Cellular Host Factors in Somatic Cells. Front Cell Dev Biol 2016; 4: 14. [http://dx.doi.org/10.3389/fcell.2016.00014] [PMID: 27014690]

[66] Hancks DC, Kazazian HH Jr. Roles for retrotransposon insertions in human disease. Mob DNA 2016; 7: 9. [http://dx.doi.org/10.1186/s13100-016-0065-9] [PMID: 27158268]

[67] Feng G, Leem YE, Levin HL. Transposon integration enhances expression of stress response genes. Nucleic Acids Res 2013; 41(2): 775-89. [http://dx.doi.org/10.1093/nar/gks1185] [PMID: 23193295]

[68] Sehgal A, Lee CY, Espenshade PJ. SREBP controls oxygen-dependent mobilization of retrotransposons in fission yeast. PLoS Genet 2007; 3(8)e131 [http://dx.doi.org/10.1371/journal.pgen.0030131] [PMID: 17696611]

[69] Miousse IR, Chalbot MC, Lumen A, Ferguson A, Kavouras IG, Koturbash I. Response of transposable elements to environmental stressors. Mutat Res Rev Mutat Res 2015; 765: 19-39. [http://dx.doi.org/10.1016/j.mrrev.2015.05.003] [PMID: 26281766]

[70] Smalla K, Jechalke S, Top EM. Plasmid Detection, Characterization, and Ecology Microbiology spectrum 2015; 3(1): Plas-0038.

[71] Wang H, Peng N, Shah SA, Huang L, She Q. Archaeal extrachromosomal genetic elements. Microbiol Mol Biol Rev 2015; 79(1): 117-52. [http://dx.doi.org/10.1128/MMBR.00042-14] [PMID: 25694123]

[72] Carroll AC, Wong A. Plasmid persistence: costs, benefits, and the plasmid paradox. Can J Microbiol 2018; 64(5): 293-304. [http://dx.doi.org/10.1139/cjm-2017-0609] [PMID: 29562144]

[73] Hausner G, Hafez M, Edgell DR. Bacterial group I introns: mobile RNA catalysts. Mob DNA 2014; 5(1): 8. [http://dx.doi.org/10.1186/1759-8753-5-8] [PMID: 24612670]

[74] Del Hoyo A, Álvarez R, Gasulla F, Casano LM, Del Campo EM. Origin and evolution of chloroplast group I introns in lichen algae. J Phycol 2018; 54(1): 66-78. [http://dx.doi.org/10.1111/jpy.12600] [PMID: 29057470]

[75] Nawrocki EP, Jones TA, Eddy SR. Group I introns are widespread in archaea. Nucleic Acids Res 2018; 46(15): 7970-6. [http://dx.doi.org/10.1093/nar/gky414] [PMID: 29788499]

[76] Novikova O, Belfort M. Mobile Group II Introns as Ancestral Eukaryotic Elements. Trends Genet

2017; 33(11): 773-83.
[http://dx.doi.org/10.1016/j.tig.2017.07.009] [PMID: 28818345]

[77] Qu G, Kaushal PS, Wang J, *et al.* Structure of a group II intron in complex with its reverse transcriptase. Nat Struct Mol Biol 2016; 23(6): 549-57.
[http://dx.doi.org/10.1038/nsmb.3220] [PMID: 27136327]

[78] Stoddard BL. Homing endonucleases from mobile group I introns: discovery to genome engineering. Mob DNA 2014; 5(1): 7.
[http://dx.doi.org/10.1186/1759-8753-5-7] [PMID: 24589358]

[79] Novick RP, Christie GE, Penadés JR. The phage-related chromosomal islands of Gram-positive bacteria. Nat Rev Microbiol 2010; 8(8): 541-51.
[http://dx.doi.org/10.1038/nrmicro2393] [PMID: 20634809]

[80] Hendrickson C, Euler CW, Nguyen SV, *et al.* Elimination of Chromosomal Island SpyCIM1 from Streptococcus pyogenes Strain SF370 Reverses the Mutator Phenotype and Alters Global Transcription. PLoS One 2015; 10(12)e0145884
[http://dx.doi.org/10.1371/journal.pone.0145884] [PMID: 26701803]

[81] Escudero JA, Loot C, Nivina A, Mazel D. The Integron: Adaptation On Demand. Microbiology spectrum. 2015; 3(2): Mdna3-0019-.

[82] Krupovic M, Prangishvili D, Hendrix RW, Bamford DH. Genomics of bacterial and archaeal viruses: dynamics within the prokaryotic virosphere. Microbiol Mol Biol Rev 2011; 75(4): 610-35.
[http://dx.doi.org/10.1128/MMBR.00011-11] [PMID: 22126996]

[83] Yasui R, Washizaki A, Furihata Y, Yonesaki T, Otsuka Y. AbpA and AbpB provide anti-phage activity in *Escherichia coli*. Genes & genetic systems 2014; 89(2): 51-60.
[http://dx.doi.org/10.1266/ggs.89.51]

[84] Gilbert C, Feschotte C. Horizontal acquisition of transposable elements and viral sequences: patterns and consequences. Curr Opin Genet Dev 2018; 49: 15-24.
[http://dx.doi.org/10.1016/j.gde.2018.02.007] [PMID: 29505963]

[85] Hosaka A, Kakutani T. Transposable elements, genome evolution and transgenerational epigenetic variation. Curr Opin Genet Dev 2018; 49: 43-8.
[http://dx.doi.org/10.1016/j.gde.2018.02.012] [PMID: 29525544]

[86] Metcalf JA, Funkhouser-Jones LJ, Brileya K, Reysenbach AL, Bordenstein SR. Antibacterial gene transfer across the tree of life. eLife 2014; 3
[http://dx.doi.org/10.7554/eLife.04266] [PMID: 25422936]

[87] Teng G, Schatz DG. Regulation and Evolution of the RAG Recombinase. Adv Immunol 2015; 128: 1-39.
[http://dx.doi.org/10.1016/bs.ai.2015.07.002] [PMID: 26477364]

[88] Kapitonov VV, Koonin EV. Evolution of the RAG1-RAG2 locus: both proteins came from the same transposon. Biol Direct 2015; 10: 20.
[http://dx.doi.org/10.1186/s13062-015-0055-8] [PMID: 25928409]

[89] Koonin EV, Makarova KS. Mobile Genetic Elements and Evolution of CRISPR-Cas Systems: All the Way There and Back. Genome Biol Evol 2017; 9(10): 2812-25.
[http://dx.doi.org/10.1093/gbe/evx192] [PMID: 28985291]

[90] Bilen M, Dufour JC, Lagier JC, *et al.* The contribution of culturomics to the repertoire of isolated human bacterial and archaeal species. Microbiome 2018; 6(1): 94.
[http://dx.doi.org/10.1186/s40168-018-0485-5] [PMID: 29793532]

[91] Woese CR, Kandler O, Wheelis ML. Towards a natural system of organisms: proposal for the domains Archaea, Bacteria, and Eucarya. Proc Natl Acad Sci USA 1990; 87(12): 4576-9.
[http://dx.doi.org/10.1073/pnas.87.12.4576] [PMID: 2112744]

[92] Mahato NK, Gupta V, Singh P, *et al.* Microbial taxonomy in the era of OMICS: application of DNA sequences, computational tools and techniques. Antonie van Leeuwenhoek 2017; 110(10): 1357-71. [http://dx.doi.org/10.1007/s10482-017-0928-1] [PMID: 28831610]

[93] Cano RJ, Borucki MK. Revival and identification of bacterial spores in 25- to 40-million-year-old Dominican amber. Science 1995; 268(5213): 1060-4. [http://dx.doi.org/10.1126/science.7538699] [PMID: 7538699]

[94] Vreeland RH, Rosenzweig WD, Powers DW. Isolation of a 250 million-year-old halotolerant bacterium from a primary salt crystal. Nature 2000; 407(6806): 897-900. [http://dx.doi.org/10.1038/35038060] [PMID: 11057666]

[95] Dieser M, Battista JR, Christner BC. DNA double-strand break repair at-15{degrees}C. Appl Environ Microbiol 2013; 79(24): 7662-8. [http://dx.doi.org/10.1128/AEM.02845-13] [PMID: 24077718]

[96] Soucy SM, Huang J, Gogarten JP. Horizontal gene transfer: building the web of life. Nat Rev Genet 2015; 16(8): 472-82. [http://dx.doi.org/10.1038/nrg3962] [PMID: 26184597]

[97] Penadés JR, Chen J, Quiles-Puchalt N, Carpena N, Novick RP. Bacteriophage-mediated spread of bacterial virulence genes. Curr Opin Microbiol 2015; 23: 171-8. [http://dx.doi.org/10.1016/j.mib.2014.11.019] [PMID: 25528295]

[98] Górski A, Jończyk-Matysiak E, Międzybrodzki R, *et al.* Phage Therapy: Beyond Antibacterial Action. Front Med (Lausanne) 2018; 5: 146. [http://dx.doi.org/10.3389/fmed.2018.00146] [PMID: 29876350]

[99] Johnston C, Martin B, Fichant G, Polard P, Claverys JP. Bacterial transformation: distribution, shared mechanisms and divergent control. Nat Rev Microbiol 2014; 12(3): 181-96. [http://dx.doi.org/10.1038/nrmicro3199] [PMID: 24509783]

[100] Ackermann M. A functional perspective on phenotypic heterogeneity in microorganisms. Nat Rev Microbiol 2015; 13(8): 497-508. [http://dx.doi.org/10.1038/nrmicro3491] [PMID: 26145732]

[101] Pollitt EJG, Diggle SP. Defining motility in the Staphylococci. Cell Mol Life Sci 2017; 74(16): 2943-58. [http://dx.doi.org/10.1007/s00018-017-2507-z] [PMID: 28378043]

[102] Lefèvre CT, Bennet M, Landau L, *et al.* Diversity of magneto-aerotactic behaviors and oxygen sensing mechanisms in cultured magnetotactic bacteria. Biophys J 2014; 107(2): 527-38. [http://dx.doi.org/10.1016/j.bpj.2014.05.043] [PMID: 25028894]

[103] Lee EJ, Groisman EA. Control of a Salmonella virulence locus by an ATP-sensing leader messenger RNA. Nature 2012; 486(7402): 271-5. [http://dx.doi.org/10.1038/nature11090] [PMID: 22699622]

[104] Tong SY, Davis JS, Eichenberger E, Holland TL, Fowler VG Jr. Staphylococcus aureus infections: epidemiology, pathophysiology, clinical manifestations, and management. Clin Microbiol Rev 2015; 28(3): 603-61. [http://dx.doi.org/10.1128/CMR.00134-14] [PMID: 26016486]

[105] Baker S, Thomson N, Weill FX, Holt KE. Genomic insights into the emergence and spread of antimicrobial-resistant bacterial pathogens. Science 2018; 360(6390): 733-8. [http://dx.doi.org/10.1126/science.aar3777] [PMID: 29773743]

[106] Otto M. Staphylococcus aureus toxins. Curr Opin Microbiol 2014; 17: 32-7. [http://dx.doi.org/10.1016/j.mib.2013.11.004] [PMID: 24581690]

[107] Pietrocola G, Nobile G, Rindi S, Speziale P. *Staphylococcus aureus* Manipulates Innate Immunity through Own and Host-Expressed Proteases. Front Cell Infect Microbiol 2017; 7: 166.

[http://dx.doi.org/10.3389/fcimb.2017.00166] [PMID: 28529927]

[108] Josse J, Laurent F, Diot A. Staphylococcal Adhesion and Host Cell Invasion: Fibronectin-Binding and Other Mechanisms. Front Microbiol 2017; 8: 2433. [http://dx.doi.org/10.3389/fmicb.2017.02433] [PMID: 29259603]

[109] Seilie ES, Bubeck Wardenburg J. Staphylococcus aureus pore-forming toxins: The interface of pathogen and host complexity. Semin Cell Dev Biol 2017; 72: 101-16. [http://dx.doi.org/10.1016/j.semcdb.2017.04.003] [PMID: 28445785]

[110] Blake KJ, Baral P, Voisin T, *et al.* Staphylococcus aureus produces pain through pore-forming toxins and neuronal TRPV1 that is silenced by QX-314. Nat Commun 2018; 9(1): 37. [http://dx.doi.org/10.1038/s41467-017-02448-6] [PMID: 29295977]

[111] Tuffs SW, Haeryfar SMM, McCormick JK. Manipulation of Innate and Adaptive Immunity by Staphylococcal Superantigens. Pathogens 2018; 7(2)E53 [http://dx.doi.org/10.3390/pathogens7020053] [PMID: 29843476]

[112] Lee RJ, Hariri BM, McMahon DB, *et al.* Bacterial d-amino acids suppress sinonasal innate immunity through sweet taste receptors in solitary chemosensory cells. Sci Signal 2017; 10(495)eaam7703 [http://dx.doi.org/10.1126/scisignal.aam7703] [PMID: 28874606]

[113] Guerra FE, Borgogna TR, Patel DM, Sward EW, Voyich JM. Epic Immune Battles of History: Neutrophils *vs.Staphylococcus aureus*. Front Cell Infect Microbiol 2017; 7: 286. [http://dx.doi.org/10.3389/fcimb.2017.00286] [PMID: 28713774]

[114] Lacoma A, Cano V, Moranta D, *et al.* Investigating intracellular persistence of Staphylococcus aureus within a murine alveolar macrophage cell line. Virulence 2017; 8(8): 1761-75. [http://dx.doi.org/10.1080/21505594.2017.1361089] [PMID: 28762868]

[115] Martinez E, Siadous FA, Bonazzi M. Tiny architects: biogenesis of intracellular replicative niches by bacterial pathogens. FEMS Microbiol Rev 2018; 42(4): 425-47. [http://dx.doi.org/10.1093/femsre/fuy013] [PMID: 29596635]

[116] Weinert LA, Welch JJ. Why Might Bacterial Pathogens Have Small Genomes? Trends in ecology & evolution 2017; 32(12): 936-47. [http://dx.doi.org/10.1016/j.tree.2017.09.006]

[117] Cervin J, Wands AM, Casselbrant A, *et al.* GM1 ganglioside-independent intoxication by Cholera toxin. PLoS Pathog 2018; 14(2)e1006862 [http://dx.doi.org/10.1371/journal.ppat.1006862] [PMID: 29432456]

[118] Pasternak Z, Njagi M, Shani Y, *et al.* In and out: an analysis of epibiotic *vs.* periplasmic bacterial predators. ISME J 2014; 8(3): 625-35. [http://dx.doi.org/10.1038/ismej.2013.164] [PMID: 24088628]

[119] Pérez J, Jiménez-Zurdo JI, Martínez-Abarca F, Millán V, Shimkets LJ, Muñoz-Dorado J. Rhizobial galactoglucan determines the predatory pattern of Myxococcus xanthus and protects Sinorhizobium meliloti from predation. Environ Microbiol 2014; 16(7): 2341-50. [http://dx.doi.org/10.1111/1462-2920.12477] [PMID: 24707988]

[120] Berleman JE, Allen S, Danielewicz MA, *et al.* The lethal cargo of Myxococcus xanthus outer membrane vesicles. Front Microbiol 2014; 5: 474. [http://dx.doi.org/10.3389/fmicb.2014.00474] [PMID: 25250022]

[121] Kumbhar C, Mudliar P, Bhatia L, Kshirsagar A, Watve M. Widespread predatory abilities in the genus Streptomyces. Arch Microbiol 2014; 196(4): 235-48. [http://dx.doi.org/10.1007/s00203-014-0961-7] [PMID: 24535490]

[122] Papenfort K, Bassler BL. Quorum sensing signal-response systems in Gram-negative bacteria. Nat Rev Microbiol 2016; 14(9): 576-88. [http://dx.doi.org/10.1038/nrmicro.2016.89] [PMID: 27510864]

[123] Abisado RG, Benomar S, Klaus JR, Dandekar AA, Chandler JR. Bacterial Quorum Sensing and Microbial Community Interactions. MBio 2018; 9(3): e02331-17. [http://dx.doi.org/10.1128/mBio.02331-17] [PMID: 29789364]

[124] Schluter J, Schoech AP, Foster KR, Mitri S. The Evolution of Quorum Sensing as a Mechanism to Infer Kinship. PLOS Comput Biol 2016; 12(4)e1004848 [http://dx.doi.org/10.1371/journal.pcbi.1004848] [PMID: 27120081]

[125] Alayande AB, Aung MM, Kim IS. Correlation Between Quorum Sensing Signal Molecules and Pseudomonas aeruginosa's Biofilm Development and Virulency. Curr Microbiol 2018; 75(7): 787-93. [http://dx.doi.org/10.1007/s00284-018-1449-5] [PMID: 29427006]

[126] Liu J, Prindle A, Humphries J, *et al.* Metabolic co-dependence gives rise to collective oscillations within biofilms. Nature 2015; 523(7562): 550-4. [http://dx.doi.org/10.1038/nature14660] [PMID: 26200335]

[127] Majumdar S, Pal S. Information transmission in microbial and fungal communication: from classical to quantum. J Cell Commun Signal 2018; 12(2): 491-502. [http://dx.doi.org/10.1007/s12079-018-0462-6] [PMID: 29476316]

[128] Prindle A, Liu J, Asally M, Ly S, Garcia-Ojalvo J, Süel GM. Ion channels enable electrical communication in bacterial communities. Nature 2015; 527(7576): 59-63. [http://dx.doi.org/10.1038/nature15709] [PMID: 26503040]

[129] Liu J, Martinez-Corral R, Prindle A, *et al.* Coupling between distant biofilms and emergence of nutrient time-sharing. Science 2017; 356(6338): 638-42. [http://dx.doi.org/10.1126/science.aah4204] [PMID: 28386026]

[130] Humphries J, Xiong L, Liu J, *et al.* Species-Independent Attraction to Biofilms through Electrical Signaling. Cell 2017; 168(1-2): 200-209.e12. [http://dx.doi.org/10.1016/j.cell.2016.12.014] [PMID: 28086091]

[131] Cao P, Dey A, Vassallo CN, Wall D. How Myxobacteria Cooperate. J Mol Biol 2015; 427(23): 3709-21. [http://dx.doi.org/10.1016/j.jmb.2015.07.022] [PMID: 26254571]

[132] Sato N, Katsumata Y, Sato K, Tajima N. Cellular Dynamics Drives the Emergence of Supracellular Structure in the Cyanobacterium, Phormidium sp. KS. Life (Basel) 2014; 4(4): 819-36. [http://dx.doi.org/10.3390/life4040819] [PMID: 25460162]

[133] Koskinen K, Pausan MR, Perras AK, *et al.* First Insights into the Diverse Human Archaeome: Specific Detection of Archaea in the Gastrointestinal Tract, Lung, and Nose and on Skin. MBio 2017; 8(6): e00824-17. [http://dx.doi.org/10.1128/mBio.00824-17] [PMID: 29138298]

[134] Ghoshal U, Shukla R, Srivastava D, Ghoshal UC. Irritable Bowel Syndrome, Particularly the Constipation-Predominant Form, Involves an Increase in *Methanobrevibacter smithii*, Which Is Associated with Higher Methane Production. Gut Liver 2016; 10(6): 932-8. [http://dx.doi.org/10.5009/gnl15588] [PMID: 27458176]

[135] Hawksworth DL, Lücking R. Fungal Diversity Revisited: 2.2 to 3.8 Million Species. Microbiol Spectr 2017; 5(4) [http://dx.doi.org/10.1128/microbiolspec.FUNK-0052-2016] [PMID: 28752818]

[136] Spatafora JW, Aime MC, Grigoriev IV, Martin F, Stajich JE, Blackwell M. The Fungal Tree of Life: from Molecular Systematics to Genome-Scale Phylogenies. Microbiol Spectr 2017; 5(5) [http://dx.doi.org/10.1128/microbiolspec.FUNK-0053-2016] [PMID: 28917057]

[137] Köhler JR, Hube B, Puccia R, Casadevall A, Perfect JR. Fungi that Infect Humans. Microbiol Spectr 2017; 5(3) [http://dx.doi.org/10.1128/microbiolspec.FUNK-0014-2016] [PMID: 28597822]

[138] Johnson L, Gaab EM, Sanchez J, Bui PQ, Nobile CJ, Hoyer KK, *et al.* Valley fever: danger lurking in a dust cloud. Microbes and infection / Institut Pasteur 2014; 16(8): 591-600. [http://dx.doi.org/10.1016/j.micinf.2014.06.011]

[139] Moreno LF, Vicente VA, de Hoog S. Black yeasts in the omics era: Achievements and challenges Medical mycology 2018; 56(suppl_1): 32-41.

[140] Fisher MC, Hawkins NJ, Sanglard D, Gurr SJ. Worldwide emergence of resistance to antifungal drugs challenges human health and food security. Science 2018; 360(6390): 739-42. [http://dx.doi.org/10.1126/science.aap7999] [PMID: 29773744]

[141] Wisecaver JH, Rokas A. Fungal metabolic gene clusters-caravans traveling across genomes and environments. Front Microbiol 2015; 6: 161. [http://dx.doi.org/10.3389/fmicb.2015.00161] [PMID: 25784900]

[142] Milner DA Jr. Malaria Pathogenesis. Cold Spring Harb Perspect Med 2018; 8(1)a025569 [http://dx.doi.org/10.1101/cshperspect.a025569] [PMID: 28533315]

[143] Pierron D, Heiske M, Razafindrazaka H, *et al.* Strong selection during the last millennium for African ancestry in the admixed population of Madagascar. Nat Commun 2018; 9(1): 932. [http://dx.doi.org/10.1038/s41467-018-03342-5] [PMID: 29500350]

[144] Nagaraja S, Ankri S. Utilization of Different Omic Approaches to Unravel Stress Response Mechanisms in the Parasite *Entamoeba histolytica.* Front Cell Infect Microbiol 2018; 8: 19. [http://dx.doi.org/10.3389/fcimb.2018.00019] [PMID: 29473019]

[145] Grace E, Asbill S, Virga K. Naegleria fowleri: pathogenesis, diagnosis, and treatment options. Antimicrob Agents Chemother 2015; 59(11): 6677-81. [http://dx.doi.org/10.1128/AAC.01293-15] [PMID: 26259797]

[146] Logan J, Navarro S, Loukas A, Giacomin P. Helminth-induced regulatory T cells and suppression of allergic responses. Curr Opin Immunol 2018; 54: 1-6. [http://dx.doi.org/10.1016/j.coi.2018.05.007] [PMID: 29852470]

[147] Zakeri A, Hansen EP, Andersen SD, Williams AR, Nejsum P. Immunomodulation by Helminths: Intracellular Pathways and Extracellular Vesicles. Front Immunol 2018; 9: 2349. [http://dx.doi.org/10.3389/fimmu.2018.02349] [PMID: 30369927]

[148] Wu D, Molofsky AB, Liang HE, *et al.* Eosinophils sustain adipose alternatively activated macrophages associated with glucose homeostasis. Science 2011; 332(6026): 243-7. [http://dx.doi.org/10.1126/science.1201475] [PMID: 21436399]

[149] Doonan J, Tarafdar A, Pineda MA, *et al.* The parasitic worm product ES-62 normalises the gut microbiota bone marrow axis in inflammatory arthritis. Nat Commun 2019; 10(1): 1554. [http://dx.doi.org/10.1038/s41467-019-09361-0] [PMID: 30952846]

[150] Marzano V, Mancinelli L, Bracaglia G, *et al.* "Omic" investigations of protozoa and worms for a deeper understanding of the human gut "parasitome". PLoS Negl Trop Dis 2017; 11(11)e0005916 [http://dx.doi.org/10.1371/journal.pntd.0005916] [PMID: 29095820]

[151] Bäckhed F, Ley RE, Sonnenburg JL, Peterson DA, Gordon JI. Host-bacterial mutualism in the human intestine. Science 2005; 307(5717): 1915-20. [http://dx.doi.org/10.1126/science.1104816] [PMID: 15790844]

[152] de Muinck EJ, Trosvik P. Individuality and convergence of the infant gut microbiota during the first year of life. Nat Commun 2018; 9(1): 2233. [http://dx.doi.org/10.1038/s41467-018-04641-7] [PMID: 29884786]

[153] Yu ZT, Chen C, Kling DE, *et al.* The principal fucosylated oligosaccharides of human milk exhibit prebiotic properties on cultured infant microbiota. Glycobiology 2013; 23(2): 169-77. [http://dx.doi.org/10.1093/glycob/cws138] [PMID: 23028202]

[154] Durack J, Kimes NE, Lin DL, *et al.* Delayed gut microbiota development in high-risk for asthma

infants is temporarily modifiable by Lactobacillus supplementation. Nat Commun 2018; 9(1): 707. [http://dx.doi.org/10.1038/s41467-018-03157-4] [PMID: 29453431]

[155] Donaldson GP, Lee SM, Mazmanian SK. Gut biogeography of the bacterial microbiota. Nat Rev Microbiol 2016; 14(1): 20-32. [http://dx.doi.org/10.1038/nrmicro3552] [PMID: 26499895]

[156] Begon M, Townsend CR, Harper JL. Ecology, individuals, populations and communities. 4th ed., Blackwell Publishing 2006.

[157] Cabral DJ, Penumutchu S, Norris C, Morones-Ramirez JR, Belenky P. Microbial competition between *Escherichia coli* and *Candida albicans* reveals a soluble fungicidal factor. Microbial cell (Graz, Austria) 2018; 5(5): 249-55. [http://dx.doi.org/10.15698/mic2018.05.631]

[158] Zhao W, Caro F, Robins W, Mekalanos JJ. Antagonism toward the intestinal microbiota and its effect on *Vibrio cholerae* virulence. Science 2018; 359(6372): 210-3. [http://dx.doi.org/10.1126/science.aap8775] [PMID: 29326272]

[159] Lurie-Weinberger MN, Gophna U. Archaea in and on the Human Body: Health Implications and Future Directions. PLoS Pathog 2015; 11(6)e1004833 [http://dx.doi.org/10.1371/journal.ppat.1004833] [PMID: 26066650]

[160] Rakoff-Nahoum S, Foster KR, Comstock LE. The evolution of cooperation within the gut microbiota. Nature 2016; 533(7602): 255-9. [http://dx.doi.org/10.1038/nature17626] [PMID: 27111508]

[161] Lefcheck JS, Byrnes JEK, Isbell F, *et al.* Biodiversity enhances ecosystem multifunctionality across trophic levels and habitats. Nat Commun 2015; 6: 6936. [http://dx.doi.org/10.1038/ncomms7936] [PMID: 25907115]

[162] McNally L, Brown SP. Building the microbiome in health and disease: niche construction and social conflict in bacteria. Philos Trans R Soc Lond B Biol Sci 2015; 370(1675): 370. [http://dx.doi.org/10.1098/rstb.2014.0298] [PMID: 26150664]

[163] Mirzaei MK, Maurice CF. Ménage à trois in the human gut: interactions between host, bacteria and phages. Nat Rev Microbiol 2017; 15(7): 397-408. [http://dx.doi.org/10.1038/nrmicro.2017.30] [PMID: 28461690]

[164] Yutin N, Makarova KS, Gussow AB, *et al.* Discovery of an expansive bacteriophage family that includes the most abundant viruses from the human gut. Nat Microbiol 2018; 3(1): 38-46. [http://dx.doi.org/10.1038/s41564-017-0053-y] [PMID: 29133882]

[165] Barr JJ. A bacteriophages journey through the human body. Immunol Rev 2017; 279(1): 106-22. [http://dx.doi.org/10.1111/imr.12565] [PMID: 28856733]

[166] Norman JM, Handley SA, Baldridge MT, *et al.* Disease-specific alterations in the enteric virome in inflammatory bowel disease. Cell 2015; 160(3): 447-60. [http://dx.doi.org/10.1016/j.cell.2015.01.002] [PMID: 25619688]

[167] Blaser MJ. The theory of disappearing microbiota and the epidemics of chronic diseases. Nat Rev Immunol 2017; 17(8): 461-3. [http://dx.doi.org/10.1038/nri.2017.77] [PMID: 28749457]

[168] Berer K, Gerdes LA, Cekanaviciute E, *et al.* Gut microbiota from multiple sclerosis patients enables spontaneous autoimmune encephalomyelitis in mice. Proc Natl Acad Sci USA 2017; 114(40): 10719-24. [http://dx.doi.org/10.1073/pnas.1711233114] [PMID: 28893994]

[169] Cekanaviciute E, Yoo BB, Runia TF, *et al.* Gut bacteria from multiple sclerosis patients modulate human T cells and exacerbate symptoms in mouse models. Proc Natl Acad Sci USA 2017; 114(40): 10713-8. [http://dx.doi.org/10.1073/pnas.1711235114] [PMID: 28893978]

[170] Castrillo G, Teixeira PJ, Paredes SH, *et al.* Root microbiota drive direct integration of phosphate stress and immunity. Nature 2017; 543(7646): 513-8.
[http://dx.doi.org/10.1038/nature21417] [PMID: 28297714]

[171] Afshinnekoo E, Meydan C, Chowdhury S, *et al.* Geospatial Resolution of Human and Bacterial Diversity with City-Scale Metagenomics. Cell Syst 2015; 1(1): 72-87.
[http://dx.doi.org/10.1016/j.cels.2015.01.001] [PMID: 26594662]

CHAPTER 3

How Bacteria Cope With Stress

Abstract: Most of the information about stress response in prokaryotes comes from studies in bacteria. Bacteria developed a set of measures to maintain homeostasis in unstable environments. In response to a particular stress, bacteria switch their transcription to an alternative mode that better serves survival in the changed environment. They do it by using alternative sigma subunits of RNA polymerase (RNAP). Some sigma factors induce a transcription switch to a particular program that adjusts bacteria to specific stress, such as heat. Other sigma factors are responsible for the general stress response that protects the bacteria from various stresses. In addition to alternative transcriptional modes, bacteria use other mechanisms to cope with stress including DNA damage stress response and programmed cell death in response to severe stress.

Keywords: Bacteria, Cold shock response, Envelope stress, General stress response, Heat shock response, (p)ppGpp, RNA thermometer, Stress, Sigma factor, Toxin-antitoxin.

INTRODUCTION

Prokaryotes live in unstable environments and face many stresses including a deficit of nutrients, non-optimal temperature, acids, alkali, salts, oxygen-derived radicals, toxic compounds, and radiation. In addition, they face competition for resources with other prokaryotes from their own and other species, and predation by viruses, bacteria, and protozoa. Pathogenic bacteria face additional stress associated with the host's immune response.

Despite their small size, prokaryotes are complex creatures, many of them encode >7000 genes with a current champion *Sorangium cellulosum* that has 11,599 genes [1]. Therefore, prokaryotes are able to build complex responses to stress. A brief description of bacterial stress responses is presented here; readers can find comprehensive information on this subject in recent reviews and books [2].

3.1. Sensing Environment

Prokaryotes have developed multiple mechanisms to monitor their internal and

Nadia Danilova

external environment and dynamically adjust their metabolic and regulatory networks to environmental changes. They use several classes of sensor proteins that include chemoreceptors, histidine kinases, Ser/Thr/Tyr protein kinases among others [3]. Bacteria achieve fast stress response through two major mechanisms, dynamic remodeling of transcriptional networks and control of protein degradation [4]. For fast and direct response they use two-component regulatory systems (TCSs) that in their simplest form contain two proteins: a sensor histidine kinase and a cognate response regulator [5, 6]. The kinase autophosphorylates in response to a specific environmental signal and passes a phosphate group to the response regulator. Phosphorylation of a response regulator induces its conformational change, activation, and regulation of its targets. Numerous variations provide adaptations for the response to a specific stressor. Nutrient sensing and uptake, chemotaxis, biofilm formation, and other processes are regulated by these systems.

Human pathogen *Staphylococcus aureus* has 16 TCSs [7]. It does not need these systems under growth arrest conditions and needs only one system, WalRK, under stable replicative conditions. Therefore, *S. aureus* TCSs are devoted almost exclusively to response to stressful changes in the environment. Each system is autonomous and self-sufficient although cross-regulation between systems also takes place. This organization allows a fast and flexible response. The TCSs can cooperate and cross-talk with other mechanisms, such as eukaryotic-like Ser/Thr kinases and phosphatases (eSTK/Ps) [8].

Bacteria can also globally switch their transcription to a different mode more appropriate for the changing environment. Therefore, they use alternative sigma subunits of RNA polymerase (RNAP) [9]. Alternative sigma factors are associated with RNAP, replacing the primary housekeeping σ factor, and directly activating or repressing subsets of genes. Different sigma factors respond to different sets of environmental signals, allowing bacteria to quickly adjust to specific environments. *Escherichia coli* has a primary housekeeping sigma factor σ70 and six alternative σ factors. They include a general stress response regulator σS and σ factors responsible for more specialized responses associated with heat shock, cell envelope stresses, flagellar motility and chemotaxis, and iron-citrate transport. Other bacteria have anywhere from 2 to more than 50 σ factors [10]. Several sigma factors can respond to the same signal, and different sigma factors can regulate the same genes in response to dissimilar signals, creating a complex network of regulation.

The availability of σ factors is regulated at various levels [11]. For example, synthesis of *E. coli* σS is increased during stress and its level is regulated by proteolysis in dependence on stress level. Partner-switching systems (PSS) common in Gram-positive bacteria are based on the interaction between the anti-sigma factor and its antagonist. The latter can be phosphorylated, in which state it does not bind the anti-sigma. Stress induces dephosphorylation of the antagonist, which then can bind anti-sigma. It leads to the release of the sigma factor from sequestration by the anti-sigma [12]. Environmental signals can be delivered to PSS through a two-component system or from specialized chemosensory systems [13].

3.2. Heat and Cold Shock Responses

An example of a specialized σ factor is σ32 of *E. coli*, which regulates heat shock response (HSR). *rpoH* mRNA encoding σ32 is a thermosensor. It has a secondary structure that prevents its translation at normal temperature. This structure melts at elevated temperatures allowing translation and accumulation of σ32 protein even before damage to other macromolecules occurs [14]. In addition, activity and stability of σ32 is controlled by protein chaperones that sense the presence of unfolded proteins in the cytoplasm and membranes [15]. σ32 enhances the production of chaperones that assist protein folding and can refold misfolded proteins. It also stimulates the production of ATP-dependent proteases that degrade proteins that cannot be restored. In addition, cells increase the production of low-molecular-weight compounds such as trehalose that protect macromolecules. σ32 regulon includes proteins that function in metabolism, membrane, DNA repair, RNA processing, transcription, translation, and transport. Therefore, in addition to protecting protein homeostasis, s32 regulon protects DNA, ribosomes, and membranes. The fluidity of membranes is maintained relatively constant at various temperatures, which is achieved by modification of lipid composition [16].

Some mRNAs have built-in sensory elements in the 5'-UTRs that regulate mRNA translation in dependence on temperature. They are called RNA thermometers (RNATs). In pathogenic bacteria, they enable a rapid upregulation of translation of some virulence factors after the bacterium enters a mammalian host where the environmental temperature rises [17]. Pathogenic bacteria have another mechanism to adjust to host temperature, which is called high-temperature response (HTR). It is different from the stress response and results in a permanent upregulation of genes needed for the new environment [18]. Similarly, the low-temperature response induces a constant expression of the appropriate genes.

Cold shock response (CSR) is not regulated by sigma factors but by several protein families. Proteins of Csp family bind mRNAs and regulate their secondary structure to promote translation [19]. CspA also regulates ribonucleotide and carbohydrate metabolism [20]. Induction of the CspA during CSR is attributed to the higher stability of its mRNA at low temperatures (10–20 s at 37°C and >30 min at 16°C). It suggests that thermosensor mRNAs are core to both HSR and CSR. Analogously, in animal cells, cold shock response is associated with decreased proliferation, which is mediated by cold-inducible RNA-binding protein (CIRP) that upregulates transcription of the cell cycle inhibitor p27Kip1 [21]. Cold shock response also involves increased levels of several ribosome-binding proteins, such as initiation factors that improve translation [22]. Overexpression of trehalose is also a common response to cold. Changes in the physical state of the membrane induced by low temperature are sensed and transmitted to the transcription of genes that encode enzymes catalyzing desaturation of fatty acid chains, which increases the membrane's fluidity [16, 23]. Cis-unsaturated fatty acids increase fluidity more efficiently than trans-forms and are specifically produced during cold stress. The cells transition back to transforms with increased temperature.

3.3. General Stress Response

In a media rich with nutrients, bacteria grow exponentially; with depletion of nutrients, they stop growing and enter a stationary phase when they do not divide, do not move and basically in a survival mode. In that state, they are less sensitive to any stress. It is part of the general stress response that is regulated by sigma factor σS encoded by rpoS gene [24]. σS mediates resistance to multiple stress including high cell density, energy shortage, starvation, cold, heat, high osmolarity, acidic pH, dehydration, and UV/DNA damage [25, 26]. Diverse stresses affect σS in different ways; they can increase its transcription, translation, or stability. For example, σS transcription is regulated by cAMP and by alarmones that are accumulated in cells experiencing starvation. High osmolarity, low pH, and cold activate rpoS translation. A complex protein network regulates σS proteolysis and input from various stresses can lead to σS stabilization. In general, changes in σS transcription provide long-term changes, whereas the inhibition of its proteolysis is responsible for the fast response. σS binding to RNAP core is another target of σS regulation; it is increased in the presence of alarmones and accessory proteins. σS is similar to the basic sigma factor σ70 in structure and RNAP holoenzymes containing both sigma factors can *in vitro* transcribe the same DNA templates. σS selectivity *in vivo* is achieved by different affinity to various promoters and usage of additional regulatory proteins.

Switching from the primary housekeeping sigma factor σ70 to σS leads to a change in the expression of 23% of the *E. coli* genes [27]. Some of these genes are transcriptional factors suggesting that some σS effects are indirect. σS targets include metabolism, storage compounds like glycogen, protective compounds like trehalose, changes in cell morphology (small ovoid cells instead of rod-shaped), membrane alterations, adhesive extracellular matrix, and control of biofilm formation. Collectively, these changes provide a transition from motile lifestyle to the stationary phase characterized by maintenance metabolism, stress resistance, high cell density, and cell adherence. One of the σS targets, Dps (DNA-binding protein from starved cells) protein contributes to survival by two mechanisms: binding to and protecting DNA and oxidizing Fe^{2+} ions to Fe^{3+} thus reducing the formation of damaging hydroxyl free radicals *via* the Fenton reaction [28]. There are variations for the use of σS factor in other gram-negative bacteria, and in some species, a general stress response is not mediated by sigma factors.

In gram-positive *Bacillus subtilis,* general stress response is regulated by σB [29]. σB activity is induced by two phosphatases: RsbP phosphatase mediates energy (nutritional) stress and RsbU mediates such environmental stresses as acid, osmotic, heat, and UV light. In the absence of stress, σB is bound by an anti-sigma factor RsbW, which phosphorylates its antagonist the anti-anti-sigma factor RsbV preventing it from binding RsbW. Stress activates RsbP or RsbU phosphatases, which dephosphorylate RsbV allowing it to bind RsbW; this switch of partners releases σB. In addition, a decrease in ATP due to energy stress decreases the ability of RsbW to phosphorylate RsbV, therefore RsbW is an energy stress sensor itself. Relative stoichiometry of the biosynthesis rates of σB and its regulators provides an ultrasensitive negative feedback loop for this network [30]. Activation of RsbU phosphatase that conveys environmental stress signals is more complex; it involves several upstream regulators and takes place in a 1.8 MDa stressosome complex [31]. σB is modulated by several sets of genes that respond to different combinations of stress signals [32].

3.4. Envelope Stress Response

In *E. coli* and other gram-negative species, the envelope is made of the outer membrane, the inner membrane, and the periplasmic space in between that contains the peptidoglycan cell wall (Fig. **3**). The outer membrane is an asymmetric bilayer with the lipopolysaccharide (LPS) layer exposed to the surface and the phospholipid layer facing the periplasm. The outer membrane is connected to the peptidoglycan wall through Lpp protein [33]. It has channels that permit the entrance of hydrophobic compounds into the periplasm; that is how b-lactam antibiotics enter periplasm to inhibit peptidoglycan wall synthesis. The inner membrane is a phospholipid bilayer with the electron transport chain proteins embedded into it. Periplasm also contains flagella motors and various sensors and chaperones that deliver proteins to the outer membrane during biosynthesis. Thus, cell envelope is a complex structure, and its correct biosynthesis and function are vital for survival. It is vigilantly monitored. Among the most important *E. coli* envelope stress response systems are an alternative sigma factor E that maintains the homeostasis of the outer membrane and Cpx two-component system that monitors the inner membrane [34].

σE is encoded in an operon which also includes σE negative regulators *rseA* and *rseB*. RseA is a protein crossing the inner membrane; its periplasmic domain binds to RseB and its cytoplasmic domain binds to σE, preventing its inclusion into RNA polymerase. A well-studied example of this system's activation is through outer membrane porins, proteins that form channels. When they are misfolded, they expose a normally buried hydrophobic sequence. This sequence then binds DegS protease, which removes the periplasmic domain of RseA making it sensitive to other proteases that degrade the remaining RseA and release σE. Different porins activate cascade differently [35]. Additional fine-tuning of this system is provided by RseB, which recognizes mistakes in intermediates in lipopolysaccharide transport and assembly and integrates this signal into RseA pathway [36]. σE regulates about 60 genes involved mostly in maintaining cell envelope including periplasmic chaperones. After resolution of stress, the response is repressed by feedback mechanisms that include sRNAs [37, 38]. sRNAs are responsible for the suppression of outer membrane biosynthesis during stress [34]. σE systems vary between bacterial species.

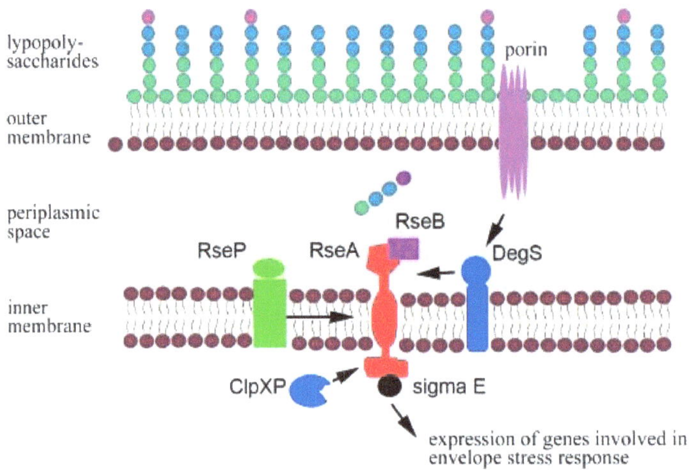

Fig. (3). σE is activated by stress signals from the outer membrane. In absence if stress, σE is bound by RseA. Signals from outer membrane activate DegS protease that removes periplasmic domain of RseA initiating its destruction by RseP and ClpXP. RseB binds lipopolysaccharide intermediates and regulates RseA.

Cpx that monitors the inner membrane is a typical two-component system [34].

CpxA is an inner membrane histidine kinase whose periplasmic domain is responsible for signal detection. Cpx is activated by misfolded proteins in the periplasm, alkaline pH, high salt, alterations in membrane lipids, and defects in peptidoglycan wall. Upon induction, CpxA autophosphorylates and passes a phosphate to the response regulator CpxR, which binds cognate sequences on the DNA to regulate gene expression. Cpx regulon includes genes for proteins that decrease cell envelope stress such as periplasmic foldases and proteases that degrade misfolded and damaged proteins. NipE is an outer membrane lipoprotein that activates CpxA upon cell attachment to hydrophobic surfaces. CpxP is a negative regulator. Cpx activation also results in suppression of the outer membrane biosynthesis through upregulation of regulatory sRNAs [34].

Rcs (regulator of capsule synthesis) monitors the outer membrane and peptidoglycan wall in enterobacteria [33, 39, 40]. A major sensor of this system is outer membrane lipoprotein RcsF, which after induction interacts with the inner membrane signaling protein IgaA. Rcs is activated by defects in synthesis and assembly of the outer membrane, increased osmolarity, desiccation, and growth on a solid surface. It also responds both to antimicrobial peptides that usually bind the negatively charged surface LPS and to beta-lactam antibiotics that interfere with peptidoglycan biosynthesis. After activation, Rcs induces stress-related mechanisms including the synthesis of extracellular polysaccharides that form a protective capsule; it is also used in biofilm formation [41].

The phage shock protein (Psp) system was identified as a response to filamentous phage infection, although later studies revealed that Psp senses not the bacteriophage itself but changes in the inner membrane permeability caused by the phage [42]. Psp is induced by phages, ionophores, ethanol, hyperosmotic shock, and mislocalization of outer membrane proteins secretins that form export channels [43]. The key component of this system is PspA, which directly binds membrane in dependence on its anionic lipid content and stored curvature elastic stress [44]. Similar systems are found in Gram-negative and Gram-positive bacteria and chloroplasts [44, 45]. Psp response is important for the survival of pathogenic bacteria in macrophages, biofilm formation and antibiotic resistance.

Gram-positive bacteria such as *B. subtilis* have several responses to the envelope stress [46]. Several σ factors such as σW, σM, and σX are involved in envelope stress response caused by antibiotics and membrane perturbations. Inducing conditions for these systems partially overlap. σM regulates genes involved in cell wall biosynthesis and cell division. In addition, it regulates genes involved in detoxification and DNA repair. σM is activated by heat shock, high salinity, superoxides, antibiotics, and low pH. In contrast, high pH activates σW. It also regulates the detoxification of antibiotics and production of bacteriocins targeting other bacteria. σX regulates enzymes that decrease the net negative charge of the membrane; this can decrease susceptibility to cationic antimicrobial peptides. Sigma V factor responds to lysozyme [47]. Each sigma factor is held in an inactive state by a specific membrane-bound anti-sigma factor.

E. coli σE is a prototype of the big family of extracytoplasmic function (ECF) sigma factors. They represent the most abundant regulatory systems in bacteria. Currently, in 7679 bacterial genomes, there are 38 000 predicted ECF sigma factors [48]. Many ECF pathways have important functions in bacterial physiology at non-stressful conditions as well.

3.5. Starvation

Bacteria size and growth rate depends on nutrients availability and can vary several folds [49]. Lack of nutrients is a major stress for any species. In most bacteria, the first line of defense against starvation is the so-called stringent response mediated by the alarmones guanosine 3'-diphosphate 5'-triphosphate (pppGpp) and guanosine 3'-diphosphate 5'-diphosphate (ppGpp) [50]. (p)ppGpp are synthesized and hydrolyzed by homologs of *E. coli* enzymes RelA and SpoT. Uncharged tRNAs accumulate during amino acid starvation and if such tRNA binds ribosome A site, it blocks translation. RelA binds to an uncharged tRNA without being activated. However, if this complex binds ribosome, RelA is activated and catalyzes the synthesis of (p)ppGpp [51]. Rel enzyme activity is also regulated by branched-chain amino acids such as valine and isoleucine that bind its carboxyl end [52].

Increase in (p)ppGpp nucleotides induces a global physiological response leading to arrest of processes associated with cell growths and to activation of stress response genes. It includes inhibition of synthesis of all major macromolecules and arrest of ribosome and membrane biogenesis and of DNA replication. (p)ppGpp can bind RNA polymerase modulating its activity, or influence it indirectly. For example, binding of alternative sigma factors to RNAP is increased in the presence of (p)ppGpp [50]. (p)ppGpp can also regulate the production of branched-chain amino acids by binding to riboswitch regions in mRNAs [53]. (p)ppGpp increases survival during various stresses and contributes to the virulence of pathogenic bacteria [54, 55]. Among (p)ppGpp effects is increased production of antibiotics such as bacilysin in *B. subtilis* [56]. (p)ppGpp mediates the phenomenon of persistence when a few cells in a population of antibiotic-sensitive bacteria become multidrug tolerant [57].

RelA/SpoT homologs (RSH) are ubiquitous throughout the bacterial kingdom. They are classified as the long RSH enzymes, the small alarmone synthetases, and the small alarmone hydrolases that differ in their activation and regulation [58]. (p)ppGpp is an important regulator of chloroplasts in plants that affects their growth and development [59].

3.6. Oxidative Stress

Reactive oxygen species (ROS) are continuously produced in aerobic bacteria during normal metabolism and their production can increase during many stresses. Some antibiotics kill bacteria by inducing oxidative stress [60]. ROS and other reactive species can also originate in the environment. Similar to other organisms, bacteria cope with oxidative stress using multiple mechanisms, the main is increased production of compounds such as glutathione that protect protein thiol groups and upregulation of enzymes that turn reactive species into harmless compounds. Bacteria also activate repair of oxidized macromolecules and increase the activity of proteasomes that degrade damaged proteins [61, 62].

In *E. coli* major regulators that respond to oxidative stress are OxyR, SoxRS, and RpoS [63]. Cysteine residues of OxyR are oxidized in the presence of hydrogen peroxide leading to the formation of intra-protein disulfide bonds and to a conformation change allowing DNA binding. OxyR controls a regulon of almost 40 genes that protect cells from hydrogen peroxide toxicity. This regulon includes a catalase that catalyzes the decomposition of hydrogen peroxide into water and oxygen. This response is terminated by thioredoxin reductase, which inactivates the activated form of OxyR. The reduced OxyR then represses *oxyR* gene.

Another important *E. coli* oxidative stress response system is SoxRS (superoxide response). SoxR, contains [2Fe-2S]-clusters, which are oxidized in the presence of superoxide radicals leading to a conformation change, DNA binding, and activation of SoxS expression. That leads to changes in expression of ~100 of cognate genes including superoxide dismutases.

Oxidative stress in *E. coli* leads to induction of RpoS (σS), which mediates general stress response [64]. In absence of stress, MqsA represses *rpoS*, whereas upon oxidative stress, MqsA is degraded by Lon protease, resulting in activation of *rpoS*.

OxyS is a small RNA that promotes cell cycle arrest in response to oxidative stress in an unusual way. It inhibits expression of the transcription termination factor *nusG*, which leads to a read-through transcription into the cryptic prophage encoding KilR protein that interferes with cell division [65].

E. coli oxidative stress response controllers are functionally conserved in a wide range of bacteria. Other sensors likely exist that use the same principle of reversible oxidation of sensitive groups. One recently discovered sensor in *Streptococcus pneumoniae* relies on single cysteine oxidation, which leads to a conformation change in the sensor's DNA binding region [66].

The preferred method of reactive species inactivation depends on the bacterium's lifestyle. For example, reactive nitrogen species (RNS) are removed by an oxidative process aerobically, but in anaerobic conditions, they are removed by reduction [67, 68].

Pathogenic bacteria need an efficient oxidative stress response to colonize their hosts and avoid host immune mechanisms that rely on ROS [69 - 72]. The oxidative stress response can be activated not only in response to oxidative stress but also in other adverse conditions [73].

3.7. Osmotic Stress

Change in the outside osmotic pressure beyond the optimal range is stress for bacteria. *E. coli* respond to hyperosmotic stress by a slowdown of translational elongation due to a reduced rate of tRNA binding to the ribosome [74]. In addition, bacteria increase the intracellular concentration of K^+ or compatible solutes. Most common compatible solutes include mannitol, trehalose, proline, glycine betaine, ectoine, and b-dimethylsulfoniopropionate [75]. Osmoprotectant transporters that bring compatible solutes inside the cell sense and respond to osmotic stress without the help of other regulators. They vary in specificity; for example, BetP is specific to glycine betaine whereas ProP accepts glycine betaine, proline betaine, ectoine, and other compounds. Initial responses are very fast and based on already active transporters and enzymes. In addition, bacteria upregulate expression of genes that encode transporters and enzymes producing compatible solutes. This provides a long-term adjustment and may involve more regulators and fine-tuning to the environment.

To survive osmotic downshift, cells limit water uptake by decreasing the inside concentration of solutes. Bacteria do so by opening membrane mechanosensitive channels to release solutes into the environment. Osmotic homeostasis in Group B Streptococcus is regulated by cyclic di-AMP [76]. It controls BusR transcription factor that regulates potassium and osmolyte transporters.

3.8. pH Stress

The optimum cytoplasmic pH for most bacteria is near neutral (7.0) while the pH of the environment varies. Non-optimal pH can compromise enzymatic activity, energy production, and can cause damage to macromolecules. The ability to withstand acidic and alkaline stress is very important as illustrated by the fact that pH is the main factor that determines the composition and distribution of bacterial and archaeal communities in various soils [77]. It is also very important for pathogenic bacteria that need to survive pH between 1.5 and 3 in the human stomach and the low pH in phagolysosomes after being phagocytosed by macrophages. *E. coli* and other enteric bacteria have a high tolerance of acidic stress [78, 79].

Bacteria often sense pH using various two-component systems, whose sensors respond to pH by conformational changes [79]. In bacteria that have a high tolerance of acids, several different systems sense and respond to acidic stress. In *H. Pylori*, two sensor kinases respond to acidity; medium acidity activates ArsS, and high acidity – FlgS.

Adjustment to an acidic condition is associated with changes in expression of many genes. For example, alpha-proteobacteria *Rhizobium tropici* respond to a shift from pH 6.8 to pH 4.5 by activation of 351 genes and repression of 32 [80]. Several regulatory pathways are involved in acidic stress response, including an alternative sigma factor S, two-component signal transduction systems (PhoP/Q, BvgAS), and main iron regulatory protein Fur. Among other factors involved in acidic stress response are ClpXP protease and nucleic acid-stabilizing proteins such as Dps.

Gram-positive *B. subtilis* also respond to acidification by changes in expression of multiple genes [81]. They include the general stress response genes and SigM regulon. *B. subtilis* also upregulate an ECF-like atypical two-subunit σ factor consisting of subunits SigO and RsoA [82].

Bacteria use several mechanisms to adjust to acidic stress [78, 79]. First, they remove extra protons from the cytoplasm by increasing their export. They do that by upregulation of components of the electron transport chain and by decreasing the expression of ATPase that brings protons into the cell in the process of ATP synthesis. In addition, alkalization of the cytoplasm is achieved by activation of amino acid decarboxylases that consume protons at low pH. The main protection against low pH is provided by glutamate decarboxylase that consumes one proton per catalytic cycle during the conversion of glutamate into gamma-aminobutyrate and carbon dioxide; then a membrane antiporter exchanges the product for more glutamate [83]. Arginine and lysine are used in an analogous way. Several systems generate ammonia from various substrates including arginine, glutamine, and adenosine. Ammonia combines with protons to generate the ammonium ions thereby increasing cytoplasmic pH [79]. In anaerobic condition, protons can be converted to hydrogen gas by hydrogenases [84]. *H. Pylori* uses urease to alkalize the periplasm [79].

Decreasing membrane fluidity and thus permeability to protons restricts the influx of protons to the cytoplasm. Membrane fluidity is decreased through a change in surface proteins and lipids *e.g.* by increasing the concentration of saturated lipids and cyclopropane fatty acids [85]. In some species, unsaturated fatty acids are modified by the addition of a methyl group to the double bonds [79]. Blocking outer membrane porins by polyphosphate or cadaverine further reduces proton influx into the periplasm. *Francisella tularensis* adjust their membranes to acidic pH by adding an additional hexose to the diglucosamine backbone of lipid A [86].

In periplasm and cytoplasm, chaperones manage the effects of acid damage. Periplasm is especially susceptible to acidic damage due to proton leakage through the outer membrane. It has specialized non ATP-dependent chaperones HdeA and HdeB that are activated by low pH and bind acid-denatured proteins [87]. They then stabilize their clients and release them, when pH normalizes. Cytoplasmic chaperones DnaK and GroEL are also induced by acidic stress [79]. Cytoplasmic chaperone Hsp31 can function as a holdase; it binds to and stabilizes unfolded proteins until pH stress is over; the proteins are then released and can refold spontaneously or with the help of ATP-dependent chaperone systems [88]. Hsp31 is also activated by heat stress. Some bacteria respond to low pH by a regulated growth arrest [89].

The alkaline stress response in bacteria is mediated by several mechanisms. Active transport of protons into the cell is achieved by upregulation of cation/proton antiporters such as Na^+/H^+ antiporter NhaA of *E. coli* [90]. Proton-pumping respiratory chain's activity decreases at alkaline pH [91]. Bacteria also decrease the expression of respiratory chain components while increasing the expression of non-proton pumping cytochrome and ATP synthase. Upregulation of amino acid deaminases and catabolic pathways that produce organic acids also helps neutralize the alkalinity. These mechanisms allow bacteria to maintain neutral or near neutral cytoplasmic pH in environments with variable pH [92].

3.9. Metals

Many metals are essential components of cells. The deficit of metals can compromise cellular functions and excess of even essential metals can be toxic. Bacteria have developed mechanisms to sense important metals, to acquire and store them, and to regulate their intracellular concentrations. Bacteria also have mechanisms for detoxification and export of toxic metals. Metal metabolism is especially critical for symbiotic and pathogenic bacteria. Host immune system tries to deprive invaders of metals especially iron, zinc, and manganese [93, 94]. After phagocytosis of pathogens, the immune system uses the opposite strategy and tries to poison them with an excess of metals [95].

Uptake and usage of essential metals is regulated by several mechanisms. Usually, metalloregulators directly bind their effector metal. For example, Fur (ferric uptake regulator) family proteins bind iron [96]. Iron levels can also be monitored through iron-sulfur clusters. Mg (II) and several other metals bind riboswitches. Riboswitch upstream from the gene, which encodes Mg uptake protein, adopts a compact conformation after Mg binding, which favors transcription termination [97]. This stops production of Mg uptake protein when enough Mg is present and helps to avoid Mg toxicity.

Iron is the most abundant metal in bacterial cells. To acquire iron, bacteria produce low-molecular-weight compounds, known as siderophores that have an extremely high affinity for Fe(III). The more soluble Fe(II) is acquired through permeases. When iron is depleted, bacteria increase the synthesis of these compounds. When iron is in excess, bacteria upregulate mechanisms of iron storage and efflux.

Iron's binding by Fur leads to a conformational change that makes Fur competent to bind DNA. The fully active form of Fur is a dimer with two Zn(II) and up to four Fe(II) atoms [97]. Fur directly suppresses promoters of up to 100 genes including those involved in iron uptake. At the same time, it upregulates the synthesis of such iron-binding proteins as bacterioferritin, superoxide dismutases, and fumarase. In *B. subtilis*, Fur regulon is derepressed gradually in three waves with the decrease of iron level [98]. First, uptake systems are induced including transporters for Fe^{2+}, ferric citrate, and siderophore petrobactin. Next, *B. subtilis* induces synthesis of its own siderophores and flavodoxins. If iron level declines further, the bacterium induces an "iron-sparing" response that blocks the translation of abundant iron-using proteins and restricts iron use to only essential iron-dependent enzymes.

Many virulence genes are regulated by Fur. For example, *Listeria monocytogenes* FrvA virulence protein is a Fe(II) exporter that protects the bacterium from iron intoxication [99].

Fur family also regulates genes involved in oxidative stress response and energy metabolism. Pro-oxidative and anti-oxidative properties of various metals explain their involvement in oxidative stress. A member of the Fur family PerR, which is present in most gram-positive bacteria, evolved to sense peroxide through iron-mediated histidine oxidation [100].

Zinc level is regulated by a paralog of Fur zinc-sensing zinc uptake regulator (Zur) [101]. Zur is also a dimeric protein with both a structural Zn(II) site and a zinc-sensing site that activates and binds DNA in Zn(II)'s presence. When zinc levels drop, operons regulated by Zur get derepressed, which first leads to mobilization of internal Zn(II) resources and then to stimulation of its import [102]. An excess of zinc in *B. subtilis* is sensed by a separate regulatory protein CzrA [97]. It is a repressor that dissociates from DNA in the presence of Zn(II) inducing two Zn efflux systems.

3.10. DNA Damage Response

Most bacterial responses to DNA damage are coordinated by "SOS response", which involves more than forty genes [103 - 105]. SOS systems vary significantly among bacteria species but are based on the same principle; in the absence of DNA damage, LexA represses promoters of genes that mediate SOS response. The repression is relieved during DNA damage. Evelin Witkin hypothesized the existence of such repressor in 1967 [106].

N-terminal domain of Lex A is responsible for DNA binding whereas the C-terminal domain is responsible for dimerization and self-protease activity. LexA self-cleavage is ineffective in the absence of DNA damage. SOS response is initiated by RecA, a DNA damage sensor that binds single-stranded DNA. The emergence of ssDNA during DNA damage can be caused by incomplete DNA replication or processing of double-strand breaks. RecA binds ssDNA and forms filaments that act as a signal of DNA damage for LexA. LexA binds to the RecA filament and assumes a conformation that favors self-cleavage [107]. Autoproteolysis of LexA derepresses the downstream genes that mediate SOS response. The time of their transcription initiation depends on the kinetics of LexA binding to target promoters. SOS genes are divided into "early", "middle", and "late" [108]. The first (early) genes to be induced belong to the nucleotide excision repair pathway, a high fidelity repair mechanism that accurately fixes a broad spectrum of lesions. Stronger damage leads to induction of "middle", genes involved in homologous recombination repair and cell cycle arrest. If SOS signaling persists for more than 40 minutes, it activates error-prone DNA polymerases that are capable to repair DNA bypassing lesions, while at the same time, introducing mutations [109, 110]. In this way, SOS response is graded to the extent of DNA damage; little damage results in increased repair whereas a prolonged extensive damage results in a switch to a survival strategy that stimulates bacterial evolution in the face of danger. In addition, SOS induces bacteriophages and increases lateral gene transfer [111, 112]. Antibiotics can trigger SOS response, which can lead to the development of antibiotic resistance by stimulating gene transfer and recombination events [113]. ssDNA that arise through several mechanisms of lateral gene transfer such as conjugation, transformation, and transduction also can induce SOS response [114].

In the absence of ssDNA, LexA will quickly restore the repression of its target genes. Additional tuning of SOS response is achieved at the stage of RecA filament formation by proteins that help to unload RecA on the ssDNA or inhibit RecA binding to ssDNA. There are also SOS-independent mechanisms of response to DNA damage [115].

Survival of bacterial species strongly depends on the effectiveness of DNA replication and repair. Radiation and desiccations-resistant *Deinococcus radiodurans* can reassemble its genome from shattered fragments [116, 117]. DNA repair can protect the organism from stress [105]. Stress tolerance also depends on the effectiveness of repair systems for RNA [118] and proteins [119], Mn(II) metabolism [120], epigenetic mechanisms [121], and other factors.

3.11. Multiple Stresses

Prokaryotes frequently experience several stresses simultaneously. For example, pathogenic bacteria such as *Listeria monocytogenes*, which contaminates food and causes listeriosis, experience several different stresses before and after entering its human host [122]. Before infection, this bacterium needs to survive food sterilization by heat or irradiation and then storage at low temperatures. When it enters the human body, it needs to survive the increased temperature, low pH of the stomach, starvation, and oxidative stress in phagocytes.

Bacteria have developed a complex system of integration of diverse signals. For example, starving cells increase *rpoS* transcription through cAMP and ppGpp, whereas high osmolarity activates *rpoS* translation [123]. In *C. crescentus*, replication initiation factor DnaA is a common point of stress response control. Proteotoxic stress induces degradation of DnaA, whereas starvation reduces *dnaA* translation [124]. DnaA level thus serves as an indicator of cumulative stress, and its low level leads to cell cycle arrest. In archaea *Halobacterium salinarum*, regulatory networks responsible for adjustment to different stressors, show significant overlap [125].

Bacteria respond to many stresses by attachment to solid surfaces and producing substances that form the extracellular matrix that encloses and protects the cells. This results in biofilm formation [37, 41, 126].

Often, exposure to one stress increases tolerance to others through changes in the general stress response mechanisms, metabolism, and shared stress-response pathways such as the production of chaperones and small compounds protecting macromolecule and increase in molecular repair [22, 127].

3.12. Programmed Cell Death in Bacteria

The concept of programmed cell death (PCD) has been developed for multicellular organisms where suicidal death of stressed, defective, damaged, and infected cells benefit the organism as a whole [128, 129]. In these organisms, PCD is also used for killing healthy cells after they served their purpose during normal development, organ sculpturing, and immune response [130]. Since bacteria were considered to be solitary organisms, initial reports about characteristic markers of PCD in bacteria exposed to antibiotics were met with skepticism [131]. Now, the idea that PCD originated at the prokaryotic stage of evolution, is commonly accepted [132, 133]. Although bacteria are unicellular organisms, they often act cooperatively guided by quorum sensing, electric signaling, and other mechanisms. In addition, they can form multicellular structures such as biofilms, filaments, and fruiting bodies where cells have different roles depending on their position in the structure. This may explain the occurrence of PCD in bacterial communities. Presumably, the death of compromised cells increases the fitness of the prokaryotic community. In addition, the community benefits from nutrients released by dying cells. Several types of PCD are suggested to operate in bacteria.

PCD is often mediated by toxin-antitoxin systems. TA consists of a stable toxin and unstable antitoxin encoded and co-transcribed off the same operon [134]. Toxins interfere with basic cellular functions such as energy production, DNA replication, translation, cell wall synthesis, and cell division, which interference leads to growth arrest or cell death [135 - 137]. Antitoxins neutralize their cognate toxins by inhibition of their synthesis, activity, by blocking their binding to the target, or by decreasing their stability. Antitoxins should be produced continuously to prevent toxin activity. TA systems are involved in translational regulation of stress response, biofilm formation, and genesis of persister cells [134]. Stress can activate specific mechanisms that degrade antitoxin, for example by proteolysis letting the toxin kill the cell. TA can also kill cells non-specifically. Any stress causing a decrease of translation can lead to cell death since the concentration of unstable anti-toxins would drop faster than that of their more stable toxins.

TA systems are classified by their mode of action and the nature of antitoxin into five types. The most common is type II TA where inhibition of protein toxin is achieved by protein antitoxin [138]. Usually, prokaryotic genomes encode multiple TAs. Therefore more than one pathway may lead from stress to growth arrest or cell death. In *E. coli*, a synchronized activation of several TA systems mediated by (p)ppGpp has been observed [139].

One of the best-characterized TA systems is MazE-MazF of *E. coli* [140]. MazF toxin inhibits translation by cleaving both mRNAs and ribosomal RNA [141]. MazE antitoxin is degraded by stress-activated cellular proteases Lon and Clp. This system responds to starvation, DNA damage, and other stresses.

Some antibiotics induce apoptosis-like death (ALD), which depends on SOS response [142, 143]. In addition to this death pathway, *Mycobacterium smegnatis* has a pathway independent of traditional LexA/RelA-mediated SOS response. It involves DNA repair genes and RelA [115]. Restriction endonucleases also can mediate ALD [144]. ALD is induced by starvation in a subpopulation of cells and survived cells benefit from nutrients released from dead cells. Interestingly, the starving cells do not use all the nutrients released by dying cells, instead, the growth is limited both by available resources and the population density [145]. Another form of regulated cell death, ferroptosis, is caused by iron-dependent lipid peroxidation [146].

3.13. Phenotypic Diversity and Persisters

Some bacteria increase their stress tolerance by generating phenotypic variations in the growth, metabolic activity, morphology, gene expression, and death patterns. They can do this by several mechanisms. Bacterial genomes can encode several variants of the same protein. One of the best-studied examples is D1 protein encoded by *psbA* gene in cyanobacteria and chloroplasts [147]. This protein functions in photosystem II enzyme complex that mediates light-induced oxidation of water into molecular oxygen. Cyanobacteria have multiple *psbA* genes that are expressed in specific conditions, including a variant that is expressed at nighttime. When this D1 variant is incorporated into photosystem II, the resulting complex is incapable to function in water oxidation thus allowing bacteria to switch to nitrogen fixation at night.

Bacterial pathogens often generate phenotypic variations by using simple DNA repeats to randomly switch genes on and off. This process is called phase-variation. Switching N6-adenine DNA methyltransferase gene ON and OFF changes methylation of the genome, leading to changes in gene expression [148]. Diverse populations can survive particular stress better. For example, some bacteria have an ability to switch to L-phase, a wall-free state [149]. This form survives better stress associated with inhibition of the peptidoglycan cell wall synthesis [150].

Antibiotic treatment often kills the majority of bacteria except for a small population, called persisters [151]. Those are characterized by slow growth rates at the time of treatment, which is thought to originate from stochastic variations in biochemical processes in individual bacteria [152]. Phenotype switching to dormancy or slow growth at high density protects such subpopulation against many stresses including antibiotic treatment [153]. TA systems play a role in the formation of persister cells through induction of dormancy [154]. Persisters are thought to be responsible for antibiotic treatment failure and relapsing infection; they also contribute to the development of antibiotic tolerance [155]. Persisters may be especially important for survival in cases of fast, severe, and unusual stress.

3.14. Stress of Living Together

Bacterial species use both cooperation and competition as their survival strategies. Some species form stable alliances especially when one species consumes metabolites produced by the other [156]. Such cross feeding often serves as a base for diversity observed in microbial communities. In contrast, bacteria that use the same resources are often involved in fierce competition [157]. In passive competition two strains compete by restricting the competitor's access to common resources, for example, by secreting digestive enzymes and siderophores, exploiting products produced by the competitor, and producing compounds that affect its adhesion. In active competition, strains directly damage each other by producing antibiotics, other toxins, and quorum signaling inhibitors. It is suggested that competition has played a dominant role in the evolution of bacterial stress response networks [158]. Several bacterial stress responses detect nutrient limitation (that can be caused by exploitative competition) or cell damage (that can be caused by active competition). Bacterial antibiotics and bacteriocins are frequently directed against competing and related organisms. They are frequently upregulated by the stress response to nutrient limitation or cell damage, but very rarely by the stress response to heat or osmotic shock, which are typically not competition related. Therefore, one of the purposes of stress response networks in bacteria may be to enable them to detect the presence of ecological competition and respond to it [158]. Antibiotics increase the formation of biofilms; strain mixing also increases biofilm formation and leads to a strong competition where one strain dominates and eventually excludes the other from the biofilm [159]. One of the beneficial functions of human commensal gut microbiota is that it competitively excludes pathogens [160].

3.15. Archaea Stress Responses

Most studies of the stress response and stress tolerance in the archaea have been performed on extremophiles that have various permanent adaptations to extreme environments. These adaptations will be discussed in the next chapter.

Archaea resemble bacteria in that they lack nuclei and typically contain a singular circular genome with operonal organization. However structural and functional features of their transcription and translation are similar to those of eukaryotes and so are their proteome maintenance and DNA damage response [161 - 163]. Both archaeal and bacterial stress response involves chaperones, heat shock and cold shock proteins, thermoprotectants, membrane and proteasome adjustments, production of antibiotics, and biofilm formation [164]. Some archaeal transcriptional factors resemble those of bacteria whereas others are similar to eukaryotic ones. For example, the oxidative stress response in archaea is mediated by bacteria-like RosR and five eukaryote-like TFIIB proteins that combinatorially regulate over 100 genes [165]. Similar to bacteria, archaea use the two-component systems to respond to stress, but although their sensory components are similar to bacterial ones, the output is different in that it relies mostly not on transcriptional regulation but on protein-protein interaction [166]. Many specific traits are different between bacteria and archaea, *e.g.* response to starvation. In bacteria, such response depends on the presence of non-loaded tRNA and is mediated by (p)ppGpp. In archaea, no (p)ppGpp was ever detected and although crenoarchaeal species seems to recognize non-loaded tRNA as a starvation signal, euryarchaeal species do not recognize it [167].

CONCLUDING REMARKS

Prokaryotes are remarkable vigorous organisms. They adjust to changes in their environment by switching to alternative transcription programs, modification of their metabolism, changing the composition of their membranes, and producing compounds that protect their macromolecules.

CONSENT FOR PUBLICATION

Not applicable.

CONFLICT OF INTEREST

The author confirms that this chapter contents have no conflict of interest.

ACKNOWLEDGEMENTS

Declare none.

REFERENCES

[1] Han K, Li ZF, Peng R, *et al.* Extraordinary expansion of a Sorangium cellulosum genome from an alkaline milieu. Sci Rep 2013; 3: 2101.
[http://dx.doi.org/10.1038/srep02101] [PMID: 23812535]

[2] Bacterial Stress Responses 2ed. Storz G, Hengge R, editors: American Society for Microbiology Press. 2011.

[3] Galperin MY. What bacteria want. Environ Microbiol 2018; 20(12): 4221-9.
[http://dx.doi.org/10.1111/1462-2920.14398] [PMID: 30187651]

[4] Guo MS, Gross CA. Stress-induced remodeling of the bacterial proteome. Curr Biol 2014; 24(10): R424-34.
[http://dx.doi.org/10.1016/j.cub.2014.03.023] [PMID: 24845675]

[5] Stock AM, Robinson VL, Goudreau PN. Two-component signal transduction. Annu Rev Biochem 2000; 69: 183-215.
[http://dx.doi.org/10.1146/annurev.biochem.69.1.183] [PMID: 10966457]

[6] Stock AM, Zhulin IB. Two-Component Signal Transduction: A Special Issue in the *Journal of Bacteriology*. J Bacteriol 2017; 199(18): e00443-17.
[http://dx.doi.org/10.1128/JB.00443-17] [PMID: 28831031]

[7] Villanueva M, García B, Valle J, *et al.* Sensory deprivation in Staphylococcus aureus. Nat Commun 2018; 9(1): 523.
[http://dx.doi.org/10.1038/s41467-018-02949-y] [PMID: 29410457]

[8] Kellogg SL, Kristich CJ. Convergence of PASTA kinase and two-component signaling in response to cell wall stress in Enterococcus faecalis. Journal of bacteriology 2018; JB.00086-.
[http://dx.doi.org/10.1128/JB.00086-18]

[9] Feklístov A, Sharon BD, Darst SA, Gross CA. Bacterial sigma factors: a historical, structural, and genomic perspective. Annu Rev Microbiol 2014; 68: 357-76.
[http://dx.doi.org/10.1146/annurev-micro-092412-155737] [PMID: 25002089]

[10] Kill K, Binnewies TT, Sicheritz-Pontén T, *et al.* Genome update: sigma factors in 240 bacterial genomes. Microbiology 2005; 151(Pt 10): 3147-50.
[http://dx.doi.org/10.1099/mic.0.28339-0] [PMID: 16207898]

[11] Paget MS. Bacterial Sigma Factors and Anti-Sigma Factors: Structure, Function and Distribution. Biomolecules 2015; 5(3): 1245-65.
[http://dx.doi.org/10.3390/biom5031245] [PMID: 26131973]

[12] Bouillet S, Arabet D, Jourlin-Castelli C, Méjean V, Iobbi-Nivol C. Regulation of σ factors by conserved partner switches controlled by divergent signalling systems. Environ Microbiol Rep 2018; 10(2): 127-39.
[http://dx.doi.org/10.1111/1758-2229.12620] [PMID: 29393573]

[13] Pané-Farré J, Quin MB, Lewis RJ, Marles-Wright J. Structure and function of the stressosome signalling hub. Subcell Biochem 2017; 83: 1-41.
[http://dx.doi.org/10.1007/978-3-319-46503-6_1] [PMID: 28271471]

[14] Morita MT, Tanaka Y, Kodama TS, Kyogoku Y, Yanagi H, Yura T. Translational induction of heat shock transcription factor sigma32: evidence for a built-in RNA thermosensor. Genes Dev 1999; 13(6): 655-65.
[http://dx.doi.org/10.1101/gad.13.6.655] [PMID: 10090722]

[15] Lim B, Miyazaki R, Neher S, *et al.* Heat shock transcription factor σ32 co-opts the signal recognition particle to regulate protein homeostasis in *E. coli*. PLoS Biol 2013; 11(12)e1001735
[http://dx.doi.org/10.1371/journal.pbio.1001735] [PMID: 24358019]

[16] Siliakus MF, van der Oost J, Kengen SWM. Adaptations of archaeal and bacterial membranes to

variations in temperature, pH and pressure. Extremophiles: Life under extreme conditions 2017; 21(4): 651-70.
[http://dx.doi.org/10.1007/s00792-017-0939-x]

[17] Loh E, Righetti F, Eichner H, Twittenhoff C, Narberhaus F. RNA Thermometers in Bacterial Pathogens. Microbiol Spectr 2018; 6(2)
[http://dx.doi.org/10.1128/microbiolspec.RWR-0012-2017] [PMID: 29623874]

[18] Schumann W. Thermosensor systems in eubacteria. Adv Exp Med Biol 2012; 739: 1-16.
[http://dx.doi.org/10.1007/978-1-4614-1704-0_1] [PMID: 22399392]

[19] Zhang Y, Burkhardt DH, Rouskin S, Li GW, Weissman JS, Gross CA. A Stress Response that Monitors and Regulates mRNA Structure Is Central to Cold Shock Adaptation. Mol Cell 2018; 70(2): 274-286.e7.
[http://dx.doi.org/10.1016/j.molcel.2018.02.035] [PMID: 29628307]

[20] Caballero CJ, Menendez-Gil P, Catalan-Moreno A, *et al.* The regulon of the RNA chaperone CspA and its auto-regulation in Staphylococcus aureus. Nucleic Acids Res 2018; 46(3): 1345-61.
[http://dx.doi.org/10.1093/nar/gkx1284] [PMID: 29309682]

[21] Roilo M, Kullmann MK, Hengst L. Cold-inducible RNA-binding protein (CIRP) induces translation of the cell-cycle inhibitor p27Kip1. Nucleic Acids Res 2018; 46(6): 3198-210.
[http://dx.doi.org/10.1093/nar/gkx1317] [PMID: 29361038]

[22] Keto-Timonen R, Hietala N, Palonen E, Hakakorpi A, Lindström M, Korkeala H. Cold Shock Proteins: A Minireview with Special Emphasis on Csp-family of Enteropathogenic Yersinia. Front Microbiol 2016; 7: 1151.
[http://dx.doi.org/10.3389/fmicb.2016.01151] [PMID: 27499753]

[23] Sengupta P, Garrity P. Sensing temperature. Curr Biol 2013; 23(8): R304-7.
[http://dx.doi.org/10.1016/j.cub.2013.03.009] [PMID: 23618661]

[24] Lange R, Hengge-Aronis R. Identification of a central regulator of stationary-phase gene expression in *Escherichia coli*. Mol Microbiol 1991; 5(1): 49-59.
[http://dx.doi.org/10.1111/j.1365-2958.1991.tb01825.x] [PMID: 1849609]

[25] Battesti A, Majdalani N, Gottesman S. The RpoS-mediated general stress response in *Escherichia coli*. Annu Rev Microbiol 2011; 65: 189-213.
[http://dx.doi.org/10.1146/annurev-micro-090110-102946] [PMID: 21639793]

[26] Chen AI, Goulian M. A network of regulators promotes dehydration tolerance in *Escherichia coli*. Environ Microbiol 2018; 20(3): 1283-95.
[http://dx.doi.org/10.1111/1462-2920.14074] [PMID: 29457688]

[27] Wong GT, Bonocora RP, Schep AN, *et al.* Genome-Wide Transcriptional Response to Varying RpoS Levels in *Escherichia coli* K-12. J Bacteriol 2017; 199(7): e00755-16.
[http://dx.doi.org/10.1128/JB.00755-16] [PMID: 28115545]

[28] Karas VO, Westerlaken I, Meyer AS. The DNA-Binding Protein from Starved Cells (Dps) Utilizes Dual Functions To Defend Cells against Multiple Stresses. J Bacteriol 2015; 197(19): 3206-15.
[http://dx.doi.org/10.1128/JB.00475-15] [PMID: 26216848]

[29] Hecker M, Pané-Farré J, Völker U. SigB-dependent general stress response in Bacillus subtilis and related gram-positive bacteria. Annu Rev Microbiol 2007; 61: 215-36.
[http://dx.doi.org/10.1146/annurev.micro.61.080706.093445] [PMID: 18035607]

[30] Narula J, Tiwari A, Igoshin OA. Role of Autoregulation and Relative Synthesis of Operon Partners in Alternative Sigma Factor Networks. PLOS Comput Biol 2016; 12(12)e1005267
[http://dx.doi.org/10.1371/journal.pcbi.1005267] [PMID: 27977677]

[31] Gaidenko TA, Price CW. Genetic evidence for a phosphorylation-independent signal transduction mechanism within the Bacillus subtilis stressosome. PLoS One 2014; 9(3)e90741
[http://dx.doi.org/10.1371/journal.pone.0090741] [PMID: 24599254]

[32] Hecker M, Reder A, Fuchs S, Pagels M, Engelmann S. Physiological proteomics and stress/starvation responses in Bacillus subtilis and Staphylococcus aureus. Res Microbiol 2009; 160(4): 245-58. [http://dx.doi.org/10.1016/j.resmic.2009.03.008] [PMID: 19403106]

[33] Asmar AT, Ferreira JL, Cohen EJ, *et al.* Communication across the bacterial cell envelope depends on the size of the periplasm. PLoS Biol 2017; 15(12)e2004303 [http://dx.doi.org/10.1371/journal.pbio.2004303] [PMID: 29257832]

[34] Grabowicz M, Silhavy TJ. Envelope Stress Responses: An Interconnected Safety Net. Trends Biochem Sci 2017; 42(3): 232-42. [http://dx.doi.org/10.1016/j.tibs.2016.10.002] [PMID: 27839654]

[35] Sohn J, Sauer RT. OMP peptides modulate the activity of DegS protease by differential binding to active and inactive conformations. Mol Cell 2009; 33(1): 64-74. [http://dx.doi.org/10.1016/j.molcel.2008.12.017] [PMID: 19150428]

[36] Lima S, Guo MS, Chaba R, Gross CA, Sauer RT. Dual molecular signals mediate the bacterial response to outer-membrane stress. Science 2013; 340(6134): 837-41. [http://dx.doi.org/10.1126/science.1235358] [PMID: 23687042]

[37] Yakhnin H, Aichele R, Ades SE, Romeo T, Babitzke P. Circuitry linking the global Csr and sigma(E)-dependent cell envelope stress response systems. Journal of bacteriology 2017; JB.00484-.

[38] Klein G, Raina S. Small regulatory bacterial RNAs regulating the envelope stress response. Biochem Soc Trans 2017; 45(2): 417-25. [http://dx.doi.org/10.1042/BST20160367] [PMID: 28408482]

[39] Hussein NA, Cho SH, Laloux G, Siam R, Collet JF. Distinct domains of *Escherichia coli* IgaA connect envelope stress sensing and down-regulation of the Rcs phosphorelay across subcellular compartments. PLoS Genet 2018; 14(5)e1007398 [http://dx.doi.org/10.1371/journal.pgen.1007398] [PMID: 29852010]

[40] Laloux G, Collet JF. Major Tom to Ground Control: How Lipoproteins Communicate Extracytoplasmic Stress to the Decision Center of the Cell. J Bacteriol 2017; 199(21): e00216-7. [http://dx.doi.org/10.1128/JB.00216-17] [PMID: 28674071]

[41] Wall E, Majdalani N, Gottesman S. The Complex Rcs Regulatory Cascade. Annu Rev Microbiol 2018; 72: 111-39. [http://dx.doi.org/10.1146/annurev-micro-090817-062640] [PMID: 29897834]

[42] Flores-Kim J, Darwin AJ. The Phage Shock Protein Response. Annu Rev Microbiol 2016; 70: 83-101. [http://dx.doi.org/10.1146/annurev-micro-102215-095359] [PMID: 27297125]

[43] Srivastava D, Moumene A, Flores-Kim J, Darwin AJ. Psp Stress Response Proteins Form a Complex with Mislocalized Secretins in the *Yersinia enterocolitica* Cytoplasmic Membrane. MBio 2017; 8(5): e01088-17. [http://dx.doi.org/10.1128/mBio.01088-17] [PMID: 28900025]

[44] McDonald C, Jovanovic G, Wallace BA, Ces O, Buck M. Structure and function of PspA and Vipp1 N-terminal peptides: Insights into the membrane stress sensing and mitigation. Biochim Biophys Acta Biomembr 2017; 1859(1): 28-39. [http://dx.doi.org/10.1016/j.bbamem.2016.10.018] [PMID: 27806910]

[45] Manganelli R, Gennaro ML. Protecting from Envelope Stress: Variations on the Phage-Shock-Protein Theme. Trends Microbiol 2017; 25(3): 205-16. [http://dx.doi.org/10.1016/j.tim.2016.10.001] [PMID: 27865622]

[46] Radeck J, Fritz G, Mascher T. The cell envelope stress response of Bacillus subtilis: from static signaling devices to dynamic regulatory network. Curr Genet 2017; 63(1): 79-90. [http://dx.doi.org/10.1007/s00294-016-0624-0] [PMID: 27344142]

[47] Lewerke LT, Kies PJ, Müh U, Ellermeier CD. Bacterial sensing: A putative amphipathic helix in RsiV

is the switch for activating σV in response to lysozyme. PLoS Genet 2018; 14(7)e1007527 [http://dx.doi.org/10.1371/journal.pgen.1007527] [PMID: 30020925]

[48] Sineva E, Savkina M, Ades SE. Themes and variations in gene regulation by extracytoplasmic function (ECF) sigma factors. Curr Opin Microbiol 2017; 36: 128-37. [http://dx.doi.org/10.1016/j.mib.2017.05.004] [PMID: 28575802]

[49] Westfall CS, Levin PA. Comprehensive analysis of central carbon metabolism illuminates connections between nutrient availability, growth rate, and cell morphology in *Escherichia coli*. PLoS Genet 2018; 14(2)e1007205 [http://dx.doi.org/10.1371/journal.pgen.1007205] [PMID: 29432413]

[50] Hauryliuk V, Atkinson GC, Murakami KS, Tenson T, Gerdes K. Recent functional insights into the role of (p)ppGpp in bacterial physiology. Nat Rev Microbiol 2015; 13(5): 298-309. [http://dx.doi.org/10.1038/nrmicro3448] [PMID: 25853779]

[51] Winther KS, Roghanian M, Gerdes K. Activation of the Stringent Response by Loading of RelA-tRNA Complexes at the Ribosomal A-Site. Mol Cell 2018; 70(1): 95-105.e4. [http://dx.doi.org/10.1016/j.molcel.2018.02.033] [PMID: 29625042]

[52] Fang M, Bauer CE. Regulation of stringent factor by branched-chain amino acids. Proc Natl Acad Sci USA 2018; 115(25): 6446-51. [http://dx.doi.org/10.1073/pnas.1803220115] [PMID: 29866825]

[53] Sherlock ME, Sudarsan N, Breaker RR. Riboswitches for the alarmone ppGpp expand the collection of RNA-based signaling systems. Proc Natl Acad Sci USA 2018; 115(23): 6052-7. [http://dx.doi.org/10.1073/pnas.1720406115] [PMID: 29784782]

[54] Murch AL, Skipp PJ, Roach PL, Oyston PCF. Whole genome transcriptomics reveals global effects including up-regulation of Francisella pathogenicity island gene expression during active stringent response in the highly virulent Francisella tularensis subsp. tularensis SCHU S4. Microbiology 2017; 163(11): 1664-79. [http://dx.doi.org/10.1099/mic.0.000550] [PMID: 29034854]

[55] Hüttener M, Prieto A, Espelt J, Bernabeu M, Juárez A. Stringent Response and AggR-Dependent Virulence Regulation in the Enteroaggregative *Escherichia coli* Strain 042. Front Microbiol 2018; 9: 717. [http://dx.doi.org/10.3389/fmicb.2018.00717] [PMID: 29692772]

[56] Inaoka T, Takahashi K, Ohnishi-Kameyama M, Yoshida M, Ochi K. Guanine nucleotides guanosine 5′-diphosphate 3′-diphosphate and GTP co-operatively regulate the production of an antibiotic bacilysin in Bacillus subtilis. J Biol Chem 2003; 278(4): 2169-76. [http://dx.doi.org/10.1074/jbc.M208722200] [PMID: 12372825]

[57] Maisonneuve E, Castro-Camargo M, Gerdes K. (p)ppGpp controls bacterial persistence by stochastic induction of toxin-antitoxin activity. Cell 2013; 154(5): 1140-50. [http://dx.doi.org/10.1016/j.cell.2013.07.048] [PMID: 23993101]

[58] Irving SE, Corrigan RM. Triggering the stringent response: signals responsible for activating (p)ppGpp synthesis in bacteria. Microbiology 2018; 164(3): 268-76. [http://dx.doi.org/10.1099/mic.0.000621] [PMID: 29493495]

[59] Field B. Green magic: regulation of the chloroplast stress response by (p)ppGpp in plants and algae. J Exp Bot 2018; 69(11): 2797-807. [http://dx.doi.org/10.1093/jxb/erx485] [PMID: 29281108]

[60] Dwyer DJ, Collins JJ, Walker GC. Unraveling the physiological complexities of antibiotic lethality. Annu Rev Pharmacol Toxicol 2015; 55: 313-32. [http://dx.doi.org/10.1146/annurev-pharmtox-010814-124712] [PMID: 25251995]

[61] Ezraty B, Gennaris A, Barras F, Collet JF. Oxidative stress, protein damage and repair in bacteria. Nat Rev Microbiol 2017; 15(7): 385-96.

[http://dx.doi.org/10.1038/nrmicro.2017.26] [PMID: 28420885]

[62] Reichmann D, Voth W, Jakob U. Maintaining a Healthy Proteome during Oxidative Stress. Mol Cell 2018; 69(2): 203-13.
[http://dx.doi.org/10.1016/j.molcel.2017.12.021] [PMID: 29351842]

[63] Imlay JA. The molecular mechanisms and physiological consequences of oxidative stress: lessons from a model bacterium. Nat Rev Microbiol 2013; 11(7): 443-54.
[http://dx.doi.org/10.1038/nrmicro3032] [PMID: 23712352]

[64] Wang X, Kim Y, Hong SH, *et al.* Antitoxin MqsA helps mediate the bacterial general stress response. Nat Chem Biol 2011; 7(6): 359-66.
[http://dx.doi.org/10.1038/nchembio.560] [PMID: 21516113]

[65] Barshishat S, Elgrably-Weiss M, Edelstein J, *et al.* OxyS small RNA induces cell cycle arrest to allow DNA damage repair. EMBO J 2018; 37(3): 413-26.
[http://dx.doi.org/10.15252/embj.201797651] [PMID: 29237698]

[66] Glanville DG, Han L, Maule AF, *et al.* RitR is an archetype for a novel family of redox sensors in the streptococci that has evolved from two-component response regulators and is required for pneumococcal colonization. PLoS Pathog 2018; 14(5)e1007052
[http://dx.doi.org/10.1371/journal.ppat.1007052] [PMID: 29750817]

[67] Tharmalingam S, Alhasawi A, Appanna VP, Lemire J, Appanna VD. Reactive nitrogen species (RNS)-resistant microbes: adaptation and medical implications. Biol Chem 2017; 398(11): 1193-208.
[http://dx.doi.org/10.1515/hsz-2017-0152] [PMID: 28622140]

[68] Cole JA. Anaerobic Bacterial Response to Nitrosative Stress. Adv Microb Physiol 2018; 72: 193-237.
[http://dx.doi.org/10.1016/bs.ampbs.2018.01.001] [PMID: 29778215]

[69] Flint A, Stintzi A, Saraiva LM. Oxidative and nitrosative stress defences of Helicobacter and Campylobacter species that counteract mammalian immunity. FEMS Microbiol Rev 2016; 40(6): 938-60.
[http://dx.doi.org/10.1093/femsre/fuw025] [PMID: 28201757]

[70] Rockwood N, Costa DL, Amaral EP, *et al. Mycobacterium tuberculosis* Induction of Heme Oxygenase-1 Expression Is Dependent on Oxidative Stress and Reflects Treatment Outcomes. Front Immunol 2017; 8: 542.
[http://dx.doi.org/10.3389/fimmu.2017.00542] [PMID: 28553288]

[71] Akgul A, Akgul A, Lawrence ML, Karsi A. Stress-related genes promote Edwardsiella ictaluri pathogenesis. PLoS One 2018; 13(3)e0194669
[http://dx.doi.org/10.1371/journal.pone.0194669] [PMID: 29554143]

[72] Schatzman SS, Culotta VC. Chemical Warfare at the Microorganismal Level: A Closer Look at the Superoxide Dismutase Enzymes of Pathogens. ACS Infect Dis 2018; 4(6): 893-903.
[http://dx.doi.org/10.1021/acsinfecdis.8b00026] [PMID: 29517910]

[73] Aunins TR, Erickson KE, Prasad N, *et al.* Spaceflight Modifies *Escherichia coli* Gene Expression in Response to Antibiotic Exposure and Reveals Role of Oxidative Stress Response. Front Microbiol 2018; 9: 310.
[http://dx.doi.org/10.3389/fmicb.2018.00310] [PMID: 29615983]

[74] Dai X, Zhu M, Warren M, *et al.* Slowdown of Translational Elongation in *Escherichia coli* under Hyperosmotic Stress. MBio 2018; 9(1): e02375-17.
[http://dx.doi.org/10.1128/mBio.02375-17] [PMID: 29440576]

[75] Gunde-Cimerman N, Plemenitaš A, Oren A. Strategies of adaptation of microorganisms of the three domains of life to high salt concentrations. FEMS Microbiol Rev 2018; 42(3): 353-75.
[http://dx.doi.org/10.1093/femsre/fuy009] [PMID: 29529204]

[76] Devaux L, Sleiman D, Mazzuoli MV, *et al.* Cyclic di-AMP regulation of osmotic homeostasis is essential in Group B Streptococcus. PLoS Genet 2018; 14(4)e1007342

[http://dx.doi.org/10.1371/journal.pgen.1007342] [PMID: 29659565]

[77] Karimi B, Terrat S, Dequiedt S, Saby NPA, Horrigue W, Lelievre M, *et al.* Biogeography of soil bacteria and archaea across France. Science advances 2018; 47: eaat1808. [http://dx.doi.org/10.1126/sciadv.aat1808]

[78] De Biase D, Lund PA. The *Escherichia coli* Acid Stress Response and Its Significance for Pathogenesis. Adv Appl Microbiol 2015; 92: 49-88. [http://dx.doi.org/10.1016/bs.aambs.2015.03.002] [PMID: 26003933]

[79] Lund P, Tramonti A, De Biase D. Coping with low pH: molecular strategies in neutralophilic bacteria. FEMS Microbiol Rev 2014; 38(6): 1091-125. [http://dx.doi.org/10.1111/1574-6976.12076] [PMID: 24898062]

[80] Guerrero-Castro J, Lozano L, Sohlenkamp C. Dissecting the Acid Stress Response of *Rhizobium tropici* CIAT 899. Front Microbiol 2018; 9: 846. [http://dx.doi.org/10.3389/fmicb.2018.00846] [PMID: 29760688]

[81] Ter Beek A, Wijman JG, Zakrzewska A, Orij R, Smits GJ, Brul S. Comparative physiological and transcriptional analysis of weak organic acid stress in Bacillus subtilis. Food microbiology 2015; 45(Pt A): 71-82. [http://dx.doi.org/10.1016/j.fm.2014.02.013]

[82] Davis MC, Smith LK, MacLellan SR. The atypical two-subunit σ factor from Bacillus subtilis is regulated by an integral membrane protein and acid stress. Microbiology 2016; 162(2): 398-407. [http://dx.doi.org/10.1099/mic.0.000223] [PMID: 26651345]

[83] Giovannercole F, Mérigoux C, Zamparelli C, *et al.* On the effect of alkaline pH and cofactor availability in the conformational and oligomeric state of *Escherichia coli* glutamate decarboxylase. Protein Eng Des Sel 2017; 30(3): 235-44. [PMID: 28062647]

[84] Noguchi K, Riggins DP, Eldahan KC, Kitko RD, Slonczewski JL. Hydrogenase-3 contributes to anaerobic acid resistance of *Escherichia coli.* PLoS One 2010; 5(4)e10132 [http://dx.doi.org/10.1371/journal.pone.0010132] [PMID: 20405029]

[85] Kanjee U, Houry WA. Mechanisms of acid resistance in *Escherichia coli.* Annu Rev Microbiol 2013; 67: 65-81. [http://dx.doi.org/10.1146/annurev-micro-092412-155708] [PMID: 23701194]

[86] Robert CB, Thomson M, Vercellone A, *et al.* Mass spectrometry analysis of intact Francisella bacteria identifies lipid A structure remodeling in response to acidic pH stress. Biochimie 2017; 141: 16-20. [http://dx.doi.org/10.1016/j.biochi.2017.08.008] [PMID: 28807561]

[87] Salmon L, Stull F, Sayle S, *et al.* The Mechanism of HdeA Unfolding and Chaperone Activation. J Mol Biol 2018; 430(1): 33-40. [http://dx.doi.org/10.1016/j.jmb.2017.11.002] [PMID: 29138002]

[88] Kim J, Choi D, Cha SY, *et al.* Zinc-mediated Reversible Multimerization of Hsp31 Enhances the Activity of Holding Chaperone. J Mol Biol 2018; 430(12): 1760-72. [http://dx.doi.org/10.1016/j.jmb.2018.04.029] [PMID: 29709570]

[89] Baker JJ, Abramovitch RB. Genetic and metabolic regulation of Mycobacterium tuberculosis acid growth arrest. Sci Rep 2018; 8(1): 4168. [http://dx.doi.org/10.1038/s41598-018-22343-4] [PMID: 29520087]

[90] Rimon A, Dwivedi M, Friedler A, Padan E. Asp133 Residue in NhaA Na^+/H^+ Antiporter Is Required for Stability Cation Binding and Transport. J Mol Biol 2018; 430(6): 867-80. [http://dx.doi.org/10.1016/j.jmb.2018.01.014] [PMID: 29410365]

[91] Belevich G, Belevich N, Knuuti J, Verkhovskaya M. Ca2+ stabilization of respiratory complex I from *Escherichia coli.* FEMS Microbiol Lett 2018; 365(12) [http://dx.doi.org/10.1093/femsle/fny097] [PMID: 29668960]

[92] Nyanga-Koumou AP, Ouoba LI, Kobawila SC, Louembe D. Response mechanisms of lactic acid bacteria to alkaline environments: a review. Crit Rev Microbiol 2012; 38(3): 185-90. [http://dx.doi.org/10.3109/1040841X.2011.640978] [PMID: 22168378]

[93] Lopez CA, Skaar EP. The Impact of Dietary Transition Metals on Host-Bacterial Interactions. Cell Host Microbe 2018; 23(6): 737-48. [http://dx.doi.org/10.1016/j.chom.2018.05.008] [PMID: 29902439]

[94] Ong CY, Berking O, Walker MJ, McEwan AG. New Insights into the Role of Zinc Acquisition and Zinc Tolerance in Group A Streptococcal Infection. Infect Immun 2018; 86(6): e00048-18. [http://dx.doi.org/10.1128/IAI.00048-18] [PMID: 29581188]

[95] Djoko KY, Ong CL, Walker MJ, McEwan AG. The Role of Copper and Zinc Toxicity in Innate Immune Defense against Bacterial Pathogens. J Biol Chem 2015; 290(31): 18954-61. [http://dx.doi.org/10.1074/jbc.R115.647099] [PMID: 26055706]

[96] Fillat MF. The FUR (ferric uptake regulator) superfamily: Diversity and versatility of key transcriptional regulators. Arch Biochem Biophys 2014; 546: 41-52. [http://dx.doi.org/10.1016/j.abb.2014.01.029] [PMID: 24513162]

[97] Chandrangsu P, Rensing C, Helmann JD. Metal homeostasis and resistance in bacteria. Nat Rev Microbiol 2017; 15(6): 338-50. [http://dx.doi.org/10.1038/nrmicro.2017.15] [PMID: 28344348]

[98] Pi H, Helmann JD. Sequential induction of Fur-regulated genes in response to iron limitation in *Bacillus subtilis*. Proc Natl Acad Sci USA 2017; 114(48): 12785-90. [http://dx.doi.org/10.1073/pnas.1713008114] [PMID: 29133393]

[99] Pi H, Patel SJ, Argüello JM, Helmann JD. The Listeria monocytogenes Fur-regulated virulence protein FrvA is an Fe(II) efflux P1B4 -type ATPase. Mol Microbiol 2016; 100(6): 1066-79. [http://dx.doi.org/10.1111/mmi.13368] [PMID: 26946370]

[100] Lee JW, Helmann JD. The PerR transcription factor senses H2O2 by metal-catalysed histidine oxidation. Nature 2006; 440(7082): 363-7. [http://dx.doi.org/10.1038/nature04537] [PMID: 16541078]

[101] Gilston BA, Wang S, Marcus MD, *et al.* Structural and mechanistic basis of zinc regulation across the *E. coli* Zur regulon. PLoS Biol 2014; 12(11)e1001987 [http://dx.doi.org/10.1371/journal.pbio.1001987] [PMID: 25369000]

[102] Shin JH, Helmann JD. Molecular logic of the Zur-regulated zinc deprivation response in Bacillus subtilis. Nat Commun 2016; 7: 12612. [http://dx.doi.org/10.1038/ncomms12612] [PMID: 27561249]

[103] Radman M. SOS repair hypothesis: phenomenology of an inducible DNA repair which is accompanied by mutagenesis. Basic Life Sci 1975; 5A: 355-67. [http://dx.doi.org/10.1007/978-1-4684-2895-7_48] [PMID: 1103845]

[104] Kreuzer KN. DNA damage responses in prokaryotes: regulating gene expression, modulating growth patterns, and manipulating replication forks. Cold Spring Harb Perspect Biol 2013; 5(11)a012674 [http://dx.doi.org/10.1101/cshperspect.a012674] [PMID: 24097899]

[105] Williams AB, Schumacher B. DNA damage responses and stress resistance: Concepts from bacterial SOS to metazoan immunity. Mech Ageing Dev 2017; 165Pt A: 27-32.

[106] Witkin EM. The radiation sensitivity of *Escherichia coli* B: a hypothesis relating filament formation and prophage induction. Proc Natl Acad Sci USA 1967; 57(5): 1275-9. [http://dx.doi.org/10.1073/pnas.57.5.1275] [PMID: 5341236]

[107] Luo Y, Pfuetzner RA, Mosimann S, *et al.* Crystal structure of LexA: a conformational switch for regulation of self-cleavage. Cell 2001; 106(5): 585-94. [http://dx.doi.org/10.1016/S0092-8674(01)00479-2] [PMID: 11551506]

[108] Culyba MJ, Kubiak JM, Mo CY, Goulian M, Kohli RM. Non-equilibrium repressor binding kinetics link DNA damage dose to transcriptional timing within the SOS gene network. PLoS Genet 2018; 14(6)e1007405
[http://dx.doi.org/10.1371/journal.pgen.1007405] [PMID: 29856734]

[109] Fitzgerald DM, Hastings PJ, Rosenberg SM. Stress-Induced Mutagenesis: Implications in Cancer and Drug Resistance. Annual review of cancer biology 2017; 1: 119-40.

[110] Henrikus SS, van Oijen AM, Robinson A. Specialised DNA polymerases in *Escherichia coli*: roles within multiple pathways. Curr Genet 2018; 64(6): 1189-96.
[http://dx.doi.org/10.1007/s00294-018-0840-x] [PMID: 29700578]

[111] Gillings MR. Lateral gene transfer, bacterial genome evolution, and the Anthropocene. Ann N Y Acad Sci 2017; 1389(1): 20-36.
[http://dx.doi.org/10.1111/nyas.13213] [PMID: 27706829]

[112] Fornelos N, Browning DF, Butala M. The Use and Abuse of LexA by Mobile Genetic Elements. Trends Microbiol 2016; 24(5): 391-401.
[http://dx.doi.org/10.1016/j.tim.2016.02.009] [PMID: 26970840]

[113] Miller C, Thomsen LE, Gaggero C, Mosseri R, Ingmer H, Cohen SN. SOS response induction by beta-lactams and bacterial defense against antibiotic lethality. Science 2004; 305(5690): 1629-31.
[http://dx.doi.org/10.1126/science.1101630] [PMID: 15308764]

[114] Baharoglu Z, Bikard D, Mazel D. Conjugative DNA transfer induces the bacterial SOS response and promotes antibiotic resistance development through integron activation. PLoS Genet 2010; 6(10)e1001165
[http://dx.doi.org/10.1371/journal.pgen.1001165] [PMID: 20975940]

[115] Müller AU, Imkamp F, Weber-Ban E. The Mycobacterial LexA/RecA-Independent DNA Damage Response Is Controlled by PafBC and the Pup-Proteasome System. Cell Rep 2018; 23(12): 3551-64.
[http://dx.doi.org/10.1016/j.celrep.2018.05.073] [PMID: 29924998]

[116] Zahradka K, Slade D, Bailone A, *et al.* Reassembly of shattered chromosomes in Deinococcus radiodurans. Nature 2006; 443(7111): 569-73.
[http://dx.doi.org/10.1038/nature05160] [PMID: 17006450]

[117] Pobegalov G, Cherevatenko G, Alekseev A, *et al.* Deinococcus radiodurans RecA nucleoprotein filaments characterized at the single-molecule level with optical tweezers. Biochem Biophys Res Commun 2015; 466(3): 426-30.
[http://dx.doi.org/10.1016/j.bbrc.2015.09.042] [PMID: 26365346]

[118] Engl C, Schaefer J, Kotta-Loizou I, Buck M. Cellular and molecular phenotypes depending upon the RNA repair system RtcAB of *Escherichia coli*. Nucleic Acids Res 2016; 44(20): 9933-41.
[PMID: 27402162]

[119] Radman M. Protein damage, radiation sensitivity and aging. DNA Repair (Amst) 2016; 44: 186-92.
[http://dx.doi.org/10.1016/j.dnarep.2016.05.025] [PMID: 27264559]

[120] Sharma A, Gaidamakova EK, Grichenko O, *et al.* Across the tree of life, radiation resistance is governed by antioxidant Mn^{2+}, gauged by paramagnetic resonance. Proc Natl Acad Sci USA 2017; 114(44): E9253-60.
[http://dx.doi.org/10.1073/pnas.1713608114] [PMID: 29042516]

[121] Cohen NR, Ross CA, Jain S, *et al.* A role for the bacterial GATC methylome in antibiotic stress survival. Nat Genet 2016; 48(5): 581-6.
[http://dx.doi.org/10.1038/ng.3530] [PMID: 26998690]

[122] NicAogáin K, O'Byrne CP. The Role of Stress and Stress Adaptations in Determining the Fate of the Bacterial Pathogen *Listeria monocytogenes* in the Food Chain. Front Microbiol 2016; 7: 1865.
[http://dx.doi.org/10.3389/fmicb.2016.01865] [PMID: 27933042]

[123] Madhugiri R, Basineni SR, Klug G. Turn-over of the small non-coding RNA RprA in *E. coli* is influenced by osmolarity. Mol Genet Genomics 2010; 284(4): 307-18. [http://dx.doi.org/10.1007/s00438-010-0568-x] [PMID: 20717695]

[124] Jonas K, Liu J, Chien P, Laub MT. Proteotoxic stress induces a cell-cycle arrest by stimulating Lon to degrade the replication initiator DnaA. Cell 2013; 154(3): 623-36. [http://dx.doi.org/10.1016/j.cell.2013.06.034] [PMID: 23911325]

[125] Darnell CL, Tonner PD, Gulli JG, Schmidler SC, Schmid AK. Systematic Discovery of Archaeal Transcription Factor Functions in Regulatory Networks through Quantitative Phenotyping Analysis. mSystems 2017; 2(5): e00032-17. [http://dx.doi.org/10.1128/mSystems.00032-17] [PMID: 28951888]

[126] Lee CK, de Anda J, Baker AE, *et al.* Multigenerational memory and adaptive adhesion in early bacterial biofilm communities. Proc Natl Acad Sci USA 2018; 115(17): 4471-6. [http://dx.doi.org/10.1073/pnas.1720071115] [PMID: 29559526]

[127] Zorraquino V, Kim M, Rai N, Tagkopoulos I. The genetic and transcriptional basis of short and long term adaptation across multiple stresses in *Escherichia coli*. Mol Biol Evol 2017; 34(3): 707-17. [PMID: 28007978]

[128] Jorgensen I, Rayamajhi M, Miao EA. Programmed cell death as a defence against infection. Nat Rev Immunol 2017; 17(3): 151-64. [http://dx.doi.org/10.1038/nri.2016.147] [PMID: 28138137]

[129] Weinlich R, Oberst A, Beere HM, Green DR. Necroptosis in development, inflammation and disease. Nat Rev Mol Cell Biol 2017; 18(2): 127-36. [http://dx.doi.org/10.1038/nrm.2016.149] [PMID: 27999438]

[130] Mosser CA, Baptista S, Arnoux I, Audinat E. Microglia in CNS development: Shaping the brain for the future. Prog Neurobiol 2017; 149-150: 1-20. [http://dx.doi.org/10.1016/j.pneurobio.2017.01.002] [PMID: 28143732]

[131] Dwyer DJ, Camacho DM, Kohanski MA, Callura JM, Collins JJ. Antibiotic-induced bacterial cell death exhibits physiological and biochemical hallmarks of apoptosis. Mol Cell 2012; 46(5): 561-72. [http://dx.doi.org/10.1016/j.molcel.2012.04.027] [PMID: 22633370]

[132] Bayles KW. Bacterial programmed cell death: making sense of a paradox. Nat Rev Microbiol 2014; 12(1): 63-9. [http://dx.doi.org/10.1038/nrmicro3136] [PMID: 24336185]

[133] Peeters SH, de Jonge MI. For the greater good: Programmed cell death in bacterial communities. Microbiol Res 2018; 207: 161-9. [http://dx.doi.org/10.1016/j.micres.2017.11.016] [PMID: 29458850]

[134] Harms A, Brodersen DE, Mitarai N, Gerdes K. Toxins, Targets, and Triggers: An Overview of Toxin-Antitoxin Biology. Mol Cell 2018; 70(5): 768-84. [http://dx.doi.org/10.1016/j.molcel.2018.01.003] [PMID: 29398446]

[135] Hõrak R, Tamman H. Desperate times call for desperate measures: benefits and costs of toxin-antitoxin systems. Curr Genet 2017; 63(1): 69-74. [http://dx.doi.org/10.1007/s00294-016-0622-2] [PMID: 27276988]

[136] Díaz-Orejas R, Espinosa M, Yeo CC. The Importance of the Expendable: Toxin-Antitoxin Genes in Plasmids and Chromosomes. Front Microbiol 2017; 8: 1479. [http://dx.doi.org/10.3389/fmicb.2017.01479] [PMID: 28824602]

[137] Rocker A, Peschke M, Kittilä T, Sakson R, Brieke C, Meinhart A. The ng_zeta1 toxin of the gonococcal epsilon/zeta toxin/antitoxin system drains precursors for cell wall synthesis. Nat Commun 2018; 9(1): 1686. [http://dx.doi.org/10.1038/s41467-018-03652-8] [PMID: 29703974]

[138] Rocker A, Meinhart A. Type II toxin: antitoxin systems. More than small selfish entities? Curr Genet 2016; 62(2): 287-90.
[http://dx.doi.org/10.1007/s00294-015-0541-7] [PMID: 26597447]

[139] Germain E, Roghanian M, Gerdes K, Maisonneuve E. Stochastic induction of persister cells by HipA through (p)ppGpp-mediated activation of mRNA endonucleases. Proc Natl Acad Sci USA 2015; 112(16): 5171-6.
[http://dx.doi.org/10.1073/pnas.1423536112] [PMID: 25848049]

[140] Sauert M, Wolfinger MT, Vesper O, Müller C, Byrgazov K, Moll I. The MazF-regulon: a toolbox for the post-transcriptional stress response in *Escherichia coli*. Nucleic Acids Res 2016; 44(14): 6660-75.
[http://dx.doi.org/10.1093/nar/gkw115] [PMID: 26908653]

[141] Culviner PH, Laub MT. Global Analysis of the E. coli Toxin MazF Reveals Widespread Cleavage of mRNA and the Inhibition of rRNA Maturation and Ribosome Biogenesis. Mol Cell 2018; 70(5): 868-880.e10.
[http://dx.doi.org/10.1016/j.molcel.2018.04.026] [PMID: 29861158]

[142] Erental A, Kalderon Z, Saada A, Smith Y, Engelberg-Kulka H. Apoptosis-like death, an extreme SOS response in *Escherichia coli*. MBio 2014; 5(4): e01426-14.
[http://dx.doi.org/10.1128/mBio.01426-14] [PMID: 25028428]

[143] Lee H, Lee DG. Arenicin-1-induced apoptosis-like response requires RecA activation and hydrogen peroxide against *Escherichia coli*. Curr Genet 2018.
[http://dx.doi.org/10.1007/s00294-018-0855-3] [PMID: 29881920]

[144] Nagamalleswari E, Rao S, Vasu K, Nagaraja V. Restriction endonuclease triggered bacterial apoptosis as a mechanism for long time survival. Nucleic Acids Res 2017; 45(14): 8423-34.
[http://dx.doi.org/10.1093/nar/gkx576] [PMID: 28854737]

[145] Takano S, Pawlowska BJ, Gudelj I, Yomo T, Tsuru S. Density-Dependent Recycling Promotes the Long-Term Survival of Bacterial Populations during Periods of Starvation. MBio 2017; 8(1): e02336-16.
[http://dx.doi.org/10.1128/mBio.02336-16] [PMID: 28174316]

[146] Conrad M, Kagan VE, Bayir H, *et al.* Regulation of lipid peroxidation and ferroptosis in diverse species. Genes Dev 2018; 32(9-10): 602-19.
[http://dx.doi.org/10.1101/gad.314674.118] [PMID: 29802123]

[147] Wegener KM, Nagarajan A, Pakrasi HB. An atypical psbA gene encodes a sentinel D1 protein to form a physiologically relevant inactive photosystem II complex in cyanobacteria. J Biol Chem 2015; 290(6): 3764-74.
[http://dx.doi.org/10.1074/jbc.M114.604124] [PMID: 25525275]

[148] Atack JM, Yang Y, Seib KL, Zhou Y, Jennings MP. A survey of Type III restriction-modification systems reveals numerous, novel epigenetic regulators controlling phase-variable regulons; phasevarions. Nucleic Acids Res 2018; 46(7): 3532-42.
[http://dx.doi.org/10.1093/nar/gky192] [PMID: 29554328]

[149] Leaver M, Domínguez-Cuevas P, Coxhead JM, Daniel RA, Errington J. Life without a wall or division machine in Bacillus subtilis. Nature 2009; 457(7231): 849-53.
[http://dx.doi.org/10.1038/nature07742] [PMID: 19212404]

[150] Kawai Y, Mickiewicz K, Errington J. Lysozyme Counteracts β-Lactam Antibiotics by Promoting the Emergence of L-Form Bacteria. Cell 2018; 172(5): 1038-1049.e10.
[http://dx.doi.org/10.1016/j.cell.2018.01.021] [PMID: 29456081]

[151] Stapels DAC, Hill PWS, Westermann AJ, *et al. Salmonella* persisters undermine host immune defenses during antibiotic treatment. Science 2018; 362(6419): 1156-60.
[http://dx.doi.org/10.1126/science.aat7148] [PMID: 30523110]

[152] Verstraeten N, Knapen W, Fauvart M, Michiels J. A Historical Perspective on Bacterial Persistence.

Methods Mol Biol 2016; 1333: 3-13.
[http://dx.doi.org/10.1007/978-1-4939-2854-5_1] [PMID: 26468095]

[153] Gutierrez A, Jain S, Bhargava P, Hamblin M, Lobritz MA, Collins JJ. Understanding and Sensitizing Density-Dependent Persistence to Quinolone Antibiotics. Mol Cell 2017; 68(6): 1147-1154.e3.
[http://dx.doi.org/10.1016/j.molcel.2017.11.012] [PMID: 29225037]

[154] Berghoff BA, Wagner EGH. RNA-based regulation in type I toxin-antitoxin systems and its implication for bacterial persistence. Curr Genet 2017; 63(6): 1011-6.
[http://dx.doi.org/10.1007/s00294-017-0710-y] [PMID: 28560584]

[155] Meylan S, Andrews IW, Collins JJ. Targeting Antibiotic Tolerance, Pathogen by Pathogen. Cell 2018; 172(6): 1228-38.
[http://dx.doi.org/10.1016/j.cell.2018.01.037] [PMID: 29522744]

[156] Gudelj I, Kinnersley M, Rashkov P, Schmidt K, Rosenzweig F. Stability of Cross-Feeding Polymorphisms in Microbial Communities. PLOS Comput Biol 2016; 12(12)e1005269
[http://dx.doi.org/10.1371/journal.pcbi.1005269] [PMID: 28036324]

[157] Bauer MA, Kainz K, Carmona-Gutierrez D, Madeo F. Microbial wars: Competition in ecological niches and within the microbiome. Microbial cell (Graz, Austria) 2018; 5(5): 215-9.

[158] Cornforth DM, Foster KR. Competition sensing: The social side of bacterial stress responses. Nat Rev Microbiol 2013; 11(4): 285-93.
[http://dx.doi.org/10.1038/nrmicro2977] [PMID: 23456045]

[159] Oliveira NM, Martinez-Garcia E, Xavier J, *et al.* Biofilm Formation As a Response to Ecological Competition. PLoS Biol 2015; 13(7)e1002191
[http://dx.doi.org/10.1371/journal.pbio.1002191] [PMID: 26158271]

[160] Mullineaux-Sanders C, Suez J, Elinav E, Frankel G. Sieving through gut models of colonization resistance. Nat Microbiol 2018; 3(2): 132-40.
[http://dx.doi.org/10.1038/s41564-017-0095-1] [PMID: 29358683]

[161] Hennell James R, Caceres EF, Escasinas A, *et al.* Functional reconstruction of a eukaryotic-like E1/E2/(RING) E3 ubiquitylation cascade from an uncultured archaeon. Nat Commun 2017; 8(1): 1120.
[http://dx.doi.org/10.1038/s41467-017-01162-7] [PMID: 29066714]

[162] Shin DS, Pratt AJ, Tainer JA. Archaeal genome guardians give insights into eukaryotic DNA replication and damage response proteins. Archaea 2014; in press: 206735.
[http://dx.doi.org/10.1155/2014/206735]

[163] Zaremba-Niedzwiedzka K, Caceres EF, Saw JH, *et al.* Asgard archaea illuminate the origin of eukaryotic cellular complexity. Nature 2017; 541(7637): 353-8.
[http://dx.doi.org/10.1038/nature21031] [PMID: 28077874]

[164] Fröls S. Archaeal biofilms: Widespread and complex. Biochem Soc Trans 2013; 41(1): 393-8.
[http://dx.doi.org/10.1042/BST20120304] [PMID: 23356317]

[165] Tonner PD, Pittman AM, Gulli JG, Sharma K, Schmid AK. A regulatory hierarchy controls the dynamic transcriptional response to extreme oxidative stress in archaea. PLoS Genet 2015; 11(1)e1004912
[http://dx.doi.org/10.1371/journal.pgen.1004912] [PMID: 25569531]

[166] Galperin MY, Makarova KS, Wolf YI, Koonin EV. Phyletic Distribution and Lineage-Specific Domain Architectures of Archaeal Two-Component Signal Transduction Systems. J Bacteriol 2018; 200(7): e00681-17.
[PMID: 29263101]

[167] Cellini A, Scoarughi GL, Poggiali P, *et al.* Stringent control in the archaeal genus Sulfolobus. Res Microbiol 2004; 155(2): 98-104.
[http://dx.doi.org/10.1016/j.resmic.2003.11.006] [PMID: 14990261]

CHAPTER 4

Extremophiles

Abstract: Many Earth habitats have an extreme cold, heat, salinity, and pressure conditions. Such places are populated by extremophiles. Thermophiles prefer environments with high temperatures. Psychrophiles live at very low temperatures. Halophiles can tolerate high salt, alkaliphiles tolerate high and acidophiles low pH. Barophiles adapt to live under high pressure. Extremophiles have permanent modifications in their membranes and in the amino acid composition of their proteins that allow them to maintain a stable internal environment under extreme conditions. Many environments have several extreme conditions, *e.g.* cold and high pressure on the ocean floor; such places are populated by polyextremophiles.

Keywords: Alkaliphile, Acidophile, Compatible Solutes, Extremophiles, Halophile, Halophile, Polyextremophile, Psychrophile, Thermophile.

INTRODUCTION

Extremophiles is a common name for organisms that live in extremely hot, cold, high saline, acidic, alkaline, or high metal content conditions that other life forms cannot tolerate [1, 2]. Extremophiles are found in hydrothermal vents, permafrost and Antarctica lakes, radioactive waste, mine drainage, several miles inside the Earth, at the ocean bottom, where the pressure reaches 110 MPa, and at mountaintops where low pressure is combined with cold and extreme solar radiation. Extremophiles show a wide metabolic versatility in their sources of nutrients and energy from photosynthesis to methane, sulfur, and iron. Environmental conditions optimal for their growth are often difficult to reproduce in the laboratory, and only a few extremophiles are adapted to grow in culture. Some extremophiles are discovered only from high-throughput DNA sequencing data obtained from the whole microbial population from a specific environment.

Extremophiles are classified according to the environments where they live. Thermophiles live at high temperatures, psychrophiles live at very low temperatures, halophiles live in high salt, alkaliphiles and acidophiles live at high or low pH, respectively, and barophiles live under high pressure. Most extremophiles can tolerate several extreme conditions.

Nadia Danilova

Most extremophiles are microorganisms with a high proportion of archaea among them. Archaea are often the most adapted extremophiles that can tolerate most extreme environments. For example, some *Picrophilus* species grow optimally at pH of 0.7 [3]. Among bacteria, cyanobacteria are often found in diverse extreme environments from ice to hot spring to deserts and hypersaline and alkaline lakes but not in acidic conditions. Eukaryotes are also found in extreme environments; they mostly include algae, fungi, and protozoa. Fungi adapt exceptionally well to various extreme environments except for very hot places. They are found in the acidic and metal-rich drainage in mines, hot and cold deserts, deep ocean floor, and in hypersaline lakes. They grow alone or in symbiosis with cyanobacteria or algae as lichens. Even some animals are found in extreme environments, *e.g.* representatives of rotifers, nematodes, arthropods, and tardigrades.

Many extremophiles are ancient organisms and their study is important for understanding the evolution of life on our planet. They are also of interest to astrobiology, since some extreme environments on Earth may resemble conditions on other planets. For example, Antarctic dry regions resemble Mars. Biomolecules from extremophiles and their peculiar biochemical pathways are of great interest to biotechnology. This is particularly true for enzymes that remain catalytically active under extremes of temperature, salinity, pH, and solvent conditions. Some of them are already used for diverse applications from polymerase chain reaction (PCR) used in science and medical tests to the production of laundry detergents and biodegradation of waste [4]. Other organic compounds produced by extremophiles find applications as stabilizers, osmoprotectants, skin moisturizers, and antimicrobials [5].

Extremophiles are useful organisms for studies of the stress response. They have various adaptations that maintain their homeostasis in extreme environments. Some of these adaptations are unique for extremophiles; for example, their macromolecules are often modified to provide stability under extreme conditions. Other adaptations of extremophiles are similar to stress responses of mesophilic organisms. For example, an adaptation of membrane lipid composition to keep proton permeability within the effective range for ATP synthesis despite changes in temperature, salinity, and pH are found in both mesophilic and extremophilic bacteria and archaea [6, 7]. It is called "homeoproton permeability adaptation". Increased proportion of unsaturated fatty acids in cell membranes is a common adaptation of psychrophiles and mesophiles to low temperature. In psychrophiles, it is a permanent feature. In mesophiles, desaturation of fatty acids by desaturases is induced as a rapid cold stress response. Other adaptations like the length of fatty acid chains or their branching require *de novo* synthesis and can be introduced only during growth under extreme conditions [8]. Next, we will discuss the adaptations of specific extremophiles.

4.1. Thermophiles

Thermophiles grow optimally between 50°C and 70°C, whereas hyperthermophiles can grow optimally up to 105°C [9]. Thermophiles grow in thermal vents, springs, and ponds heated by geothermal activity or by intensive solar irradiation. Geothermal environment existed on Earth throughout its entire history and was especially widespread at its early stages. It is thought that life originated in these ecosystems [10, 11]. Some anoxic geothermal environments such as at the sea floor may resemble those that existed on Earth 4 billion years ago. Organisms that live there, take advantage of natural redox couples, such as hydrogen and sulfur. In other places, thermophiles adapted to oxygen. Hyperthermophilic organisms often utilize a wide variety of unusual electron donors and acceptors; for example, the archaeon *Ferroglobus placidus* uses aromatic compounds as electron donors and Fe(III) as an electron acceptor [12, 13].

Bacteria, archaea, and eukaryotes can be thermophilic although archaea predominate. Diversity of thermophiles decreases with increasing temperature and extreme pH [14, 15].

Thermophiles adjust to life at high temperature by multiple changes in their macromolecules, membranes, and metabolism. The major problem encountered by cells at a high temperature is unfolding of proteins, which exposes their hydrophobic cores and leads to aggregation. Thermophilic proteins have several adaptations that let them maintain activity at high temperatures. They include an increased number of large hydrophobic residues, disulfide and ionic bonds [16, 17]. Thermophilic proteins typically have a tighter packing of their hydrophobic core, which decreases solvent exposure. Additional disulfide bridges increase the rigidity of thermophilic proteins, preventing alteration of their quaternary structure. Most thermophilic proteins tend to have extra salt bridges, which increase the resistance to unfolding under heat and make the folded state more stable [18]. They also have an increased number of charged residues on the surface [19]. Thermophilic enzymes are nearly inactive at room temperature as a result of their compactness and rigidity.

Molecular chaperones from thermophiles that assist protein folding also show adaptations to high temperature. For example, the Trigger factor from hyperthermophiles strongly binds proteins during folding and counteracts the fast folding rate imposed by high temperature. The same factor in psychrophiles has only weak chaperone activity [20].

Heat causes increased membrane fluidity and disorganization. Accordingly, both mesophiles and thermophiles respond to heat by increasing membrane order and stiffness. Common changes in membrane composition in thermophilic bacteria include an increase in the proportion of saturated fatty acids in their membranes, increase in the degree of fatty acids saturation, usage of long-chain fatty acids, usage of branched fatty acids, and formation of cyclopropane ring from unsaturated fatty acids [7, 8, 21]. Membranes of thermophilic archaea are made of tetraether monolayers [6]. Ether lipids are very stable, which is important in extremely hot conditions. Membranes made of ether lipids have low permeability for protons even at high temperatures, which permit thermophilic archaea to maintain proton motive force under extreme conditions. Some thermophilic bacteria have membranes with tetraether lipids that resemble archaea's membranes [8]. Archaea membrane's stability is further increased by incorporation of pentacyclic and macrocyclic archaeols that are cross-linked at the ends of isoprenoid chains.

Nucleic acid protection in thermophiles and mesophiles includes stabilization of nucleic acids by monovalent cations, mostly through increased concentration of K^+ and polyamines. Both archaea and bacteria hyperthermophiles produce unusual long linear DNA-binding polyamines such as caldopentamine and caldohexamine, and branched RNA-binding polyamines such as tetrakis(3-aminopropyl) ammonium and N(4)-aminopropyl spermidine [22]. Modification of nucleotides such as methylation of cytosine and 2'OH group of the ribose in RNA is also frequently observed and is thought to increase RNA stability [23, 24]. GC content in structural RNA in some thermophiles shows a positive correlation with their optimal growth temperatures [25]. Change in supercoiling of DNA is another example of a permanent adaptation of thermophiles. In mesophilic bacteria, DNA is negatively supercoiled, which makes the strand separation easier. In contrast, in thermophilic bacteria, DNA is positively supercoiled, which makes it more stable [26].

Similar to mesophilic organisms, thermophiles respond to various non-thermal stresses [27]. An interesting response to UV treatment or to chemically induced double-strand breaks in DNA have been observed in the thermophile *Sulfolobus* species, which grow with a preference for pH 2–3 and temperature of 75–80°C. UV induces pili formation in this species, which promotes cell aggregation. It increases DNA transfer among *Sulfolobus* cells for DNA repair *via* homologous recombination [28, 29]. Enhanced DNA repair contributes to the high-stress tolerance of this archaea.

4.2. Psychrophiles

Most of our planet is permanently cold. This includes most of the ocean, which is, except for a surface layer, permanently at 4–5°C, permafrost, polar regions, glaciers, and mountains. Organisms that live in places where the temperature ranges from –20°C to +10°C are called psychrophiles [30]. Many terrestrial organisms live at fluctuating temperatures and are capable of surviving periods of cold temperature; those are called cold-tolerant or psychrotrophs. Often, psychrophiles face additional stresses associated with cold environments, such as high pressure, desiccation, irradiation, high or low pH, high osmotic pressure, and low nutrient availability.

Cold places are populated at a surprisingly high density by organisms from all three domains of life [31]. Microorganisms are the most numerous inhabitants largely represented by bacteria that form the basis of food webs for other organisms. Some multicellular organisms such as isopods, lichens, and mosses are also common in extremely cold places [32]. A few microorganisms can live in ice; they use photoautotrophy where sunlight penetrates the ice or chemoautotrophy in dark subglacial habitats [33]. Interestingly, the abundance of bacteria and archaea in benthic deep-sea layers increases from the middle to high latitudes; especially for archaea species that belong to *Thaumarchaeota* [34].

Organisms living at low temperature face problems opposite to thermophiles. Structural flexibility of macromolecules decreases in cold leading to less membrane fluidity and lower enzymatic activity. Translation and post-translational processing are also decreased at a low temperature. To maintain life-sustaining metabolic rates, psychrophiles need to make their proteins more flexible and the membranes more fluid [17, 35].

Psychrophiles evolved several permanent adaptations to overcome the effects of low temperature. To increase the efficiency of translation, their genomes often encode multiple copies of tRNA species for all amino acids, as well as an increased variety and number of chaperones. For example, four copies of the chaperone DnaJ are encoded in the genome of *Psychromonas ingrahamii* [36].

Permanent adjustment of bacterial and archaeal membranes to low temperature includes an increased proportion of mono- and polyunsaturated fatty acids, conversion of trans- to cis-isomeric fatty acids, more lipids with methyl branched chains, and shorter fatty acids (<12 carbons) [8]. In addition, psychrophiles often have polar and non-polar carotenoids in their membranes. Yeasts adapted to cold places also have an increased proportion of polyunsaturated fatty acids in their membranes [37]. Many species encode OLE (ornate, large, extremophilic) non-coding RNAs that localize to cell membranes and increase their resistance to cold and substances such as alcohol [38].

To increase enzymes' flexibility and activity psychrophiles make use of multiple mechanisms [16, 17, 39]. They decrease the content of amino acids such as arginine and proline that form hydrogen bonds and salt bridges, thus reducing conformational flexibility. *Psychrobacter arcticus*, which grows at temperatures as low as –10°C in Siberian permafrost has a low content of these amino acids [40]. At the same time, psychrophiles increase the content of amino acids with small or neutral side chains such as glycine. In general, psychrophilic proteins have a reduced hydrophobic core and a less charged protein surface to maintain flexibility and activity under cold temperatures. Cold-adapted enzymes have less hydrophobic amino acids in their active sites, which allow them to function at subfreezing temperatures. Psychrophilic enzymes have high activity at room temperature. However, as a consequence of their structural flexibility, psychrophilic proteins are unstable and thermolabile and can have lower fidelity [41].

Another important surviving strategy of psychrophiles is to use compounds such as antifreeze proteins and compatible solutes that reduce the formation of cytoplasmic ice crystals at freezing temperatures [40, 42]. Psychrophiles constitutively express a number of the cold-shock proteins induced in response to cold exposure in mesophiles [43].

Low level of nutrients is a frequent feature of cold environments, and psychrophiles overcome it by adjusting their metabolism and using such alternative sources of energy as acetate; by secretion of hydrolytic extracellular enzymes that are active at low temperatures, and by other means [37, 40, 42].

Cold-tolerant organisms living in environments with fluctuating temperatures adjust to cold by a massive change in their gene expression. Archaeon *Methanolobus psychrophilus* grown at 4°C *versus* 18°C changes the expression of 1295 genes (944 up-regulated/351 down-regulated), which is ~40% of its genome [43]. Examples of genes upregulated at low temperature include antioxidative enzymes such as catalases and superoxide dismutases, which may be associated with increased solubility of oxygen and increased ROS formation at low temperatures. Also, upregulated are membrane transporters responsible for the uptake of nutrients and compatible solutes, and for recycling of membrane components. Biosynthesis of compounds, such as trehalose, which prevents denaturation and aggregation of proteins, scavenges free radicals and stabilizes cellular membranes, is also increased. Production of exopolysaccharides, which lower the freezing point of water, is also upregulated. Expression of genes involved in locomotion and iron uptake is generally suppressed at cold temperature.

Psychrophilic organisms often have high resistance to various stresses. For example, methanogenic archaea from Siberian permafrost are highly resistant not only to subzero temperatures, but also to starvation, desiccation, high salinity, and radiation [44].

4.3. Halophiles

Halophiles are organisms that inhabit hypersaline environments, like the Dead Sea in Israel and the Great Salt Lake in the USA. They are classified as slight halophiles that grow in 2–5% of NaCl, moderate halophiles that tolerate 5–20%, and extreme halophiles that tolerate saturating concentrations of NaCl. Representatives of all domains of life are found among halophiles. Most halophiles are archaea. *Halobacteriaceae* species such as *Haloquadratum walsbyi* are often the dominant species in hypersaline waters. A large bloom of *Halobacteriaceae* makes water reddish because of bacteriorhodopsin in their membranes. Halophiles also include some photosynthetic, lithotrophic, and heterotrophic bacteria. Eukaryotes that are found in hypersaline environment include the alga *Dunaliella salina* and some heterotrophs, such as fungi and brine shrimp. Hypersaline environments are often coupled with other stressful conditions, such as acidic or alkaline pH, high temperature, and low nutrients.

Halophiles have two major mechanisms to balance the extreme osmotic pressure, which are called "salt-out" and "salt-in". "Salt-out" strategy is also used by non-halophiles during osmotic stress response; it is associated with accumulation inside the cytoplasm of compatible solutes to increase the internal osmolarity. The solutes accumulate either by uptake from the environment or by *de novo* synthesis. This strategy is accompanied by the export of inorganic ions to keep cytoplasmic concentration of ions low. Various species use different compatible solutes. For example, freshwater strains of cyanobacteria accumulate the sugars trehalose and/or sucrose in response to osmotic shock; the moderately halotolerant species often accumulate glucosylglycerol, whereas halophilic species use glycine betaine and ectoine [45 - 47]. Algae of the *Dunatiella* species accumulate glycerol in the cytoplasm in molar concentrations proportionally to the concentration of NaCl in the media [48]. Halophilic organisms adapted to life in hot environments produce compatible solutes that typically have a negative charge associated either with a carboxylic group or a phosphodiester. One example is mannosylglycerate, which is produced in response to osmotic stress in several thermophilic bacteria and archaea [49, 50]

The excessive osmotic pressure can also be balanced by the "salt-in" strategy. It is found in extremely halophilic archaea of the family *Halobacteriaceae* and the bacteria of the order *Haloanaerobiales.* It involves the accumulation of inorganic ions, typically K^+. KCl can accumulate in the cells of these organisms at concentrations of 1.2 to 4.5M, and their proteins are adapted to function in such high salt. Potassium accrual and sodium expulsion during osmotic upshock are achieved *via* the secondary transport with the proton gradient as an energy source. Potassium loss during downshock is achieved through both the secondary transport and non-specific ion loss through mechanosensitive channels [51].

At high salt concentrations, water is bound by the salt's ions and is less available to proteins, which can cause hydrophobic amino acids to lose hydration and lead to protein denaturation. Halophiles developed special adaptations that make proteins stable and functional at very high salinity [17, 52, 53]. Genomic and structural analyses of halophilic enzymes have revealed an increased number of the acidic amino acids Gln and Asn and of short polar side chains amino acids, while the numbers of lysines and large hydrophobic residues are decreased, which enhances the enzymes' solubility. Halophilic proteins often have a low pI, which at the normal cellular pH results in a negative charge and increased number of salt bridges that keeps the water shell around the proteins. Haloenzymes are often polyextremophilic and can withstand high temperature, organic solvents, and chaotropic agents; salt is required for their stability [54].

Halophiles adapted their membranes to have a low proton and sodium permeability in a high salt environment. The destabilizing effect of salt is compensated by an increase of the ratio of phosphatidylglycerol to phosphatidylethanolamine. Extremely halophilic archaea have archaetidylglycerol methylphosphate as the polar lipid instead of phospholipids with choline, ethanolamine, and serine head groups, which makes their membranes very stable [55]. Glycosylation of S-layers also increases the stability of archaeal membranes at high salinity [56].

Extremely halophilic archaea are often resistant to other stresses such as extreme starvation, desiccation, and radiation. They protect themselves from UV by pigmentation with carotenoids, by polyploidy, and by efficient DNA repair [57]. Polyploidy was found in several haloarchaea; hypothetically it provides advantages for growth and survival due to the low rate of mutations and usage of extra DNA copies as a source of phosphate [58]. Several survival strategies seem to be specific to *Haloarchaea*, for example, secretion of an extremely large protein halomucin, which surrounds the cells forming a protecting cloud, gas vesicles that promote cell flotation, and production of spherical dormant stages [59]. Halophilic archaea have been reported to survive for millions of years within brine inclusions in salt crystals.

Exposure of the halophilic archaea *Halolamina* to 5.5M NaCl induces up-regulation of 2149 genes and down-regulation of 1638 genes [60]. Upregulated are membrane transporters, CRISPR-Cas systems, oxidative stress proteins, iron metabolism proteins, and the production of osmoprotectant solutes. Downregulated are genes involved in DNA replication, transcription, translation, mismatch and nucleotide excision repair.

Microscopic halophilic eukaryotes use strategies similar to prokaryotes to adapt to high salt environments including a shift in protein amino acid composition, lipid remodeling, changes in ion transport, and changes in signal transduction [61]. Accumulation of compatible solutes inside cells is also their common tactic [62]. Interestingly, bacterivorous protists such as *Halocafeteria seosinensis* that use the "salt-out" strategy have intron-containing genes for biosynthesis of ectoine and hydroxyectoine that have been acquired through lateral gene transfer from their food — ectoine/hydroxyectoine-producing bacteria [63].

Many halophilic species have variable modes of anaerobic metabolism, and those that have bacteriorhodopsin can use light as an energy source. Some species can even use perchlorate as an electron acceptor, which is of interest to astrobiology since perchlorate is present in Martian soils [64]. Carbon monoxide can support halophilic organisms in putative brines found on Mars. The ability of halophiles to oxidize CO under oxic, suboxic (0.2% oxygen), and anoxic conditions has been demonstrated in several species [65].

Hypersaline environments are often associated with additional stressors such as intense UV irradiation that leads to direct DNA damage or oxidative damage, extreme temperatures and pH. Organisms need complex adaptations to such environments including adjustments to the cell wall structure, plasma membrane lipid composition, membrane transport systems, bioenergetics, and osmoregulation [66].

4.4. Alkaliphiles

Alkaliphiles are organisms that typically grow well at pH 9 and extremophiles can grow at pH as high as 12–13. Alkaline environments often originate by serpentinization, which involves the reaction of water with minerals like olivine. In this process, new minerals are produced along with hydrogen, methane, and highly alkaline fluids. Serpentinization is a common process at oceanic tectonic plate boundaries whereas terrestrial sites are rare. One such site is in the Cedars area in northern California. Cedars springs have pH 11–12, a low redox potential, and low ionic concentrations, making it an exceptionally challenging environment for life. However, representatives of both bacteria and archaea are found in these springs [67]. Some alkaline environments are formed by water evaporation as in lakes without outflow, which are usually highly saline such as Mono Lake in California. Its prokaryotic community is dominated by bacteria from the phyla *Proteobacteria*, *Firmicutes*, and *Bacteroidetes* [68].

To protect cytoplasmic macromolecules, alkaliphiles must maintain the cytoplasmic pH significantly lower than the pH on the outside. This is achieved through both passive and active approaches. Alkaliphiles passively maintain the cytoplasmic pH through the high buffering capacity of the cytoplasm and membrane modifications [8, 69]. Buffering capacity of cytoplasm depends on the presence of polyamines (spermine, spermidine, and putrescine) and proteins rich in acidic amino acids. Membranes of alkaliphiles contain acidic compounds such as galacturonic acid, glutamic acid, and teichuronic acid. The acidic nature of the membrane prevents the entry of hydroxide ions but permits uptake of sodium. In addition, the peptidoglycan layer of the cell wall of alkaliphiles contains an increased number of hexosamines and amino acids. There is a correlation between lipid characteristics and pH optimum of an alkaliphile: those growing at higher pHs have an increase in short-chain saturated fatty acids and a decrease in long-chain unsaturated fatty acids [70]. Some species produce extracellular flocs composed of a complex mixture of polymeric substances that cover the cells and let them maintain a lower pH inside the flocs [71].

pH homeostasis is also maintained actively through the action of cation/proton antiporters. They are located in membranes of all cells and catalyze an electrogenic exchange of outwardly moving sodium ions (or in some instances potassium ions) for a greater number of entering protons [72]. The multi-subunit Mrp-type antiporter and a large repertoire of additional antiporters help alkaliphiles to sustain a cytoplasmic pH that is often more than two pH units below the pH of the medium [73, 74].

In addition, macromolecules of alkaliphiles can tolerate higher cytoplasmic pHs than those of neutrophiles. This is why enzymes from alkaliphiles that are adapted to high pH are commonly used in the manufacturing of detergents and other biotechnological applications [75].

When cytoplasm is more acidic than the outside, this creates a pH gradient across the cytoplasmic membrane that is in reverse to the orientation of trans-membrane proton-motive force necessary for ATP synthases. Alkaliphiles apply multiple strategies to circumvent this problem. Some anaerobic alkaliphiles pump sodium instead of protons, and use sodium based ATPases to generate ATP [76]. However most aerobic alkaliphiles couple ATP synthesis to protons. Since the ΔpH in alkaliphiles is reversed (the inside is more acidic than outside) and the cytoplasm has a high negative ion capacity, this creates a high enough transmembrane electrical potential $\Delta\Psi$ (inside negative) to maintains a sufficient proton-motive force [77] (Fig. **4**).

In addition, the altered composition of alkaliphile membranes let them bind protons thus increasing proton concentration near the surface. The surface proteins of alkaliphiles such as *Bacillus pseudofirmus* OF4 usually have a low pI, which can contribute to proton retention [69]. Moreover, ATP synthases of alkaliphiles are modified at several motifs that may increase the enzyme activity at elevated pH [78, 79].

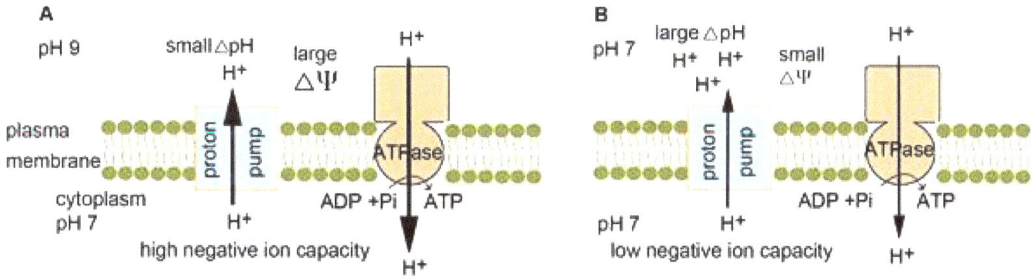

Fig. (4). Large transmembrane electrical potential ΔΨ of alkaliphiles compensates for the low difference in H+ concentration across the membrane at high pH. (**A**) A proton motive force at alkaline environment is composed from a small ΔpH and large Δψ. (**B**) At neutral pH, opposed is observed with large ΔpH and small Δψ.

ATP synthase might be present in the genome of the last common ancestor, although no proteins of proton-pumping chains have been traced to LUCA. It is hypothesized that LUCA lived in alkaline hydrothermal vents that were more alkaline then the ocean outside and might use this naturally existing pH gradient to synthesize ATP [80].

Alkaline waters are often low in nutrients and their inhabitants need to diversify sources of energy and nutrients. A haloalkaliphilic bacterium *Halanaerobium hydrogeniformans* utilize a wide range of carbon sources from simple sugars like glucose to hemicellulose and cellulose [81]. It converts them to various products including hydrogen. Betaproteobacteria isolated from highly alkaline (pH 11.6) serpentinizing springs are capable of autotrophic growth with hydrogen, calcium carbonate and oxygen [82].

4.5. Acidophiles

Highly acidic environments are relatively scarce and are generally associated with volcanic activity or mining [83]. Often, sulfate is the main anion in such an environment. Low pH facilitates metal solubility, and therefore, acidic waters can have increased concentrations of heavy metals. An important bacterial acidophile is *Acidithiobacillus ferrooxidans*, a chemolithoautotrophic gram-negative bacteria that grow at pH1.3–4.5, derives energy from the oxidation of iron using oxygen as an electron acceptor, and builds its carbon molecules using the air CO_2 [84, 85]. *A. ferrooxidans* is used in metal extraction from sulfide ores by bioleaching. Another well-studied bacterial acidophile is the human pathogen *Helicobacter pylori* that live in an extremely acidic environment of the stomach with pH 1.5–3.5 [86]. Acidophilic iron-oxidizing wall-lacking archaea from the family *Ferro- plasmaceae* grow in slime streamers and sediments at pH 0.5 [87, 88]. Eukaryotes are also found among acidophiles including alga *Dunaliella acidophila* and ciliate *Urotricha*.

Most acidophiles can maintain their cytoplasmic pH close to neutral *via* alkalization and buffering. Their internal pH is also protected by a special composition of the membranes that make them highly impermeable. Thermo-acidophilic sulfobacilli and alicyclobacilli such as *Alicyclobacillus acidocaldarius* that cause spoilage of fruit juices have as major membrane components fatty acids containing a terminal cyclic group such as cyclohexyl and cycloheptyl. These so-called ω-alicyclic fatty acids may protect the membrane from high-temperature and low-pH through restriction of lipid movement and of transmembrane diffusion of molecules [89]. Polar hopanoids-based lipids found in these bacteria may play the same role. Archaea *Ferroplasma* cell membranes contain caldarchaetidyl glycerol tetraether lipids that have very low proton permeability because of the isoprenoid core, which could be a major contributor to the extreme acid tolerance of these cell wall-less organisms [88, 90]. In addition, some acidophiles form a protective polysaccharide coating on the cell's outside and *A. ferrooxidans* can adjust the composition of membrane lipids and porins to pH [69].

Membrane ΔpH of extreme acidophiles is high because it is alkaline on the inside relative to the outside. High ΔpH results in a large proton motive force, which has to be reduced to normal value by a reversed ΔΨ (positive on the inside). In spite of such compensation, the PMF of acidophiles is still high, around −200 mV [69, 77].

H. pylori survives in the stomach by maintaining its periplasmic pH at ~6.1, which allows it to maintain the cytoplasmic pH ~ 7.0 and Δψ ~ 100 mV [69]. This is in contrast to *E. coli*, whose periplasmic pH is in equilibrium with the medium. *H. pylori* uses urease to buffer its periplasm. In response to acidic pH, its UreI channels open and associate with urease, permitting rapid access of urea to the enzyme. NH_3 and H_2CO_3 produced by urease are transported through the same UreI channels to the periplasm thereby avoiding over-alkalization of the cytoplasm while buffering the periplasm. Because of the large ΔpH, acidophilic bacteria are sensitive to organic acids. In addition, the inside-positive Δψ makes them sensitive to toxic anions.

Some acidophiles such as bacteria that oxidize ethanol to acetic acid have acidified cytoplasm. Several metabolic processes inside the cells of these bacteria serve to reduce acetic acid damage including assimilation of acetate through the tricarboxylic acid (TCA) cycle and export of acetate by various transporters [91, 92]. Proteins of these acidophiles attain acid stability through several mechanisms. Comparison of proteomes from bacteria with acidic and neutral cytoplasm shows a reduction of basic residues with acid-liable amides, increased usage of smaller and more compact residues, and an increased number of aromatic and aliphatic residues [93]. Therefore the numbers of Ala, Pro, and Thr is increased in proteins from bacteria with acidic cytoplasm, whereas the numbers of Asn, Gln, Ile, and Tyr are reduced. Relocation of salt-bridges also can increase acid resistance.

The secreted acidophilic and thermostable enzyme endo-β-glucanase from *Sulfolobus solfataricus* works optimally at pH 1.8 and temperature 80°C [94]. The modeling of this enzyme yields a mostly negatively charged surface with an excess of glutamic and aspartic acid residues, which minimizes destabilization caused by low pH.

Acidophilic organisms have high resistance to metals. *A. ferrooxidans* resistance to copper is mediated by multiple factors including ATPases, efflux pumps, a copper-binding chaperone, and the periplasmic copper-binding proteins rusticyanin and AcoP [95, 96]. Some factors may contribute to the resistance of this organism to uranium *via* thiol homeostasis and uranium binding [97].

4.6. Piezophiles

Humans and other organisms are adjusted to a certain pressure and its change in pressure can lead to stress and even death. It is well known that scuba divers that experience steep changes in water pressure suffer considerable injury and morbidity. Piezophiles are organisms that are adapted to live under high hydrostatic or lithostatic pressure. They colonize environments in the depth of the ocean and deep under the earth surface. For each kilometer of water depth hydrostatic pressure increases by approximately 100 atmospheres or 10 megapascal (MPa), whereas the lithostatic pressure of the continental crust increases by 27–30 MPa for each kilometer. However, life is found even 11 km in the ocean depth in the Mariana trench [98] and underground at the depth of 5 km [99]. Deep sediments and oceanic crust also harbor life. The so-called deep biosphere is typically defined as all oceanic and continental biotopes living under pressure of over 10 MPa.

The deep ocean is cold, 2–3°C on average except for hydrothermal vents where the temperature of hydrothermal fluids can reach 460°C and pH is acidic. Temperature is also increased deep underground. Because high-pressure environments are so dissimilar, piezophiles are very diverse organisms. Most archaeal piezophiles, such as *Sulfolobus solfataricus*, are thermophilic and belong to *Thermococcales,* whereas γ-proteobacteria are abundant among psychrophilic piezophiles [100, 101]. There are also eukaryotic piezophiles including fungi, invertebrates such as echinoderms and isopods, and fish from which Myxini, Chondrichthyes and Actinopterygii are found at depths of 2743, 4156 and 8370 m, respectively [102, 103]. Various eukaryotic piezophiles have been found in a South African gold mine at the depth of 1.3 km [104]. Deep biosphere harbors the largest number of unknown species on our planet. For example, one study of deep-sea prokaryotic community shows the presence of at least 3600 taxa, from which ~50% are new species [105].

High pressure impacts the conformation of macromolecules. Lipids respond to compression by changing their conformation and packing, which decreases membrane fluidity; this effect is similar to the effect of low temperature [8, 100]. High pressure induces dissociation of ribosomes. Proteins under pressure change their volume and folding [106].

Piezophiles have multiple adaptations in their proteins, membranes, and metabolism that allow them to live under pressure. In order to increase membrane fluidity, piezophiles use the same tactics as other organisms use to overcome the effect of low temperature [8]. One way to increase fluidity is to increase the proportion of unsaturated fatty acids in cellular membranes. Another frequent tactic is an increase in the polar lipids phosphatidylglycerol and phosphatidylcholine instead of phosphatidylethanolamine.

Proteostasis of piezophiles is also adapted to high pressure [107]. Adaptations involve an increase in protein flexibility and reduced hydration [108]. It is thought that a tightly packed hydrophobic core contributes to protein resistance against pressure [100]. Piezophilic adaptations that allow for tighter packing include the prevalence of small hydrogen-bonding amino acids and reduction in glycine and proline. Piezophilic proteins also developed their more compact structure by changes in the degree of their multimerization. Some enzymes of piezophiles increased their flexibility at the expense of stability [109].

High levels of heat shock proteins have been found in piezophilic bacteria such as *Shewanella violacea* [110]. Some of them show a pressure-dependent increase in mRNA expression. A major target of one of these HSPs, chaperonin, is RplB, a component of the 50S ribosome. Therefore HSPs are important for the maintenance of protein production at high pressure. Piezophiles have also adaptations in ribosomal RNA. A correlation has been found between the presence of sequence extensions in loops 8, 11 and 49 of the 16S rRNA molecule and the optimal pressure for gammaproteobacteria [111].

Accumulation of compatible solutes may also play a role in the adaptation to high pressure. Increased glutamate levels have been found in cells of piezophilic *Desulfovibrio hydrothermalis* [112]. In the *Photobacterium profundum*, cells accumulate glutamate and glycine betaine at normal pressure (0.1 MPa), but at their optimal growth pressure (28 MPa) they preferentially accumulate β-hydroxybutyrate and β-hydroxybutyrate oligomers [113]. The concentration of the osmolyte trimethylamine N-oxide that stabilizes proteins against pressure increases with depth from 40 to 261 mmol/kg in fishes living from 0 to 4,850 m and up to 386 mmol/kg in fishes living at 7,000 m [103]. Piezophile *Thermococcus barophilus* responds to lowering pressure below optimal by the accumulation of osmolyte mannosylglycerate [114].

Differential analysis of the proteome of *P. profundum*, which grows optimally at 28 MPa but can also grow at atmospheric pressure, shows the most significant changes in response to changing pressure in the expression of proteins involved in metabolic pathways and oxidative phosphorylation [115].

4.7. Polyextremophiles

Many environments have several kinds of extreme conditions and organisms living in these environments are called polyextremophiles. An example of a prokaryotic polyextremophile is *Sulfolobus acidocaldarius*, an archaeon that lives in hot springs at pH 3 and 80°C. Another example is the microbial communities around hot springs at the seafloor where environment features high temperature, high-pressure, absence of light and oxygen, and scarcity of food [1, 2]. Eukaryotic polyextremophile black yeasts from the order *Chaetothyriale*s can survive in high and low pH, extreme temperatures, presence of toxins, low nutrient availability, and osmotic and mechanical stress [116]. Tardigrade, a microscopic invertebrate is also a polyextremophile. In hibernation mode, called the tun state, tardigrades survive temperatures from −272°C to 151°C, vacuum (leading to extreme dehydration), the pressure of 6,000 atm as well as exposure to X-rays and gamma-rays [2, 117]. Active tardigrades too are tolerant to many extreme conditions such as cold and radiation.

Often, similar strategies are used by different species to adjust to a specific challenge, on the other hand, a certain adjustment may help to battle several stressors. For example, diverse organisms produce trehalose in response to various stressors [118]. Trehalose is produced by a thermophilic archaeon *Sulfolobus acidocaldarius* in response to osmotic stress. Tardigrades, nematodes, and the crustacean *Artemia salina* produce it in response to desiccation. The arctic insect *Megaphorura arctica* produces it in response to freezing conditions. Trehalose preserves the structure and function of proteins. It may also help to fight infections. Lysozyme, which is a part of the innate immune system and is responsible for breaking down bacterial polysaccharide walls, is stabilized by trehalose that preserves its enzymatic activity in adverse conditions [118]. Other common strategies that provide protection against diverse stresses include increased production of chaperones and compounds that neutralize ROS.

The discovery that life can exist under extreme conditions reinforced the hypothesis that life may exist on other planets and can be transferred from one planet to another. Liquid water is present on Mars and other bodies of the solar system along with compounds that can be used as a source of energy such as H_2 and iron. Several methanogens show a high survival rate under conditions that simulate the thermal and pressure conditions on these planetary bodies [119]. Our solar system, however, may not be the easiest place for the interplanetary spread of life in comparison to the recently discovered system of seven Earth-like planets orbiting a dwarf sun [120]. Some scientists even speculate that the whole universe was once an incubator for life heated by the energy of Big Bang [121]. Theoretically, other forms of life, not based on carbon biopolymers and water as a solvent, may be possible, for example, with silicon as a carbon alternative. However, no traces of such alternative life forms have been found yet.

CONCLUDING REMARKS

In the past few decades, we have realized that organisms called extremophiles populate many Earth environments with extreme conditions. The presence of liquid water appears most critical for the existence of life, whereas other conditions may vary greatly from acidic to alkaline pH, from extremely low to high pressure, salinity, radiation, and toxic metal concentrations. Living in such conditions requires specific adaptations. They include modification of membrane and intracellular proteins and production of low molecular weight compounds that protect cellular constituents.

CONSENT FOR PUBLICATION

Not applicable.

CONFLICT OF INTEREST

The author confirms that this chapter contents have no conflict of interest.

ACKNOWLEDGEMENTS

Declare none.

REFERENCES

[1] Rothschild LJ, Mancinelli RL. Life in extreme environments. Nature 2001; 409(6823): 1092-101. [http://dx.doi.org/10.1038/35059215] [PMID: 11234023]

[2] Rampelotto PH. Extremophiles and extreme environments. Life (Basel) 2013; 3(3): 482-5. [http://dx.doi.org/10.3390/life3030482] [PMID: 25369817]

[3] Schleper C, Puehler G, Holz I, *et al.* Picrophilus gen. nov., fam. nov.: a novel aerobic, heterotrophic, thermoacidophilic genus and family comprising archaea capable of growth around pH 0. J Bacteriol 1995; 177(24): 7050-9. [http://dx.doi.org/10.1128/jb.177.24.7050-7059.1995] [PMID: 8522509]

[4] Plewniak F, Crognale S, Rossetti S, Bertin PN. A Genomic Outlook on Bioremediation: The Case of Arsenic Removal. Front Microbiol 2018; 9: 820. [http://dx.doi.org/10.3389/fmicb.2018.00820] [PMID: 29755441]

[5] Sahli R, Rivière C, Neut C, *et al.* An ecological approach to discover new bioactive extracts and products: the case of extremophile plants. J Pharm Pharmacol 2017; 69(8): 1041-55. [http://dx.doi.org/10.1111/jphp.12728] [PMID: 28444868]

[6] Albers SV, van de Vossenberg JL, Driessen AJ, Konings WN. Adaptations of the archaeal cell membrane to heat stress. Front Biosci 2000; 5: D813-20. [http://dx.doi.org/10.2741/albers] [PMID: 10966867]

[7] Al-Beloshei NE, Al-Awadhi H, Al-Khalaf RA, Afzal M. A comparative study of fatty acid profile and formation of biofilm in Geobacillus gargensis exposed to variable abiotic stress. Can J Microbiol 2015; 61(1): 48-59. [http://dx.doi.org/10.1139/cjm-2014-0615] [PMID: 25496258]

[8] Siliakus MF, van der Oost J, Kengen SWM. Adaptations of archaeal and bacterial membranes to variations in temperature, pH and pressure 2017. [http://dx.doi.org/10.1007/s00792-017-0939-x]

[9] Counts JA, Zeldes BM, Lee LL, Straub CT, Adams MWW, Kelly RM. Physiological, metabolic and biotechnological features of extremely thermophilic microorganisms. Wiley Interdiscip Rev Syst Biol Med 2017; 9(3) [http://dx.doi.org/10.1002/wsbm.1377] [PMID: 28206708]

[10] Weiss MC, Sousa FL, Mrnjavac N, *et al.* The physiology and habitat of the last universal common ancestor. Nat Microbiol 2016; 1(9): 16116. [http://dx.doi.org/10.1038/nmicrobiol.2016.116] [PMID: 27562259]

[11] Dodd MS, Papineau D, Grenne T, *et al.* Evidence for early life in Earth's oldest hydrothermal vent precipitates. Nature 2017; 543(7643): 60-4. [http://dx.doi.org/10.1038/nature21377] [PMID: 28252057]

[12] Tor JM, Lovley DR. Anaerobic degradation of aromatic compounds coupled to Fe(III) reduction by Ferroglobus placidus. Environ Microbiol 2001; 3(4): 281-7. [http://dx.doi.org/10.1046/j.1462-2920.2001.00192.x] [PMID: 11359514]

[13] Schmid G, René SB, Boll M. Enzymes of the benzoyl-coenzyme A degradation pathway in the hyperthermophilic archaeon Ferroglobus placidus. Environ Microbiol 2015; 17(9): 3289-300. [http://dx.doi.org/10.1111/1462-2920.12785] [PMID: 25630364]

[14] Mardanov AV, Gumerov VM, Beletsky AV, Ravin NV. Microbial diversity in acidic thermal pools in the Uzon Caldera, Kamchatka. Antonie van Leeuwenhoek 2018; 111(1): 35-43. [http://dx.doi.org/10.1007/s10482-017-0924-5] [PMID: 28815328]

[15] Oliverio AM, Power JF, Washburne A, Cary SC, Stott MB, Fierer N. The ecology and diversity of microbial eukaryotes in geothermal springs. ISME J 2018; 12(8): 1918-28. [http://dx.doi.org/10.1038/s41396-018-0104-2] [PMID: 29662145]

[16] Reed CJ, Lewis H, Trejo E, Winston V, Evilia C. Protein adaptations in archaeal extremophiles. Archaea 2013; 2013373275
[http://dx.doi.org/10.1155/2013/373275] [PMID: 24151449]

[17] Brininger C, Spradlin S, Cobani L, Evilia C. The more adaptive to change, the more likely you are to survive: Protein adaptation in extremophiles. Semin Cell Dev Biol 2018; 84: 158-69.
[http://dx.doi.org/10.1016/j.semcdb.2017.12.016] [PMID: 29288800]

[18] Chan CH, Yu TH, Wong KB. Stabilizing salt-bridge enhances protein thermostability by reducing the heat capacity change of unfolding. PLoS One 2011; 6(6)e21624
[http://dx.doi.org/10.1371/journal.pone.0021624] [PMID: 21720566]

[19] Fukuchi S, Nishikawa K. Protein surface amino acid compositions distinctively differ between thermophilic and mesophilic bacteria. J Mol Biol 2001; 309(4): 835-43.
[http://dx.doi.org/10.1006/jmbi.2001.4718] [PMID: 11399062]

[20] Godin-Roulling A, Schmidpeter PA, Schmid FX, Feller G. Functional adaptations of the bacterial chaperone trigger factor to extreme environmental temperatures. Environ Microbiol 2015; 17(7): 2407-20.
[http://dx.doi.org/10.1111/1462-2920.12707] [PMID: 25389111]

[21] Chen YY, Gänzle MG. Influence of cyclopropane fatty acids on heat, high pressure, acid and oxidative resistance in *Escherichia coli*. Int J Food Microbiol 2016; 222: 16-22.
[http://dx.doi.org/10.1016/j.ijfoodmicro.2016.01.017] [PMID: 26828814]

[22] Lightfoot HL, Hall J. Endogenous polyamine function-the RNA perspective. Nucleic Acids Res 2014; 42(18): 11275-90.
[http://dx.doi.org/10.1093/nar/gku837] [PMID: 25232095]

[23] Baldridge KC, Contreras LM. Functional implications of ribosomal RNA methylation in response to environmental stress. Crit Rev Biochem Mol Biol 2014; 49(1): 69-89.
[http://dx.doi.org/10.3109/10409238.2013.859229] [PMID: 24261569]

[24] Schosserer M, Minois N, Angerer TB, *et al.* Methylation of ribosomal RNA by NSUN5 is a conserved mechanism modulating organismal lifespan. Nat Commun 2015; 6: 6158.
[http://dx.doi.org/10.1038/ncomms7158] [PMID: 25635753]

[25] Miralles F. Compositional properties and thermal adaptation of SRP-RNA in bacteria and archaea. J Mol Evol 2010; 70(2): 181-9.
[http://dx.doi.org/10.1007/s00239-009-9319-1] [PMID: 20069286]

[26] Galburt EA, Tomko EJ, Stump WT, Ruiz Manzano A. Force-dependent melting of supercoiled DNA at thermophilic temperatures. Biophys Chem 2014; 187-188: 23-8.
[http://dx.doi.org/10.1016/j.bpc.2014.01.001] [PMID: 24486433]

[27] Ranawat P, Rawat S. Stress response physiology of thermophiles. Arch Microbiol 2017; 199(3): 391-414.
[http://dx.doi.org/10.1007/s00203-016-1331-4] [PMID: 28097384]

[28] Fröls S, Ajon M, Wagner M, *et al.* UV-inducible cellular aggregation of the hyperthermophilic archaeon Sulfolobus solfataricus is mediated by pili formation. Mol Microbiol 2008; 70(4): 938-52.
[http://dx.doi.org/10.1111/j.1365-2958.2008.06459.x] [PMID: 18990182]

[29] Schult F, Le TN, Albersmeier A, *et al.* Effect of UV irradiation on Sulfolobus acidocaldarius and involvement of the general transcription factor TFB3 in the early UV response. Nucleic Acids Res 2018; 46(14): 7179-92.
[http://dx.doi.org/10.1093/nar/gky527] [PMID: 29982548]

[30] Cavicchioli R. On the concept of a psychrophile. ISME J 2016; 10(4): 793-5.
[http://dx.doi.org/10.1038/ismej.2015.160] [PMID: 26371407]

[31] Cavicchioli R. Microbial ecology of Antarctic aquatic systems. Nat Rev Microbiol 2015; 13(11): 691-

706.
[http://dx.doi.org/10.1038/nrmicro3549] [PMID: 26456925]

[32] Chown SL, Clarke A, Fraser CI, Cary SC, Moon KL, McGeoch MA. The changing form of Antarctic biodiversity. Nature 2015; 522(7557): 431-8.
[http://dx.doi.org/10.1038/nature14505] [PMID: 26108852]

[33] Boetius A, Anesio AM, Deming JW, Mikucki JA, Rapp JZ. Microbial ecology of the cryosphere: sea ice and glacial habitats. Nat Rev Microbiol 2015; 13(11): 677-90.
[http://dx.doi.org/10.1038/nrmicro3522] [PMID: 26344407]

[34] Danovaro R, Corinaldesi C, Dell'Anno A, Rastelli E. Potential impact of global climate change on benthic deep-sea microbes. FEMS Microbiol Lett 2017; 364(23)
[http://dx.doi.org/10.1093/femsle/fnx214] [PMID: 29045616]

[35] Bakermans C. Adaptations to marine *versus* terrestrial low temperature environments as revealed by comparative genomic analyses of the genus Psychrobacter. FEMS Microbiol Ecol 2018; 94(7)
[http://dx.doi.org/10.1093/femsec/fiy102] [PMID: 29868789]

[36] De Maayer P, Anderson D, Cary C, Cowan DA. Some like it cold: understanding the survival strategies of psychrophiles. EMBO Rep 2014; 15(5): 508-17.
[http://dx.doi.org/10.1002/embr.201338170] [PMID: 24671034]

[37] Alcaino J, Cifuentes V, Baeza M. Physiological adaptations of yeasts living in cold environments and their potential applications 2015.
[http://dx.doi.org/10.1007/s11274-015-1900-8]

[38] Harris KA, Zhou Z, Peters ML, Wilkins SG, Breaker RR. A second RNA-binding protein is essential for ethanol tolerance provided by the bacterial OLE ribonucleoprotein complex. Proc Natl Acad Sci USA 2018; 115(27): E6319-28.
[http://dx.doi.org/10.1073/pnas.1803191115] [PMID: 29915070]

[39] Feller G. Protein stability and enzyme activity at extreme biological temperatures 2010.
[http://dx.doi.org/10.1088/0953-8984/22/32/323101]

[40] Ayala-del-Río HL, Chain PS, Grzymski JJ, *et al.* The genome sequence of Psychrobacter arcticus 273-4, a psychroactive Siberian permafrost bacterium, reveals mechanisms for adaptation to low-temperature growth. Appl Environ Microbiol 2010; 76(7): 2304-12.
[http://dx.doi.org/10.1128/AEM.02101-09] [PMID: 20154119]

[41] Ernst FGM, Erber L, Sammler J, Jühling F, Betat H, Mörl M. Cold adaptation of tRNA nucleotidyltransferases: A tradeoff in activity, stability and fidelity. RNA Biol 2018; 15(1): 144-55.
[http://dx.doi.org/10.1080/15476286.2017.1391445] [PMID: 29099323]

[42] Tribelli PM, López NI. Reporting Key Features in Cold-Adapted Bacteria. Life (Basel) 2018; 8(1)E8
[http://dx.doi.org/10.3390/life8010008] [PMID: 29534000]

[43] Rodrigues DF, Tiedje JM. Coping with our cold planet. Appl Environ Microbiol 2008; 74(6): 1677-86.
[http://dx.doi.org/10.1128/AEM.02000-07] [PMID: 18203855]

[44] Morozova D, Moeller R, Rettberg P, Wagner D. Enhanced Radiation Resistance of Methanosarcina soligelidi SMA-21, a New Methanogenic Archaeon Isolated from a Siberian Permafrost-Affected Soil in Direct Comparison to Methanosarcina barkeri. Astrobiology 2015; 15(11): 951-60.
[http://dx.doi.org/10.1089/ast.2015.1319] [PMID: 26544020]

[45] Pade N, Hagemann M. Salt acclimation of cyanobacteria and their application in biotechnology. Life (Basel) 2014; 5(1): 25-49.
[http://dx.doi.org/10.3390/life5010025] [PMID: 25551682]

[46] Czech L, Hermann L, Stöveken N, *et al.* Role of the Extremolytes Ectoine and Hydroxyectoine as Stress Protectants and Nutrients: Genetics, Phylogenomics, Biochemistry, and Structural Analysis. Genes (Basel) 2018; 9(4)E177
[http://dx.doi.org/10.3390/genes9040177] [PMID: 29565833]

[47] León MJ, Hoffmann T, Sánchez-Porro C, Heider J, Ventosa A, Bremer E. Compatible Solute Synthesis and Import by the Moderate Halophile *Spiribacter salinus*: Physiology and Genomics. Front Microbiol 2018; 9: 108.
[http://dx.doi.org/10.3389/fmicb.2018.00108] [PMID: 29497403]

[48] Ben-Amotz A, Avron M. The Role of Glycerol in the Osmotic Regulation of the Halophilic Alga Dunaliella parva. Plant Physiol 1973; 51(5): 875-8.
[http://dx.doi.org/10.1104/pp.51.5.875] [PMID: 16658431]

[49] Empadinhas N, da Costa MS. 2006.

[50] Borges N, Jorge CD, Goncalves LG, Goncalves S, Matias PM, Santos H. Mannosylglycerate: structural analysis of biosynthesis and evolutionary history 2014.
[http://dx.doi.org/10.1007/s00792-014-0661-x]

[51] Becker EA, Seitzer PM, Tritt A, *et al.* Phylogenetically driven sequencing of extremely halophilic archaea reveals strategies for static and dynamic osmo-response. PLoS Genet 2014; 10(11)e1004784
[http://dx.doi.org/10.1371/journal.pgen.1004784] [PMID: 25393412]

[52] Mevarech M, Frolow F, Gloss LM. Halophilic enzymes: proteins with a grain of salt. Biophys Chem 2000; 86(2-3): 155-64.
[http://dx.doi.org/10.1016/S0301-4622(00)00126-5] [PMID: 11026680]

[53] DasSarma S, DasSarma P. Halophiles and their enzymes: negativity put to good use. Curr Opin Microbiol 2015; 25: 120-6.
[http://dx.doi.org/10.1016/j.mib.2015.05.009] [PMID: 26066288]

[54] Sinha R, Khare SK. Protective role of salt in catalysis and maintaining structure of halophilic proteins against denaturation. Front Microbiol 2014; 5: 165.
[http://dx.doi.org/10.3389/fmicb.2014.00165] [PMID: 24782853]

[55] Tenchov B, Vescio EM, Sprott GD, Zeidel ML, Mathai JC. Salt tolerance of archaeal extremely halophilic lipid membranes. J Biol Chem 2006; 281(15): 10016-23.
[http://dx.doi.org/10.1074/jbc.M600369200] [PMID: 16484230]

[56] Guan Z, Naparstek S, Calo D, Eichler J. Protein glycosylation as an adaptive response in Archaea: growth at different salt concentrations leads to alterations in Haloferax volcanii S-layer glycoprotein N-glycosylation. Environ Microbiol 2012; 14(3): 743-53.
[http://dx.doi.org/10.1111/j.1462-2920.2011.02625.x] [PMID: 22029420]

[57] Jones DL, Baxter BK. DNA Repair and Photoprotection: Mechanisms of Overcoming Environmental Ultraviolet Radiation Exposure in Halophilic Archaea. Front Microbiol 2017; 8: 1882.
[http://dx.doi.org/10.3389/fmicb.2017.01882] [PMID: 29033920]

[58] Zerulla K, Soppa J. Polyploidy in haloarchaea: advantages for growth and survival. Front Microbiol 2014; 5: 274.
[http://dx.doi.org/10.3389/fmicb.2014.00274] [PMID: 24982654]

[59] Stan-Lotter H, Fendrihan S. Halophilic Archaea: Life with Desiccation, Radiation and Oligotrophy over Geological Times. Life (Basel) 2015; 5(3): 1487-96.
[http://dx.doi.org/10.3390/life5031487] [PMID: 26226005]

[60] Kurt-Kızıldoğan A, Abanoz B, Okay S. Global transcriptome analysis of Halolamina sp. to decipher the salt tolerance in extremely halophilic archaea. Gene 2017; 601: 56-64.
[http://dx.doi.org/10.1016/j.gene.2016.11.042] [PMID: 27919704]

[61] Harding T, Simpson AGB. Recent Advances in Halophilic Protozoa Research. J Eukaryot Microbiol 2018; 65(4): 556-70.
[http://dx.doi.org/10.1111/jeu.12495] [PMID: 29266533]

[62] Weinisch L, Kühner S, Roth R, *et al.* Identification of osmoadaptive strategies in the halophile, heterotrophic ciliate Schmidingerothrix salinarum. PLoS Biol 2018; 16(1)e2003892

[http://dx.doi.org/10.1371/journal.pbio.2003892] [PMID: 29357351]

[63] Czech L, Bremer E. With a pinch of extra salt-Did predatory protists steal genes from their food? PLoS Biol 2018; 16(2)e2005163 [http://dx.doi.org/10.1371/journal.pbio.2005163] [PMID: 29394244]

[64] Oren A. Halophilic archaea on Earth and in space: growth and survival under extreme conditions 2014. [http://dx.doi.org/10.1098/rsta.2014.0194]

[65] King GM. Carbon monoxide as a metabolic energy source for extremely halophilic microbes: implications for microbial activity in Mars regolith. Proc Natl Acad Sci USA 2015; 112(14): 4465-70. [http://dx.doi.org/10.1073/pnas.1424989112] [PMID: 25831529]

[66] Banciu HL, Muntyan MS. Adaptive strategies in the double-extremophilic prokaryotes inhabiting soda lakes. Curr Opin Microbiol 2015; 25: 73-9. [http://dx.doi.org/10.1016/j.mib.2015.05.003] [PMID: 26025020]

[67] Suzuki S, Ishii S, Wu A, *et al.* Microbial diversity in The Cedars, an ultrabasic, ultrareducing, and low salinity serpentinizing ecosystem. Proc Natl Acad Sci USA 2013; 110(38): 15336-41. [http://dx.doi.org/10.1073/pnas.1302426110] [PMID: 24003156]

[68] Edwardson CF, Hollibaugh JT. Composition and Activity of Microbial Communities along the Redox Gradient of an Alkaline, Hypersaline, Lake. Front Microbiol 2018; 9: 14. [http://dx.doi.org/10.3389/fmicb.2018.00014] [PMID: 29445359]

[69] Krulwich TA, Sachs G, Padan E. Molecular aspects of bacterial pH sensing and homeostasis. Nat Rev Microbiol 2011; 9(5): 330-43. [http://dx.doi.org/10.1038/nrmicro2549] [PMID: 21464825]

[70] Bajerski F, Wagner D, Mangelsdorf K. Cell Membrane Fatty Acid Composition of *Chryseobacterium frigidisoli* PB4^T, Isolated from Antarctic Glacier Forefield Soils, in Response to Changing Temperature and pH Conditions. Front Microbiol 2017; 8: 677. [http://dx.doi.org/10.3389/fmicb.2017.00677] [PMID: 28469614]

[71] Charles CJ, Rout SP, Patel KA, *et al.* Floc Formation Reduces the pH Stress Experienced by Microorganisms Living in Alkaline Environments. Appl Environ Microbiol 2017; 83(6)e02985-16 [http://dx.doi.org/10.1128/AEM.02985-16] [PMID: 28087527]

[72] Lee C, Kang HJ, von Ballmoos C, *et al.* A two-domain elevator mechanism for sodium/proton antiport. Nature 2013; 501(7468): 573-7. [http://dx.doi.org/10.1038/nature12484] [PMID: 23995679]

[73] Mesbah NM, Cook GM, Wiegel J. The halophilic alkalithermophile Natranaerobius thermophilus adapts to multiple environmental extremes using a large repertoire of Na(K)/H antiporters. Mol Microbiol 2009; 74(2): 270-81. [http://dx.doi.org/10.1111/j.1365-2958.2009.06845.x] [PMID: 19708921]

[74] Ito M, Morino M, Krulwich TA. Mrp Antiporters Have Important Roles in Diverse Bacteria and Archaea. Front Microbiol 2017; 8: 2325. [http://dx.doi.org/10.3389/fmicb.2017.02325] [PMID: 29218041]

[75] Raddadi N, Cherif A, Daffonchio D, Neifar M, Fava F. Biotechnological applications of extremophiles, extremozymes and extremolytes. Appl Microbiol Biotechnol 2015; 99(19): 7907-13. [http://dx.doi.org/10.1007/s00253-015-6874-9] [PMID: 26272092]

[76] Meier T, Polzer P, Diederichs K, Welte W, Dimroth P. Structure of the rotor ring of F-Type Na+-ATPase from Ilyobacter tartaricus. Science 2005; 308(5722): 659-62. [http://dx.doi.org/10.1126/science.1111199] [PMID: 15860619]

[77] Goto T, Hirabayashi T, Morimoto H, *et al.* Contribution of intracellular negative ion capacity to Donnan effect across the membrane in alkaliphilic Bacillus spp. J Bioenerg Biomembr 2016; 48(1): 87-96.

[http://dx.doi.org/10.1007/s10863-015-9641-9] [PMID: 26749514]

[78] Preiss L, Hicks DB, Suzuki S, Meier T, Krulwich TA. Alkaliphilic Bacteria with Impact on Industrial Applications, Concepts of Early Life Forms, and Bioenergetics of ATP Synthesis. Front Bioeng Biotechnol 2015; 3: 75. [http://dx.doi.org/10.3389/fbioe.2015.00075] [PMID: 26090360]

[79] McMillan DG, Watanabe R, Ueno H, Cook GM, Noji H. Biophysical Characterization of a Thermoalkaliphilic Molecular Motor with a High Stepping Torque Gives Insight into Evolutionary ATP Synthase Adaptation. J Biol Chem 2016; 291(46): 23965-77. [http://dx.doi.org/10.1074/jbc.M116.743633] [PMID: 27624936]

[80] Weiss MC, Preiner M, Xavier JC, Zimorski V, Martin WF. The last universal common ancestor between ancient Earth chemistry and the onset of genetics. PLoS Genet 2018; 14(8)e1007518 [http://dx.doi.org/10.1371/journal.pgen.1007518] [PMID: 30114187]

[81] Mormile MR. Going from microbial ecology to genome data and back: studies on a haloalkaliphilic bacterium isolated from Soap Lake, Washington State. Front Microbiol 2014; 5: 628. [http://dx.doi.org/10.3389/fmicb.2014.00628] [PMID: 25477871]

[82] Suzuki S, Kuenen JG, Schipper K, *et al.* Physiological and genomic features of highly alkaliphilic hydrogen-utilizing Betaproteobacteria from a continental serpentinizing site. Nat Commun 2014; 5: 3900. [http://dx.doi.org/10.1038/ncomms4900] [PMID: 24845058]

[83] Méndez-García C, Peláez AI, Mesa V, Sánchez J, Golyshina OV, Ferrer M. Microbial diversity and metabolic networks in acid mine drainage habitats. Front Microbiol 2015; 6: 475. [PMID: 26074887]

[84] Li TF, Painter RG, Ban B, Blake RC II. The Multicenter Aerobic Iron Respiratory Chain of Acidithiobacillus ferrooxidans Functions as an Ensemble with a Single Macroscopic Rate Constant. J Biol Chem 2015; 290(30): 18293-303. [http://dx.doi.org/10.1074/jbc.M115.657551] [PMID: 26041781]

[85] Tsai YC, Lapina MC, Bhushan S, Mueller-Cajar O. Identification and characterization of multiple rubisco activases in chemoautotrophic bacteria. Nat Commun 2015; 6: 8883. [http://dx.doi.org/10.1038/ncomms9883] [PMID: 26567524]

[86] Tomb JF, White O, Kerlavage AR, *et al.* The complete genome sequence of the gastric pathogen Helicobacter pylori. Nature 1997; 388(6642): 539-47. [http://dx.doi.org/10.1038/41483] [PMID: 9252185]

[87] Edwards KJ, Bond PL, Gihring TM, Banfield JF. An archaeal iron-oxidizing extreme acidophile important in acid mine drainage. Science 2000; 287(5459): 1796-9. [http://dx.doi.org/10.1126/science.287.5459.1796] [PMID: 10710303]

[88] Golyshina OV, Timmis KN. Ferroplasma and relatives, recently discovered cell wall-lacking archaea making a living in extremely acid, heavy metal-rich environments. Environ Microbiol 2005; 7(9): 1277-88. [http://dx.doi.org/10.1111/j.1462-2920.2005.00861.x] [PMID: 16104851]

[89] Poger D, Mark AE. Effect of Ring Size in ω-Alicyclic Fatty Acids on the Structural and Dynamical Properties Associated with Fluidity in Lipid Bilayers. Langmuir 2015; 31(42): 11574-82. [http://dx.doi.org/10.1021/acs.langmuir.5b02635] [PMID: 26444798]

[90] Macalady JL, Vestling MM, Baumler D, Boekelheide N, Kaspar CW, Banfield JF. Tetraether-linked membrane monolayers in Ferroplasma spp: a key to survival in acid 2004. [http://dx.doi.org/10.1007/s00792-004-0404-5]

[91] Mullins EA, Francois JA, Kappock TJ. A specialized citric acid cycle requiring succinyl-coenzyme A (CoA):acetate CoA-transferase (AarC) confers acetic acid resistance on the acidophile Acetobacter aceti. J Bacteriol 2008; 190(14): 4933-40.

[http://dx.doi.org/10.1128/JB.00405-08] [PMID: 18502856]

[92] Andrés-Barrao C, Saad MM, Cabello Ferrete E, *et al.* Metaproteomics and ultrastructure characterization of Komagataeibacter spp. involved in high-acid spirit vinegar production. Food Microbiol 2016; 55: 112-22.
[http://dx.doi.org/10.1016/j.fm.2015.10.012] [PMID: 26742622]

[93] Fang Y, Middaugh CR, Fang J. *In silico* classification of proteins from acidic and neutral cytoplasms. PLoS One 2012; 7(9)e45585
[http://dx.doi.org/10.1371/journal.pone.0045585] [PMID: 23049817]

[94] Huang Y, Krauss G, Cottaz S, Driguez H, Lipps G. A highly acid-stable and thermostable endo-bet--glucanase from the thermoacidophilic archaeon Sulfolobus solfataricus. Biochem J 2005; 385(Pt 2): 581-8.
[http://dx.doi.org/10.1042/BJ20041388] [PMID: 15456402]

[95] Navarro CA, von Bernath D, Martínez-Bussenius C, Castillo RA, Jerez CA. Cytoplasmic CopZ-Like Protein and Periplasmic Rusticyanin and AcoP Proteins as Possible Copper Resistance Determinants in Acidithiobacillus ferrooxidans ATCC 23270. Appl Environ Microbiol 2015; 82(4): 1015-22.
[http://dx.doi.org/10.1128/AEM.02810-15] [PMID: 26637599]

[96] Latorre M, Ehrenfeld N, Cortés MP, *et al.* Global transcriptional responses of Acidithiobacillus ferrooxidans Wenelen under different sulfide minerals. Bioresour Technol 2016; 200: 29-34.
[http://dx.doi.org/10.1016/j.biortech.2015.09.110] [PMID: 26476161]

[97] Dekker L, Arsène-Ploetze F, Santini JM. Comparative proteomics of Acidithiobacillus ferrooxidans grown in the presence and absence of uranium. Res Microbiol 2016; 167(3): 234-9.
[http://dx.doi.org/10.1016/j.resmic.2016.01.007] [PMID: 26829305]

[98] Moran MA. The global ocean microbiome. Science 2015; 350(6266)aac8455
[http://dx.doi.org/10.1126/science.aac8455] [PMID: 26659059]

[99] Itävaara M, Salavirta H, Marjamaa K, Ruskeeniemi T. Geomicrobiology and Metagenomics of Terrestrial Deep Subsurface Microbiomes. Adv Appl Microbiol 2016; 94: 1-77.
[http://dx.doi.org/10.1016/bs.aambs.2015.12.001] [PMID: 26917241]

[100] Jebbar M, Franzetti B, Girard E, Oger P. Microbial diversity and adaptation to high hydrostatic pressure in deep-sea hydrothermal vents prokaryotes 2015.
[http://dx.doi.org/10.1007/s00792-015-0760-3]

[101] Li M, Baker BJ, Anantharaman K, Jain S, Breier JA, Dick GJ. Genomic and transcriptomic evidence for scavenging of diverse organic compounds by widespread deep-sea archaea. Nat Commun 2015; 6: 8933.
[http://dx.doi.org/10.1038/ncomms9933] [PMID: 26573375]

[102] Priede IG, Froese R. Colonization of the deep sea by fishes. J Fish Biol 2013; 83(6): 1528-50.
[http://dx.doi.org/10.1111/jfb.12265] [PMID: 24298950]

[103] Yancey PH, Gerringer ME, Drazen JC, Rowden AA, Jamieson A. Marine fish may be biochemically constrained from inhabiting the deepest ocean depths. Proc Natl Acad Sci USA 2014; 111(12): 4461-5.
[http://dx.doi.org/10.1073/pnas.1322003111] [PMID: 24591588]

[104] Borgonie G, Linage-Alvarez B, Ojo AO, *et al.* Eukaryotic opportunists dominate the deep-subsurface biosphere in South Africa. Nat Commun 2015; 6: 8952.
[http://dx.doi.org/10.1038/ncomms9952] [PMID: 26597082]

[105] Salazar G, Cornejo-Castillo FM, Benítez-Barrios V, *et al.* Global diversity and biogeography of deep-sea pelagic prokaryotes. ISME J 2016; 10(3): 596-608.
[http://dx.doi.org/10.1038/ismej.2015.137] [PMID: 26251871]

[106] Chen CR, Makhatadze GI. Molecular determinant of the effects of hydrostatic pressure on protein folding stability. Nat Commun 2017; 8: 14561.
[http://dx.doi.org/10.1038/ncomms14561] [PMID: 28169271]

[107] Gayán E, Govers SK, Aertsen A. Impact of high hydrostatic pressure on bacterial proteostasis. Biophys Chem 2017; 231: 3-9.
[http://dx.doi.org/10.1016/j.bpc.2017.03.005] [PMID: 28365058]

[108] Martinez N, Michoud G, Cario A, *et al.* High protein flexibility and reduced hydration water dynamics are key pressure adaptive strategies in prokaryotes. Sci Rep 2016; 6: 32816.
[http://dx.doi.org/10.1038/srep32816] [PMID: 27595789]

[109] Ichiye T. Enzymes from piezophiles. Semin Cell Dev Biol 2018; 84: 138-46.
[http://dx.doi.org/10.1016/j.semcdb.2018.01.004] [PMID: 29331641]

[110] Sato H, Nakasone K, Yoshida T, Kato C, Maruyama T. Increases of heat shock proteins and their mRNAs at high hydrostatic pressure in a deep-sea piezophilic bacterium, Shewanella violacea. Extremophiles : life under extreme conditions 2015; 19(4): 751-62.

[111] Lauro FM, Chastain RA, Blankenship LE, Yayanos AA, Bartlett DH. The unique 16S rRNA genes of piezophiles reflect both phylogeny and adaptation. Appl Environ Microbiol 2007; 73(3): 838-45.
[http://dx.doi.org/10.1128/AEM.01726-06] [PMID: 17158629]

[112] Amrani A, Bergon A, Holota H, *et al.* Transcriptomics reveal several gene expression patterns in the piezophile Desulfovibrio hydrothermalis in response to hydrostatic pressure. PLoS One 2014; 9(9)e106831
[http://dx.doi.org/10.1371/journal.pone.0106831] [PMID: 25215865]

[113] Martin DD, Bartlett DH, Roberts MF. Solute accumulation in the deep-sea bacterium Photobacterium profundum 2002.
[http://dx.doi.org/10.1007/s00792-002-0288-1]

[114] Cario A, Jebbar M, Thiel A, Kervarec N, Oger PM. Molecular chaperone accumulation as a function of stress evidences adaptation to high hydrostatic pressure in the piezophilic archaeon Thermococcus barophilus. Sci Rep 2016; 6: 29483.
[http://dx.doi.org/10.1038/srep29483] [PMID: 27378270]

[115] Le Bihan T, Rayner J, Roy MM, Spagnolo L. Photobacterium profundum under pressure: a MS-based label-free quantitative proteomics study. PLoS One 2013; 8(5)e60897
[http://dx.doi.org/10.1371/journal.pone.0060897] [PMID: 23741291]

[116] Moreno LF, Vicente VA, de Hoog S. Black yeasts in the omics era: Achievements and challenges. Medical mycology 2018; 56(suppl_1): 32-41.

[117] Weronika E, Lukasz K. Tardigrades in Space Research - Past and Future. Origins of life and evolution of the biosphere : the journal of the International Society for the Study of the Origin of Life 2017; 47(4): 545-3.

[118] Magazù S, Migliardo F, Gonzalez MA, Mondelli C, Parker SF, Vertessy BG. Molecular mechanisms of survival strategies in extreme conditions. Life (Basel) 2012; 2(4): 364-76.
[http://dx.doi.org/10.3390/life2040364] [PMID: 25371270]

[119] Taubner RS, Schleper C, Firneis MG, Rittmann SK. Assessing the Ecophysiology of Methanogens in the Context of Recent Astrobiological and Planetological Studies. Life (Basel) 2015; 5(4): 1652-86.
[http://dx.doi.org/10.3390/life5041652] [PMID: 26703739]

[120] Lingam M, Loeb A. Enhanced interplanetary panspermia in the TRAPPIST-1 system. Proc Natl Acad Sci USA 2017; 114(26): 6689-93.
[http://dx.doi.org/10.1073/pnas.1703517114] [PMID: 28611223]

[121] Merali Z. Life possible in the early Universe. Nature 2013; 504(7479): 201.
[http://dx.doi.org/10.1038/504201a] [PMID: 24336268]

CHAPTER 5

Immunodefenses of Prokaryotes

Abstract: Prokaryotes protect themselves from viruses and mobile genetic elements at several levels. They can produce exopolysaccharide capsules around the cell or slime that covers viral receptors and prevents viral binding. Secreted extracellular enzymes can destroy pathogenic nuclear acids around the cell. After a pathogen enters the cell, it can be destroyed by sugar-nonspecific nucleases that cleave both DNA and RNA. A common defense mechanism is restriction-modification (R/M). R/M systems are based on methylation of the host DNA at specific sequences, whereas viral DNA not protected by methylation is quickly degraded by restriction endonucleases. R/M systems differ in the type of enzymes and recognition principles. Most archaea and half of the bacteria have the prokaryotic adaptive immune system CRISPR-Cas. This system consists of CRISPR (clustered regularly interspersed short palindromic repeats) and CRISPR-associated (Cas) proteins. Other defenses include Argonaute-based systems that utilize both DNA and RNA guides to destroy foreign nucleic acids and abortive infection (Abi), which is a measure of last resort, such that death of the infected cells prevents spread of infection.

Keywords: Abortive infection, Argonaute, Bacteriophage exclusion, CRISPR-Cas, Extracellular nucleases, Phase variation, Restriction-modification system, Sugar-nonspecific nucleases, Superinfection exclusion.

INTRODUCTION

Viruses and other parasitic genetic elements constantly attack prokaryotes. Most bacteriophages have double strand DNA genomes and belong to Caudovirales. Single strand DNA viruses are less common, and dsRNA and ssRNA are rare [1]. All studied archaeal viruses have DNA genomes, and although some RNA viruses have been found in archaeal habitats, their hosts have not been confirmed [2].

Prokaryotes sense infection using two major immune recognition strategies conserved in evolution. One is based on the recognition of "non-self" features, which are specific for parasites. The other strategy is based on attaching a label to "self" molecules and recognizing the absence of the "self" label as a feature of parasites. An immune response can also be activated in response to the disruption of homeostasis [3]. Immune responses are subdivided into innate that are genetically encoded and adaptive that are built *de novo* in individual organisms

Nadia Danilova

against unfamiliar pathogens [4].

The known immune defense systems constitute up to 10% of prokaryotic genomes. Thermophilic bacteria and archaea have a higher representation of immune genes in their genomes, whereas parasitic prokaryotes have less immune genes [5]. Most studies of prokaryotic immunity have been done on bacteria with few on archaea.

Diverse mechanisms are used to restrict the infection on various stages including virus attachment, adsorption, replication, integration into the host DNA, and cell exit [4]. If an infection cannot be controlled, cells activate mechanisms of abortive infection (Abi) that kill both the host cell and the invader.

Immune genes often cluster on the chromosomes in defense islands in association with transposable elements, which favors horizontal gene transfer and diversification of defense systems [6]. Viruses constantly develop mechanisms to avoid host defenses [7]. This necessitates the existence in prokaryotes of multiple immune mechanisms. A recent analysis of the unknown genes in the defense islands revealed at least nine new antiviral systems and predicted many more [8]. Co-evolution of viruses and bacteria is characterized by fluctuating selection dynamics where hosts are more resistant to their contemporary than past and future parasites [9]. Below, we discuss major known defense systems of prokaryotes.

5.1. Secreted Extracellular Molecules

Many prokaryotes secrete various molecules that can serve as a weapon against other cells and as a defense against parasites. For example, extracellular nucleases can destroy pathogenic nucleic acids in the cell's vicinity. Nucleases are also important for the survival of pathogenic bacteria inside their hosts. Human pathogenic group A Streptococcus bacteria escape the human immune system by degrading DNA of the neutrophil's extracellular traps by a secreted nuclease [10]. *Staphylococcus aureus* also uses a nuclease to degrade the extracellular traps and adenosine synthase A to produce deoxyadenosine, which blocks purine salvage pathways in phagocytes and kills them [11]. Bacterial plant pathogens also use nucleases to escape extracellular traps made by the plant root border cells [12].

Prokaryotes produce various toxins and antibiotics that can protect from predator bacteria. *Bacillus subtilis* produces bacillaene that inhibits protein synthesis and protects from predation by *Myxococcus xanthus* [13].

Polymorphic toxin systems (PTSs) include a secreted multidomain toxin and an immunity (anti-toxin) protein able to protect the producing cell from the toxin's effects [14]. C-terminal domains of the polymorphic toxins carry toxic activity, whereas the N-terminal domains are involved in delivery systems. Toxic domains encode various enzymes including nucleases targeting DNA, tRNA, rRNA, nucleic acid modifying enzymes, peptidases, and lipases. The immunity protein is encoded next to a toxin. The locus may also contain genes encoding alternative toxic domains that can recombine with the 3 end of the toxin gene to generate new toxins. Some bacterial species produce toxins that are active in both bacterial and eukaryotic cells including phagocytes [15].

The toxin-producing cell delivers toxins to nearby cells through special methods such as type VI secretion and contact-dependent growth inhibition systems [16]. Injection of a toxin can result in cell death unless the target cell expresses the corresponding immunity protein. This is why kin cells are immune to the toxin. Moreover, the delivery of a toxin to a kin cell serves as a signal that promotes biofilm formation and other community defense actions [16, 17].

Many prokaryotes can make exopolysaccharide capsules coating them in response to some stress signals. These structures are difficult to penetrate and they protect cells from antibiotics and parasites. In addition, prokaryotes often form biofilms, where they are better protected against antibiotics than free-living cells [18]. Moreover, cells living in a biofilm are protected against killing by predators including bacterial predator *M. xanthus* and a worm *Caenorhabditis elegans* [19]. Besides biofilms, cells can form other superstructures in response to a threat. For example, *B. subtilis* exposed to predation by *M. xanthus* forms complex tree-like brachiating structures filled with endospores embedded within a dense matrix [20].

5.2. Preventing Cell Entry

The important line of defense is a mechanical barrier between the prokaryotic cell and the environment. Most bacteria and archaea have cell walls outside their cytoplasmic membranes that are built of sugar polymers as in most bacteria and methanogenic archaea or of the surface (S) layers of glycosylated proteins as in most archaea. Exopolysaccharide capsules enveloping cells also serve as a mechanical barrier as do biofilms.

However, viruses evolved to use cell walls and exopolysaccharide capsules to recognize their hosts. Many components of the outer membrane serve as viral receptors including peptidoglycan, lipopolysaccharides, teichoic acids, various membrane proteins, fimbriae, flagellum, and sex pilus [21, 22]. More than one receptor can be involved in recognition. Receptor binding is a prerequisite for the infection, and specificity of viral receptor-binding proteins determines the host range. Viral receptor-binding proteins and their receptors co-evolve; host receptors evolve to avoid recognition while viral proteins evolve to overcome host changes and improve recognition [23].

Host adaptations include blocking viral receptors by other molecules and hiding them by forming a capsule around the cell or by excreting exopolysaccharides that form a slime covering the receptor. Capsule-encoding genes are among fast evolving [24]. Hence, viruses develop enzymes degrading these structures [7]. Many viruses also degrade exopolysaccharides that form biofilms [25]. Some prokaryotes allow the binding but then shed the receptors together with the bound virus in the outer membrane vesicle [26, 27]. The downside of this strategy is that membrane vesicles can promote an infection spread by transfer of the viral receptors from susceptible to resistant cells [28].

Prokaryotes use diversification of viral receptors as a common antiviral strategy. It allows a subset of the population to avoid recognition. Diversification of surface molecules is also important for pathogenic bacteria; it helps to avoid the host immune system. Various mechanisms are used to generate such diversity including mutations, glycosylation, phase variation, horizontal gene transfer, gene conversion, and diversity-generating retroelements (DGRs).

Enterococcus faecalis avoids recognition by lytic phages *via* mutations in its cell wall protein that serves as their receptor [29]. Mutations in the type IV pilus prevent phage infection in *P. aeruginosa* [30]. This mechanism of diversification has its downside since structures that serve as viral receptors are often important for cell physiology and their modification may reduce fitness [7]. Glycosylation of receptors can also prevent recognition. *P.aeruginosa* has five variants of type IV pilus. Three variants are used as receptors by several phages. Two variants are glycosylated and not recognized by phages [30]. *Klebsiella pneumoniae* has more than 80 types of polysialic acid (K-antigen); such diversity decreases the chance of bacteriophage infection and also helps this human pathogen to avoid the immune system [31].

Phase variation is a common mechanism of receptor diversification. This mechanism is also used by pathogens to diversify their surface antigens. It involves reversible site-specific recombination, which acts *via* inversion of DNA fragments or by insertion of genetic elements that disrupt gene expression. Variation in protein expression by this mechanism is often achieved by the ON-OFF switch. For example, the length of oligopolysaccharide O-antigen in *Salmonella enterica,* which serves as a receptor for several bacteriophages, depends on OpvAB operon. OpvAB (OFF) generates the standard O-antigen length, whereas OpvAB (ON) generates shorter oligopolysaccharide that phages do not bind [32, 33]. Acquisition of phage resistance comes at the price of a lower virulence of *Salmonella* with the truncated antigen. *Neisseria meningitidis* (the meningococcus) has 115 genes controlled by phase variation mechanism [34].

Phase variation can be achieved epigenetically by switching between different methylation modes of the whole genome. These epigenetic regulatory systems are called phasevarions — phase-variable regulons [35 - 37].

Gene conversion is also used for diversification. *Neisseria gonorrhoeae* uses it to change the sequence of the major subunit of type IV pilus (Tfp) that mediates adherence, motility, and defense against neutrophil, and natural transformation. Tfp is encoded by *pilE* locus. In addition to this locus, *N. gonorrhoeae* has many promoter-less, partial-coding silent copies (*pilS*) that are used in gene-conversion reactions to transfer their sequences into the *pilE* locus [38].

Diversity-generating retroelements (DGRs) are genetic elements that use reverse transcription to introduce sequence variants into the target genes [39 - 41]. Bacteria, archaea, and viruses use this diversification mechanism [42]. It involves a variable sequence inside a target gene and a homologous template region that sits next to the end of the gene. A special reverse transcriptase that is error prone at adenine bases transcribes the template region and converts its adenine residues into random nucleotides; this transcript then replaces the original variable region of the target gene. Target genes are expressed on the cell surface and have the C-type lectin fold, which provides a static scaffold for display of highly variable sequences [43]. DGR mechanism let organisms adjust quickly to environmental changes.

Viral receptor-binding proteins evolve to recognize modified host receptor or by acquiring specificity to a new receptor. Bacteriophages evolve through mutations but also through recombination with other phages [44]. An interesting example of bacteriophage λ host-recognition gene J evolution has recently been described. A mutation in this gene leads to conformational bistability of J and to the formation of multiple phenotypic forms specific to different receptors, so that one form of J binds OmpF, whereas the other binds LamB [45].

The first contact of virus with the host cell is typically reversible and is followed by an irreversible attachment to the same or a different receptor [21]. This step triggers conformational changes that facilitate the insertion of viral genetic material into the host cell. Mechanisms of insertion differ between viruses. Myoviruses use the contraction of the tail to inject the genetic material into the host cell. Podoviruses that have short non-contractile tails enzymatically degrade the host cell wall and release capsid proteins that form a channel for transperiplasmic transport of the viral genome. Siphoviruses that have long non-contractile tails also use enzymes and the ejected proteins to deliver their genetic materials. Within minutes, the cellular machinery switches to the production of viral components and viral assembly. Phages are released from cells by cell lysis, extrusion, or sometimes by budding. All stages of infection can be targeted by the host immune defense factors.

5.3. Non-specific Silencing of Foreign Genes

If a foreign DNA passes the outer membrane, it still can be degraded by periplasmic nucleases. They often belong to sugar-nonspecific nucleases that cleave DNA and RNA in a sequence-independent manner [46]. Some of them may be encoded on prophages [47].

The newly acquired foreign DNA can potentially be harmful to the host and it is usually kept repressed until the host finds a way to use it [48, 49]. This repression is called xenogenic silencing and is performed by proteins from H-NS, Lsr2, MvaT or Rok families. They bind to the foreign DNA with higher AT-content than in the resident genome and repress its expression. Both DNA acquired through horizontal gene transfer and provirus DNA can be silenced in this way. To use newly acquired genes in a controlled manner, prokaryotes apply counter-silencing mechanisms [50]. Since many genes involved in virulence and immunodefense are acquired through horizontal gene transfer, xenogenic silencing is an important regulator of these functions [51].

5.4. Restriction Modification Systems

In the host cell, viral DNA can be quickly cleaved at specific sites by restriction endonucleases from restriction-modification (R/M) systems. Each R/M system includes DNA methyltransferase (MT) and restriction endonuclease (RE) that targets a specific DNA sequence. MT methylates host DNA on these sequences, which protects them from digestion by the RE [52, 53]. Thereby, R/M system labels “self” DNA by methylation and destroys all non-labeled DNAs. R/M systems are classified into several types that differ in enzymes, recognition sites, a site of hydrolysis, and ATP/GTP requirements.

Type I R/Ms are found in over half of all known bacterial genomes. Type I system is a protein complex that in addition to RE and MT also includes specificity peptides [54]. One specificity subunit recognizes foreign DNA sequence and anchors the enzyme complex at it. Then, the complex performs ATP-dependent translocation of DNA through the stationary enzyme and cleaves it at random sites that could be a long distance from a recognition site [55].

The most common type II R/M consists of separate RE and MT typically encoded within the same operon [56]. Their activities are balanced by specialized transcription factors called C (controller) proteins [57]. Type II enzymes usually recognize short, mostly palindromic motifs and cleave DNA within or close to the recognition site. These enzymes are widely used in molecular biology.

Type III MT contains a DNA binding specificity domain. It works as a multisubunit complex with helicase activity. It slides along the DNA using ATP hydrolysis; one MT subunit recognizes DNA sequence, while the other methylates adenine on one of the DNA strands [58]. Type III RE needs two unmethylated DNA sequences in a head-to-head orientation for cleavage at 25-27 bp downstream of the recognition site.

Type IV systems cleave modified, typically methylated, hydroxymethylated, or glucosyl-hydroxymethylated DNA, and have no MT [59]. A DISARM (defense island system associated with restriction-modification) system that involved several enzymes has recently been described [60]. DISARM methylase modifies CCWGG sequences in host DNA as a marker of self.

Another variant of R/M system uses adenine DNA glycosylase instead of the endonuclease. Glycosylase produces two opposing apurinic/apyrimidinic sites that are processed to double-strand break by other enzymes [61]. Other kinds of DNA modification may also serve as restriction modification mechanisms. *Salmonella enterica* restricts DNA from *Escherichia coli*, while protecting its own DNA by phosphorothioation [62, 63]. Recent studies suggest that modification of DNA with 7-deazaguanine derivatives also serves as a restriction-modification mechanism [64].

R/M enzymes are often encoded not by bacterial genome but by plasmids or prophages and may behave like selfish genetic elements [65]. Off-target cleavage of host DNA by a R/M restriction endonuclease may promote homologous recombination and boost genetic adaptations [66]. In addition, genetic exchange within and between lineages with cognate R/M systems contributes to the horizontal gene flow [67].

Phase variation in R/M system produces variations in methylation of specific sequences and reduces the emergence of resistant phages [68, 69]. At the same time, phase variation allows some cells to acquire phage DNA. Many genes that encode Type III N6-adenine DNA-methyltransferases are phase variable [70].

Most bacteria have several R/M systems so that if a phage escapes one system, another can still destroy it. The importance of R/M as a crucial defense mechanism is evidenced by diverse adaptations developed by viruses to avoid these systems [7]. Changes in phage DNA protecting against restriction include a decrease of the number of recognition sites, difference in their position, or orientation, and incorporation of unusual bases. Some modifications provide protection against multiple R/M systems. Examples include enzymes that convert adenine to N(6)-methyladenine or N(6)-(1-acetamido) adenine [7, 71]. 5-(2-aminoethoxy)methyluridine is an example of an enzymatically hypermodified pyrimidine in phage DNA [72]. Viruses also produce antirestriction proteins that inhibit R/M systems. ArdA from ColiB-P9 phage, Ocr from T7 phage inhibit type I R/M activity whereas Arn from T4 phage inhibits type IV R/M [73]. These proteins belong to the family of DNA-mimicking proteins since their structure resembles the double-helical B-form DNA. They also inhibit nucleoid protein H-NS that is involved in xenogeneic silencing of foreign DNA [74]. Host R/M system can be subverted by stimulation of host methylases to modify phage DNA or by the destruction of cofactors required for R/M activity [7]. To fight these phages, some bacteria developed restriction systems that recognize and cut only modified DNA. In turn, phages developed means to block such enzymes. For example, phage T4 protects its DNA from *E. coli* gmrSD restriction endonuclease that specifically targets and cleaves glucosylated hydroxymethylcytosine modified DNA by injecting 76-amino-acid-residue nuclease inhibitor into the infected host [75]. Mrr restriction endonuclease of *E. coli* that cuts methylated DNA, is suggested to prevent epigenetic modifications by foreign DNA methyltransferases [76].

R/M systems are not used in organisms more complex than bacteria. However, the principle of recognition of DNA methylation as a means of "self"/"non-self" discrimination was maintained throughout evolution. Immune cells of vertebrates recognize unmethylated CpG dinucleotides in bacterial DNA as a "non-self" marker.

5.5. Bacteriophage Exclusion (BREX) Systems

An immune mechanism called bacteriophage exclusion (BREX) system that confers resistance to a broad range of phages has been found in *Bacillus cereus* [77]. It includes a putative Lon-like protease, a protein with alkaline phosphatase domain, a putative RNA-binding protein, a DNA methylase, a protein with ATPase domain, and a protein of unknown function. BREX allows for phage adsorption but blocks replication of the phage DNA. Methylation of non-palindromic TAGGAG sequences may guide self/non-self discrimination, but in contrast to R/M systems, no cleavage of phage DNA has been observed. A related system called phage growth limitation (Pgl) has been found in several bacterial species [78]. BREX-like systems may be present in ~10% of microbial species.

5.6. CRISPR-Cas Immunity

CRISPR-Cas is a prokaryotic adaptive immune system found in 85% of archaea and 45% of bacteria [79 - 83]. Giant viruses, double-stranded DNA viruses with extremely large genomes use a similar system to block infection by virophages [84]. CRISPR-Cas consists of CRISPR (clustered regularly interspersed short palindromic repeats) and CRISPR-associated (Cas) proteins. CRISPR are organized as arrays of short, partially palindromic sequences interspaced by short (typically 30-40 bp) "spacer" sequences derived from viruses and other parasitic genetic elements [85 - 88]. Spacers represent the genetic memory of previously encountered invaders and are used to direct sequence-specific destruction of foreign genetic elements. They let prokaryotes rapidly eradicate repeated infections and protect the progeny from the same infection. CRISPR-Cas immunity is an example of immunity based on specific recognition of "non-self" contrary to R/M systems that indiscriminately destroy all DNAs except protected "self".

CRISPR-associated (Cas) proteins are typically encoded next to one of the arrays. Some Cas proteins mediate acquisition of new spacers upon entry of a genetic element into the cell. Other Cas proteins participate in the generation of guide RNAs from the spacers and in the destruction of intruder nucleic acids. CRISPR-Cas systems are currently subdivided into two classes, six types, and more than 20 subtypes [89]. They are extremely variable in their composition, effectors, and regulations. The class I includes the most common type I systems and type III. Class II includes type II systems. The same tasks in different systems are performed by single enzymes or by enzyme complexes. For example, in type II, a single protein Cas9 is responsible for foreign DNA destruction, whereas in type I systems the role of Cas9 is played by a system of several enzymes called Cascade. Most systems target DNA although some of them such as type III can target RNA. Techniques based on type II CRISPR are becoming indispensable tools in research, biotechnology, and medicine [90, 91].

CRISPR-Cas action includes three stages: adaptation, expression and maturation, and interference (Fig. **5**). At the adaptation stage, Cas proteins acquire spacers that become integrated into the CRISPR array [92, 93]. Acquisition of completely new spacers is called naïve CRISPR adaptation. When the cell already has spacers specific to the invading genetic element, the existing spacers are used to enhance the acquisition of new spacers in a process called primed CRISPR adaptation [94].

Adaptation starts with the selection and extraction of spacers from a foreign genome. Prokaryotes need to avoid acquisition of spacers from their own DNA since that would lead to autoimmunity. Therefore they use various mechanisms to bias spacer acquisition to foreign nucleic acids [92, 93]. Spacers are acquired during viral injection preferentially from a region of the viral genome that enters the cell first [95]. In *E. coli*, the DNA repair enzyme nuclease RecBCD is suggested to be involved in spacer acquisition [96]. RecBCD binds the free ends of dsDNA to perform an end resection to repair double-strand breaks. Sequences generated in this process can serve as spacers. RecBCD does not discriminate between host and viral DNA but it stops at a sequence called a Chi site, which is common in *E. coli* genome but not in viral DNA. Therefore, much more "non-self" than "self" degradation products are generated. Low rate of spacer acquisition also can protect against autoimmunity. Only one in 10^7 cells acquires a spacer from an infecting virus and the rates for spacers from "self" DNA are even lower [93]. In addition, CRISPR-Cas activity is induced only in response to infection.

Moreover, only a subset of sequences can be used as spacers. In type I and type II, proto-spacer sequences need to have a short protospacer-adjacent motif (PAM) immediately downstream from them. PAM motif is specific for a particular CRISPR-Cas and is recognized during spacer acquisition by the Cas enzymes [93, 97]. Type III acquires spacers only from the transcribed DNA.

The presence of a spacer matching the invading DNA can stimulate the acquisition of new spacers from this DNA through a mechanism named "priming". For example, when type I Cascade complex directed by the existing spacer binds viral DNA, it recruits Cas3, whose nuclease and helicase activity destroys the foreign genome. At the same time, new spacers are generated [98]. Partial match also promotes increased spacer acquisition, which enables prokaryotes to defend themselves against rapidly evolving viruses. Recent data suggest that priming is used in other types of CRISPR-Cas systems as well [94].

At the next stage, suitable spacers are processed and incorporated into CRISPR arrays. In type I systems, Cas4 recognizes PAM motif, processes protospacers to the correct size, and prevents integration of unprocessed spacers into CRISPR [99, 100]. New spacers are integrated into the junction between the leader sequence and the first repeat of the CRISPR array. A complex of endonucleases Cas1 and Cas2 catalyzes spacer acquisition in nearly all systems with involvement of other Cas proteins and various accessory factors [101, 102]. This complex captures the incoming protospacer, docks on the leader-repeat site of the CRISPR array, and catalyzes the integration [93, 103]. Therefore the position of the spacer in an array reflects the order of its acquisition with spacers from more recent infections placed closer to the leader. The expression of leader-adjacent spacers is higher in comparison to older spacers, thus directing the immune response at the latest offenders.

After the array is transcribed into RNA, it is processed into small CRISPR RNAs (crRNAs), each containing a single spacer, that serve as antisense guides for nucleases to locate and cleave the cognate foreign nucleic acids. Target DNA needs to be flanked by a PAM motif [104]. The best-studied types I, II, and III use different Cas proteins for processing of crRNAs and for interference with invading nucleic acids. In type I, crRNA precursors are processed by Cas6 or Cas5d, and interference is handled by Cascade-Cas3 [105 - 107]. In type II, pre-crRNA is transcribed from the same locus as tracrRNA, which encodes anti-repeat sequence tracrRNA that base pairs with repeats from crRNA (Fig. **5**). tracrRNA-crRNA duplexes are cleaved by RNase III in the presence of Cas9 and are used as guides for Cas9 to cleave the target DNA [108, 109]. Cas9 can degrade not only DNA but also RNAs [110]. An effector Cas12a from type VA type CRISPR-Cas is RNA-guided DNA endonuclease finds its target sequence by one-dimensional diffusion along DNA and cleaves at both strands [111].

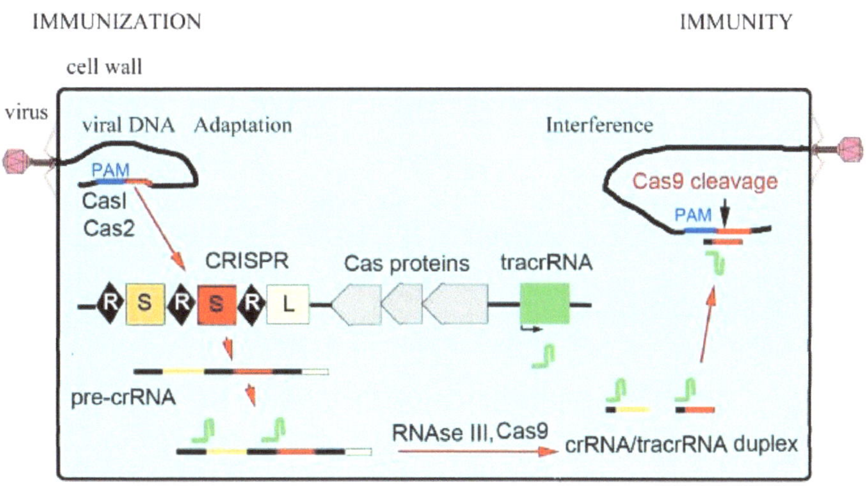

Fig. (5). CRISPR-Cas immunity. A fragment from an incoming genetic element such as a virus is incorporated into a CRISPR array as a spacer (red). RNA is transcribed from the array and form a hybrid with complimentary tracrRNA transcribed from the same locus. The hybrid is cleaved into short crRNAs/tracrRNA duplexes that serve as guides for Cas enzymes to cleave the intruder DNA. Schematic shows a simplified type II mechanism. L – leader, R – repeat, S – spacer, PAM – protospacer-adjacent motif.

In type III, crRNA-guided nucleases perform co-transcriptional cleavage of target DNA and its transcripts [112, 113]. Csm6 RNase that degrades RNA transcripts is activated through a second messenger, cyclic oligoadenylate product [114]. The messenger is produced from ATP by Cas10 subunit of the interference complex. Type VI also target RNA [115].

CRISPR-Cas systems are regulated at multiple levels [116]. Upon viral infection, general host transcription, translation, and metabolism are diminished while genes of CRISPR-Cas systems are upregulated in both bacteria and archaea. This upregulation depends on the nature of the virus; of two viruses infecting *Sulfolobus solfataricus*, only one induces CRISPR-Cas [117]. Diverse CRISPR systems are upregulated during infection in different ways, suggesting variations in their functions and regulations [118]. Mechanisms of CRISPR-Cas regulation are not completely understood and may include quorum sensing regulators, envelope stress, cAMP levels, metabolic state, and other factors. The long evolutionary history of these systems and their origin in some species through horizontal gene transfer suggest species may differ substantially in how they regulate their CRISP-Cas systems.

Quorum sensing coordinates bacterial responses. The chance of infection increases when bacteria grow to high density and can come in contact with a larger number of viruses and plasmid. Studies on several species show that quorum sensing regulates CRISPR-Cas. In *Serratia*, CRISPR-Cas systems are activated only at high cell density and a corresponding concentration of quorum sensing molecules [116]. In *P. aeruginosa*, CRISPR-Cas is also induced *via* quorum sensing activators [119].

In the absence of stress, histone-like nucleoid-structuring (H-NS) proteins suppress *E. coli* CRISPR-Cas systems [120]. H-NS is known to bind and silence horizontally acquired genes with GC content lower than the host genome [121]. It has been suggested that H-NS binding to DNA from a virus or a plasmid titrates out the H-NS available for CRISPR-Cas leading to its de-repression [122].

LeuO, the regulator of leucine biosynthesis operon of *E. coli*, is a major antagonist of the H-NS [123]. LeuO has also been implicated in CRISPR–Cas regulation [124]. LeuO is activated by the membrane stress system Bsc in *E. coli* [125]. In *M. xanthus* extracytoplasmic stress is regulated by sigma factor DdvS, which also regulates the type IIIB CRISPR-Cas system in this species [126].

Viral infection changes host metabolism, which typically induces upregulation of glucose and of nucleotide metabolism [127]. Many metabolic pathways in bacteria are controlled by cAMP receptor protein (CRP). CRP is also involved in CRISPR–Cas regulation [116, 128].

Fast response of CRISPR-Cas is hypothesized to depend on low expression of

CRISPR arrays in the absence of infection while cas operon stays silent [129]. In the type I system of the archaeon *Sulfolobus*, interference complex Cascade and the transcriptional regulator Csa3b repress the interference gene cassette. During viral infection, Cascade competitively binds crRNA-matching protospacers relieving transcriptional repression. Csa3a is also involved in CRISPR–Cas regulation in this species [130]. Most known regulators target Cas proteins. Cbp1 from *Sulfolobus islandicus*, on the contrary, regulates expression of the DNA repeats [131].

Phages developed multiple mechanisms to avoid CRISPR-Cas. Study of virus-host co-evolution in *Flavobacterium columnare* shows some general tendencies [132]. The CRISPR spacers usually target the terminal end of the phage genome. Correspondently, phages accumulate most mutations in protospacers and PAMs from this region. Phages with substitutions close to PAM have a higher probability to avoid CRISPR-Cas. However, in a prokaryotic population, individual cells acquire different spacers from the same pathogen. It means pathogens cannot overcome CRISPR-Cas immunity only through mutations in their genomes since some of the host cells will survive and pass immunity to the next generations [133]. Therefore, pathogens developed an extensive arsenal of anti-CRISPR (Acr) proteins that includes more than 20 protein families [134 - 138]. Acr proteins typically inhibit Cas enzymes; for example AcrIIA6 from *Streptococcus thermophilus* inhibits Cas9 [139] and archaeal AcrID1 inhibits Cas10d protein [136]. Mechanisms of Acr action are very diverse. For example, AcrF3 form a complex with Cas3 that locks Cas3 in an ADP-bound form that prevents the entrance of DNA into the helicase and Cascade recruitment [140].

DNA modification such as replacement of cytosine with glucosyl-hydroxymethylcytosine also protects T4 phage from attack by the CRISPR-Cas9 system [141]. In addition, phages can encode their own CRISPR-Cas systems that can be used to attack the antiviral chromosomal island of a bacterial host [142]. To create a sufficient concentration of Acr protein the number of phages that enter the cell needs to exceed a particular threshold. If the phage fails to replicate, it leaves the cell immunosuppressed, which predisposes it to a successful infection by other phages. Therefore phages often cooperate in order to overcome CRISPR resistance [143, 144].

Evolution of human pathogens in the era of antibiotics depends on their ability to acquire antibiotic resistance genes. Some studies show the restrictive effect of CRISPR-Cas on horizontal gene transfer. In *E. faecalis* there is an inverse correlation between the presence of CRISPR loci and number of drug resistance genes [145]. Other studies suggest that CRISPR-Cas immunity enhances horizontal gene transfer by transduction [146]. It is possible that CRISPR-Cas can discriminate between different types of gene transfer. CRISPR-Cas systems contribute to other cell functions besides adaptive immunity such as DNA repair, signal transduction, and programmed cell death [147].

CRISPR-Cas and R/M systems both cleave foreign DNA. These systems complement each other, they are compatible since CRISPR-Cas acquisition or interference activities are not affected by the methylation status of DNA [148, 149].

5.7. Argonaute-based Nucleic Acid Interference

Argonaute (Ago)-like proteins harbor an RNaseH-like domain that has endonucleolytic activity. In eukaryotes, Agos are the core of RNA interference machinery. They use small RNAs as guides to destroy or silence RNA targets. They also participate in immunodefense. Agos are found in all life domains including ~10% of bacteria and one-third of archaea [150, 151]. In prokaryotes, Agos are involved in protection against parasitic genetic elements mostly DNAs.

Contrary to eukaryotes, prokaryotic Agos utilize both DNA and RNA guides to destroy foreign nucleic acids [152, 153]. For example, Ago from the bacterium *Thermus thermophilus* (TtAgo) uses 5'-phosphorylated DNA guides to destroy plasmid DNA [154]. DNA guides from *T. thermophilus* are 13-25 nucleotides in length and have a strong bias for 5'-end deoxycytidine. Ago proteins from other species show diverse preferences or none at all [153]. DNA guides are thought to be generated through a guide-independent Ago nuclease activity [155, 156].

Ago from *Rhodobacter sphaeroides* binds both small RNAs and DNAs and uses small DNA guides to target foreign plasmid DNA [157]. *Marinitoga piezophila* Ago encoded within CRISPR-Cas operon binds 5'-hydroxylated guide RNAs and cleaves ssRNAs and ssDNAs [158].

A subset of Ago-associated proteins in prokaryotes have TIR (Toll/interleukin-1 receptor) domain [159]. TIR domain is common for immune proteins from plants and animals that are involved in pathogen recognition and immune signaling. The presence of TIR domain in prokaryotic immune proteins suggests conservation of some immune principles across all life domains.

5.8. Abortive Infection

Prokaryotes can defend themselves against infection not only on an individual level but on the level of the population as well. Altruistic suicide upon infection by lytic phages known as abortive infection (Abi) is common in bacterial species and is an effective strategy to limit phage spread [26, 160]. Often Abi systems are encoded by plasmids, which may promote their spread [161]. Abi genes constitute approximately 0.1% of all genes in archaea and bacteria [162].

Many different mechanisms can lead to Abi. *Lactococcus lactis* has over 20 different Abi systems (from AbiA to AbiZ), whose actions vary. Some systems such as AbiA and AbiF inhibit phage DNA replication and transcription along with Abi promotion, whereas others such as AbiT and AbiZ, have no noticeable effect on the phage but cause lysis of phage-infected cells [163].

Some of *L. lactis* Abi systems function as toxin/antitoxin (TA) pairs [164]. TA systems protect against phage spread in many bacteria [165 - 167]. The main mechanism of Abi effect of TA systems may depend on a general decrease in transcription and translation of bacterial genes, which is usually associated with a viral infection. Decrease of translation may disrupt the equilibrium between toxin and antitoxin since the concentration of unstable antitoxin would decrease faster. This would free the toxin. Phages develop means to inhibit the activity of TA systems [168, 169]. For example, immediately after infection with bacteriophage T4, MazF toxin is ADP-ribosylated by ADP-ribosyltransferase encoded by T4. This results in a decrease of MazF RNA cleavage activity, suggesting that the phage inactivates MazF during infection [170]. Rex system of lambda phage inhibits cell death induced by several TA systems by inhibiting cleavage of their antitoxins [7, 171]. In this way, viruses promote host cell survival until they complete their reproduction cycle.

Immunity genes typically co-localize with PCD modules, such as TAs. This is suggestive of immunity-PCD coupling likely mediated by shared proteins that sense damage and "predict" the outcome of infections. In type VI CRISPR-Cas systems, the same enzyme that inactivates the target RNA can execute cell suicide, representing a case of the ultimate integration of immunity and PCD [172].

5.9. Superinfection Exclusion

Infection by many viruses prevents superinfection by a homologous second virus [26, 173]. An exclusion system is often encoded in a resident plasmid or a provirus. Established temperate phages inhibit infection by other phages through diverse mechanisms. Some of them cause surface modifications, whereas others function inside the cell [174].

A number of viruses mediate superinfection exclusion at the stage of virus entry. It is achieved by downregulation of the cell surface receptor, its modification, prevention of membrane fusion with the viral envelope, production of lipoproteins that upon insertion into cytoplasmic membrane bind the virus and block its insertion into the membrane, and other mechanisms [175, 176]. Modification of cellular surface mediated by *P. aeruginosa* resident phages causes a very efficient exclusion [174].

Superinfection exclusion can also take place after the secondary virus has already entered the cell. Some phages produce repressor proteins that inhibit virus replication [174]. Often, more than one mechanism is used for exclusion. For example, T-even coliphage protein Imm blocks DNA transfer across the plasma membrane and partially inhibits the release of DNA from the superinfecting virion, and Sp inhibits degradation of bacterial cell-wall peptidoglycan by a phage-associated lysozyme [177]. Bovine viral diarrhea virus inhibits superinfection by blocking viral entry and by inhibition of viral RNA replication [178].

Some phage superinfection exclusion systems can interfere with essential components of the cellular machinery such as translation, which may result in cell death. For example, early in lytic growth, mycobacteriophage *Fruitloop* produces a protein that binds and inactivates a mycobacterial protein involved in cell wall biosynthesis. This excludes superinfection by phages such as *Rosebush* that depend on recognition of cell wall [179].

Viral competition can potentially be used to prevent human diseases. For example, co-infection with flaviviruses that are specific to mosquito and cannot infect humans can potentially be used to block flavivisuses associated with human diseases such as West Nile virus [180, 181].

5.10. New Immune Systems

Known immunodefense systems are typically encoded in the defense islands, which also contain many uncharacterized genes [6]. Analysis of these genes suggests that they encode new antiviral systems. Recently, 28 systems have been predicted and expressed in *E. coli* and *B. subtilis* [8]. 9 systems showed protection from infection by at least one phage. The mechanism of action has been partially characterized for several systems.

The Zorya system, named after a Slavic deity, is a potential Abi system present in ~3% of bacteria. It includes a proton channel that is also used in the flagella motors. This channel may be used to depolarize the membrane and kill the cell if phage inactivation fails.

The Thoeris system, named after an Egyptian god, is found in ~4% of bacteria and archaea. It includes a protein with nicotinamide adenine dinucleotide (NAD)-binding domain and a protein with TIR domain. A recent study demonstrated that TIR domain proteins cleave NAD(+) into nicotinamide and ADP-ribose (ADPR) [182]. Some TIR domains produce a cyclic ADPR (cADPR) molecule. Cyclic di- and linear oligonucleotides are used in immune signaling by all species and a large number of proteins with specificity to nucleotides are encoded in prokaryotic genomes [183]. NAD+ binding activity is essential for the antiviral activity of Thoeris system [8].

Recently discovered phage-inducible chromosomal islands contribute to immunodefense in many bacteria. *Vibrio cholerae* phage-inducible chromosomal island-like element protects from ICP1 phage. It encodes a serine recombinase that uses ICP1 protein as a recombination directionality factor to excise the island in response to phage infection. That reduces phage genome replication and accelerates lysis of infected cells [6].

The number of sequenced genomes of prokaryotes continues to increase. Most systems discovered by Doron *et al.* [8] contain a large number of genes with unknown function. One may predict that many new unexpected mechanisms of immunodefense of prokaryotes will be discovered in the near future.

CONCLUDING REMARKS

Prokaryotes have surprisingly variable and sophisticated immunodefense mechanisms. They include mechanisms based on direct recognition of pathogens. They also use the "missing self" strategy in their R/M systems. CRISPR-Cas is an adaptive immune system of prokaryotes that provides a memory response during repeated infection and also protects the progeny. The suicide of infected cells is also a common strategy of immunodefense that benefits the rest of the population.

CONSENT FOR PUBLICATION

Not applicable.

CONFLICT OF INTEREST

The author confirms that this chapter contents have no conflict of interest.

ACKNOWLEDGEMENTS

Declare none.

REFERENCES

[1] Ofir G, Sorek R. Contemporary Phage Biology: From Classic Models to New Insights. Cell 2018; 172(6): 1260-70.
[http://dx.doi.org/10.1016/j.cell.2017.10.045] [PMID: 29522746]

[2] Krupovic M, Cvirkaite-Krupovic V, Iranzo J, Prangishvili D, Koonin EV. Viruses of archaea: Structural, functional, environmental and evolutionary genomics. Virus Res 2018; 244: 181-93.
[http://dx.doi.org/10.1016/j.virusres.2017.11.025] [PMID: 29175107]

[3] Colaço HG, Moita LF. Initiation of innate immune responses by surveillance of homeostasis perturbations. FEBS J 2016; 283(13): 2448-57.
[http://dx.doi.org/10.1111/febs.13730] [PMID: 27037950]

[4] Labrie SJ, Samson JE, Moineau S. Bacteriophage resistance mechanisms. Nat Rev Microbiol 2010; 8(5): 317-27.
[http://dx.doi.org/10.1038/nrmicro2315] [PMID: 20348932]

[5] Koonin EV, Makarova KS, Wolf YI. Evolutionary Genomics of Defense Systems in Archaea and Bacteria. Annu Rev Microbiol 2017; 71: 233-61.
[http://dx.doi.org/10.1146/annurev-micro-090816-093830] [PMID: 28657885]

[6] McKitterick AC, Seed KD. Anti-phage islands force their target phage to directly mediate island excision and spread. Nat Commun 2018; 9(1): 2348.
[http://dx.doi.org/10.1038/s41467-018-04786-5] [PMID: 29904071]

[7] Samson JE, Magadán AH, Sabri M, Moineau S. Revenge of the phages: Defeating bacterial defences. Nat Rev Microbiol 2013; 11(10): 675-87.
[http://dx.doi.org/10.1038/nrmicro3096] [PMID: 23979432]

[8] Doron S, Melamed S, Ofir G, *et al.* Systematic discovery of antiphage defense systems in the microbial pangenome. Science 2018; 359(6379)eaar4120
[http://dx.doi.org/10.1126/science.aar4120] [PMID: 29371424]

[9] Gómez P, Buckling A. Bacteria-phage antagonistic coevolution in soil. Science 2011; 332(6025): 106-9.
[http://dx.doi.org/10.1126/science.1198767] [PMID: 21454789]

[10] Sumby P, Barbian KD, Gardner DJ, *et al.* Extracellular deoxyribonuclease made by group A Streptococcus assists pathogenesis by enhancing evasion of the innate immune response. Proc Natl Acad Sci USA 2005; 102(5): 1679-84.
[http://dx.doi.org/10.1073/pnas.0406641102] [PMID: 15668390]

[11] Winstel V, Missiakas D, Schneewind O. *Staphylococcus aureus* targets the purine salvage pathway to kill phagocytes. Proc Natl Acad Sci USA 2018; 115(26): 6846-51.
[http://dx.doi.org/10.1073/pnas.1805622115] [PMID: 29891696]

[12] Tran TM, MacIntyre A, Hawes M, Allen C. Escaping Underground Nets: Extracellular DNases Degrade Plant Extracellular Traps and Contribute to Virulence of the Plant Pathogenic Bacterium Ralstonia solanacearum. PLoS Pathog 2016; 12(6)e1005686
[http://dx.doi.org/10.1371/journal.ppat.1005686] [PMID: 27336156]

[13] Müller S, Strack SN, Hoefler BC, Straight PD, Kearns DB, Kirby JR. Bacillaene and sporulation protect Bacillus subtilis from predation by Myxococcus xanthus. Appl Environ Microbiol 2014; 80(18): 5603-10.
[http://dx.doi.org/10.1128/AEM.01621-14] [PMID: 25002419]

[14] Jamet A, Nassif X. New players in the toxin field: Polymorphic toxin systems in bacteria. MBio 2015; 6(3): e00285-15.
[http://dx.doi.org/10.1128/mBio.00285-15] [PMID: 25944858]

[15] Ray A, Schwartz N, de Souza Santos M, Zhang J, Orth K, Salomon D. Type VI secretion system MIX-effectors carry both antibacterial and anti-eukaryotic activities. EMBO Rep 2017; 18(11): 1978-

90.
[http://dx.doi.org/10.15252/embr.201744226] [PMID: 28912123]

[16] Garcia EC. Contact-dependent interbacterial toxins deliver a message. Curr Opin Microbiol 2018; 42: 40-6.
[http://dx.doi.org/10.1016/j.mib.2017.09.011] [PMID: 29078204]

[17] Garcia EC, Perault AI, Marlatt SA, Cotter PA. Interbacterial signaling *via* Burkholderia contact-dependent growth inhibition system proteins. Proc Natl Acad Sci USA 2016; 113(29): 8296-301.
[http://dx.doi.org/10.1073/pnas.1606323113] [PMID: 27335458]

[18] Gunn JS, Bakaletz LO, Wozniak DJ. What's on the Outside Matters: The Role of the Extracellular Polymeric Substance of Gram-negative Biofilms in Evading Host Immunity and as a Target for Therapeutic Intervention. J Biol Chem 2016; 291(24): 12538-46.
[http://dx.doi.org/10.1074/jbc.R115.707547] [PMID: 27129225]

[19] DePas WH, Syed AK, Sifuentes M, *et al.* Biofilm formation protects *Escherichia coli* against killing by Caenorhabditis elegans and Myxococcus xanthus. Appl Environ Microbiol 2014; 80(22): 7079-87.
[http://dx.doi.org/10.1128/AEM.02464-14] [PMID: 25192998]

[20] Müller S, Strack SN, Ryan SE, Kearns DB, Kirby JR. Predation by Myxococcus xanthus induces Bacillus subtilis to form spore-filled megastructures. Appl Environ Microbiol 2015; 81(1): 203-10.
[http://dx.doi.org/10.1128/AEM.02448-14] [PMID: 25326308]

[21] Bertozzi Silva J, Storms Z, Sauvageau D. Host receptors for bacteriophage adsorption. FEMS Microbiol Lett 2016; 363(4)fnw002
[http://dx.doi.org/10.1093/femsle/fnw002] [PMID: 26755501]

[22] Dunne M, Hupfeld M, Klumpp J, Loessner MJ. Molecular Basis of Bacterial Host Interactions by Gram-Positive Targeting Bacteriophages. Viruses 2018; 10(8)E397
[http://dx.doi.org/10.3390/v10080397] [PMID: 30060549]

[23] Oechslin F. Resistance Development to Bacteriophages Occurring during Bacteriophage Therapy. Viruses 2018; 10(7)E351
[http://dx.doi.org/10.3390/v10070351] [PMID: 29966329]

[24] Zhang R, Sun Q, Chen Y, *et al.* Ts-Hsp70 induces protective immunity against Trichinella spiralis infection in mouse by activating dendritic cells through TLR2 and TLR4. PLoS Negl Trop Dis 2018; 12(5)e0006502
[http://dx.doi.org/10.1371/journal.pntd.0006502] [PMID: 29775453]

[25] Asija K, Teschke CM. Lessons from bacteriophages part 2: A saga of scientific breakthroughs and prospects for their use in human health. PLoS Pathog 2018; 14(5)e1006970
[http://dx.doi.org/10.1371/journal.ppat.1006970] [PMID: 29772006]

[26] Seed KD. Battling Phages: How Bacteria Defend against Viral Attack. PLoS Pathog 2015; 11(6)e1004847
[http://dx.doi.org/10.1371/journal.ppat.1004847] [PMID: 26066799]

[27] Reyes-Robles T, Dillard RS, Cairns LS, Silva-Valenzuela CA, Housman M, Ali A, *et al.* Vibrio cholerae outer membrane vesicles inhibit bacteriophage infection. Journal of bacteriology 2018; 00792-17.
[http://dx.doi.org/10.1128/JB.00792-17]

[28] Tzipilevich E, Habusha M, Ben-Yehuda S. Acquisition of Phage Sensitivity by Bacteria through Exchange of Phage Receptors. Cell 2017; 168(1-2): 186-199.e12.
[http://dx.doi.org/10.1016/j.cell.2016.12.003] [PMID: 28041851]

[29] Duerkop BA, Huo W, Bhardwaj P, Palmer KL, Hooper LV. Molecular Basis for Lytic Bacteriophage Resistance in Enterococci. MBio 2016; 7(4): e01304-16.
[http://dx.doi.org/10.1128/mBio.01304-16] [PMID: 27578757]

[30] Harvey H, Bondy-Denomy J, Marquis H, Sztanko KM, Davidson AR, Burrows LL. Pseudomonas

aeruginosa defends against phages through type IV pilus glycosylation. Nat Microbiol 2018; 3(1): 47-52.
[http://dx.doi.org/10.1038/s41564-017-0061-y] [PMID: 29133883]

[31] Letarov AV, Kulikov EE. Adsorption of Bacteriophages on Bacterial Cells. Biochemistry (Mosc) 2017; 82(13): 1632-58.
[http://dx.doi.org/10.1134/S0006297917130053] [PMID: 29523063]

[32] Cota I, Sánchez-Romero MA, Hernández SB, Pucciarelli MG, García-Del Portillo F, Casadesús J. Epigenetic Control of Salmonella enterica O-Antigen Chain Length: A Tradeoff between Virulence and Bacteriophage Resistance. PLoS Genet 2015; 11(11)e1005667
[http://dx.doi.org/10.1371/journal.pgen.1005667] [PMID: 26583926]

[33] Cota I, Bunk B, Spröer C, Overmann J, König C, Casadesús J. OxyR-dependent formation of DNA methylation patterns in OpvABOFF and OpvABON cell lineages of Salmonella enterica. Nucleic Acids Res 2016; 44(8): 3595-609.
[http://dx.doi.org/10.1093/nar/gkv1483] [PMID: 26687718]

[34] Siena E, D'Aurizio R, Riley D, *et al.* In-silico prediction and deep-DNA sequencing validation indicate phase variation in 115 Neisseria meningitidis genes. BMC Genomics 2016; 17(1): 843.
[http://dx.doi.org/10.1186/s12864-016-3185-1] [PMID: 27793092]

[35] De Ste Croix M, Vacca I, Kwun MJ, Ralph JD, Bentley SD, Haigh R, *et al.* Phase-variable methylation and epigenetic regulation by type I restriction-modification systems. FEMS microbiology reviews 2017; 41: S3-s15.
[http://dx.doi.org/10.1093/femsre/fux025]

[36] Atack JM, Tan A, Bakaletz LO, Jennings MP, Seib KL. Phasevarions of Bacterial Pathogens: Methylomics Sheds New Light on Old Enemies. Trends Microbiol 2018; 26(8): 715-26.
[http://dx.doi.org/10.1016/j.tim.2018.01.008] [PMID: 29452952]

[37] Phillips ZN, Husna AU, Jennings MP, Seib KL, Atack JM. 2019.

[38] Rotman E, Webber DM, Seifert HS. Analyzing Neisseria gonorrhoeae pilin antigenic variation using 454 sequencing technology. J Bacteriol 2016; 198(18): 2470-82.
[http://dx.doi.org/10.1128/JB.00330-16] [PMID: 27381912]

[39] Liu M, Deora R, Doulatov SR, *et al.* Reverse transcriptase-mediated tropism switching in Bordetella bacteriophage. Science 2002; 295(5562): 2091-4.
[http://dx.doi.org/10.1126/science.1067467] [PMID: 11896279]

[40] Doulatov S, Hodes A, Dai L, *et al.* Tropism switching in Bordetella bacteriophage defines a family of diversity-generating retroelements. Nature 2004; 431(7007): 476-81.
[http://dx.doi.org/10.1038/nature02833] [PMID: 15386016]

[41] Wu L, Gingery M, Abebe M, *et al.* Diversity-generating retroelements: natural variation, classification and evolution inferred from a large-scale genomic survey. Nucleic Acids Res 2018; 46(1): 11-24.
[http://dx.doi.org/10.1093/nar/gkx1150] [PMID: 29186518]

[42] Paul BG, Burstein D, Castelle CJ, *et al.* Retroelement-guided protein diversification abounds in vast lineages of Bacteria and Archaea. Nat Microbiol 2017; 2: 17045.
[http://dx.doi.org/10.1038/nmicrobiol.2017.45] [PMID: 28368387]

[43] McMahon SA, Miller JL, Lawton JA, *et al.* The C-type lectin fold as an evolutionary solution for massive sequence variation. Nat Struct Mol Biol 2005; 12(10): 886-92.
[http://dx.doi.org/10.1038/nsmb992] [PMID: 16170324]

[44] Kupczok A, Neve H, Huang KD, *et al.* Rates of Mutation and Recombination in Siphoviridae Phage Genome Evolution over Three Decades. Mol Biol Evol 2018; 35(5): 1147-59.
[http://dx.doi.org/10.1093/molbev/msy027] [PMID: 29688542]

[45] Petrie KL, Palmer ND, Johnson DT, *et al.* Destabilizing mutations encode nongenetic variation that drives evolutionary innovation. Science 2018; 359(6383): 1542-5.

[http://dx.doi.org/10.1126/science.aar1954] [PMID: 29599247]

[46] Hsia KC, Li CL, Yuan HS. Structural and functional insight into sugar-nonspecific nucleases in host defense. Curr Opin Struct Biol 2005; 15(1): 126-34. [http://dx.doi.org/10.1016/j.sbi.2005.01.015] [PMID: 15718143]

[47] Gaasbeek EJ, Wagenaar JA, Guilhabert MR, van Putten JP, Parker CT, van der Wal FJ. Nucleases encoded by the integrated elements CJIE2 and CJIE4 inhibit natural transformation of Campylobacter jejuni. J Bacteriol 2010; 192(4): 936-41. [http://dx.doi.org/10.1128/JB.00867-09] [PMID: 20023031]

[48] Singh K, Milstein JN, Navarre WW. Xenogeneic Silencing and Its Impact on Bacterial Genomes. Annu Rev Microbiol 2016; 70: 199-213. [http://dx.doi.org/10.1146/annurev-micro-102215-095301] [PMID: 27359215]

[49] Lamberte LE, Baniulyte G, Singh SS, *et al.* Horizontally acquired AT-rich genes in *Escherichia coli* cause toxicity by sequestering RNA polymerase. Nat Microbiol 2017; 2: 16249. [http://dx.doi.org/10.1038/nmicrobiol.2016.249] [PMID: 28067866]

[50] Will WR, Navarre WW, Fang FC. Integrated circuits: How transcriptional silencing and counter-silencing facilitate bacterial evolution. Curr Opin Microbiol 2015; 23: 8-13. [http://dx.doi.org/10.1016/j.mib.2014.10.005] [PMID: 25461567]

[51] Platenkamp A, Mellies JL. Environment Controls LEE Regulation in Enteropathogenic *Escherichia coli*. Front Microbiol 2018; 9: 1694. [http://dx.doi.org/10.3389/fmicb.2018.01694] [PMID: 30140259]

[52] Luria SE, Human ML. A nonhereditary, host-induced variation of bacterial viruses. J Bacteriol 1952; 64(4): 557-69. [PMID: 12999684]

[53] Loenen WA, Dryden DT, Raleigh EA, Wilson GG, Murray NE. Highlights of the DNA cutters: A short history of the restriction enzymes. Nucleic Acids Res 2014; 42(1): 3-19. [http://dx.doi.org/10.1093/nar/gkt990] [PMID: 24141096]

[54] Loenen WA, Dryden DT, Raleigh EA, Wilson GG. Type I restriction enzymes and their relatives. Nucleic Acids Res 2014; 42(1): 20-44. [http://dx.doi.org/10.1093/nar/gkt847] [PMID: 24068554]

[55] Grinkevich P, Sinha D, Iermak I, *et al.* Crystal structure of a novel domain of the motor subunit of the Type I restriction enzyme EcoR124 involved in complex assembly and DNA binding. J Biol Chem 2018; 293(39): 15043-54. [http://dx.doi.org/10.1074/jbc.RA118.003978] [PMID: 30054276]

[56] Pingoud A, Wilson GG, Wende W. Type II restriction endonucleases-a historical perspective and more. Nucleic Acids Res 2014; 42(12): 7489-527. [http://dx.doi.org/10.1093/nar/gku447] [PMID: 24878924]

[57] Tao T, Bourne JC, Blumenthal RM. A family of regulatory genes associated with type II restriction-modification systems. J Bacteriol 1991; 173(4): 1367-75. [http://dx.doi.org/10.1128/jb.173.4.1367-1375.1991] [PMID: 1995588]

[58] Gupta YK, Chan SH, Xu SY, Aggarwal AK. Structural basis of asymmetric DNA methylation and ATP-triggered long-range diffusion by EcoP15I. Nat Commun 2015; 6: 7363. [http://dx.doi.org/10.1038/ncomms8363] [PMID: 26067164]

[59] Machnicka MA, Kaminska KH, Dunin-Horkawicz S, Bujnicki JM. Phylogenomics and sequence-structure-function relationships in the GmrSD family of Type IV restriction enzymes. BMC Bioinformatics 2015; 16: 336. [http://dx.doi.org/10.1186/s12859-015-0773-z] [PMID: 26493560]

[60] Ofir G, Melamed S, Sberro H, *et al.* DISARM is a widespread bacterial defence system with broad anti-phage activities. Nat Microbiol 2018; 3(1): 90-8.

[http://dx.doi.org/10.1038/s41564-017-0051-0] [PMID: 29085076]

[61] Miyazono K, Furuta Y, Watanabe-Matsui M, *et al.* A sequence-specific DNA glycosylase mediates restriction-modification in Pyrococcus abyssi. Nat Commun 2014; 5: 3178. [http://dx.doi.org/10.1038/ncomms4178] [PMID: 24458096]

[62] Xu T, Yao F, Zhou X, Deng Z, You D. A novel host-specific restriction system associated with DNA backbone S-modification in Salmonella. Nucleic Acids Res 2010; 38(20): 7133-41. [http://dx.doi.org/10.1093/nar/gkq610] [PMID: 20627870]

[63] Cao B, Zheng X, Cheng Q, *et al. in vitro* analysis of phosphorothioate modification of DNA reveals substrate recognition by a multiprotein complex. Sci Rep 2015; 5: 12513. [http://dx.doi.org/10.1038/srep12513] [PMID: 26213215]

[64] Thiaville JJ, Kellner SM, Yuan Y, *et al.* Novel genomic island modifies DNA with 7-deazaguanine derivatives. Proc Natl Acad Sci USA 2016; 113(11): E1452-9. [http://dx.doi.org/10.1073/pnas.1518570113] [PMID: 26929322]

[65] Kobayashi I. Behavior of restriction-modification systems as selfish mobile elements and their impact on genome evolution. Nucleic Acids Res 2001; 29(18): 3742-56. [http://dx.doi.org/10.1093/nar/29.18.3742] [PMID: 11557807]

[66] Pleška M, Qian L, Okura R, *et al.* Bacterial Autoimmunity Due to a Restriction-Modification System. Curr Biol 2016; 26(3): 404-9. [http://dx.doi.org/10.1016/j.cub.2015.12.041] [PMID: 26804559]

[67] Oliveira PH, Touchon M, Rocha EP. Regulation of genetic flux between bacteria by restriction-modification systems. Proc Natl Acad Sci USA 2016; 113(20): 5658-63. [http://dx.doi.org/10.1073/pnas.1603257113] [PMID: 27140615]

[68] Srikhanta YN, Dowideit SJ, Edwards JL, *et al.* Phasevarions mediate random switching of gene expression in pathogenic Neisseria. PLoS Pathog 2009; 5(4)e1000400 [http://dx.doi.org/10.1371/journal.ppat.1000400] [PMID: 19390608]

[69] Anjum A, Brathwaite KJ, Aidley J, *et al.* Phase variation of a Type IIG restriction-modification enzyme alters site-specific methylation patterns and gene expression in *Campylobacter jejuni* strain NCTC11168. Nucleic Acids Res 2016; 44(10): 4581-94. [http://dx.doi.org/10.1093/nar/gkw019] [PMID: 26786317]

[70] Atack JM, Yang Y, Seib KL, Zhou Y, Jennings MP. A survey of Type III restriction-modification systems reveals numerous, novel epigenetic regulators controlling phase-variable regulons; phasevarions. Nucleic Acids Res 2018; 46(7): 3532-42. [http://dx.doi.org/10.1093/nar/gky192] [PMID: 29554328]

[71] Drozdz M, Piekarowicz A, Bujnicki JM, Radlinska M. Novel non-specific DNA adenine methyltransferases. Nucleic Acids Res 2012; 40(5): 2119-30. [http://dx.doi.org/10.1093/nar/gkr1039] [PMID: 22102579]

[72] Lee YJ, Dai N, Walsh SE, *et al.* Identification and biosynthesis of thymidine hypermodifications in the genomic DNA of widespread bacterial viruses. Proc Natl Acad Sci USA 2018; 115(14): E3116-25. [http://dx.doi.org/10.1073/pnas.1714812115] [PMID: 29555775]

[73] Chen K, Reuter M, Sanghvi B, *et al.* ArdA proteins from different mobile genetic elements can bind to the EcoKI Type I DNA methyltransferase of *E. coli* K12. Biochim Biophys Acta 2014; 1844(3): 505-11. [http://dx.doi.org/10.1016/j.bbapap.2013.12.008] [PMID: 24368349]

[74] Melkina OE, Goryanin II, Zavilgelsky GB. The DNA-mimic antirestriction proteins ArdA ColIB-P9, Arn T4, and Ocr T7 as activators of H-NS-dependent gene transcription. Microbiol Res 2016; 192: 283-91. [http://dx.doi.org/10.1016/j.micres.2016.07.008] [PMID: 27664747]

[75] Rifat D, Wright NT, Varney KM, Weber DJ, Black LW. Restriction endonuclease inhibitor IPI* of

bacteriophage T4: A novel structure for a dedicated target. J Mol Biol 2008; 375(3): 720-34. [http://dx.doi.org/10.1016/j.jmb.2007.10.064] [PMID: 18037438]

[76] Tesfazgi Mebrhatu M, Wywial E, Ghosh A, *et al.* Evidence for an evolutionary antagonism between Mrr and Type III modification systems. Nucleic Acids Res 2011; 39(14): 5991-6001. [http://dx.doi.org/10.1093/nar/gkr219] [PMID: 21504983]

[77] Goldfarb T, Sberro H, Weinstock E, *et al.* BREX is a novel phage resistance system widespread in microbial genomes. EMBO J 2015; 34(2): 169-83. [http://dx.doi.org/10.15252/embj.201489455] [PMID: 25452498]

[78] Hoskisson PA, Sumby P, Smith MCM. The phage growth limitation system in *Streptomyces coelicolor* A(3)2 is a toxin/antitoxin system, comprising enzymes with DNA methyltransferase, protein kinase and ATPase activity. Virology 2015; 477: 100-9. [http://dx.doi.org/10.1016/j.virol.2014.12.036] [PMID: 25592393]

[79] Barrangou R, Fremaux C, Deveau H, *et al.* CRISPR provides acquired resistance against viruses in prokaryotes. Science 2007; 315(5819): 1709-12. [http://dx.doi.org/10.1126/science.1138140] [PMID: 17379808]

[80] Brouns SJ, Jore MM, Lundgren M, *et al.* Small CRISPR RNAs guide antiviral defense in prokaryotes. Science 2008; 321(5891): 960-4. [http://dx.doi.org/10.1126/science.1159689] [PMID: 18703739]

[81] Jinek M, Chylinski K, Fonfara I, Hauer M, Doudna JA, Charpentier E. A programmable dual-RN- -guided DNA endonuclease in adaptive bacterial immunity. Science 2012; 337(6096): 816-21. [http://dx.doi.org/10.1126/science.1225829] [PMID: 22745249]

[82] Lander ES. The Heroes of CRISPR. Cell 2016; 164(1-2): 18-28. [http://dx.doi.org/10.1016/j.cell.2015.12.041] [PMID: 26771483]

[83] Hille F, Richter H, Wong SP, Bratovič M, Ressel S, Charpentier E. The Biology of CRISPR-Cas: Backward and Forward. Cell 2018; 172(6): 1239-59. [http://dx.doi.org/10.1016/j.cell.2017.11.032] [PMID: 29522745]

[84] Levasseur A, Bekliz M, Chabrière E, Pontarotti P, La Scola B, Raoult D. MIMIVIRE is a defence system in mimivirus that confers resistance to virophage. Nature 2016; 531(7593): 249-52. [http://dx.doi.org/10.1038/nature17146] [PMID: 26934229]

[85] Bolotin A, Quinquis B, Sorokin A, Ehrlich SD. Clustered regularly interspaced short palindrome repeats (CRISPRs) have spacers of extrachromosomal origin. Microbiology 2005; 151(Pt 8): 2551-61. [http://dx.doi.org/10.1099/mic.0.28048-0] [PMID: 16079334]

[86] Mojica FJ, Díez-Villaseñor C, García-Martínez J, Soria E. Intervening sequences of regularly spaced prokaryotic repeats derive from foreign genetic elements. J Mol Evol 2005; 60(2): 174-82. [http://dx.doi.org/10.1007/s00239-004-0046-3] [PMID: 15791728]

[87] Pourcel C, Salvignol G, Vergnaud G. CRISPR elements in Yersinia pestis acquire new repeats by preferential uptake of bacteriophage DNA, and provide additional tools for evolutionary studies. Microbiology 2005; 151(Pt 3): 653-63. [http://dx.doi.org/10.1099/mic.0.27437-0] [PMID: 15758212]

[88] Marraffini LA, Sontheimer EJ. CRISPR interference limits horizontal gene transfer in staphylococci by targeting DNA. Science 2008; 322(5909): 1843-5. [http://dx.doi.org/10.1126/science.1165771] [PMID: 19095942]

[89] Koonin EV, Makarova KS, Zhang F. Diversity, classification and evolution of CRISPR-Cas systems. Curr Opin Microbiol 2017; 37: 67-78. [http://dx.doi.org/10.1016/j.mib.2017.05.008] [PMID: 28605718]

[90] Adli M. The CRISPR tool kit for genome editing and beyond. Nat Commun 2018; 9(1): 1911. [http://dx.doi.org/10.1038/s41467-018-04252-2] [PMID: 29765029]

[91] Knott GJ, Doudna JA. CRISPR-Cas guides the future of genetic engineering. Science 2018; 361(6405): 866-9.
[http://dx.doi.org/10.1126/science.aat5011] [PMID: 30166482]

[92] Jackson SA, McKenzie RE, Fagerlund RD, Kieper SN, Fineran PC, Brouns SJ. CRISPR-Cas: Adapting to change. Science 2017; 356(6333)eaal5056
[http://dx.doi.org/10.1126/science.aal5056] [PMID: 28385959]

[93] McGinn J, Marraffini LA. Molecular mechanisms of CRISPR-Cas spacer acquisition. Nat Rev Microbiol 2019; 17(1): 7-12.
[http://dx.doi.org/10.1038/s41579-018-0071-7] [PMID: 30171202]

[94] Nicholson TJ, Jackson SA, Croft BI, Staals RHJ, Fineran PC, Brown CM. Bioinformatic evidence of widespread priming in type I and II CRISPR-Cas systems. RNA Biol 2019; 16(4): 566-76.
[http://dx.doi.org/10.1080/15476286.2018.1509662] [PMID: 30157725]

[95] Modell JW, Jiang W, Marraffini LA. CRISPR-Cas systems exploit viral DNA injection to establish and maintain adaptive immunity. Nature 2017; 544(7648): 101-4.
[http://dx.doi.org/10.1038/nature21719] [PMID: 28355179]

[96] Amitai G, Sorek R. CRISPR-Cas adaptation: Insights into the mechanism of action. Nat Rev Microbiol 2016; 14(2): 67-76.
[http://dx.doi.org/10.1038/nrmicro.2015.14] [PMID: 26751509]

[97] Heler R, Samai P, Modell JW, *et al.* Cas9 specifies functional viral targets during CRISPR-Cas adaptation. Nature 2015; 519(7542): 199-202.
[http://dx.doi.org/10.1038/nature14245] [PMID: 25707807]

[98] Künne T, Kieper SN, Bannenberg JW, *et al.* Cas3-Derived Target DNA Degradation Fragments Fuel Primed CRISPR Adaptation. Mol Cell 2016; 63(5): 852-64.
[http://dx.doi.org/10.1016/j.molcel.2016.07.011] [PMID: 27546790]

[99] Lee H, Zhou Y, Taylor DW, Sashital DG. Cas4-Dependent Prespacer Processing Ensures High-Fidelity Programming of CRISPR Arrays. Mol Cell 2018; 70(1): 48-59.e5.
[http://dx.doi.org/10.1016/j.molcel.2018.03.003] [PMID: 29602742]

[100] Shiimori M, Garrett SC, Graveley BR, Terns MP. Cas4 Nucleases Define the PAM, Length, and Orientation of DNA Fragments Integrated at CRISPR Loci. Mol Cell 2018; 70(5): 814-824.e6.
[http://dx.doi.org/10.1016/j.molcel.2018.05.002] [PMID: 29883605]

[101] Nuñez JK, Lee AS, Engelman A, Doudna JA. Integrase-mediated spacer acquisition during CRISPR-Cas adaptive immunity. Nature 2015; 519(7542): 193-8.
[http://dx.doi.org/10.1038/nature14237] [PMID: 25707795]

[102] Shipman SL, Nivala J, Macklis JD, Church GM. Molecular recordings by directed CRISPR spacer acquisition. Science 2016; 353(6298)aaf1175
[http://dx.doi.org/10.1126/science.aaf1175] [PMID: 27284167]

[103] Rollie C, Graham S, Rouillon C, White MF. Prespacer processing and specific integration in a Type I-A CRISPR system. Nucleic Acids Res 2018; 46(3): 1007-20.
[http://dx.doi.org/10.1093/nar/gkx1232] [PMID: 29228332]

[104] Westra ER, van Erp PB, Künne T, *et al.* CRISPR immunity relies on the consecutive binding and degradation of negatively supercoiled invader DNA by Cascade and Cas3. Mol Cell 2012; 46(5): 595-605.
[http://dx.doi.org/10.1016/j.molcel.2012.03.018] [PMID: 22521689]

[105] Cooper LA, Stringer AM, Wade JT. Determining the Specificity of Cascade Binding, Interference, and Primed Adaptation *in vivo* in the *Escherichia coli* Type I-E CRISPR-Cas System. MBio 2018; 9(2): e02100-17.
[http://dx.doi.org/10.1128/mBio.02100-17] [PMID: 29666291]

[106] van Erp PBG, Patterson A, Kant R, *et al.* Conformational Dynamics of DNA Binding and Cas3 Recruitment by the CRISPR RNA-Guided Cascade Complex. ACS Chem Biol 2018; 13(2): 481-90. [http://dx.doi.org/10.1021/acschembio.7b00649] [PMID: 29035497]

[107] Xiao Y, Luo M, Dolan AE, Liao M, Ke A. Structure basis for RNA-guided DNA degradation by Cascade and Cas3. Science 2018; 361(6397)eaat0839 [http://dx.doi.org/10.1126/science.aat0839] [PMID: 29880725]

[108] Faure G, Shmakov SA, Makarova KS, *et al.* Comparative genomics and evolution of trans-activating RNAs in Class 2 CRISPR-Cas systems. RNA Biol 2019; 16(4): 435-48. [http://dx.doi.org/10.1080/15476286.2018.1493331] [PMID: 30103650]

[109] Jiang F, Doudna JA. CRISPR-Cas9 Structures and Mechanisms. Annu Rev Biophys 2017; 46: 505-29. [http://dx.doi.org/10.1146/annurev-biophys-062215-010822] [PMID: 28375731]

[110] Strutt SC, Torrez RM, Kaya E, Negrete OA, Doudna JA. RNA-dependent RNA targeting by CRISPR-Cas9. eLife 2018; 7e32724 [http://dx.doi.org/10.7554/eLife.32724] [PMID: 29303478]

[111] Jeon Y, Choi YH, Jang Y, *et al.* Direct observation of DNA target searching and cleavage by CRISPR-Cas12a. Nat Commun 2018; 9(1): 2777. [http://dx.doi.org/10.1038/s41467-018-05245-x] [PMID: 30018371]

[112] Jiang W, Samai P, Marraffini LA. Degradation of Phage Transcripts by CRISPR-Associated RNases Enables Type III CRISPR-Cas Immunity. Cell 2016; 164(4): 710-21. [http://dx.doi.org/10.1016/j.cell.2015.12.053] [PMID: 26853474]

[113] Pyenson NC, Marraffini LA. Type III CRISPR-Cas systems: When DNA cleavage just isn't enough. Curr Opin Microbiol 2017; 37: 150-4. [http://dx.doi.org/10.1016/j.mib.2017.08.003] [PMID: 28865392]

[114] Niewoehner O, Garcia-Doval C, Rostøl JT, *et al.* Type III CRISPR-Cas systems produce cyclic oligoadenylate second messengers. Nature 2017; 548(7669): 543-8. [http://dx.doi.org/10.1038/nature23467] [PMID: 28722012]

[115] O'Connell MR. Molecular Mechanisms of RNA Targeting by Cas13-containing Type VI CRISPR-Cas Systems. J Mol Biol 2019; 431(1): 66-87. [http://dx.doi.org/10.1016/j.jmb.2018.06.029] [PMID: 29940185]

[116] Patterson AG, Yevstigneyeva MS, Fineran PC. Regulation of CRISPR-Cas adaptive immune systems. Curr Opin Microbiol 2017; 37: 1-7. [http://dx.doi.org/10.1016/j.mib.2017.02.004] [PMID: 28359988]

[117] Fusco S, Liguori R, Limauro D, Bartolucci S, She Q, Contursi P. Transcriptome analysis of Sulfolobus solfataricus infected with two related fuselloviruses reveals novel insights into the regulation of CRISPR-Cas system. Biochimie 2015; 118: 322-32. [http://dx.doi.org/10.1016/j.biochi.2015.04.006] [PMID: 25891845]

[118] Quax TE, Voet M, Sismeiro O, *et al.* Massive activation of archaeal defense genes during viral infection. J Virol 2013; 87(15): 8419-28. [http://dx.doi.org/10.1128/JVI.01020-13] [PMID: 23698312]

[119] Høyland-Kroghsbo NM, Paczkowski J, Mukherjee S, *et al.* Quorum sensing controls the *Pseudomonas aeruginosa* CRISPR-Cas adaptive immune system. Proc Natl Acad Sci USA 2017; 114(1): 131-5. [http://dx.doi.org/10.1073/pnas.1617415113] [PMID: 27849583]

[120] Pul U, Wurm R, Arslan Z, Geissen R, Hofmann N, Wagner R. Identification and characterization of *E. coli* CRISPR-cas promoters and their silencing by H-NS. Mol Microbiol 2010; 75(6): 1495-512. [http://dx.doi.org/10.1111/j.1365-2958.2010.07073.x] [PMID: 20132443]

[121] Navarre WW, Porwollik S, Wang Y, *et al.* Selective silencing of foreign DNA with low GC content by the H-NS protein in Salmonella. Science 2006; 313(5784): 236-8.

[http://dx.doi.org/10.1126/science.1128794] [PMID: 16763111]

[122] Raivio T. Identifying your enemies-could envelope stress trigger microbial immunity? Mol Microbiol 2011; 79(3): 557-61.
[http://dx.doi.org/10.1111/j.1365-2958.2010.07485.x] [PMID: 21255103]

[123] Shimada T, Bridier A, Briandet R, Ishihama A. Novel roles of LeuO in transcription regulation of *E.coli* genome: Antagonistic interplay with the universal silencer H-NS. Mol Microbiol 2011; 82(2): 378-97.
[http://dx.doi.org/10.1111/j.1365-2958.2011.07818.x] [PMID: 21883529]

[124] Westra ER, Pul U, Heidrich N, *et al.* H-NS-mediated repression of CRISPR-based immunity in *Escherichia coli* K12 can be relieved by the transcription activator LeuO. Mol Microbiol 2010; 77(6): 1380-93.
[http://dx.doi.org/10.1111/j.1365-2958.2010.07315.x] [PMID: 20659289]

[125] Stratmann T, Pul Ü, Wurm R, Wagner R, Schnetz K. RcsB-BglJ activates the *Escherichia coli* leuO gene, encoding an H-NS antagonist and pleiotropic regulator of virulence determinants. Mol Microbiol 2012; 83(6): 1109-23.
[http://dx.doi.org/10.1111/j.1365-2958.2012.07993.x] [PMID: 22295907]

[126] Bernal-Bernal D, Abellón-Ruiz J, Iniesta AA, *et al.* Multifactorial control of the expression of a CRISPR-Cas system by an extracytoplasmic function σ/anti-σ pair and a global regulatory complex. Nucleic Acids Res 2018; 46(13): 6726-45.
[http://dx.doi.org/10.1093/nar/gky475] [PMID: 29893914]

[127] De Smet J, Zimmermann M, Kogadeeva M, *et al.* High coverage metabolomics analysis reveals phage-specific alterations to *Pseudomonas aeruginosa* physiology during infection. ISME J 2016; 10(8): 1823-35.
[http://dx.doi.org/10.1038/ismej.2016.3] [PMID: 26882266]

[128] Yang CD, Chen YH, Huang HY, Huang HD, Tseng CP. CRP represses the CRISPR/Cas system in *Escherichia coli*: Evidence that endogenous CRISPR spacers impede phage P1 replication. Mol Microbiol 2014; 92(5): 1072-91.
[http://dx.doi.org/10.1111/mmi.12614] [PMID: 24720807]

[129] He F, Vestergaard G, Peng W, She Q, Peng X. CRISPR-Cas type I-A Cascade complex couples viral infection surveillance to host transcriptional regulation in the dependence of Csa3b. Nucleic Acids Res 2017; 45(4): 1902-13.
[PMID: 27980065]

[130] Liu T, Liu Z, Ye Q, *et al.* Coupling transcriptional activation of CRISPR-Cas system and DNA repair genes by Csa3a in *Sulfolobus islandicus.* Nucleic Acids Res 2017; 45(15): 8978-92.
[http://dx.doi.org/10.1093/nar/gkx612] [PMID: 28911114]

[131] Deng L, Kenchappa CS, Peng X, She Q, Garrett RA. Modulation of CRISPR locus transcription by the repeat-binding protein Cbp1 in Sulfolobus. Nucleic Acids Res 2012; 40(6): 2470-80.
[http://dx.doi.org/10.1093/nar/gkr1111] [PMID: 22139923]

[132] Laanto E, Hoikkala V, Ravantti J, Sundberg LR. Long-term genomic coevolution of host-parasite interaction in the natural environment. Nat Commun 2017; 8(1): 111.
[http://dx.doi.org/10.1038/s41467-017-00158-7] [PMID: 28740072]

[133] van Houte S, Ekroth AK, Broniewski JM, *et al.* The diversity-generating benefits of a prokaryotic adaptive immune system. Nature 2016; 532(7599): 385-8.
[http://dx.doi.org/10.1038/nature17436] [PMID: 27074511]

[134] Bondy-Denomy J, Garcia B, Strum S, *et al.* Multiple mechanisms for CRISPR-Cas inhibition by anti-CRISPR proteins. Nature 2015; 526(7571): 136-9.
[http://dx.doi.org/10.1038/nature15254] [PMID: 26416740]

[135] Borges AL, Davidson AR, Bondy-Denomy J. The Discovery, Mechanisms, and Evolutionary Impact

of Anti-CRISPRs. Annu Rev Virol 2017; 4(1): 37-59. [http://dx.doi.org/10.1146/annurev-virology-101416-041616] [PMID: 28749735]

[136] He F, Bhoobalan-Chitty Y, Van LB, *et al.* Anti-CRISPR proteins encoded by archaeal lytic viruses inhibit subtype I-D immunity. Nat Microbiol 2018; 3(4): 461-9. [http://dx.doi.org/10.1038/s41564-018-0120-z] [PMID: 29507349]

[137] Marino ND, Zhang JY, Borges AL, *et al.* Discovery of widespread type I and type V CRISPR-Cas inhibitors. Science 2018; 362(6411): 240-2. [http://dx.doi.org/10.1126/science.aau5174] [PMID: 30190308]

[138] Pawluk A, Davidson AR, Maxwell KL. Anti-CRISPR: Discovery, mechanism and function. Nat Rev Microbiol 2018; 16(1): 12-7. [http://dx.doi.org/10.1038/nrmicro.2017.120] [PMID: 29062071]

[139] Hynes AP, Rousseau GM, Agudelo D, *et al.* Widespread anti-CRISPR proteins in virulent bacteriophages inhibit a range of Cas9 proteins. Nat Commun 2018; 9(1): 2919. [http://dx.doi.org/10.1038/s41467-018-05092-w] [PMID: 30046034]

[140] Wang X, Yao D, Xu JG, *et al.* Structural basis of Cas3 inhibition by the bacteriophage protein AcrF3. Nat Struct Mol Biol 2016; 23(9): 868-70. [http://dx.doi.org/10.1038/nsmb.3269] [PMID: 27455460]

[141] Bryson AL, Hwang Y, Sherrill-Mix S, *et al.* Covalent Modification of Bacteriophage T4 DNA Inhibits CRISPR-Cas9. MBio 2015; 6(3)e00648 [http://dx.doi.org/10.1128/mBio.00648-15] [PMID: 26081634]

[142] Seed KD, Lazinski DW, Calderwood SB, Camilli A. A bacteriophage encodes its own CRISPR/Cas adaptive response to evade host innate immunity. Nature 2013; 494(7438): 489-91. [http://dx.doi.org/10.1038/nature11927] [PMID: 23446421]

[143] Borges AL, Zhang JY, Rollins MF, Osuna BA, Wiedenheft B, Bondy-Denomy J. Bacteriophage Cooperation Suppresses CRISPR-Cas3 and Cas9 Immunity. Cell 2018; 174(4): 917-925.e10. [http://dx.doi.org/10.1016/j.cell.2018.06.013] [PMID: 30033364]

[144] Landsberger M, Gandon S, Meaden S, *et al.* Anti-CRISPR Phages Cooperate to Overcome CRISPR-Cas Immunity. Cell 2018; 174(4): 908-916.e12. [http://dx.doi.org/10.1016/j.cell.2018.05.058] [PMID: 30033365]

[145] Hullahalli K, Rodrigues M, Nguyen UT, Palmer K. An Attenuated CRISPR-Cas System in Enterococcus faecalis Permits DNA Acquisition. MBio 2018; 9(3): e00414-8. [http://dx.doi.org/10.1128/mBio.00414-18] [PMID: 29717009]

[146] Watson BNJ, Staals RHJ, Fineran PC. CRISPR-Cas-Mediated Phage Resistance Enhances Horizontal Gene Transfer by Transduction. MBio 2018; 9(1): e02406-17. [http://dx.doi.org/10.1128/mBio.02406-17] [PMID: 29440578]

[147] Faure G, Makarova KS, Koonin EV. CRISPR-Cas: Complex Functional Networks and Multiple Roles beyond Adaptive Immunity. J Mol Biol 2019; 431(1): 3-20. [http://dx.doi.org/10.1016/j.jmb.2018.08.030] [PMID: 30193985]

[148] Dupuis ME, Villion M, Magadán AH, Moineau S. CRISPR-Cas and restriction-modification systems are compatible and increase phage resistance. Nat Commun 2013; 4: 2087. [http://dx.doi.org/10.1038/ncomms3087] [PMID: 23820428]

[149] Price VJ, Huo W, Sharifi A, Palmer KL. CRISPR-Cas and Restriction-Modification Act Additively against Conjugative Antibiotic Resistance Plasmid Transfer in *Enterococcus faecalis.* MSphere 2016; 1(3): e00064-16. [http://dx.doi.org/10.1128/mSphere.00064-16] [PMID: 27303749]

[150] Koonin EV. Evolution of RNA- and DNA-guided antivirus defense systems in prokaryotes and eukaryotes: Common ancestry *vs* convergence. Biol Direct 2017; 12(1): 5. [http://dx.doi.org/10.1186/s13062-017-0177-2] [PMID: 28187792]

[151] Ryazansky S, Kulbachinskiy A, Aravin AA. The Expanded Universe of Prokaryotic Argonaute Proteins. MBio 2018; 9(6): e01935-18.
[http://dx.doi.org/10.1128/mBio.01935-18] [PMID: 30563906]

[152] Lisitskaya L, Aravin AA, Kulbachinskiy A. DNA interference and beyond: Structure and functions of prokaryotic Argonaute proteins. Nat Commun 2018; 9(1): 5165.
[http://dx.doi.org/10.1038/s41467-018-07449-7] [PMID: 30514832]

[153] Willkomm S, Makarova KS, Grohmann D. DNA silencing by prokaryotic Argonaute proteins adds a new layer of defense against invading nucleic acids. FEMS Microbiol Rev 2018; 42(3): 376-87.
[http://dx.doi.org/10.1093/femsre/fuy010] [PMID: 29579258]

[154] Swarts DC, Jore MM, Westra ER, *et al.* DNA-guided DNA interference by a prokaryotic Argonaute. Nature 2014; 507(7491): 258-61.
[http://dx.doi.org/10.1038/nature12971] [PMID: 24531762]

[155] Swarts DC, Szczepaniak M, Sheng G, *et al.* Autonomous Generation and Loading of DNA Guides by Bacterial Argonaute. Mol Cell 2017; 65(6): 985-998.e6.
[http://dx.doi.org/10.1016/j.molcel.2017.01.033] [PMID: 28262506]

[156] Zander A, Willkomm S, Ofer S, *et al.* Guide-independent DNA cleavage by archaeal Argonaute from *Methanocaldococcus jannaschii.* Nat Microbiol 2017; 2: 17034.
[http://dx.doi.org/10.1038/nmicrobiol.2017.34] [PMID: 28319081]

[157] Olovnikov I, Chan K, Sachidanandam R, Newman DK, Aravin AA. Bacterial argonaute samples the transcriptome to identify foreign DNA. Mol Cell 2013; 51(5): 594-605.
[http://dx.doi.org/10.1016/j.molcel.2013.08.014] [PMID: 24034694]

[158] Kaya E, Doxzen KW, Knoll KR, *et al.* A bacterial Argonaute with noncanonical guide RNA specificity. Proc Natl Acad Sci USA 2016; 113(15): 4057-62.
[http://dx.doi.org/10.1073/pnas.1524385113] [PMID: 27035975]

[159] Makarova KS, Wolf YI, van der Oost J, Koonin EV. Prokaryotic homologs of Argonaute proteins are predicted to function as key components of a novel system of defense against mobile genetic elements. Biol Direct 2009; 4: 29.
[http://dx.doi.org/10.1186/1745-6150-4-29] [PMID: 19706170]

[160] Berngruber TW, Lion S, Gandon S. Evolution of suicide as a defence strategy against pathogens in a spatially structured environment. Ecol Lett 2013; 16(4): 446-53.
[http://dx.doi.org/10.1111/ele.12064] [PMID: 23331662]

[161] Ainsworth S, Mahony J, van Sinderen D. The plasmid complement of *Lactococcus lactis* UC509.9 encodes multiple bacteriophage resistance systems. Appl Environ Microbiol 2014; 80(14): 4341-9.
[http://dx.doi.org/10.1128/AEM.01070-14] [PMID: 24814781]

[162] Makarova KS, Wolf YI, Koonin EV. Comparative genomics of defense systems in archaea and bacteria. Nucleic Acids Res 2013; 41(8): 4360-77.
[http://dx.doi.org/10.1093/nar/gkt157] [PMID: 23470997]

[163] Chopin MC, Chopin A, Bidnenko E. Phage abortive infection in lactococci: Variations on a theme. Curr Opin Microbiol 2005; 8(4): 473-9.
[http://dx.doi.org/10.1016/j.mib.2005.06.006] [PMID: 15979388]

[164] Dy RL, Przybilski R, Semeijn K, Salmond GP, Fineran PC. A widespread bacteriophage abortive infection system functions through a Type IV toxin-antitoxin mechanism. Nucleic Acids Res 2014; 42(7): 4590-605.
[http://dx.doi.org/10.1093/nar/gkt1419] [PMID: 24465005]

[165] Fineran PC, Blower TR, Foulds IJ, Humphreys DP, Lilley KS, Salmond GP. The phage abortive infection system, ToxIN, functions as a protein-RNA toxin-antitoxin pair. Proc Natl Acad Sci USA 2009; 106(3): 894-9.
[http://dx.doi.org/10.1073/pnas.0808832106] [PMID: 19124776]

[166] Sberro H, Leavitt A, Kiro R, *et al.* Discovery of functional toxin/antitoxin systems in bacteria by shotgun cloning. Mol Cell 2013; 50(1): 136-48. [http://dx.doi.org/10.1016/j.molcel.2013.02.002] [PMID: 23478446]

[167] Short FL, Akusobi C, Broadhurst WR, Salmond GPC. The bacterial Type III toxin-antitoxin system, ToxIN, is a dynamic protein-RNA complex with stability-dependent antiviral abortive infection activity. Sci Rep 2018; 8(1): 1013. [http://dx.doi.org/10.1038/s41598-017-18696-x] [PMID: 29343718]

[168] Blower TR, Chai R, Przybilski R, *et al.* Evolution of Pectobacterium Bacteriophage PhiM1 To Escape Two Bifunctional Type III Toxin-Antitoxin and Abortive Infection Systems through Mutations in a Single Viral Gene. Appl Environ Microbiol 2017; 83(8): e03229-16. [http://dx.doi.org/10.1128/AEM.03229-16] [PMID: 28159786]

[169] Chen B, Akusobi C, Fang X, Salmond GPC. Environmental T4-Family Bacteriophages Evolve to Escape Abortive Infection *via* Multiple Routes in a Bacterial Host Employing "Altruistic Suicide" through Type III Toxin-Antitoxin Systems. Front Microbiol 2017; 8: 1006. [http://dx.doi.org/10.3389/fmicb.2017.01006] [PMID: 28620370]

[170] Alawneh AM, Qi D, Yonesaki T, Otsuka Y. An ADP-ribosyltransferase Alt of bacteriophage T4 negatively regulates the *Escherichia coli* MazF toxin of a toxin-antitoxin module. Mol Microbiol 2016; 99(1): 188-98. [http://dx.doi.org/10.1111/mmi.13225] [PMID: 26395283]

[171] Engelberg-Kulka H, Reches M, Narasimhan S, *et al.* rexB of bacteriophage lambda is an anti-cell death gene. Proc Natl Acad Sci USA 1998; 95(26): 15481-6. [http://dx.doi.org/10.1073/pnas.95.26.15481] [PMID: 9860994]

[172] Koonin EV, Zhang F. Coupling immunity and programmed cell suicide in prokaryotes: Life-or-death choices. BioEssays 2017; 39(1): 1-9. [http://dx.doi.org/10.1002/bies.201600186] [PMID: 27896818]

[173] Díaz-Muñoz SL. Viral coinfection is shaped by host ecology and virus-virus interactions across diverse microbial taxa and environments. Virus Evol 2017; 3(1)vex011 [http://dx.doi.org/10.1093/ve/vex011] [PMID: 28469939]

[174] Bondy-Denomy J, Qian J, Westra ER, *et al.* Prophages mediate defense against phage infection through diverse mechanisms. ISME J 2016; 10(12): 2854-66. [http://dx.doi.org/10.1038/ismej.2016.79] [PMID: 27258950]

[175] Bebeacua C, Lorenzo Fajardo JC, Blangy S, *et al.* X-ray structure of a superinfection exclusion lipoprotein from phage TP-J34 and identification of the tape measure protein as its target. Mol Microbiol 2013; 89(1): 152-65. [http://dx.doi.org/10.1111/mmi.12267] [PMID: 23692331]

[176] Laliberte JP, Moss B. A novel mode of poxvirus superinfection exclusion that prevents fusion of the lipid bilayers of viral and cellular membranes. J Virol 2014; 88(17): 9751-68. [http://dx.doi.org/10.1128/JVI.00816-14] [PMID: 24920806]

[177] Lu MJ, Henning U. Superinfection exclusion by T-even-type coliphages. Trends Microbiol 1994; 2(4): 137-9. [http://dx.doi.org/10.1016/0966-842X(94)90601-7] [PMID: 8012757]

[178] Lee YM, Tscherne DM, Yun SI, Frolov I, Rice CM. Dual mechanisms of pestiviral superinfection exclusion at entry and RNA replication. J Virol 2005; 79(6): 3231-42. [http://dx.doi.org/10.1128/JVI.79.6.3231-3242.2005] [PMID: 15731218]

[179] Ko CC, Hatfull GF. Mycobacteriophage Fruitloop gp52 inactivates Wag31 (DivIVA) to prevent heterotypic superinfection. Mol Microbiol 2018; 108(4): 443-60. [http://dx.doi.org/10.1111/mmi.13946] [PMID: 29488662]

[180] Goenaga S, Kenney JL, Duggal NK, *et al.* Potential for Co-Infection of a Mosquito-Specific

Flavivirus, Nhumirim Virus, to Block West Nile Virus Transmission in Mosquitoes. Viruses 2015; 7(11): 5801-12.
[http://dx.doi.org/10.3390/v7112911] [PMID: 26569286]

[181] Nasar F, Erasmus JH, Haddow AD, Tesh RB, Weaver SC. Eilat virus induces both homologous and heterologous interference. Virology 2015; 484: 51-8.
[http://dx.doi.org/10.1016/j.virol.2015.05.009] [PMID: 26068885]

[182] Essuman K, Summers DW, Sasaki Y, *et al.* TIR Domain Proteins Are an Ancient Family of NAD^+-Consuming Enzymes. Curr Biol 2018; 28(3): 421-430.e4.
[http://dx.doi.org/10.1016/j.cub.2017.12.024] [PMID: 29395922]

[183] Burroughs AM, Zhang D, Schäffer DE, Iyer LM, Aravind L. Comparative genomic analyses reveal a vast, novel network of nucleotide-centric systems in biological conflicts, immunity and signaling. Nucleic Acids Res 2015; 43(22): 10633-54.
[http://dx.doi.org/10.1093/nar/gkv1267] [PMID: 26590262]

CHAPTER 6

Interaction of Stress and Immune Responses in Prokaryotes

Abstract: In prokaryote cells, stress response interacts with immune response. Stress caused by diverse factors from DNA damage to heat, low pH, heavy metals, antibiotics, and toxins leads to the induction of proviruses and movement of mobile genetic elements. This can lead to horizontal gene transfer of immune mechanisms encoded by such elements and stimulate the evolution of bacterial immune defenses. The viral infection is associated with changes in host membranes and metabolism, which can be sensed as stress. Such stress then affects immune mechanisms. Stress can activate CRISPR-Cas systems preventively before they encounter an intruder. During DNA damage and associated DNA repair, unmodified R/M recognition sites may appear in the prokaryotic genome. To avoid targeting such sites by R/M systems, their activity is downregulated during stress.

Keywords: Bacteriophage-encoded toxins, CRISPR-Cas, Envelope stress response, Immunity, Prokaryotes, Provirus induction, Phage shock protein, R/M autoimmunity, Restriction alleviation, Stress, Stress response.

INTRODUCTION

Often, prokaryotes face both stress and pathogens. Immunodefense and stress response pathways can compete for cellular resources or cooperate. In archaeon *Sulfolobus islandicus* and bacterium *Campylobacter jejuni,* viral infection downregulates stress response genes such as HSPs [1, 2]. Moreover, viruses can exploit stress response for their advantage. For example, for their propagation, nuclear-replicating viruses need a stress-induced protein that exports mRNAs from the cell nucleus [3]

On the other hand, some reports show that bacteria exposed to heat, chlorine, or UV stress develop increased phage resistance [4]. Oxidative stress decreases phage infection [5]. Stress may lead to cellular dormancy through a stringent response, SOS response, toxin-antitoxin systems, and other pathways [6]. Dormancy does not prevent viral infection but restricts the number of progeny and delays their release [7]. In some cases, immune genes and stress response genes can directly regulate each other [8, 9].

Nadia Danilova

Latent infection is associated with a long-term presence in prokaryotic cells of DNA from viruses and other mobile genetic elements (MGEs). It can lead to changes in host metabolism, stress resistance, and immune response that are beneficial to the host [10]. Stress typically stimulates induction and propagation of MGEs through horizontal gene transfer. This can promote the spread of useful traits such as immune resistance. Specific interactions of stress and immune mechanisms are discussed below.

6.1. Mobile Genetic Elements are Induced by Stress

In bacteria and archaea, cellular stress caused by diverse noxious factors including heat, irradiation, low pH, heavy metals, ROS, antibiotics, and toxins leads to induction of proviruses and other MGEs [11 - 14]. Provirus induction is often noted during the SOS response associated with DNA damage. DNA damage is sensed by RecA protein; activated RecA promotes self-cleavage of the transcription repressor LexA, which unleashes the SOS response [15]. Some bacteriophages, such as bacteriophage λ in *E. coli*, encode CI repressor that undergoes self-cleavage in response to DNA damage and RecA activation [16, 17]. Cl cleavage starts a lytic cycle. Other phages do not encode SOS-sensitive repressor but rely only on the host factors such as LexA. Often, LexA regulates genes of a mobile genetic element together with an additional repressor that can be encoded by the cell or by MGEs. For example, LexA together with a cellular repressor controls CTXphi phage of *Vibrio cholerae* that encodes cholera toxin [18]. Bacteriophage GIL01 of *Bacillus thuringiensis* encodes a LexA homolog that activates the phage's lytic gene expression [19]. Repressors of *Salmonella* lambdoid prophages are not cleaved upon induction but inactivated by antirepressor proteins produced during SOS response [20]. The same antirepressor can bind repressors from different prophages, which leads to a coordinated induction of multiple prophages.

One way for a prophage to sense cellular stress is hypothesized based on the organization of CI repressor [21]. Cl mediates DNA looping that serves both to suppress promoters of lytic genes and to prevent overexpression of repressor. Some data show that the looping depends on supercoiling of DNA [22, 23]. Supercoiling is an indicator of cell health; it modulates the accessibility of chromatin to regulatory proteins and gene expression. The authors suggest that regulatory loops in DNA such as that mediated by Cl repressor may sense supercoiling of the genome in which they are embedded. In this way, a provirus sense any condition that adversely affects cellular health and can respond to it by its induction.

Stress factors that do not cause DNA damage and SOS response can still induce prophages. They include low pH, high temperature, heavy metals, and other adverse conditions [24]. Lytic promoters pL and pR of prophage λ in *E. coli* are activated when cells are exposed to the uncoupler of oxidative phosphorylation carbonyl cyanide m-chlorophenyl hydrazine, which shuts down ATP synthesis [14]. Induction of the prophage requires host RecA and CI repressor.

In the bacterial population, prophages are induced spontaneously in a small proportion of cells [12, 25]. A study of *E. coli* infected with λ phage shows that 1 to 2% of the prophage is induced in monoxenic mouse intestine through DNA damage-mediated response. Sporadic DNA damage may occur during normal cell function, such as replication, or in response to such stressors as ROS, temperature, or UV radiation. For example, uptake of iron by *Shewanella oneidensis* during biofilm formation triggers prophage induction and cell lysis likely through iron-induced DNA damage [11]. Simultaneously acting stressors increase the frequency of prophage induction. Phage induction may also depend on variations in the gene expression among individual cells [26]. Strains of *E. coli* that have shiga toxin-encoding prophages have increased frequency of spontaneous induction, which is explained by decreased levels of the cellular repressor [27].

For some phages, their induction in the presence of antibiotics potentiates antibiotic lethality through phage-encoded proteins. Ref is a phage P1 encoded endonuclease that cleaves damaged DNA tagged by RecA protein during SOS response. This increases the lethality during the lytic cycle of the phage. Ref also inhibits cell division independently of the SOS response [28].

Acquisition of foreign DNA including drug resistance genes is also potentiated by stress both in bacteria and archaea [29, 30]. In *Rhodobacter capsulatus*, SOS response activates gene transfer agent that packages random ~4 kb fragments of the genome and delivers it to surrounding cells [31].

Virulence of pathogenic bacteria evolves within the infected host by transfer of bacteriophages that encode virulence factors (lysogenic conversion). Bacteriophage transfer is strongly promoted by inflammation. In the murine *Salmonella typhimurium* diarrhea model, reactive oxygen and nitrogen species associated with inflammation trigger the bacterial SOS response and upregulate the expression of the phage antirepressor Tum, which activates phage production and transfer [32].

Quorum sensing communication molecules also can induce prophages. Bacterial universal communication molecule AI-2 induces virulence genes and promotes their spread *via* phage release [33].

Movement of other MGEs such as plasmids and transposons is also increased by stress. Transposon movement leads to increased intrachromosomal rearrangements, whereas plasmids are involved in gene transfer between different cells. Plasmids are exchanged even in the absence of stress providing steady gene flow between cells [34].

6.2. Mobile Genetic Elements can Improve the Host's Fitness

Infections with MGEs are detrimental to the prokaryotic host, which has necessitated the evolution of elaborate immune mechanisms protecting against foreign nucleic acids. However, in some cases, MGEs are beneficial to the host. Recent findings show that a provirus integrated into the prokaryotic genome can encode genes that do not participate in the viral life cycle but instead contribute to the host's fitness making the virus-host relationship mutualistic [35 - 38]. Such genes are often encoded within moron (from "more on") elements. These additional genes may function in the prevention of infection by other viruses through superinfection exclusion mechanisms, in the enhancement of the host's metabolism through the use of new nutrients or new electron acceptors, in antibiotic resistance, and in an increase of the host's stress tolerance.

Some lysogens such as *E. coli* infected with phage M13 grow faster and to a higher density than uninfected bacteria [37, 39]. Viruses often disseminate antibiotic resistance genes [40]. Multiple studies show that proviruses can bring to their hosts an increased stress resistance. Lysogenization of *Staphylococcus aureus* RN450 by phages ϕ11 and ϕ80α leads to activation of the SigB regulon responsible for the general stress response [41]. *S. aureus* episomal prophage varphiSa4ms confers increased survival after heat stress [42]. Enterohemorrhagic *E.coli* strain owes its acid resistance to transcriptional regulators encoded by a prophage [43].

Some prokaryotic hosts benefit from metabolic traits encoded by prophages. Temperate phage SopEF of *Salmonella enterica* serotype *Typhimurium* encodes an inducible nitric oxide synthase that produces precursors to the electron acceptor nitrate [44]. This can provide the host with a superior electron acceptor for anaerobic respiration in an anoxic inflamed gut. Another phage, CD38-2, improves sugar uptake and metabolism in its host *Clostridium difficile* [45].

Even prophages that lost the ability to reproduce can increase the fitness of the host. *E. coli* prophage CPS-53 proteins YfdK, YfdO, and YfdS enhance resistance to oxidative stress; prophages e14, CPS-53, and CP4-57 increase resistance to acids; and prophages e14 and rac increase biofilm formation [46]. Cryptic prophages are involved in inhibition of *E. coli* replication under stressful environmental conditions including antibiotic treatments [46, 47].

Proviruses can also carry genes that mediate superinfection exclusion, thus increasing resistance to infection by other viruses. Superinfection exclusion works through diverse mechanisms discussed in the previous chapter [48].

At the population level, lysogeny can influence biofilm formation [25, 36, 49]. *Clostridium difficile* cell wall protein CwpV is expressed in phase variation mode and promotes cell aggregation. 95% of lysogenic cells carrying CD38-2 prophage express CwpV, as opposed to only 5% of wild-type cells. This may facilitate host cells aggregation and formation of biofilms [45]. An important human pathogen *Pseudomonas aeruginosa* forms biofilms at sites of chronic infection in the lungs. Filamentous Pf phage increases *P. aeruginosa* virulence. It promotes bacterial adhesion to peptidoglycan; limits bacteria spread, and reduce the lung injury and inflammatory response [50]. A filamentous phage MDAΦ from *Neisseria meningitides* also increases its virulence by promoting bacteria aggregation on epithelial cells [51].

A small number of cells lysed by the activated proviruses may benefit the rest of the population due to the release of nutrients, bacteriocins, and DNA [36]. DNA released during lysis of cells that reside in biofilm strengthens the biofilm matrix [52].

Any provirus or other MGE that can be beneficial for the host needs to pass the barrier of prokaryotic immune defenses such as R/M and CRISPR-Cas systems. Immunocompromised hosts acquire new genes much easier. For example, multidrug-resistant *Enterococcus faecalis* strains that lack R/M and CRISPR-Cas have a 4-log increase in acquisition of plasmids [53]. Individual cells in a bacterial population that lost their immunity due to mutations can acquire foreign DNA including beneficial phages, creating populational heterogeneity and contributing to bacterial evolution. In addition, current data suggest that prokaryotes can have selective immunity. For example, the *Staphylococcus epidermidis* CRISPR-Cas type III system prevents infection by lytic phages but tolerates lysogenization by temperate phages [54]. Such conditional tolerance results from differences in transcription of the target. Type III CRISPR-Cas system targets only transcribed DNAs. Since temperate viruses are dormant, they do not induce an immune response. If the same provirus is induced and starts transcription and a lytic cycle, it would be targeted. The ability of CRISPR-Cas immune systems to discriminate between lytic and temperate viruses enables prokaryotes to selectively permit the entrance of temperate viruses. Other mechanisms such as BREX system and suppression of the newly acquired genes may also help to keep a balance between immune defense and genomic evolvability.

6.3. Infection Induces Envelope Stress Response

Prokaryotic envelopes are complex structures serving multiple purposes from physical fortification of cells to exchange of information with the environment, delivery of nutrients, excretion of waste and toxins, and production of energy. Multiple systems monitor their function and sense envelope stress. The viral infection is associated with several changes in the host and some findings suggest that they can be sensed as stress under certain conditions [55]. Absorption of the virus and its extrusion from the cell may be associated with insertion into the membrane of viral proteins that differ in their characteristics from cellular proteins. Furthermore, binding of a virus to its receptor and penetration of exogenous DNA induce a transient depolarization of the host membrane [56, 57]. The degree of depolarization depends on the viral load, the number of viral receptors on the membrane and concentration of Ca^{2+}.

In *E. coli* and other Gram-negative bacteria, there are at least five systems monitoring envelope function [58 - 61]. Alternative sigma factor E and Cpx system respond to misfolded proteins in membrane and in periplasm, respectively. Bae system responds to noxious substances including antibiotics and other toxins and regulates their efflux. Rcs system monitors the quality of membrane components, and the phage shock protein (Psp) system monitors membrane integrity.

Psp system was named so because it is induced during filamentous phage infection of *E. coli* [62]. Psp is also induced by heat, ionophores, ethanol, hyperosmotic shock, as well as by mislocalization to the inner membrane of secretins that form export channels in the outer membrane [63]. In general, Psp senses the inner membrane's permeability.

In *E. coli*, Psp system consists of seven proteins. PspA is a membrane protein that functions as a negative regulator of the system at non-stressful conditions, and as an effector upon stress. In non-inducing conditions, it binds and inhibits the transcriptional activator of Psp system, PspF. PspA's N-terminal region contains two amphipathic helixes: one is used for membrane binding, the other — for PspF binding [64]. Lipids are amphipathic molecules that have various degrees of spontaneous curvature in a monolayer. Interaction of lipids and proteins in the membrane results in stored curvature elastic (SCE) stress. Changes in membrane characteristic can lead to a PspA conformational change and its dissociation from PspF. PspF uses ATP hydrolysis to activate the alternative sigma factor σ54. Activated σ54 replaces the general sigma factor in RNA polymerase, which leads to transcription of stress response genes. PspA preferentially binds to membranes with lipid-packing defects and an increased SCE stress, which means that at some conditions PspA can sense envelope stress directly [64]. PspA may reduce SCE stress by inserting its amphipathic helix into the membrane and forming higher-order oligomers that have a scaffolding effect preventing the membrane transition into a porous phase.

Many inducers of Psp system have an effect on SCE stress. Specifically, mislocalization of outer membrane proteins secretins and insertion of viral proteins can cause a hydrophobic mismatch that would change the SCE stress profile of the surrounding bilayer. Temperature, high pressure, fatty acids, and alcohol can cause lipid-packing defects.

Besides Psp, other stress responses have been reported during viral infection. The reaction of *Yersinia enterocolitica* to a viral infection is characterized by up-regulation of several genes involved in stress response, including the Cpx pathway, ABC transporters, phage-, osmotic -, and cold-shock proteins [65].

The early response of *Lactococcus lactis* to infection with the lytic phage c2 prior to its DNA replication aims at biogenesis, modification, and regulatory processes of the cell envelope [66]. The response includes induction of membrane stress proteins, D-alanylation of the cell wall lipoteichoic acids, maintenance of the proton motive force, and energy conservation. The authors suggest that the response is modulated by the Psp system, the global regulator SpxB, and the two-component system CesSR. These data demonstrate how viral infection of prokaryotic cells may cause a membrane disturbance that sensed by the cell as envelope stress.

6.4. CRISPR-Cas Activation in Response to Stress

Envelope stress caused by a viral infection can, at least in some cases, be signaled to immune mechanisms and activate them preventively before they encounter the intruder. CRISPR-Cas activation by envelope stress is the most studied example of cross talk between stress- and immune responses (Fig. **6**). One of the first indications of communication between envelope stress response and CRISPR-Cas systems was found during the experimental induction of envelope stress by targeting the twin-arginine translocation (Tat) pathway by a plasmid-encoded fusion protein. It resulted in upregulation of Cas enzymes and plasmid destruction [9]. BaeSR was necessary for this effect but the exact mechanism of CRISPR-Cas activation was not uncovered.

In *Myxococcus xanthus,* an extracytoplasmic function (ECF) sigma factor DdvS and its membrane-bound anti-sigma DdvA control CRISPR-Cas system expression [67]. DdvA senses membrane signal whereas DdvS binds and regulates promoters of Cas cluster. Many membrane genes are encoded in CRISPR-Cas type III loci and may be involved in the function of these systems [68].

E. coli envelope stress can also stimulate CRISPR-Cas indirectly through upregulation of LeuO by the membrane stress system Bsc [69]. LeuO is a major antagonist of histone-like nucleoid-structuring (H-NS) proteins that suppress CRISPR-Cas systems [70, 71]. LeuO binds to two sites flanking the casA promoter, which H-NS also binds. Antagonistic regulation of CRISPR-Cas by LeuO and H-NS suggests that when H-NS binds exogenous DNA, the amount of LeuO free to bind casA promoter increases [72].

CRISPR-Cas systems can be induced not only by envelope stress. The alternative sigma factor RpoN of *Salmonella typhimurium* binds to *cas1* promoter [73]. RpoN is induced by many stresses including nitrogen starvation, heavy metals *etc.* These stresses, therefore, would lead to upregulation of CRISPR-Cas.

In *Streptococcus mutans,* the two-component regulator VicRK involved in bacteriocin production, cell death, transformation, and biofilm formation upregulates CRISPR-Cas systems in response to temperature, oxidative, and envelope stress [74]. Type I and type II systems respond differently to various stresses; VicRK represses type IIA and activates type IC systems in response to oxidative stress.

In *E. coli*, high temperature can cause denaturation of membrane proteins leading to envelope stress. In response, cells induce protein G, a homolog of the eukaryotic chaperone heat-shock protein 90 [75]. This protein is involved in the maintenance of functional levels of *E. coli* Cas3 protein and it is necessary for CRISPR-Cas system function [75, 76].

The above data show that CRISPR-Cas systems in various species respond to envelope stress. In addition, evidence from some species suggests that these systems may also respond to other stresses.

6.5. CRISPR-Cas Responds to Metabolic Changes

The viral infection leads to profound changes in host metabolism [77, 78]. It creates an increased demand for intermediates for producing new viruses and building viral maturation compartments inside their host. Viruses typically induce upregulation of glucose and nucleotide metabolism. They switch aerobic respiration to glycolysis, which provides an increased amount of glucose metabolites that can be used in biosynthetic pathways. In tumor cells it is known as the Warburg effect.

Bacteria respond to low glucose by an increase in cAMP levels. Recent studies show that a complex of cAMP and cAMP receptor protein (CRP) activates the *cas* operon in *Pectobacterium atrosepticum*, increasing the expression of genes encoding Cas1, Cas2, and the Csy complex [79, 80] (Fig. **6**). Viral infection of *Thermus thermophilus* also results in cAMP-CRP-dependent upregulation of some *cas* genes whereas upregulation of CRISPR array is CRP independent [81].

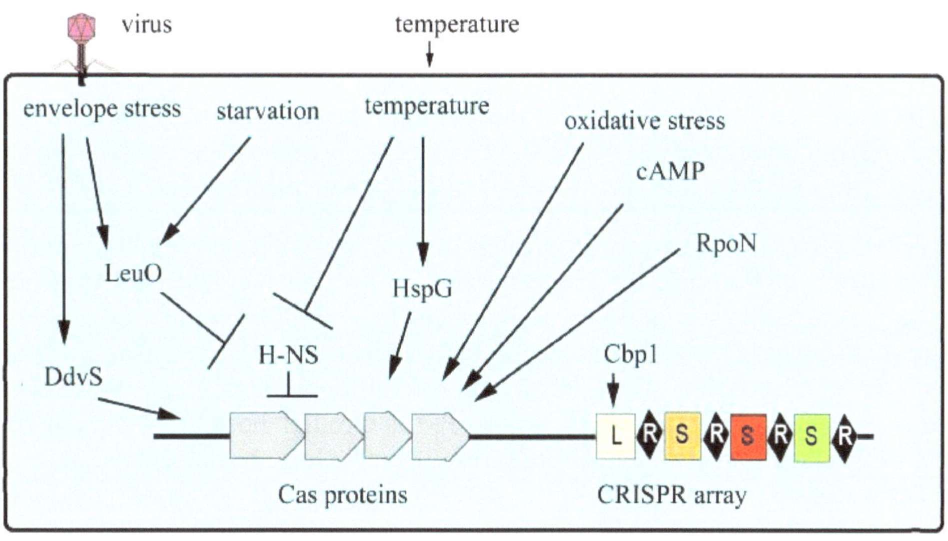

Fig. (6). Expression of CRISPR-Cas system depends on cellular conditions. It can be affected by various factors such as envelope stress, temperature, oxidative stress, and disturbances of metabolism.

LeuO, H-NS antagonist and CRISPR-Cas system activator, is induced not only in response to envelope stress but also in response to starvation for branched-chain amino acids [82]. These studies suggest that perturbation of normal metabolic state of the cell caused by the acquisition and replication of parasitic DNA, such as phages and plasmids, influences the defense function of CRISPR-Cas through multiple pathways.

6.6. Effect of Stress on R/M Systems

R/M systems usually work by modifying cellular DNA by methylation and destroying unmodified foreign DNA by a restriction endonuclease. In some conditions, unmodified R/M recognition sites can arise in the cell's own genome. It can happen during repair of DNA damage, genome rearrangement, or recombination. R/M systems can target such sites [83]. Horizontal transfer of a new R/M system can also be dangerous for the unmodified host genome. To avoid self-targeting, bacteria have evolved a mechanism that controls endonuclease expression or activity until a methyltransferase finishes genome modification. This is called restriction alleviation because it reduces the level of restriction of unmethylated DNA. It was discovered for the type I R/M of *E. coli* infected with λ phage and exposed to UV irradiation [84]. Restriction alleviation decreases but does not completely abolish the restriction of foreign DNA. For example, following DNA damage caused by UV, EcoK1 restriction is reduced 2 to 3 orders of magnitude [85]. Other stresses that result in DNA damage also lead to restriction alleviation.

In type I systems such as EcoKI, the primary pathway is degradation of the restriction endonuclease by ClpXP [83]. Type I endonucleases can also be inhibited by regulatory proteins, some of which can be plasmid-encoded [86, 87].

Higher efficiency of restriction alleviation for the bacterial chromosome was attributed to the structural differences between foreign and host DNAs. Foreign DNA enters the cell as a random coil, whereas host DNA is condensed and bound by various ligands, which makes it less accessible to nucleases [88]. The efficiency of restriction alleviation may also depend on DNA translocation lifetime in type I systems, which is longer for circular host DNA *versus* linear DNA. Nucleases associated with such long-lived DNA translocation states are targets for ClpXP protease [89].

Type II systems have a family of regulatory proteins that delay the expression of endonuclease relative to the methylase, which is especially important when an R/M system is acquired by horizontal gene transfer [90].

In these ways, bacterial genomes are protected against self-targeting by restriction endonucleases during DNA damage or recombination and after a horizontal transfer of a new R/M system. Protection against self-targeting by R/M systems is not absolute and can fail under stress. 100 MPa hydrostatic pressure shock in *E. coli* causes SOS response, which is due to DNA damage by the type IV restriction endonuclease Mrr [91]. Starvation also can increase self-targeting. At normal *E. coli* growth condition, self-restriction by EcoRI is rare but it is increased when cells grow in a minimal medium [92].

6.7. MGEs and Pathogenic Bacteria

The environmental niches occupied by some prokaryotes are within other species. Thousands of prokaryotic species populate human gut and skin. Their interactions with the human fall into general categories of symbiosis/mutualism, commensalism, or parasitism. Successful colonization of a host organism by a pathogen depends on the pathogen's virulence. Virulence factors function in diverse processes associated with the infection including adhesion, host colonization, persistence, and immune evasion. The opportunistic pathogen *Legionella pneumophila* uses ~300 different effector proteins to survive inside macrophages [93].

In some pathogens, virulence factors are encoded by MGEs. Bacteriophages can carry genes that turn harmless bacteria into a pathogen [94]. One of the most studied examples of virulence dependent on a bacteriophage is the cholera toxin genes encoded by the filamentous bacteriophage CTXφ [95]. The toxin promotes adenosine diphosphate (ADP)-ribosylation of the signaling protein GSalpha leading to high levels of cAMP, water loss by epithelial cells of the gut, and diarrhea [96]. CTXφ phage infection converts the non-toxigenic *Vibrio cholerae* strains into toxigenic ones. CTXφ genome can be chromosomally integrated or replicated as a plasmid. A cellular repressor, whose activity is regulated by SOS response to DNA damage, controls CTXφ production [18]. Cholera toxin production is also induced by starvation in dependence on ppGpp [97].

The Shiga toxin genes are also encoded in lambdoid prophages. They perform the conversion of commensal *E. coli* into a dangerous pathogen [27, 98]. In target cells, the toxin depurinates a single adenine residue in 28S ribosomal RNA, which prevents binding of amino-acyl-tRNA to the ribosome and leads to arrest of protein synthesis and cell death. The symptoms range from mild diarrhea to hemolytic uremic syndrome. Shiga toxin-encoding prophages are induced by many types of stress. Hydrogen peroxide induces Shiga toxin phages in RecA and SOS response-dependent manner [99]. Stringent response and acid stress also induce Shiga toxin phages [100].

In order to play a role in virulence of pathogenic bacteria, bacteriophage-encoded toxins need to be expressed and released, which happens only after prophage induction, lytic cycle, and death of the host cell. Studies of Shiga toxin-producing *E. coli* suggest that prophage induction takes place in a small fraction of bacterial cells that undergo "altruistic death" so that the rest of the population benefits from Shiga toxin killing of the host immune cells [101]. Virulence factors can also be encoded inside phage-related chromosomal islands, which use the phage reproduction cycle for their own transduction [102]. Besides toxins, bacteriophages and plasmids that infect pathogenic bacteria may encode additional genes for traits beneficial to their hosts such as antimicrobial resistance [94, 103].

6.8. The Role of Stress Response in Bacterial Pathogenesis

Pathogenic bacteria encounter various unfavorable environmental conditions inside their hosts. To infect a human, they need to survive contact with skin or mucous membranes that are rich in lysozymes, antibacterial peptides, and other antibacterial substances. Pathogens that enter through the gastrointestinal tract need to survive low pH in the stomach, bile salts and other surfactants, digestive enzymes, and anaerobic condition. Inside the body, the human immune system attacks the bacteria in multiple ways including immune cells, antibacterial peptides, ROS, antibodies, and the complement.

Intracellular pathogens face low pH, lysosomal enzymes, antimicrobial proteins, and ROS among other factors. In addition, the host organism tries to deprive pathogens of essential factors such as iron, leading to nutritional stress.

Therefore, pathogenic bacteria need a highly functional stress response to successfully infect the host. RpoS, a general stress response sigma factor, is necessary for efficient infection by *E. coli* [104]. The alternative sigma factor SigB of *Staphylococcus aureus* is required for intracellular growth of this pathogen [105]. Following internalization by human cells, SigB undergoes fast activation and switches *Staphylococcu*s transcription to mobilize the pathogen for a fight with the human immune system. This increases its resistance to antibiotics, oxidative stress, and damage to macromolecules. In SigB absence, all *S. aureus* cells are cleared by the host phagocytes within a few days [106]. Presence of certain prophages further increases *S. aureus* stress resistance [41, 42].

When phagocytes ingest bacteria they produce various oxidants in a process called 'respiratory burst'. It is generated in phagosomes by a NADPH oxidase multi-protein complex (NOX2) that has a catalytic core consisting of CYBB and CYBA and several cytosolic components [107]. NOX2 produces the superoxide anion O_2^-, which gives rise to other oxidants. Superoxide dismutase turns superoxide into hydrogen peroxide. Nitric oxide synthase is also induced; it produces nitric oxide (·NO), which reacts with superoxide to produce peroxynitrite and other RNSs. Neutrophil myeloperoxidase uses hydrogen peroxide to generate hypochlorous acid (HOCl), which is also a strong oxidant. Neutrophil extracellular traps, which entangle extracellular pathogens, also use ROS to kill them. The importance of NOX2 in immunodefense is underscored by severe immunodeficiency called the chronic granulomatous disease, which is caused by mutations in NOX2 components [108]. When pathogens escape from phagolysosomes into the cytosol they encounter ROS and RNS species produced by the mitochondria. Therefore it is vital for pathogens to have mechanisms that can mitigate oxidative stress [109]. These mechanisms include the production of enzymes and low-molecular-weight compounds detoxifying ROS, sequestration of heavy metals, and repair of macromolecules. Most genes of *Listeria monocytogenes* necessary for adjustment to intracellular pathogenesis play important roles in redox regulation [110].

For a pathogen, maintenance of the integrity and function of its cell envelope is also of great importance, and factors responsible for envelope stress response contribute to bacterial virulence. Specifically, the phage shock protein response supports virulence by preserving membrane PMF and other membrane functions in adverse conditions and by promoting biofilm formation [55]. CpxR/A system involved in envelope stress monitoring also regulates virulence [111, 112]. Dimethyl adenosine transferase KsgA is required for cell-envelope strength and virulence in *Salmonella enteritidis* [113]

Many other factors involved in bacterial stress response also contribute to virulence. For example, the global ferric uptake regulator (Fur) is important for the virulence of *E. coli* [114]. Nutritional immunity is associated with host-mediated restriction of a pathogen's access to important nutrients such as metals iron, manganese, and zinc. Pathogenic bacteria have evolved special adaptations to deal with nutritional immunity. For example, non-pathogenic *Staphylococci* have a superoxide dismutase that needs manganese for its activity. In contrast, the human pathogen *S. aureus* has an additional superoxide dismutase that is capable to function with both iron and manganese [115]. Hence, *S. aureus* can resist manganese starvation imposed by the host using iron instead.

Pathogenic bacteria cope with particular stress by a corresponding adjustment of their metabolism. The enteric pathogen *Salmonella* responds to macrophage-imposed nitrosative and oxidative stress by DksA-dependent changes in the expression of more than 400 genes including several antioxidants and oxidoreductases important in redox buffering, NO radical detoxification, damage repair, and cysteine biosynthesis [116]. DNA methylation landscape of *Mycoplasma hyorhinis* changes after the switch from a free-living lifestyle to life inside human cells [117]. *E. coli* adjusts its metabolism after moving from the lower gastrointestinal tract into the urinary tract with different stressors and nutritional environment [118].

Cooperation between pathogenic bacteria is important to establish infection. *Pseudomonas aeruginosa* uses quorum sensing to coordinately switch on the production of virulence factors, to form a biofilm, or to secrete various factors [119].

6.9. Role of Stress in Evolution of Prokaryotic Immunity

Co-evolution of parasites with their hosts is often viewed as a major evolutionary force that shaped evolutionary transitions at least up to and including the origin of multicellularity [120 - 122]. Prokaryotes co-evolve in a persistent arms race with their parasites, in which prokaryotes are perfecting the existing defense mechanisms and inventing new ones while parasites are developing ways to bypass them. Parasites maintain diversity inside the host population and drive divergence between the subpopulations. Parasite diversity affects the host population dynamics and evolution since every parasite drives the selection of a unique resistance mechanism [123]. In some cases, hosts and parasites evolve to form a mutualistic relationship where the host benefits from superinfection exclusion and uses genes encoded by MGEs to enhance its metabolism [35, 36, 124, 125].

Immune mechanisms alone are not sufficient to protect prokaryotic organisms from parasites. Abortive infection evolved to clean populations from parasitic elements through the sacrifice of the infected cells by Abi, a prokaryotic version of programmed cell death [126 - 128]. The existence of multiple independent Abi systems points to vital importance of this mechanism in protecting a prokaryotic population as a whole from parasites.

In addition to its role in immune defense, programmed cell death is common in prokaryotes exposed to extreme stress from the environment or from the malfunction of the cellular apparatus [129]. TA systems may have evolved to clean population from defective cells and maintain population fitness [130 - 133].

Stress drives evolution in part by activating mechanisms that increase the ability of cells to evolve including the increased movement of MGEs and increased rate of mutation [134]. Genes involved in immune mechanisms and PCD are often encoded by MGEs and can be distributed by horizontal gene transfer. This distribution is promoted by stress. Therefore, MGEs are essential elements in the evolution of both stress resistance and immunity in prokaryotes. MGEs movement can change regulatory pathways, create new functions and introduce mutations. Integration of foreign DNA and intrachromosomal gene shuffling by homologous recombination correlate with lifestyle of bacteria and are higher among pathogens compared to commensal strains because of an additional stress from host defenses [135].

A 93-gene network that promotes mutagenesis exists in *E. coli*. Most of it senses stress and operates upstream from the key stress response hubs RpoS, RpoE, and SOS [136]. One of the responses to stress is switching the repair of DNA double-strand breaks to the error-prone DNA polymerases, which introduce mutations during repair [137]. Other factors may be involved in the regulation of mutagenesis. Antibiotic treatment-induced mutagenesis is associated with increased expression of error-prone DNA polymerases, downregulation of mismatch repair caused by oxidative stress, and transcription-replication conflicts [138, 139]. A massive RpoS-dependent diversification in aging colonies of *E. coli* has been observed. It resulted mostly from mutations; some mutants formed biofilm more easily whereas others gained the ability to use different nutrients or to tolerate various antibiotics or oxidative stress [140]. Mutations that increase stress tolerance may decrease the growth rate at normal condition, so various conflicting factors may affect the evolutional outcome [141].

Epigenetic modification of genome is a way of fast response to environmental stressors, which can also produce heritable phenotypes with different levels of transmission fidelity [142]. Recent studies show that epigenetics can significantly contribute to the evolution of eukaryotes [143]. Epigenetic regulation is also important for prokaryotes. For example, in the absence of adenine methylation at GATC sites the survival of *E. coli* treated by antibiotics is reduced because of DNA repair malfunction [144]. Prokaryotes can acquire heritable epigenetic changes after exposure to a particular stress as illustrates by the example of *E. coli* that, when cultured under increasing concentration of HOCl, evolved resistance through epigenetic changes leading to a constitutively expressed OxyR regulon [145]. Epigenetic mechanisms can also affect the expression and movement of MGEs [146].

Many stresses encountered by bacterial pathogens inside their hosts from nutrient limitation to oxidative stress imposed by immune cells and to exposure to antimicrobial agents have been linked to the evolution of more resilient pathogens. Increased resilience results from the protective changes in bacterial physiology, activation of stress-response mechanisms, promotion of resistant lifestyles such as biofilms, and induction of resistant mutations [147]. In addition, the stress imposed by the host can activate MGEs in pathogenic bacteria, which may increase their mobility. Re-assortment of virulence genes and acquisition of new ones can lead to pathogens with new features. For example, a scarlet fever outbreak was associated with acquisition by *Streptococcus pyogenes* of transposable elements encoding multidrug resistance and of a phage encoding the superantigen SSA [148]. For genes encoded by prophages, gene transfer depends on prophage induction, and since different prophages respond differentially to various stresses, pathogens with unique features can emerge [149]. Antibiotic usage also may provoke prophage induction and promote gene dissemination among isolates. For example, antibiotic-induced SOS response was shown to increase the horizontal dissemination of pathogenicity island-encoded virulence factors in *Staphylococc*i [150]. SOS response was reported to induce recombination of an integron that captures and spreads antibiotic resistance genes among Gram-negative pathogens [151].

Recent data suggest that stress promotes the selection of the fittest viral mutants [152]. The mechanism behind this phenomenon may be the phenotypic heterogeneity of the viral population. A study of a human gut virome shows that temperate phages have lower mutation rates in comparison to the lytic phages *Microviridae*, whose substitution rates are so high that in 2.5 years a new viral species could emerge [153]. Altogether, these data illustrate the centrality of MGEs and stress in the evolution of prokaryotes' pathogens and immunity.

CONCLUDING REMARKS

Immunodefense and stress response pathways are tightly interconnected in prokaryotes. Stress decreases their immune fitness by inducing movement of various mobile genetic elements and activating lytic gene expression in proviruses. On the other hand, stress promotes the evolution of prokaryotes by inducing mutagenesis, which may result in improved immunity. Envelope stress and metabolic imbalance caused by infection can be sensed by immune mechanisms and activate them preventively. In such cases stress response potentiates immunity.

CONSENT FOR PUBLICATION

Not applicable.

CONFLICT OF INTEREST

The author confirms that this chapter contents have no conflict of interest.

ACKNOWLEDGEMENTS

Declared none.

REFERENCES

[1] Quax TE, Voet M, Sismeiro O, *et al.* Massive activation of archaeal defense genes during viral infection. J Virol 2013; 87(15): 8419-28.
[http://dx.doi.org/10.1128/JVI.01020-13] [PMID: 23698312]

[2] Brathwaite KJ, Siringan P, Connerton PL, Connerton IF. Host adaption to the bacteriophage carrier state of Campylobacter jejuni. Res Microbiol 2015; 166(6): 504-15.
[http://dx.doi.org/10.1016/j.resmic.2015.05.003] [PMID: 26004283]

[3] Younis S, Kamel W, Falkeborn T, *et al.* Multiple nuclear-replicating viruses require the stress-induced protein ZC3H11A for efficient growth. Proc Natl Acad Sci USA 2018; 115(16): E3808-16.
[http://dx.doi.org/10.1073/pnas.1722333115] [PMID: 29610341]

[4] Ameh EM, Tyrrel S, Harris J, Ignatiou A, Orlova E, Nocker A. The absence or presence of a lytic coliphage affects the response of *Escherichia coli* to heat, chlorine, or UV exposure. Folia Microbiol (Praha) 2018; 63(5): 599-606.
[http://dx.doi.org/10.1007/s12223-018-0600-9] [PMID: 29546616]

[5] Sacher JC, Flint A, Butcher J, *et al.* Transcriptomic Analysis of the *Campylobacter jejuni* Response to T4-Like Phage NCTC 12673 Infection. Viruses 2018; 10(6)E332
[http://dx.doi.org/10.3390/v10060332] [PMID: 29914170]

[6] Gutierrez A, Jain S, Bhargava P, Hamblin M, Lobritz MA, Collins JJ. Understanding and Sensitizing Density-Dependent Persistence to Quinolone Antibiotics. Mol Cell 2017; 68(6): 1147-1154.e3.
[http://dx.doi.org/10.1016/j.molcel.2017.11.012] [PMID: 29225037]

[7] Kutter E, Bryan D, Ray G, Brewster E, Blasdel B, Guttman B. From Host to Phage Metabolism: Hot Tales of Phage T4's Takeover of *E. coli*. Viruses 2018; 10(7)E387
[http://dx.doi.org/10.3390/v10070387] [PMID: 30037085]

[8] Huang Q, Luo H, Liu M, *et al.* The effect of Mycobacterium tuberculosis CRISPR-associated Cas2 (Rv2816c) on stress response genes expression, morphology and macrophage survival of Mycobacterium smegmatis. Infect Genet Evol 2016; 40: 295-301.
[http://dx.doi.org/10.1016/j.meegid.2015.10.019] [PMID: 26498723]

[9] Perez-Rodriguez R, Haitjema C, Huang Q, *et al.* Envelope stress is a trigger of CRISPR RNA-mediated DNA silencing in *Escherichia coli*. Mol Microbiol 2011; 79(3): 584-99.
[http://dx.doi.org/10.1111/j.1365-2958.2010.07482.x] [PMID: 21255106]

[10] Fernández L, Rodríguez A, García P. Phage or foe: an insight into the impact of viral predation on microbial communities. ISME J 2018; 12(5): 1171-9.
[http://dx.doi.org/10.1038/s41396-018-0049-5] [PMID: 29371652]

[11] Binnenkade L, Teichmann L, Thormann KM. Iron triggers λSo prophage induction and release of extracellular DNA in Shewanella oneidensis MR-1 biofilms. Appl Environ Microbiol 2014; 80(17): 5304-16.
[http://dx.doi.org/10.1128/AEM.01480-14] [PMID: 24951794]

[12] De Paepe M, Tournier L, Moncaut E, Son O, Langella P, Petit MA. Carriage of λ Latent Virus Is Costly for Its Bacterial Host due to Frequent Reactivation in Monoxenic Mouse Intestine. PLoS Genet 2016; 12(2)e1005861
[http://dx.doi.org/10.1371/journal.pgen.1005861] [PMID: 26871586]

[13] Górski A, Jończyk-Matysiak E, Międzybrodzki R, *et al.* Phage Therapy: Beyond Antibacterial Action. Front Med (Lausanne) 2018; 5: 146.
[http://dx.doi.org/10.3389/fmed.2018.00146] [PMID: 29876350]

[14] Thomason LC, Court DL. Evidence that bacteriophage λ lysogens may induce in response to the proton motive force uncoupler CCCP. FEMS Microbiol Lett 2016; 363(3)fnv244
[http://dx.doi.org/10.1093/femsle/fnv244] [PMID: 26705574]

[15] Kreuzer KN. DNA damage responses in prokaryotes: regulating gene expression, modulating growth patterns, and manipulating replication forks. Cold Spring Harb Perspect Biol 2013; 5(11)a012674
[http://dx.doi.org/10.1101/cshperspect.a012674] [PMID: 24097899]

[16] Roberts JW, Roberts CW, Craig NL. *Escherichia coli* recA gene product inactivates phage lambda repressor. Proc Natl Acad Sci USA 1978; 75(10): 4714-8.
[http://dx.doi.org/10.1073/pnas.75.10.4714] [PMID: 368796]

[17] Fornelos N, Browning DF, Butala M. The Use and Abuse of LexA by Mobile Genetic Elements. Trends Microbiol 2016; 24(5): 391-401.
[http://dx.doi.org/10.1016/j.tim.2016.02.009] [PMID: 26970840]

[18] Quinones M, Kimsey HH, Waldor MK. LexA cleavage is required for CTX prophage induction. Mol Cell 2005; 17(2): 291-300.

[http://dx.doi.org/10.1016/j.molcel.2004.11.046] [PMID: 15664197]

[19] Fornelos N, Browning DF, Pavlin A, *et al.* Lytic gene expression in the temperate bacteriophage GIL01 is activated by a phage-encoded LexA homologue. Nucleic Acids Res 2018; 46(18): 9432-43. [http://dx.doi.org/10.1093/nar/gky646] [PMID: 30053203]

[20] Lemire S, Figueroa-Bossi N, Bossi L. Bacteriophage crosstalk: coordination of prophage induction by trans-acting antirepressors. PLoS Genet 2011; 7(6)e1002149 [http://dx.doi.org/10.1371/journal.pgen.1002149] [PMID: 21731505]

[21] Michalowski CB, Little JW. Role of cis-acting sites in stimulation of the phage λ P(RM) promoter by CI-mediated looping. J Bacteriol 2013; 195(15): 3401-11. [http://dx.doi.org/10.1128/JB.02148-12] [PMID: 23708136]

[22] Ding Y, Manzo C, Fulcrand G, Leng F, Dunlap D, Finzi L. DNA supercoiling: a regulatory signal for the λ repressor. Proc Natl Acad Sci USA 2014; 111(43): 15402-7. [http://dx.doi.org/10.1073/pnas.1320644111] [PMID: 25319264]

[23] Yan Y, Leng F, Finzi L, Dunlap D. Protein-mediated looping of DNA under tension requires supercoiling. Nucleic Acids Res 2018; 46(5): 2370-9. [http://dx.doi.org/10.1093/nar/gky021] [PMID: 29365152]

[24] Choi J, Kotay SM, Goel R. Various physico-chemical stress factors cause prophage induction in Nitrosospira multiformis 25196--an ammonia oxidizing bacteria. Water Res 2010; 44(15): 4550-8. [http://dx.doi.org/10.1016/j.watres.2010.04.040] [PMID: 20630557]

[25] Nanda AM, Thormann K, Frunzke J. Impact of spontaneous prophage induction on the fitness of bacterial populations and host-microbe interactions. J Bacteriol 2015; 197(3): 410-9. [http://dx.doi.org/10.1128/JB.02230-14] [PMID: 25404701]

[26] Miyazaki R, Minoia M, Pradervand N, Sulser S, Reinhard F, van der Meer JR. Cellular variability of RpoS expression underlies subpopulation activation of an integrative and conjugative element. PLoS Genet 2012; 8(7)e1002818 [http://dx.doi.org/10.1371/journal.pgen.1002818] [PMID: 22807690]

[27] Chakraborty D, Clark E, Mauro SA, Koudelka GB. Molecular Mechanisms Governing "Hair-Trigger" Induction of Shiga Toxin-Encoding Prophages. Viruses 2018; 10(5)E228 [http://dx.doi.org/10.3390/v10050228] [PMID: 29710828]

[28] Ronayne EA, Wan YC, Boudreau BA, Landick R, Cox MM. P1 Ref Endonuclease: A Molecular Mechanism for Phage-Enhanced Antibiotic Lethality. PLoS Genet 2016; 12(1)e1005797 [http://dx.doi.org/10.1371/journal.pgen.1005797] [PMID: 26765929]

[29] Ajon M, Fröls S, van Wolferen M, *et al.* UV-inducible DNA exchange in hyperthermophilic archaea mediated by type IV pili. Mol Microbiol 2011; 82(4): 807-17. [http://dx.doi.org/10.1111/j.1365-2958.2011.07861.x] [PMID: 21999488]

[30] Soucy SM, Huang J, Gogarten JP. Horizontal gene transfer: building the web of life. Nat Rev Genet 2015; 16(8): 472-82. [http://dx.doi.org/10.1038/nrg3962] [PMID: 26184597]

[31] Kuchinski KS, Brimacombe CA, Westbye AB, Ding H, Beatty JT. The SOS Response Master Regulator LexA Regulates the Gene Transfer Agent of Rhodobacter capsulatus and Represses Transcription of the Signal Transduction Protein CckA. J Bacteriol 2016; 198(7): 1137-48. [http://dx.doi.org/10.1128/JB.00839-15] [PMID: 26833411]

[32] Diard M, Bakkeren E, Cornuault JK, *et al.* Inflammation boosts bacteriophage transfer between *Salmonella* spp. Science 2017; 355(6330): 1211-5. [http://dx.doi.org/10.1126/science.aaf8451] [PMID: 28302859]

[33] Rossmann FS, Racek T, Wobser D, *et al.* Phage-mediated dispersal of biofilm and distribution of bacterial virulence genes is induced by quorum sensing. PLoS Pathog 2015; 11(2)e1004653 [http://dx.doi.org/10.1371/journal.ppat.1004653] [PMID: 25706310]

[34] Lopatkin AJ, Meredith HR, Srimani JK, Pfeiffer C, Durrett R, You L. Persistence and reversal of plasmid-mediated antibiotic resistance. Nat Commun 2017; 8(1): 1689. [http://dx.doi.org/10.1038/s41467-017-01532-1] [PMID: 29162798]

[35] Roossinck MJ. Move over, bacteria! Viruses make their mark as mutualistic microbial symbionts. J Virol 2015; 89(13): 6532-5. [http://dx.doi.org/10.1128/JVI.02974-14] [PMID: 25903335]

[36] Obeng N, Pratama AA, Elsas JDV. The Significance of Mutualistic Phages for Bacterial Ecology and Evolution. Trends Microbiol 2016; 24(6): 440-9. [http://dx.doi.org/10.1016/j.tim.2015.12.009] [PMID: 26826796]

[37] Shapiro JW, Williams ES, Turner PE. Evolution of parasitism and mutualism between filamentous phage M13 and *Escherichia coli*. PeerJ 2016; 4e2060 [http://dx.doi.org/10.7717/peerj.2060] [PMID: 27257543]

[38] Tsao YF, Taylor VL, Kala S, *et al.* Phage morons play important roles in controlling Pseudomonas aeruginosa phenotypes. J Bacteriol 2018; 200(22): e00189-18. [http://dx.doi.org/10.1128/JB.00189-18] [PMID: 30150232]

[39] Edlin G, Lin L, Kudrna R. Lambda lysogens of *E. coli* reproduce more rapidly than non-lysogens. Nature 1975; 255(5511): 735-7. [http://dx.doi.org/10.1038/255735a0] [PMID: 1094307]

[40] Adams-Sapper S, Gayoso A, Riley LW. Stress-Adaptive Responses Associated with High-Level Carbapenem Resistance in KPC-Producing *Klebsiella pneumoniae.* J Pathogens 2018; 20183028290 [http://dx.doi.org/10.1155/2018/3028290] [PMID: 29657865]

[41] Fernández L, González S, Quiles-Puchalt N, *et al.* Lysogenization of Staphylococcus aureus RN450 by phages ϕ11 and ϕ80α leads to the activation of the SigB regulon. Sci Rep 2018; 8(1): 12662. [http://dx.doi.org/10.1038/s41598-018-31107-z] [PMID: 30139986]

[42] Deutsch DR, Utter B, Verratti KJ, Sichtig H, Tallon LJ, Fischetti VA. Extra-Chromosomal DNA Sequencing Reveals Episomal Prophages Capable of Impacting Virulence Factor Expression in *Staphylococcus aureus.* Front Microbiol 2018; 9: 1406. [http://dx.doi.org/10.3389/fmicb.2018.01406] [PMID: 30013526]

[43] Yang J, Russell TW, Hocking DM, *et al.* Control of acid resistance pathways of enterohemorrhagic *Escherichia coli* strain EDL933 by PsrB, a prophage-encoded AraC-like regulator. Infect Immun 2015; 83(1): 346-53. [http://dx.doi.org/10.1128/IAI.02758-14] [PMID: 25368119]

[44] Lopez CA, Winter SE, Rivera-Chávez F, *et al.* Phage-mediated acquisition of a type III secreted effector protein boosts growth of salmonella by nitrate respiration. MBio 2012; 3(3)e00143-12 [http://dx.doi.org/10.1128/mBio.00143-12] [PMID: 22691391]

[45] Sekulovic O, Fortier LC. Global transcriptional response of Clostridium difficile carrying the CD38 prophage. Appl Environ Microbiol 2015; 81(4): 1364-74. [http://dx.doi.org/10.1128/AEM.03656-14] [PMID: 25501487]

[46] Wang X, Kim Y, Ma Q, *et al.* Cryptic prophages help bacteria cope with adverse environments. Nat Commun 2010; 1: 147. [http://dx.doi.org/10.1038/ncomms1146] [PMID: 21266997]

[47] Noguchi Y, Katayama T. The *Escherichia coli* Cryptic Prophage Protein YfdR Binds to DnaA and Initiation of Chromosomal Replication Is Inhibited by Overexpression of the Gene Cluster yfdQ-yfd--yfdS-yfdT. Front Microbiol 2016; 7: 239. [http://dx.doi.org/10.3389/fmicb.2016.00239] [PMID: 26973617]

[48] Bondy-Denomy J, Qian J, Westra ER, *et al.* Prophages mediate defense against phage infection through diverse mechanisms. ISME J 2016; 10(12): 2854-66. [http://dx.doi.org/10.1038/ismej.2016.79] [PMID: 27258950]

[49] Gödeke J, Paul K, Lassak J, Thormann KM. Phage-induced lysis enhances biofilm formation in Shewanella oneidensis MR-1. ISME J 2011; 5(4): 613-26. [http://dx.doi.org/10.1038/ismej.2010.153] [PMID: 20962878]

[50] Secor PR, Michaels LA, Smigiel KS, *et al.* Filamentous Bacteriophage Produced by Pseudomonas aeruginosa Alters the Inflammatory Response and Promotes Noninvasive Infection *in vivo*. Infect Immun 2016; 85(1): e00648-16. [PMID: 27795361]

[51] Bille E, Meyer J, Jamet A, *et al.* A virulence-associated filamentous bacteriophage of Neisseria meningitidis increases host-cell colonisation. PLoS Pathog 2017; 13(7)e1006495 [http://dx.doi.org/10.1371/journal.ppat.1006495] [PMID: 28704569]

[52] Whitchurch CB, Tolker-Nielsen T, Ragas PC, Mattick JS. Extracellular DNA required for bacterial biofilm formation. Science 2002; 295(5559): 1487. [http://dx.doi.org/10.1126/science.295.5559.1487] [PMID: 11859186]

[53] Price VJ, Huo W, Sharifi A, Palmer KL. CRISPR-Cas and Restriction-Modification Act Additively against Conjugative Antibiotic Resistance Plasmid Transfer in Enterococcus faecalis. MSphere 2016; 1(3): e00064-16. [http://dx.doi.org/10.1128/mSphere.00064-16] [PMID: 27303749]

[54] Goldberg GW, Jiang W, Bikard D, Marraffini LA. Conditional tolerance of temperate phages *via* transcription-dependent CRISPR-Cas targeting. Nature 2014; 514(7524): 633-7. [http://dx.doi.org/10.1038/nature13637] [PMID: 25174707]

[55] Darwin AJ. Stress relief during host infection: The phage shock protein response supports bacterial virulence in various ways. PLoS Pathog 2013; 9(7)e1003388 [http://dx.doi.org/10.1371/journal.ppat.1003388] [PMID: 23853578]

[56] Labedan B, Letellier L. Membrane potential changes during the first steps of coliphage infection. Proc Natl Acad Sci USA 1981; 78(1): 215-9. [http://dx.doi.org/10.1073/pnas.78.1.215] [PMID: 7017710]

[57] Jakutytė L, Lurz R, Baptista C, *et al.* First steps of bacteriophage SPP1 entry into Bacillus subtilis. Virology 2012; 422(2): 425-34. [http://dx.doi.org/10.1016/j.virol.2011.11.010] [PMID: 22154073]

[58] Bury-Moné S, Nomane Y, Reymond N, *et al.* Global analysis of extracytoplasmic stress signaling in *Escherichia coli*. PLoS Genet 2009; 5(9)e1000651 [http://dx.doi.org/10.1371/journal.pgen.1000651] [PMID: 19763168]

[59] Ades SE, Hayden JD. M.E. L. Envelope stress.Bacterial stress responses. 2nd ed. Washington, DC: ASM press 2011; pp. 115-31.

[60] Delhaye A, Collet JF, Laloux G. Fine-Tuning of the Cpx Envelope Stress Response Is Required for Cell Wall Homeostasis in *Escherichia coli*. MBio 2016; 7(1): e00047-16. [http://dx.doi.org/10.1128/mBio.00047-16] [PMID: 26908573]

[61] Grabowicz M, Silhavy TJ. Envelope Stress Responses: An Interconnected Safety Net. Trends Biochem Sci 2017; 42(3): 232-42. [http://dx.doi.org/10.1016/j.tibs.2016.10.002] [PMID: 27839654]

[62] Brissette JL, Russel M, Weiner L, Model P. Phage shock protein, a stress protein of *Escherichia coli*. Proc Natl Acad Sci USA 1990; 87(3): 862-6. [http://dx.doi.org/10.1073/pnas.87.3.862] [PMID: 2105503]

[63] Flores-Kim J, Darwin AJ. The Phage Shock Protein Response. Annu Rev Microbiol 2016; 70: 83-101. [http://dx.doi.org/10.1146/annurev-micro-102215-095359] [PMID: 27297125]

[64] McDonald C, Jovanovic G, Wallace BA, Ces O, Buck M. Structure and function of PspA and Vipp1 N-terminal peptides: Insights into the membrane stress sensing and mitigation. Biochim Biophys Acta

Biomembr 2017; 1859(1): 28-39.
[http://dx.doi.org/10.1016/j.bbamem.2016.10.018] [PMID: 27806910]

[65] Leskinen K, Blasdel BG, Lavigne R, Skurnik M. RNA-Sequencing Reveals the Progression of Phage-Host Interactions between φR1-37 and Yersinia enterocolitica. Viruses 2016; 8(4): 111.
[http://dx.doi.org/10.3390/v8040111] [PMID: 27110815]

[66] Fallico V, Ross RP, Fitzgerald GF, McAuliffe O. Genetic response to bacteriophage infection in Lactococcus lactis reveals a four-strand approach involving induction of membrane stress proteins, D-alanylation of the cell wall, maintenance of proton motive force, and energy conservation. J Virol 2011; 85(22): 12032-42.
[http://dx.doi.org/10.1128/JVI.00275-11] [PMID: 21880765]

[67] Bernal-Bernal D, Abellón-Ruiz J, Iniesta AA, *et al.* Multifactorial control of the expression of a CRISPR-Cas system by an extracytoplasmic function σ/anti-σ pair and a global regulatory complex. Nucleic Acids Res 2018; 46(13): 6726-45.
[http://dx.doi.org/10.1093/nar/gky475] [PMID: 29893914]

[68] Shmakov SA, Makarova KS, Wolf YI, Severinov KV, Koonin EV. Systematic prediction of genes functionally linked to CRISPR-Cas systems by gene neighborhood analysis. Proc Natl Acad Sci USA 2018; 115(23): E5307-16.
[http://dx.doi.org/10.1073/pnas.1803440115] [PMID: 29784811]

[69] Stratmann T, Pul Ü, Wurm R, Wagner R, Schnetz K. RcsB-BglJ activates the *Escherichia coli* leuO gene, encoding an H-NS antagonist and pleiotropic regulator of virulence determinants. Mol Microbiol 2012; 83(6): 1109-23.
[http://dx.doi.org/10.1111/j.1365-2958.2012.07993.x] [PMID: 22295907]

[70] Pul U, Wurm R, Arslan Z, Geissen R, Hofmann N, Wagner R. Identification and characterization of *E. coli* CRISPR-cas promoters and their silencing by H-NS. Mol Microbiol 2010; 75(6): 1495-512.
[http://dx.doi.org/10.1111/j.1365-2958.2010.07073.x] [PMID: 20132443]

[71] Westra ER, Pul U, Heidrich N, *et al.* H-NS-mediated repression of CRISPR-based immunity in *Escherichia coli* K12 can be relieved by the transcription activator LeuO. Mol Microbiol 2010; 77(6): 1380-93.
[http://dx.doi.org/10.1111/j.1365-2958.2010.07315.x] [PMID: 20659289]

[72] Ratner HK, Sampson TR, Weiss DS. I can see CRISPR now, even when phage are gone: a view on alternative CRISPR-Cas functions from the prokaryotic envelope. Curr Opin Infect Dis 2015; 28(3): 267-74.
[http://dx.doi.org/10.1097/QCO.0000000000000154] [PMID: 25887612]

[73] Samuels DJ, Frye JG, Porwollik S, *et al.* Use of a promiscuous, constitutively-active bacterial enhancer-binding protein to define the σ^{54} (RpoN) regulon of Salmonella Typhimurium LT2. BMC Genomics 2013; 14: 602.
[http://dx.doi.org/10.1186/1471-2164-14-602] [PMID: 24007446]

[74] Serbanescu MA, Cordova M, Krastel K, *et al.* Role of the Streptococcus mutans CRISPR-Cas systems in immunity and cell physiology. J Bacteriol 2015; 197(4): 749-61.
[http://dx.doi.org/10.1128/JB.02333-14] [PMID: 25488301]

[75] Yosef I, Goren MG, Kiro R, Edgar R, Qimron U. High-temperature protein G is essential for activity of the *Escherichia coli* clustered regularly interspaced short palindromic repeats (CRISPR)/Cas system. Proc Natl Acad Sci USA 2011; 108(50): 20136-41.
[http://dx.doi.org/10.1073/pnas.1113519108] [PMID: 22114197]

[76] Majsec K, Bolt EL, Ivančić-Baće I. Cas3 is a limiting factor for CRISPR-Cas immunity in *Escherichia coli* cells lacking H-NS. BMC Microbiol 2016; 16: 28.
[http://dx.doi.org/10.1186/s12866-016-0643-5] [PMID: 26956996]

[77] Goodwin CM, Xu S, Munger J. Stealing the Keys to the Kitchen: Viral Manipulation of the Host Cell Metabolic Network. Trends Microbiol 2015; 23(12): 789-98.

[http://dx.doi.org/10.1016/j.tim.2015.08.007] [PMID: 26439298]

[78] De Smet J, Zimmermann M, Kogadeeva M, *et al.* High coverage metabolomics analysis reveals phage-specific alterations to Pseudomonas aeruginosa physiology during infection. ISME J 2016; 10(8): 1823-35.
[http://dx.doi.org/10.1038/ismej.2016.3] [PMID: 26882266]

[79] Patterson AG, Chang JT, Taylor C, Fineran PC. Regulation of the Type I-F CRISPR-Cas system by CRP-cAMP and GalM controls spacer acquisition and interference. Nucleic Acids Res 2015; 43(12): 6038-48.
[http://dx.doi.org/10.1093/nar/gkv517] [PMID: 26007654]

[80] Patterson AG, Yevstigneyeva MS, Fineran PC. Regulation of CRISPR-Cas adaptive immune systems. Curr Opin Microbiol 2017; 37: 1-7.
[http://dx.doi.org/10.1016/j.mib.2017.02.004] [PMID: 28359988]

[81] Agari Y, Sakamoto K, Tamakoshi M, Oshima T, Kuramitsu S, Shinkai A. Transcription profile of Thermus thermophilus CRISPR systems after phage infection. J Mol Biol 2010; 395(2): 270-81.
[http://dx.doi.org/10.1016/j.jmb.2009.10.057] [PMID: 19891975]

[82] Majumder A, Fang M, Tsai KJ, Ueguchi C, Mizuno T, Wu HY. LeuO expression in response to starvation for branched-chain amino acids. J Biol Chem 2001; 276(22): 19046-51.
[http://dx.doi.org/10.1074/jbc.M100945200] [PMID: 11376008]

[83] Blakely GW, Murray NE. Control of the endonuclease activity of type I restriction-modification systems is required to maintain chromosome integrity following homologous recombination. Mol Microbiol 2006; 60(4): 883-93.
[http://dx.doi.org/10.1111/j.1365-2958.2006.05144.x] [PMID: 16677300]

[84] Thoms B, Wackernagel W. UV-induced alleviation of lambda restriction in *Escherichia coli* K-12: kinetics of induction and specificity of this SOS function. Molecular & general genetics. MGG 1982; 186(1): 111-7.
[PMID: 6213835]

[85] Kelleher JE, Raleigh EA. Response to UV damage by four *Escherichia coli* K-12 restriction systems. J Bacteriol 1994; 176(19): 5888-96.
[http://dx.doi.org/10.1128/jb.176.19.5888-5896.1994] [PMID: 7928948]

[86] Rezulak M, Borsuk I, Mruk I. Natural C-independent expression of restriction endonuclease in a C protein-associated restriction-modification system. Nucleic Acids Res 2016; 44(6): 2646-60.
[http://dx.doi.org/10.1093/nar/gkv1331] [PMID: 26656489]

[87] Goryanin II, Kudryavtseva AA, Balabanov VP, Biryukova VS, Manukhov IV, Zavilgelsky GB. Antirestriction activities of KlcA (RP4) and ArdB (R64) proteins. FEMS Microbiol Lett 2018; 365(23)
[http://dx.doi.org/10.1093/femsle/fny227] [PMID: 30239714]

[88] Keatch SA, Su TJ, Dryden DT. Alleviation of restriction by DNA condensation and non-specific DNA binding ligands. Nucleic Acids Res 2004; 32(19): 5841-50.
[http://dx.doi.org/10.1093/nar/gkh918] [PMID: 15520467]

[89] Simons M, Diffin FM, Szczelkun MD. ClpXP protease targets long-lived DNA translocation states of a helicase-like motor to cause restriction alleviation. Nucleic Acids Res 2014; 42(19): 12082-91.
[http://dx.doi.org/10.1093/nar/gku851] [PMID: 25260590]

[90] Morozova N, Sabantsev A, Bogdanova E, *et al.* Temporal dynamics of methyltransferase and restriction endonuclease accumulation in individual cells after introducing a restriction-modification system. Nucleic Acids Res 2016; 44(2): 790-800.
[http://dx.doi.org/10.1093/nar/gkv1490] [PMID: 26687717]

[91] Bourges AC, Torres Montaguth OE, Ghosh A, *et al.* High pressure activation of the Mrr restriction endonuclease in *Escherichia coli* involves tetramer dissociation. Nucleic Acids Res 2017; 45(9): 5323-32.

[http://dx.doi.org/10.1093/nar/gkx192] [PMID: 28369499]

[92] Pleška M, Qian L, Okura R, *et al.* Bacterial Autoimmunity Due to a Restriction-Modification System. Curr Biol 2016; 26(3): 404-9.
[http://dx.doi.org/10.1016/j.cub.2015.12.041] [PMID: 26804559]

[93] Hoffmann C, Finsel I, Otto A, *et al.* Functional analysis of novel Rab GTPases identified in the proteome of purified Legionella-containing vacuoles from macrophages. Cell Microbiol 2014; 16(7): 1034-52.
[PMID: 24373249]

[94] Touchon M, Moura de Sousa JA, Rocha EP. Embracing the enemy: the diversification of microbial gene repertoires by phage-mediated horizontal gene transfer. Curr Opin Microbiol 2017; 38: 66-73.
[http://dx.doi.org/10.1016/j.mib.2017.04.010] [PMID: 28527384]

[95] Waldor MK, Mekalanos JJ. Lysogenic conversion by a filamentous phage encoding cholera toxin. Science 1996; 272(5270): 1910-4.
[http://dx.doi.org/10.1126/science.272.5270.1910] [PMID: 8658163]

[96] O'Neal CJ, Jobling MG, Holmes RK, Hol WG. Structural basis for the activation of cholera toxin by human ARF6-GTP. Science 2005; 309(5737): 1093-6.
[http://dx.doi.org/10.1126/science.1113398] [PMID: 16099990]

[97] Oh YT, Park Y, Yoon MY, *et al.* Cholera toxin production during anaerobic trimethylamine N-oxide respiration is mediated by stringent response in Vibrio cholerae. J Biol Chem 2014; 289(19): 13232-42.
[http://dx.doi.org/10.1074/jbc.M113.540088] [PMID: 24648517]

[98] Tyler JS, Beeri K, Reynolds JL, *et al.* Prophage induction is enhanced and required for renal disease and lethality in an EHEC mouse model. PLoS Pathog 2013; 9(3)e1003236
[http://dx.doi.org/10.1371/journal.ppat.1003236] [PMID: 23555250]

[99] Licznerska K, Dydecka A, Bloch S, *et al.* The Role of the Exo-Xis Region in Oxidative Stress-Mediated Induction of Shiga Toxin-Converting Prophages. Oxid Med Cell Longev 2016; 20168453135
[http://dx.doi.org/10.1155/2016/8453135] [PMID: 26798427]

[100] Fang Y, Mercer RG, McMullen LM, Gänzle MG. Induction of Shiga Toxin-Encoding Prophage by Abiotic Environmental Stress in Food. Appl Environ Microbiol 2017; 83(19): e01378-17.
[http://dx.doi.org/10.1128/AEM.01378-17] [PMID: 28778890]

[101] Łoś JM, Łoś M, Węgrzyn A, Węgrzyn G. Altruism of Shiga toxin-producing *Escherichia coli*: recent hypothesis *versus* experimental results. Front Cell Infect Microbiol 2013; 2: 166.
[http://dx.doi.org/10.3389/fcimb.2012.00166] [PMID: 23316482]

[102] Novick RP, Christie GE, Penadés JR. The phage-related chromosomal islands of Gram-positive bacteria. Nat Rev Microbiol 2010; 8(8): 541-51.
[http://dx.doi.org/10.1038/nrmicro2393] [PMID: 20634809]

[103] Carroll AC, Wong A. Plasmid persistence: costs, benefits, and the plasmid paradox. Can J Microbiol 2018; 64(5): 293-304.
[http://dx.doi.org/10.1139/cjm-2017-0609] [PMID: 29562144]

[104] Hryckowian AJ, Welch RA. RpoS contributes to phagocyte oxidase-mediated stress resistance during urinary tract infection by *Escherichia coli* CFT073. MBio 2013; 4(1): e00023-13.
[http://dx.doi.org/10.1128/mBio.00023-13] [PMID: 23404396]

[105] Pförtner H, Burian MS, Michalik S, *et al.* Activation of the alternative sigma factor SigB of Staphylococcus aureus following internalization by epithelial cells - an *in vivo* proteomics perspective. Int J Med Microbiol 2014; 304(2): 177-87.
[http://dx.doi.org/10.1016/j.ijmm.2013.11.014] [PMID: 24480029]

[106] Tuchscherr L, Bischoff M, Lattar SM, *et al.* Sigma Factor SigB Is Crucial to Mediate Staphylococcus

aureus Adaptation during Chronic Infections. PLoS Pathog 2015; 11(4)e1004870 [http://dx.doi.org/10.1371/journal.ppat.1004870] [PMID: 25923704]

[107] Thomas DC. The phagocyte respiratory burst: Historical perspectives and recent advances. Immunol Lett 2017; 192: 88-96. [http://dx.doi.org/10.1016/j.imlet.2017.08.016] [PMID: 28864335]

[108] Arnold DE, Heimall JR. A Review of Chronic Granulomatous Disease. Adv Ther 2017; 34(12): 2543-57. [http://dx.doi.org/10.1007/s12325-017-0636-2] [PMID: 29168144]

[109] Reniere ML. Reduce, Induce, Thrive: Bacterial Redox Sensing during Pathogenesis. J Bacteriol 2018; 200(17): e00128-18. [http://dx.doi.org/10.1128/JB.00128-18] [PMID: 29891640]

[110] Reniere ML, Whiteley AT, Portnoy DA. An *in vivo* Selection Identifies Listeria monocytogenes Genes Required to Sense the Intracellular Environment and Activate Virulence Factor Expression. PLoS Pathog 2016; 12(7)e1005741 [http://dx.doi.org/10.1371/journal.ppat.1005741] [PMID: 27414028]

[111] López C, Checa SK, Soncini FC. CpxR/CpxA Controls *scsABCD* Transcription To Counteract Copper and Oxidative Stress in Salmonella enterica Serovar Typhimurium. J Bacteriol 2018; 200(16): e00126-18. [http://dx.doi.org/10.1128/JB.00126-18] [PMID: 29866803]

[112] Matter LB, Ares MA, Abundes-Gallegos J, *et al.* The CpxRA stress response system regulates virulence features of avian pathogenic *Escherichia coli.* Environ Microbiol 2018; 20(9): 3363-77. [http://dx.doi.org/10.1111/1462-2920.14368] [PMID: 30062827]

[113] Chiok KL, Paul NC, Adekanmbi EO, Srivastava SK, Shah DH. Dimethyl adenosine transferase (KsgA) contributes to cell-envelope fitness in Salmonella Enteritidis. Microbiol Res 2018; 216: 108-19. [http://dx.doi.org/10.1016/j.micres.2018.08.009] [PMID: 30269850]

[114] Porcheron G, Dozois CM. Interplay between iron homeostasis and virulence: Fur and RyhB as major regulators of bacterial pathogenicity. Vet Microbiol 2015; 179(1-2): 2-14. [http://dx.doi.org/10.1016/j.vetmic.2015.03.024] [PMID: 25888312]

[115] Garcia YM, Barwinska-Sendra A, Tarrant E, Skaar EP, Waldron KJ, Kehl-Fie TE. A Superoxide Dismutase Capable of Functioning with Iron or Manganese Promotes the Resistance of Staphylococcus aureus to Calprotectin and Nutritional Immunity. PLoS Pathog 2017; 13(1)e1006125 [http://dx.doi.org/10.1371/journal.ppat.1006125] [PMID: 28103306]

[116] Crawford MA, Henard CA, Tapscott T, Porwollik S, McClelland M, Vázquez-Torres A. DksA-Dependent Transcriptional Regulation in Salmonella Experiencing Nitrosative Stress. Front Microbiol 2016; 7: 444. [http://dx.doi.org/10.3389/fmicb.2016.00444] [PMID: 27065993]

[117] Chernov AV, Reyes L, Peterson S, Strongin AY. Depletion of CG-Specific Methylation in Mycoplasma hyorhinis Genomic DNA after Host Cell Invasion. PLoS One 2015; 10(11)e0142529 [http://dx.doi.org/10.1371/journal.pone.0142529] [PMID: 26544880]

[118] Mann R, Mediati DG, Duggin IG, Harry EJ, Bottomley AL. Metabolic Adaptations of Uropathogenic *E. coli* in the Urinary Tract. Front Cell Infect Microbiol 2017; 7: 241. [http://dx.doi.org/10.3389/fcimb.2017.00241] [PMID: 28642845]

[119] Holm A, Vikström E. Quorum sensing communication between bacteria and human cells: signals, targets, and functions. Front Plant Sci 2014; 5: 309. [http://dx.doi.org/10.3389/fpls.2014.00309] [PMID: 25018766]

[120] Koskella B, Brockhurst MA. Bacteria-phage coevolution as a driver of ecological and evolutionary processes in microbial communities. FEMS Microbiol Rev 2014; 38(5): 916-31.

[http://dx.doi.org/10.1111/1574-6976.12072] [PMID: 24617569]

[121] Szathmáry E. Toward major evolutionary transitions theory 2.0. Proc Natl Acad Sci USA 2015; 112(33): 10104-11.
[http://dx.doi.org/10.1073/pnas.1421398112] [PMID: 25838283]

[122] Koonin EV. Viruses and mobile elements as drivers of evolutionary transitions. Philos Trans R Soc Lond B Biol Sci 2016; 371(1701): 20150442..
[http://dx.doi.org/10.1098/rstb.2015.0442]

[123] Betts A, Gifford DR, MacLean RC, King KC. Parasite diversity drives rapid host dynamics and evolution of resistance in a bacteria-phage system. Evolution 2016; 70(5): 969-78.
[http://dx.doi.org/10.1111/evo.12909] [PMID: 27005577]

[124] Gómez P, Buckling A. Bacteria-phage antagonistic coevolution in soil. Science 2011; 332(6025): 106-9.
[http://dx.doi.org/10.1126/science.1198767] [PMID: 21454789]

[125] Ogilvie LA, Jones BV. The human gut virome: a multifaceted majority. Front Microbiol 2015; 6: 918.
[http://dx.doi.org/10.3389/fmicb.2015.00918] [PMID: 26441861]

[126] Chopin MC, Chopin A, Bidnenko E. Phage abortive infection in lactococci: variations on a theme. Curr Opin Microbiol 2005; 8(4): 473-9.
[http://dx.doi.org/10.1016/j.mib.2005.06.006] [PMID: 15979388]

[127] Iranzo J, Lobkovsky AE, Wolf YI, Koonin EV. Virus-host arms race at the joint origin of multicellularity and programmed cell death. Cell Cycle 2014; 13(19): 3083-8.
[http://dx.doi.org/10.4161/15384101.2014.949496] [PMID: 25486567]

[128] Koonin EV, Zhang F. Coupling immunity and programmed cell suicide in prokaryotes: Life-or-death choices. BioEssays 2017; 39(1): 1-9.
[http://dx.doi.org/10.1002/bies.201600186] [PMID: 27896818]

[129] Nagamalleswari E, Rao S, Vasu K, Nagaraja V. Restriction endonuclease triggered bacterial apoptosis as a mechanism for long time survival. Nucleic Acids Res 2017; 45(14): 8423-34.
[http://dx.doi.org/10.1093/nar/gkx576] [PMID: 28854737]

[130] Díaz-Orejas R, Espinosa M, Yeo CC. The Importance of the Expendable: Toxin-Antitoxin Genes in Plasmids and Chromosomes. Front Microbiol 2017; 8: 1479.
[http://dx.doi.org/10.3389/fmicb.2017.01479] [PMID: 28824602]

[131] Hõrak R, Tamman H. Desperate times call for desperate measures: benefits and costs of toxin-antitoxin systems. Curr Genet 2017; 63(1): 69-74.
[http://dx.doi.org/10.1007/s00294-016-0622-2] [PMID: 27276988]

[132] Otsuka Y. Prokaryotic toxin-antitoxin systems: novel regulations of the toxins. Curr Genet 2016; 62(2): 379-82.
[http://dx.doi.org/10.1007/s00294-015-0557-z] [PMID: 26780368]

[133] Tamman H, Ainelo A, Tagel M, Hõrak R. Stability of the GraA Antitoxin Depends on Growth Phase, ATP Level, and Global Regulator MexT. J Bacteriol 2015; 198(5): 787-96.
[http://dx.doi.org/10.1128/JB.00684-15] [PMID: 26668267]

[134] Nevo E. Evolution under environmental stress at macro- and microscales. Genome Biol Evol 2011; 3: 1039-52.
[http://dx.doi.org/10.1093/gbe/evr052] [PMID: 21979157]

[135] Rodríguez-Beltrán J, Tourret J, Tenaillon O, *et al.* High Recombinant Frequency in Extraintestinal Pathogenic *Escherichia coli* Strains. Mol Biol Evol 2015; 32(7): 1708-16.
[http://dx.doi.org/10.1093/molbev/msv072] [PMID: 25804522]

[136] Al Mamun AA, Lombardo MJ, Shee C, *et al.* Identity and function of a large gene network underlying mutagenic repair of DNA breaks. Science 2012; 338(6112): 1344-8.

[http://dx.doi.org/10.1126/science.1226683] [PMID: 23224554]

[137] Barreto B, Rogers E, Xia J, *et al.* Small RNA GcvB promotes mutagenic break repair by opposing the membrane stress response. J Bacteriol 2016; 198(24): 3296-308. [http://dx.doi.org/10.1128/JB.00555-16] [PMID: 27698081]

[138] Jee J, Rasouly A, Shamovsky I, *et al.* Rates and mechanisms of bacterial mutagenesis from maximum-depth sequencing. Nature 2016; 534(7609): 693-6. [http://dx.doi.org/10.1038/nature18313] [PMID: 27338792]

[139] Long H, Miller SF, Strauss C, *et al.* Antibiotic treatment enhances the genome-wide mutation rate of target cells. Proc Natl Acad Sci USA 2016; 113(18): E2498-505. [http://dx.doi.org/10.1073/pnas.1601208113] [PMID: 27091991]

[140] Saint-Ruf C, Garfa-Traoré M, Collin V, Cordier C, Franceschi C, Matic I. Massive diversification in aging colonies of *Escherichia coli*. J Bacteriol 2014; 196(17): 3059-73. [http://dx.doi.org/10.1128/JB.01421-13] [PMID: 24982303]

[141] Utrilla J, O'Brien EJ, Chen K, *et al.* Global Rebalancing of Cellular Resources by Pleiotropic Point Mutations Illustrates a Multi-scale Mechanism of Adaptive Evolution. Cell Syst 2016; 2(4): 260-71. [http://dx.doi.org/10.1016/j.cels.2016.04.003] [PMID: 27135538]

[142] Blow MJ, Clark TA, Daum CG, *et al.* The Epigenomic Landscape of Prokaryotes. PLoS Genet 2016; 12(2)e1005854 [http://dx.doi.org/10.1371/journal.pgen.1005854] [PMID: 26870957]

[143] Rey O, Danchin E, Mirouze M, Loot C, Blanchet S. Adaptation to Global Change: A Transposable Element-Epigenetics Perspective. Trends in ecology & evolution 2016; (): -. [http://dx.doi.org/10.1016/j.tree.2016.03.013]

[144] Cohen NR, Ross CA, Jain S, *et al.* A role for the bacterial GATC methylome in antibiotic stress survival. Nat Genet 2016; 48(5): 581-6. [http://dx.doi.org/10.1038/ng.3530] [PMID: 26998690]

[145] Gundlach J, Winter J. Evolution of *Escherichia coli* for maximum HOCl resistance through constitutive expression of the OxyR regulon. Microbiology 2014; 160(Pt 8): 1690-704. [http://dx.doi.org/10.1099/mic.0.074815-0] [PMID: 24899627]

[146] Fedoroff NV. Presidential address. Transposable elements, epigenetics, and genome evolution. Science 2012; 338(6108): 758-67. [http://dx.doi.org/10.1126/science.338.6108.758] [PMID: 23145453]

[147] Poole K. Stress responses as determinants of antimicrobial resistance in Gram-negative bacteria. Trends Microbiol 2012; 20(5): 227-34. [http://dx.doi.org/10.1016/j.tim.2012.02.004] [PMID: 22424589]

[148] Davies MR, Holden MT, Coupland P, *et al.* Emergence of scarlet fever Streptococcus pyogenes emm12 clones in Hong Kong is associated with toxin acquisition and multidrug resistance. Nat Genet 2015; 47(1): 84-7. [http://dx.doi.org/10.1038/ng.3147] [PMID: 25401300]

[149] Garcia-Russell N, Elrod B, Dominguez K. Stress-induced prophage DNA replication in Salmonella enterica serovar Typhimurium. Infect Genet Evol 2009; 9(5): 889-95. [http://dx.doi.org/10.1016/j.meegid.2009.05.017] [PMID: 19501196]

[150] Ubeda C, Maiques E, Knecht E, Lasa I, Novick RP, Penadés JR. Antibiotic-induced SOS response promotes horizontal dissemination of pathogenicity island-encoded virulence factors in staphylococci. Mol Microbiol 2005; 56(3): 836-44. [http://dx.doi.org/10.1111/j.1365-2958.2005.04584.x] [PMID: 15819636]

[151] Guerin E, Cambray G, Sanchez-Alberola N, *et al.* The SOS response controls integron recombination. Science 2009; 324(5930): 1034. [http://dx.doi.org/10.1126/science.1172914] [PMID: 19460999]

[152] Yosef I, Edgar R, Qimron U. Phenotypic heterogeneity in a bacteriophage population only appears as stress-induced mutagenesis. Curr Genet 2016; 62(4): 771-3. [http://dx.doi.org/10.1007/s00294-016-0606-2] [PMID: 27126384]

[153] Minot S, Bryson A, Chehoud C, Wu GD, Lewis JD, Bushman FD. Rapid evolution of the human gut virome. Proc Natl Acad Sci USA 2013; 110(30): 12450-5. [http://dx.doi.org/10.1073/pnas.1300833110] [PMID: 23836644]

CHAPTER 7

Response of Eukaryotes to Stress

Abstract: Eukaryotic stress response involves changes in transcription, translation, and proteostasis. Some changes are similar across many stress responses. Stressors such as misfolded proteins, oxidative stress, viral RNAs, heme deficiency, heat shock, amino acid shortage, and others activate the integrated stress response (ISR) kinases PERK, PKR, HRI, and GCN2; they phosphorylate eIF2α inducing global inhibition of translation. Some stresses, *e.g.* oxidative and osmotic can activate several kinases. Transcriptional response to stress involves the suppression of genes functioning in energy-consuming processes such as protein biosynthesis and activation of genes that mitigate damage caused by stress. Many stresses interfere with the synthesis and folding of proteins. When unfolded protein level exceeds folding and clearance capacity of the cell, the unfolded protein response is triggered, which increases that capacity by producing more chaperones and proteases. Autophagy is another common response to stress that helps to cleanse the cell of aggregated proteins and dysfunctional organelles and replenish the supply of biosynthetic precursors and energy. There are also responses specific to heat, cold, osmotic, pH, oxidative, and mitochondrial stress. They involve activation of programs that mediate adjustment to specific stress.

Keywords: Autophagy, Cold stress, Endoplasmic reticulum stress, General stress response, Heat stress, Integrated stress response, Mitochondrial stress, Osmotic stress, Oxidative stress, PH stress, Unfolded protein response.

INTRODUCTION

The emergence of eukaryotic cells 1.6-2.1 billion years ago was a major evolutionary transition [1]. Eukaryotes are believed to originate from within Archaea and many processes in their cells are functionally similar to the archaeal ones [2 - 4]. Similar to prokaryotes, eukaryotes respond to stress both with a general response that leads to inhibition of cell proliferation and an increase in the repair of macromolecules and with a specific response to a particular stress.

Several features of eukaryotic cells made their stress responses different from prokaryotes. First of all, eukaryotic cells do not use their membranes to produce energy; instead, they generate ATP in the mitochondria or chloroplasts. External factors, therefore, affect their energy production to a lesser degree. Eukaryotic cells are typically much larger than prokaryotic ones with a smaller surface-to-volume ratio, slower metabolisms, and longer generation time.

Nadia Danilova

Compartmentalization of cellular processes is characteristic of eukaryotes, with many internal structures enclosed by membranes, which also affects their stress responses.

Eukaryotic stress response usually involves many levels and includes changes in transcription, translation, protein levels and modifications, and epigenetic modifications [5]. Regulation of stress response becomes especially complex in multicellular organisms, where allostatic adjustment to a stressful environment can involve coordinated changes in the expression of thousands of genes [6]. Stressed cells communicate with other cells, which induce local or systemic responses increasing an organism's resistance [7]. Despite the prosurvival purpose of the stress response, it can be harmful to the organism if excessive or out of control.

7.1. Integrated Stress Response

Eukaryotic cells respond to severe stress with the integrated stress response (ISR), which leads to inhibition of translation of most mRNA [8]. Translation is a complex process that is orchestrated by a set of eukaryotic initiation factors (eIFs). eIF2 is a heterotrimeric GTPase that recruits the initiator methionyl tRNA to the 40S ribosomal subunit. Following the start codon selection and GTP hydrolysis, eIF2-GDP is recycled by the nucleotide exchange factor eIF2B that exchanges GDP back for GTP. Phosphorylation of eIF2α subunit disrupts this process and inhibits the initiation of translation [9]. Eukaryotic cells use eIF2α phosphorylation to inhibit translation during stress. In vertebrates, four eIF2α kinases have been identified that sense distinct stresses (Fig. 7). PERK (PKB-like ER kinase) responds to the presence of misfolded proteins in the endoplasmic reticulum (ER), disturbance of the redox balance, hypoxia, and drop in the energy level. PKR (protein kinase RNA dependent) responds mostly to viral double-stranded RNA but also to oxidative and ribotoxic stress and bacterial infections. HRI (heme-regulated inhibitor kinase) is activated by heme deficiency, heavy metals, high salt, heat shock, and oxidative stress. GCN2 (general control non-derepressible-2) is mostly activated by amino acid shortage through the binding of uncharged tRNA [10]. Some stresses such as oxidative and osmotic can activate several kinases.

GCN2 is conserved in all eukaryotes and the sole eIF2α kinase — in plants. HRI is conserved in fungi but not in plants or protists. PERK is conserved in animals but not in plants and fungi; PKR is only conserved in vertebrates [10].

When stress kinases phosphorylate eIF2α, it inhibits translation initiation at the AUG start codons. However, it increases protein synthesis on some mRNAs such as ATF4 and ER chaperone BiP from the alternative start codons UUG and CUG in their 5′ UTR upstream open reading frames [11]. Increased expression of chaperone BiP helps to restore homeostasis. ATF4 transcriptional factor promotes diverse cell fates from cell survival to programmed cell death, depending on the stress intensity and the cellular context [12]. It is believed that a short-lived ISR is a pro-survival response that helps to restore homeostasis, whereas a prolonged ISR is a cell death inducer.

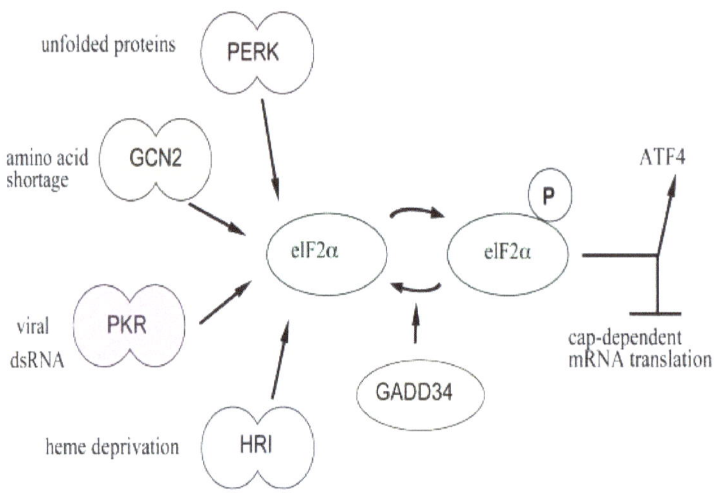

Fig. (7). Integrated stress response is sensed by four kinases. They phosphorylate eIF2α, which prevents translation initiation at the AUG start codons but promotes selective translation of some targets such as ATF4.

Global translational suppression and translational de-repression of selected targets are common adjustments to stress. Activation of all four eIF2α kinases leads to selective translation with some specificity due to phosphorylation of additional targets besides eIF2α [13]. Expression of interferons during PKR-induced translational arrest is achieved through decreasing eIF2α phosphorylation by the induction of protein phosphatase 1 cofactor GADD34 [14]. Several other mechanisms support the selective translation of mRNA during stress. They include internal ribosome entry sites (IRES)-mediated translation, eIF3-dependent cap translation, and other mechanisms [15 - 17]. ATF4 translation is regulated by mRNA methylation at adenosine N(6) [18].

An interesting example of an alternative start codon usage is stress-induced translation of the mitochondrial ribosomal protein L18 from the CUG start codon. The modified form of L18 then incorporates not into the mitochondria but into the cytosolic 80S ribosome [19]. This hybrid ribosome facilitates translation of Hsp70 chaperone protein during stress. Aberrant translation from nucleotide repeats, which is associated with many neurodegenerative diseases can also be increased by stress [20].

ISR-independent pathways can also contribute to the suppression of global translation during stress. For example, in addition to eIF2α phosphorylation through the kinase PERK, cold stress decreases mitochondrial function, thus leading to energy depletion, activation of the AMP-activated protein kinase (AMPK), and inhibition of mTOR signaling, which, in turn, suppresses translation [21].

Infections often lead to ISR. Viral double-stranded RNAs directly activate PKR kinase, whereas other kinases can sense cellular changes induced by pathogens [22, 23].

Stalled translation pre-initiation complexes that are formed during ISR are sequestrated into RNA-protein complexes called stress granules (SGs) [24]. SGs can also be assembled independently of eIF2α phosphorylation in response to a variety of stressors [25]. SGs are composed of mRNA bound to translation initiation factors, proteins that control RNA stability and metabolism, and signaling proteins. From SGs, the mRNA can be moved to processing (P) bodies for destruction [26]. Alternatively, after stress resolution, the translation of these mRNAs can resume.

SG formation prevents apoptosis by seizing some of the apoptosis-promoting regulatory factors. Therefore, SGs promote cell survival during stress. Excessive or multiple stresses disrupt SGs formation resulting in apoptosis. For example, reactive oxygen species such as H_2O_2 prevent SG assembly [27].

7.2. Transcriptional Stress Response

General stress response on the transcriptional level is most studied in yeast species where it is known as the environmental stress response (ESR). In *Saccharomyces cerevisiae,* diverse stresses including oxidative, heat, osmotic, and starvation induce similar changes in expression of approximately 900 genes [28]. Among them, the expression of approximately 600 genes is reduced. About 70% of these genes function in protein biosynthesis, which includes ribosome biogenesis and translation. This is consistent with the fact that protein biosynthesis consumes a majority of resources in actively growing cells. Therefore, the first step in reallocating cellular resources for stress response is to inhibit translation and stop the cell cycle progression [29]. Next, the cell activates the general defense mechanisms by upregulating expression of approximately 300 genes [28]. These genes function in processes that mitigate damage caused by the stress and in an adjustment of the metabolism to the adverse conditions. Common changes include increase in compounds that detoxify ROS, in compounds such as trehalose that protect macromolecules, increase in the levels of chaperones that help to refold denatured proteins, upregulation of proteases that degrade proteins that cannot be re-folded, and stimulation of processes that increase nutrient availability such as glucose transport, glycogen utilization, autophagy, and fatty acids metabolism [28, 30, 31]. The level of gene's expression is also regulated by changes in mRNA's stability and by interactions of mRNAs with regulatory proteins [32].

Yeast's ESR response is a result of cooperation between many transcription factors [33]. They include master regulators, such as Mns2 that bind the stress response element (STRE), a common sequence in the promoters of stress-responsive genes. Master regulators are engaged in response to particular stress only when the stress level is high. They cooperate with regulators that are responsible for specific stresses such as Hog1 that regulates response to osmotic stress. The specific factors respond faster and with a wider range of magnitude. Protein kinase A, MEC and TOR pathways, and multiple other factors are also involved in yeast ESR. The response can be epigenetically inherited through changes in Msn2 localization in the nucleus [34]. ESR response varies between yeast species in dependence on their habitat [35]. ESR is thought to protect homeostasis while the cells are adapting to new environmental conditions [28]. ESR response is associated with the increased mutagenesis induced by proteotoxic stress [36].

In animals, various tissues have distinct specialized cells that have different transcriptional stress response with limited overlap between the cell types [37, 38]. Responses of the same cell types to different stressors show some common features, although in comparison to yeast ESR, fewer genes are involved in response to multiple stressors [6]. In mammalian cells, modifications and compartmentalization of the chromatin along with docking of Pol II at promoters and enhancers of stress response genes allow for a fast, prewired, and directional response to stress [39].

Epigenetic regulation plays an important role in stress response and is also used to create a memory of a stressful environmental condition in order to develop a faster response in the future [40]. Epigenetic mechanisms include DNA methylation, histone modifications, relocation of a gene to generate a prolonged permissive chromatin state, and regulation by long non-coding RNAs (lncRNAs). Manipulation of DNA methylation can lead to improved heat stress response in chicken [41]. Recent studies keep uncovering more ways to regulate transcription. For example, transcription can be regulated not only by Lys acetylation of histones but also by other short-chain Lys acylations. Beta-hydroxybutyrylation is induced during starvation and leads to modulation of gene expression [42]. Stress can also induce post-transcriptional modifications of mRNAs [43]. Introns also may play an important role in the regulation of stress response. Yeast introns are rapidly excised, debranched, and degraded during fast growth. However, during starvation and other stresses, a subset of introns accumulates as linear RNAs that remain associated with spliceosomes [44]. Deletion of the introns negatively affects yeast stress response. Introns repress expression of ribosomal protein genes during starvation, which improves yeast survival [45]. miRNAs biogenesis, localization, and function also can be influenced by stress adding an additional level of regulation [46].

7.3. Endoplasmic Reticulum Stress and Unfolded Protein Response

The endoplasmic reticulum (ER) of eukaryotic cells is a tubular-reticular network separated from the cytosol by a single lipid bilayer. It is a place of synthesis and modifications of secreted and membrane proteins, lipids, steroids, and oligosaccharides. Calcium storage and detoxification of various products also takes place in the ER. The process of protein production is strictly controlled. Only properly modified and folded proteins can exit ER, those that failed quality control are destroyed through ER-associated degradation (ERAD) pathway or autophagy [47]. ERAD promotes translocation of misfolded and unfolded proteins through special channels in the ER membrane into the cytosol where they are degraded by the ubiquitin-proteasome machinery. Various mannosidases are involved in processing and sorting of misfolded glycosylated proteins from those that are properly folded. The crucial importance of ERAD for protein production is underscored by the fact that more than 60 human diseases are associated with this pathway [48].

Deficit of energy, protein overproduction, calcium dysregulation, infection, and other factors can interfere with protein biosynthesis, leading to an increase in a number of unfolded proteins. When the level of unfolded proteins exceeds ER folding and clearance capacity, it leads to cell stress and the unfolded protein response (UPR) [49, 50]. Unfolded proteins are sensed by inositol-requiring enzyme 1 (IRE1), protein kinase R-like ER kinase (PERK), and the activating transcriptional factor 6 (ATF6). Under normal conditions, the luminal domains of these sensors are associated with binding immunoglobulin protein (BiP). When unfolded proteins accumulate, BiP dissociates from the sensors, which activates the downstream signaling. IRE1 can also be activated directly by unfolded proteins [51]. Aberrant lipid composition of ER membranes is also sensed by IRE1 through its transmembrane domain [52, 53]. IRE1 may also sense protein-to-lipid ratio in the ER membrane [54]. ATF6 can be activated by sphingolipids independently of proteotoxic stress sensing [55].

The activated sensors induce UPR to restore ER homeostasis using three major mechanisms: global inhibition of protein synthesis, increase of ER folding capacity, and increase of ERAD [49, 50]. The activated PERK phosphorylates the translation initiating factor eIF2α, leading to inhibition of protein synthesis through the integrated stress response mechanism [8]. ATF6, activated by unfolded proteins, undergoes a regulated proteolysis releasing the N-terminal ATF6α transcriptional factor, which translocates to the nucleus and activates genes with ER stress response elements. They mostly encode ER resident proteins including chaperones and proteins involved in the ERAD pathway. ATF6 signaling is considered to be mostly pro-survival [56]. When IRE1 sensor is activated, it oligomerizes, which triggers its RNase activity. IRE1 cleaves XBP1 pre-mRNA at the stem-loop structures leading to its unconventional splicing [57]. The spliced mRNA translates into an XBP1 protein isoform XBP1s that serves as a transcriptional activator of UPR. It moves to the nucleus and stimulates biosynthesis of lipids for the ER membrane expansion, chaperones, and for proteins that function in ERAD and autophagy.

XBP1 mRNA not spliced by IRE1 encodes a constitutively expressed XBP1 isoform that functions as a negative feedback regulator of the transcription factor XBP1s isoform promoting its degradation. Overactivation of the UPR is also prevented by inhibition of IRE1 expression by ATF6α [58].

If homeostasis is not restored, UPR switches from the pro-survival to a pro-apoptotic mode. Integrated stress response associated with PERK activation leads to selective activation of translation of some mRNAs including ATF4 factor [11]. ATF4 increases the expression of stress response genes including the pro-apoptotic CHOP [59]. Some targets of ATF6 and IRE1 pathways are also pro-apoptotic. The choice between life and death is influenced by the duration of UPR signaling. During the early response, the cytoprotective output dominates, but a persistent ER stress attenuates the pro-survival effects while maintaining the pro-apoptotic signaling [60]. When ER stress exceeds a particular threshold, enough XBP1s accumulates to bind a conserved low-affinity site at the promoter of KLF9-encoding gene [61]. KLF9 induces expression of genes that facilitate calcium release from ER and promote cell death. IRE1 additionally contributes to apoptotic signaling by cleaving ER-associated mRNAs by a mechanism called RIDD, which is non-specific and different from the XBP1 splicing mechanism [57]. This, together with PERK suppression of translation, decreases the amount of translated proteins and promotes cell death.

ER homeostasis is regulated on many levels. ER size and activity are rapidly adjusted to cellular demands [62]. The sensitivity of ER response can be illustrated by its response to eating. Sensory food perception is sufficient to activate, through melanocortin, hepatic mTOR signaling leading to Xbp1 splicing, increased expression of ER genes and morphological ER remodeling [63]. UPR response decreases with aging, and its induction can prolong life in experimental animals [64]. However, the over-activity of UPR can cause damage. Dysregulation of UPR contributes to pathogenesis of many human diseases from diabetes to cancer and neurodegeneration [49, 65 - 68].

7.4. Autophagy

Autophagy is a conserved process of self-digestion of some cellular content in lysosomes (or vacuoles in yeast) of eukaryotic cells [69]. It is a major cytoprotective mechanism that plays a key role in response to stress. It serves to remove damaged cellular components and also as a recycling system that produces building blocks for the synthesis of macromolecules and for energy production. Many cellular processes from metabolism to immunity depend on autophagy and a wide range of human diseases is associated with the malfunction of this system [70, 71]. The importance of autophagy was recognized by the awarding of the 2016 Nobel Prize to Yoshinori Ohsumi for research in autophagy [72].

Autophagy is mediated by a set of conserved autophagy-related proteins (Atg) together with species-specific factors [73]. A major autophagy type is a macroautophagy, an engulfment of cytosolic components into double-membrane vesicles, which then fuse with lysosomes to degrade their cargo. The biogenesis of autophagosomes in yeast involves the Atg proteins assemble to form the pre-autophagosomal structure (PAS), in which the isolation membrane or phagophore forms [74, 75]. Activities of ULK1 and class III phosphatidylinositol 3 kinases are necessary for autophagosome formation. A crucial step is conjugation of Atg8/LC3 with phosphatidylethanolamine in the phagophore membrane. This lipidated form of LC3 serves as a marker of autophagosomes and it is removed before their fusion with lysosomes. In mammals, ER regions enriched for the lipid phosphatidylinositol 3-phosphate known as omegasome, serve as a platform for autophagosome formation [76]. After initiation at the PAS or omegasome, the phagophore grows as a cup-shaped membrane engulfing cellular matter [77]. Expansion of the phagophore requires ATG2, which localizes to the PI3P-enriched ER-phagophore junction and mediates the transfer of lipid membranes from the ER and/or the vesicles to the phagophore [78 - 80]. Closure of the phagophore membrane produces the autophagosome, which fuses with the lysosome to form the autolysosome. Catabolic enzymes then degrade the autophagic cargo and the products are recycled.

Macroautophagy is a non-specific process with all content of autophagosome being indiscriminately degraded. A direct engulfment of cytosolic components by lysosomal invagination is called microautophagy. Several selective types of autophagy are recognized. One type of selective autophagy is chaperone-mediated. It refers to a process of selective delivery of cytosolic proteins to lysosomes by chaperones [81]. Chaperone-bound complexes are recognized by the lysosome membrane receptors. Cells use selective autophagy to regulate glucose and lipid metabolism, remove excess of DNA damage repair proteins after they finished the repair, and regulate stress response. Autophagy is also used to degrade specific organelles [82]. Defective mitochondria are removed by mitophagy [83], whereas ribosomes are removed by ribophagy [84].

Autophagy can be induced through diverse pathways that converge on an interconnected web of key autophagy regulators that integrate all information about cell homeostasis. It is regulated at many levels from transcription to post-translational modifications [73]. In yeast cells, different types of autophagy are induced depending on changes in carbon source availability [85]. In general, autophagy is a pro survival mechanism although it can promote cell death in some conditions [70, 86, 87].

7.5. Mitochondrial Stress

Mitochondria are the power plant of the eukaryotic cell. With more than 1000 of different proteins derived from mitochondrial and nuclear genomes, they are involved in many other processes including redox homeostasis, biosynthesis of steroids, amino acids, nucleotides, and iron-sulfur clusters, calcium homeostasis, signaling, apoptosis, and autophagy. Mitochondria are also the cell's stress sensing hub [88].

Mitochondria are very important for cell health maintenance. During energy stress, they increase the capacity of their electron transport chains and induce AMPK-mediated fragmentation to produce more mitochondria [89]. They help to prevent damage from unfolded proteins; cytosolic misfolded proteins prone to aggregation are imported into mitochondria and degraded [90]. Mitochondria mitigate oxidative stress by the regulation of ROS production [91, 92].

There is a constant flow of information between the nucleus and mitochondria. Nucleus communicates with mitochondria through anterograde pathways and receives information from mitochondria through retrograde pathways [93 - 95].

Mutations, mtDNA depletion, oxidative stress, misfolded proteins, and other factors can cause mitochondrial malfunction, which signals to the nucleus through mitochondrial unfolded protein response UPR (mt), mitochondrial peptides, and damage-associated molecular pattern (DAMP) molecules [94]. This leads to the expression of nuclear genes that compensate for the defect.

UPR (mt) is an important mechanism of mitochondrial homeostasis maintenance. Most mitochondrial proteins are encoded in the genome and are produced and folded in the cytoplasm before being imported into the mitochondria. Malfunction of mitochondria reduces import and leads to UPR (mt) [96, 97]. UPR (mt) induction is well studied in *Caenorhabditis elegans*. In this species, the transcriptional factor ATFS-1 is a central sensor and regulator of UPR (mt) [98]. In the absence of stress, ATFS-1 is imported into mitochondria and degraded. If mitochondria do not function properly, such import is reduced, which causes ATFS-1 accumulation in the cytosol and its import into the nucleus where it induces transcription of mitochondrial chaperones, proteases involved in the degradation of damaged proteins, antioxidant factors, and other targets that recover mitochondrial function. In mammals, the orthologous factor ATF5 may play an analogous role [98, 99]. Mitochondrial stress can cause a global change in the chromatin structure through histone H3K9 methylation that can persist through a lifetime of *C. elegans* [100].

If changes induced by mitochondria stress response are not sufficient to restore homeostasis, removal of defective mitochondria is initiated. This involves cutting of defective mitochondrial parts by the mitochondria-derived vesicles and deletion of the whole organelle by mitophagy [83]. Inhibition of mitophagy in *C. elegans* decreases their survival [101].

Mitochondrial malfunction contributes to many human diseases [96, 102]. Many environmental and pathogenic toxins target mitochondria. Some of them interfere with the electron transport chain, *e.g.* rotenone inhibits the transfer of electrons from complex 1 to ubiquinone. This leads to the energy crisis, increase in H_2O_2 production, and oxidative stress [103]. The cytotoxic HIV-1 protein TAT damages mitochondrial membranes [104]. Certain bacterial toxins can damage mitochondria indirectly. *Pseudomonas aeruginosa* produces a siderophore pyoverdine that translocates into the host and sequesters iron. Deficit of iron prevents the normal function of the mitochondria and induces a stress response that can lead to mitophagy [105].

7.6. Starvation

Eukaryotic cells directly sense intracellular and environmental nutrient levels.

Presence of some amino acids, especially leucine and arginine, is sensed through TOR (target of rapamycin) pathway, which coordinates growth with availability of nutrients from yeast to humans [106]. Amino acids stimulate Rag GTPase, which then recruits the mTORC1 complex to lysosomes for activation by Rheb GTPase [107, 108]. Rheb itself is controlled by many means including growth factors, energy, glucose, oxygen, stress, *etc*. Inputs from both Rags and Rheb GTPases are necessary to promote anabolic growth. When amino acids are scarce, mTORC1 translocates from lysosomes to the cytoplasm and catabolic pathways are induced such as usage of stored nutrients and autophagy. In the presence of a constitutively active RagA, mouse embryos develop normally, but cannot survive periods of fasting as newborns, since their TOR pathway cannot switch from the anabolic to a catabolic mode [109]. These data show that mTORC1 is necessary to maintain homeostasis during starvation. Totally, 26 proteins from the TOR pathway are involved in nutrient sensing; among them, SLC38A9 senses lysosomal arginine, Sestrin2 sense cytosolic leucine, and CASTOR1 senses arginine [110]. mTORC1 also senses methionine [111]. The importance of this amino acid is highlighted by the fact that its restriction in a diet extends the life span [112]. Since the restriction of only a single amino acid mimics known effects of dietary restriction on life span, it can be used in the design of life-span extension diets. mTORC1 senses other nutrients such as cholesterol [113]. It is interesting that animal nutrient sensors identified so far appear to originate from prokaryotic enzymes [110].

Glucose is the primary fuel for cells. Therefore, glucose intake, storage, mobilization, and usage are tightly regulated. Glucose is sensed on many levels from cells to the central nervous system [114]. In humans, the hormones insulin and glucagon regulate glucose metabolism. Insulin lowers blood glucose and promotes its storage as glycogen, whereas glucagon promotes glycogen breakdown and gluconeogenesis to increase blood glucose. Insulin secretion by pancreatic beta cells is controlled by glucose phosphorylation by glucokinase, which is viewed as a glucose sensor, and by the activity of several ion channels [115, 116]. In neurons, mitochondrial dynamics serves as a glucose-sensing mechanism [117]. Excessive intake of nutrients disrupts the normal glucose sensing, leading to ER stress, increased Ca^{2+} levels, mitochondrial dysfunction, high ROS levels, and inflammation, and can lead to type 2 diabetes [118].

Sensing the energy levels by AMPK is also important for the regulation of nutrient homeostasis [119, 120]. AMPK contains the catalytic alpha subunit and regulatory beta and gamma subunits that form three modules: the catalytic, the carbohydrate-binding, and the nucleotide-binding. Binding of AMP to the gamma subunit promotes AMPK phosphorylation by upstream kinases such as LKB1 and activation. Multiple kinases convey additional signals such as levels of glucose and Ca2+ to AMPK by phosphorylation of various residues. AMPK subunits have multiple isoforms with a variable expression between different cell types. Isoforms are used differentially in response to various stresses [121]. Activated AMPK phosphorylates metabolic enzymes and transcriptional factors to suppress biosynthetic pathways that consume ATP and to stimulate catabolism that generates ATP. AMPK promotes the breakdown of carbohydrates and lipids, suppresses protein synthesis through mTORC1 inhibition, and promotes autophagy. Animals have additional regulators of energy metabolism such as thyroid hormones [122].

Prokaryotes can accumulate and store some nutrients during their abundance, in animals, this strategy becomes prominent with the emergence of cells specializing in nutrient storage such as fat tissue and liver that stores glycogen. In response to starvation, eukaryotes activate mechanisms that increase nutrient availability through the use of stored nutrients, use of alternative nutrients, and activation of catabolic processes such as proteasome-mediated degradation and autophagy [123, 124].

Eukaryotic unicellular organisms have an additional way to respond to starvation — by differentiation. Yeasts can sporulate, switch to pseudohyphae growth as elongated chains of small cells, or enter a non-proliferative state known as the stationary phase or quiescence [125]. Moderate deficiency of nutrients causes pseudohyphae growth, which let the yeast to spread to new areas faster than during normal growth. Sporulation is induced when starving cells switch to use of non-fermentable carbon sources. When all food sources are exhausted quiescence is induced. Ultimately, cell fate is decided by metabolic signaling [126].

Epigenetic mechanisms are also involved in response to starvation. For example, beta-hydroxybutyrate and other ketone bodies are produced from fatty acids in the mitochondria during glucose starvation. These compounds acylate Lys in histones, which changes the spectrum of expressed genes [42]. N-6 adenosine methylation of mRNAs is used by yeast in response to nutrient limitation to direct cell differentiation [127].

7.7. Oxidative Stress

Eukaryotic cells have many mechanisms that counteract oxidative stress at the level of transcription, translation, and protein turnover. Similar to bacteria, eukaryotes use reversible oxidation of cysteine residues of certain proteins to sense oxidative stress. One example is *S. cerevisiae* Yap1p. At normal ROS levels, Yap1p is exported from the nucleus. Oxidizing agents cause the formation of Yap1p intermolecular disulfide bond that masks its nuclear export sequence

[128]. This leads to Yap1p accumulation in the nucleus and upregulation of its targets. Yap1p controls the expression of over 70 genes involved in oxidative stress response including superoxide dismutase, catalase, and thioredoxin. The oxidized Yap1p is reduced by thioredoxin and reused. Other factors may activate this system as well, and it involves additional regulators to fine-tune the response.

In addition to Yap1p, other yeast pathways respond to oxidative stress by transcriptional changes. For example, overexpression of the general stress response transcriptional factor Msn2 leads to an increase in production of ROS degrading enzymes and to ROS resistance [129].

Transcriptional response plays the dominant role in yeast response to oxidative stress but other processes such as mRNA turnover also contribute to the response [130]. It includes increased decay of mRNAs encoding ribosomal proteins and stabilization of mRNAs for genes functioning in protein degradation. Changes in translation also contribute to the response. From yeasts to human and plants, tRNAs are cleaved in response to oxidative stress, which decreases the general translation [131]. Although general translation must be reduced, the cell needs to increase the translation of genes involved in oxidative stress response. One mechanism is ribosome modification through polyubiquitination of ribosomal proteins, which is increased during oxidative stress. It increases the formation and stability of polysomes thus promoting translation [132]. Decreased methylation at a single C2278 residue within a conserved region of 25S rRNA leads to structural changes and increased recruitment of oxidative stress-responsive mRNAs [133].

In animals, response to oxidative stress becomes especially complex. It depends on the intensity of the stress. Mild stress is mostly sensed by the Keap1/Nrf2 system [134]. In the absence of stress, Keap1 assists the ubiquitination of Nrf2 transcription factor targeting it for proteasomal degradation. Therefore Nrf2 activity is constitutively suppressed. Oxidation of Keap1 cysteine residues leads to a conformational change that inactivates Keap1 E3 ligase activity. This results in Nrf2 accumulation and nuclear translocation; in the nucleus, it heterodimerizes with the transcription factor Maf. This complex binds a consensus DNA sequence (the antioxidant response element, or ARE) to direct expression of more than 250 genes involved in oxidative stress response. The Keap1-Nrf2 signaling pathway is subject to multiple layers of regulation at the transcriptional, translational, and post-translational levels including epigenetic regulation by DNA methylation, histone modifications, and microRNAs [135, 136]. Moreover, diverse electrophiles oxidize different Keap1 cysteine residues and activate the Keap1/Nrf2 system differentially allowing for additional plasticity [137].

Nrf2 can also be controlled through Keap1-independent mechanisms. Nrf2 phosphorylation by glycogen synthase kinase 3 (GSK3) promotes its degradation [138]. In contrast, when phosphorylated by PERK, PKC, extracellular signal-regulated kinases (ERK), and p38 mitogen-activated protein kinase Nrf2 dissociates from Keap1 and escapes degradation [139, 140].

Nrf2 targets include canonical antioxidant genes for catalase, superoxide dismutase, glutathione peroxidase, peroxiredoxin, thioredoxin reductase, and glutathione reductase. In addition, Nrf2 also regulates genes involved in broader functions such as glucose metabolism, proteostasis, and ubiquitin-proteasomal system [141, 142]. Nrf2 suppresses inflammation by blocking transcription of proinflammatory cytokines IL-6 and IL-1b [143].

Activation of Nrf2 is a potential way of treatment for some inflammatory diseases such as neurodegeneration [144]. High Nrf2 activity, however, has a downside. Hyperactivation of Nrf2 increases stress tolerance of Drosophila at the cost of accelerated aging [145]. Nrf2 expression in cancer cells increases their resistance to treatments and their survival [146]. Activating mutations of Nrf2 lead to a multisystem disorder with immunodeficiency and neurological symptoms [147]. Keap1-null mutation that leads to a constitutive Nrf2 activation is postnatally lethal [148]. These data highlight the importance of redox balance for cell health.

High-intensity oxidative stress induces signaling pathways besides Nrf2. They include NF-kB, MAPK signaling cascades, p38, JNK, PI3K, protein kinase C, and various tyrosine kinases [140, 149]. Some of these signaling pathways also utilize reversible oxidation of a sensor protein. For example, MAPK oxidative stress pathway that includes 79 genes starts with MAP3K5/ASK1 kinase. ASK1 forms a high molecular weight complex with regulatory proteins including a reduced form of thioredoxin that inhibits ASK1's kinase activity [150]. In the presence of oxidants, the thioredoxin is oxidized and dissociates from ASK1. This starts a series of events that include ASK1 autophosphorylation on a threonine residue critical for its kinase activity, recruitment of other activating factors into the complex, and oligomerization leading to ASK1 activation. ASK1's downstream targets include JNK and p38 that play a key role in apoptosis caused by oxidative stress. Inhibition of thioredoxin dissociation from ASK1 by the HIV encoded protein Nef is proposed as a mechanism of apoptosis suppression in HIV-infected T cells [151].

Oxidative stress is thought to contribute to aging and human disorders including neurodegeneration, cancer, metabolic problems, and cardiovascular diseases. For example, frailty among elderly people is associated with low levels of antioxidants such as carotenoids beta-cryptoxanthin and zeaxanthin [152]. Many components of the oxidative stress response are of interest as therapeutic targets

[144, 153]. However, the initial view that all free radicals and oxidants are dangerous and need to be removed by antioxidants has changed to a more balanced view with the understanding that these molecules are an integral part of the redox balance required for normal cell functioning.

7.8. Heat Stress

The heat stress response (HSR) is one of the most conserved adaptive responses that protect against proteins misfolding, aggregation, and malfunction. It includes increased production of chaperones and other factors that protect macromolecules along with increased production of components of the protein degradation machinery. In eukaryotes, heat-shock transcription factors (HSFs) control this response [154]. There are six human HSFs with different roles in stress response. HSF1 is a conserved HSF activated by many proteotoxic conditions. Current models suggest that in unstressed cells HSF1 is monomeric and is associated with heat shock proteins such as HSP70 and HSP90 [155, 156]. Under heat stress, HSPs dissociate from HSF1, and the latter undergoes a temperature-dependent conformational change, moves to the nucleus, trimerizes and binds to heat shock promoter elements [154, 157]. HSF1 is regulated by several kinases that adjust the response to specific cellular conditions and is degraded by the ubiquitin proteasomal system. Promoters and enhancers of highly expressed genes such as HSF1 are often pre-loaded with general transcriptional factors. This ensures a fast response upon HSF1 binding [158]. This also ensures an immediate stop of the response as soon as HSF1 is removed [159].

Chaperones confer tolerance not only for hyperthermia but also for a variety of stresses including cold, hyperoxia, hypoxia, and other conditions that interfere with protein folding. This cross-tolerance is noted in all organisms. In humans, HSF1 is involved in many processes where protein quality is important, such as growth and differentiation, ER stress response, autophagy, immunity, cancer, and apoptosis [154, 160, 161]. HSF1 signaling is increased in cancer and decreased in neurodegenerative disorders [154].

Besides chaperones, other factors can influence sensitivity to heat shock including TOR and PKA pathways, lipid metabolism, vacuolar H(+)-ATPase, vacuolar protein sorting, and mitochondrial genome maintenance/translation [162]. mRNA levels in yeast under heat shock are affected not only by the rate of transcription but also by a change in RNA stability [163]. During replication, heat can induce DNA breaks leading to the induction of senescence in cells in the early S-phase of the cell cycle [164].

7.9. Osmotic Stress

The response of eukaryotic cells to pH and osmotic stress is well studied in budding yeast *S. cerevisiae* due to their wide use in biotechnology. During their life cycle, yeasts are exposed to drastic changes in osmolarity, pH, and nutrient sources. Yeast's preferred source of energy is glucose. Glucose is fermented into ethanol and other low molecular weight products such as acetate that are collectively called non-fermentable carbon sources [165]. These products acidify the medium up to pH 4. Ethanol produced during fermentation disrupts the water balance and causes osmotic stress. It also affects membrane fluidity. When glucose is depleted, protein kinase Snf1 is activated triggering the use of alternative sugars and non-fermentable carbon sources by aerobic respiration. During growth on these carbon sources, yeasts generate CO_2 as the end product, which makes the medium alkaline up to pH 8.

Yeast response to ethanol has common features with heat stress response such as a change in membrane lipid composition leading to its decreased fluidity [166]. The osmotic stress response in yeast depends on MAPK cascades [167, 168]. Specifically, MAP kinase Hog1 translocates transiently from cytosol to the nucleus during high external osmolarity and changes the expression of multiple genes. Hog1 induces a long non-coding RNA (lncRNA), which is in antisense orientation to a gene that encodes cell cycle controller CDK1 kinase. This lncRNA mediates a more efficient re-entry of the cell into the cell cycle after stress [169]. The most important response to high osmotic pressure in all life forms is increased synthesis and intracellular retention of compatible osmolytes [170]. Exposed to 1M NaCl *Candida albicans* downregulates 84 genes associated with translation and ribosome biogenesis, and upregulates 64 genes involved in stress response including genes functioning in glycerol synthesis, a key osmolyte for this species [171]. Ciliate *Schmidingerothrix salinarum* also responds to osmotic stress by accumulation of solutes using the "salt out" strategy [172]. Cell wall remodeling is also an important component of the response to osmotic stress [173].

Mammals sense and maintain the osmotic pressure of blood and extracellular fluids through multiple mechanisms [174]. Major control is performed by the central nervous system through sympathetic activity on contraction/dilation of vascular smooth muscles, as well as through hormonal, and behavioral adjustments.

7.10. DNA Damage Response

First insight into eukaryotic cell response to DNA damage came from the Weinert and Hartwell yeast studies [175]. They found that DNA damage causes the arrest of cell cycle in G2 phase mediated by RAD9 gene, which leads to cell death unless such damage is repaired. DNA damage response (DDR) is conserved from yeast to humans with additional levels of regulation evolved in multicellular organisms. DDR network regulates transcription and controls cell cycle, apoptosis, senescence, and DNA repair through protein modifications and epigenetics.

Eukaryotic DDR is controlled mostly by two related members of phosphatidylinositol 3-kinase-like kinase (PIKK) family [176, 177]. Ataxia-telangiectasia mutated (ATM) kinase senses DNA double-strand breaks, whereas ATM and Rad3-related (ATR) kinase senses single-stranded DNA [178]. Activated ATM and ATR phosphorylate their targets leading to a signaling cascade that adjusts cellular homeostasis to counteract the effects of DNA damage [179].

ssDNA may result from DNA damage, processing of double-strand breaks, or during stalling of a replication fork. Replication protein A complex (RPA) binds and stabilizes ssDNA and ATR is recruited to ssDNA coated with RPA [180]. ATR also responds to aberrant structures of chromatin, which can arise due to replication stress, defective duplication of centrosomes, malfunction of the nucleolus, or during osmotic and mechanical stress [181, 182].

DNA double-strand breaks (DSB) may result from DNA damage, replication fork collapse, or apoptosis-initiated chromatin fragmentation. They are recognized by ATM kinase through Mre11-Rad50-Nbs1 (MRN) complex [183]. In the absence of DNA damage, ATM exists as an inactive homodimer. When there are DSB, MRN complex promotes ATM autophosphorylation, monomerization, and recruitment to DSB sites. Some data suggest that ATM senses and responds to chromatin alterations in the context of DNA damage [184]. ATM can also be activated by transcription-blocking lesions. Pausing of RNA polymerase at such lesions causes chromatin displacement of spliceosomes and R-loop formation [185]. Formaldehyde treatment leading to histone adducts and DNA-protein crosslinks also activates ATM [186]. In the presence of oxidants, ATM forms an active disulfide-cross-linked dimer [187] and regulates a set of targets that mostly overlaps with the targets phosphorylated during the canonical DSB response [188, 189].

ATR and ATM targets include the mediator protein 53BP1, histone H2AX/H2AFX, proteins involved in replication such as DNA polymerase and MCMs (minichromosome maintenance proteins), proteins involved in transcription, splicing, and RNA metabolism [179]. Among the targets are also proteins involved in DNA repair and recombination such as MSH2, XPA, Fanconi anemia proteins, BRCA1/2, and others. Some targets are related to cell cycle regulation including cyclin E, the checkpoint kinase CHK1, and p53. Many ATM targets are common with ATR targets. Other targets are kinase-specific; they include checkpoint kinases CHK1 for ATR and CHK2 for ATM. Activation of ATR/ATM most often leads to inhibition of cell cycle, activation of DNA repair, or apoptosis depending on the cell's status. ATR/ATM activation can also induce pro-survival signaling through IGF-P13K-AKT pathway and selective autophagy [190].

p53 is a downstream target of ATM and ATR that transcriptionally regulates hundreds of genes. It decides cell fate in dependence on the degree of DNA damage and other factors [191, 192]. P53 can promote cell survival by stimulating DNA repair and autophagy or, alternatively, induce cell senescence and apoptosis.

Eukaryotic DNA is packaged with histone proteins into nucleosomes to form two types of chromatin. Euchromatin is loosely packed, contains active genes, and is easily accessible, whereas heterochromatin is tightly packed. Repair efficiency depends on the position of a DNA lesion on the chromosome, chromatin modifications of the damaged locus, and other factors [193]. DNA in active and open chromatin states undergo repair earlier than DNA from heterochromatin [194].

DNA repair requires chromatin remodeling with the involvement of many ATP-dependent enzymes. To satisfy the increased energy demand during DDR, ATP can be generated in the cell nucleus through hydrolysis of poly (ADP-ribose) to ADP-ribose, which is used by the pyrophosphatase NUDIX5 to produce ATP [195].

Since DDR response is extremely broad, it affects virtually all aspects of cellular homeostasis. Changes introduced by activation of the DDR network can confer resistance to other stresses such as proteotoxic stress [196, 197]. Deficiency of the DDR pathway underlies many human diseases including developmental disorders, neural degeneration, immune deficiency, and cancer. Some viruses exploit DDR for their propagation [198]. Pharmacological disabling of DDR is used in cancer treatment [199, 200]. When DNA damage repair is unsuccessful, the same DDR network that directed repairs induces senescence or initiates regulated cell death.

7.11. Regulated Cell Death

Regulated cell death (RCD) has been observed in prokaryotes and in single cell eukaryotes such as yeast [201]. But only in multicellular organisms where the fate of individual cells is sacrificed in the interest of the organism as a whole, regulated cell death becomes an indispensable part of organismal life. In these organisms, RCD is not only essential for the removal of stressed, damaged, transformed, and infected cells. RCD is also involved in normal development, tissue remodeling, regeneration, and turnover of immune cells, and in this role, it is better known as programmed cell death (PCD) [202 - 204]. Every day the human body produces 10 –100 billion new cells and the same number of old cells die mostly by PCD [205]. PCD defects underlie many human diseases from immune disorders to cancer.

Diverse RCD programs emerged during animal evolution, providing a flexible response to the needs of the organism. They include apoptosis, autophagic cell death, several regulated necrotic pathways such as necroptosis, pyroptosis, ferroptosis, and other death pathways [206].

Apoptosis depends on activation of a cascade of proteases called caspases that cleave various cellular components, leading to chromatin condensation, DNA fragmentation, membrane blebbing, and formation of apoptotic bodies [207]. Only a few genes are involved in apoptosis in the nematode *C. elegans,* whereas in mammals, the apoptotic network comprises more than 300 genes [208 - 210].

Apoptosis can be induced through extrinsic or intrinsic pathways. The intrinsic pathway is activated by internal stress such as DNA damage. Its activation depends on the interaction of pro-apoptotic and anti-apoptotic proteins of the Bcl2 family. Pro-apoptotic members Bax and Bak can form pores in the mitochondrial outer membrane, leading to release of pro-apoptogenic factors, including cytochrome c, and Smac from the mitochondrial intermembrane space into the cytosol. Human apoptotic peptidase activating factor APAF1 binds cytochrome *c* and forms an oligomeric apoptosome that binds and cleaves a precursor of initiator caspase-9 [211 - 213]. The initiator caspase then activates downstream effector caspases Casp3 and Casp7 that function to dismantle the cell. Cytosolic Smac promotes apoptosis by binding to the inhibitor of apoptosis (IAP) family proteins, such as XIAP that blocks caspases.

Bcl2 family includes anti-apoptotic proteins such as Bcl2, pro-apoptotic pore-formers such as Bax and Bak, and pro-apoptotic BH_3-only proteins such as Bad and Puma. Anti-apoptotic and pro-apoptotic Bcl2 family members have a similar structure with four BH domains. They can bind each other and BH_3-only members through their BH_3 domains. Some BH_3-only proteins are activators; when they bind Bax or Bak they activate them, which leads to their oligomerization and pore formation. Anti-apoptotic Bcl2 proteins bind pore-formers and BH_3-only activators through BH_3 domains and inactivate them. In turn, they are inactivated by BH_3-only sensitizer proteins. The whole system exists in equilibrium. The outcome is determined by concentrations and affinities of Bcl2 family members [214]. Their regulation is complemented by post-translational modifications; for example, phosphorylation of Bax on Ser184 by AKT has an inhibitory effect. Mitochondrial conditions also influence the likelihood of mitochondria to undergo membrane permeabilization.

The extrinsic pathway is activated by the binding of extracellular ligands to members of TNF receptor superfamily [213, 215]. For example, Fas ligand binds to Fas receptor (CD95/Apo1), leading to the formation of the death-inducing signaling complex (DISC). It includes Fas-associated death domain protein (FADD) and initiator caspase 8 [216]. Autoprocessing of caspase 8 in DISC leads to initiation of a downstream caspase cascade [217]. Caspase 8, while inducing apoptosis, suppresses an alternative form of RCD —necroptosis [218, 219]. The extrinsic pathway provides the means for specialized immune cells to trigger apoptosis in transformed and infected cells. It can also lead to NFkB activation.

Both the extrinsic and intrinsic pathways lead to activation of effector caspases-3 and -7 that cleave more than 1000 proteins and cause cell death. Apoptotic cells display a specific phenotype, which includes chromatin condensation (pyknosis), membrane blebbing, and cell shrinkage and fragmentation. Special mechanisms evolved to prevent the release of the content of apoptotic cells into surrounding tissues, which can cause inflammation. Apoptotic cells secrete and expose on the cell surface molecules that attract phagocytes and promote their engulfment without activation of inflammation [205, 220]. An important feature of apoptotic cells is caspase-dependent exposure on their surface of phosphatidylserine, an "eat me" signal, which is recognized by several phagocytic receptors [221]. Phagocytes produce anti-inflammatory compounds during clearing of apoptotic cells and cell debris. However, an extrinsic apoptotic pathway, such as Fas/CD95-induced apoptosis, can induce production of chemokines such as MCP1 and IL-8 that promote chemotaxis of phagocytes toward apoptotic cells [222].

Ferroptosis is a form of RCD associated with iron-dependent lipid peroxidation [223 - 227]. Polyunsaturated fatty acids (PUFAs) are susceptible to oxidation into lipid hydroperoxides (L-OOH) by enzymes such as lipoxygenases and by ROS [228]. In the presence of iron, L-OOH can form highly reactive lipid alkoxy radicals (L-O$^\cdot$) that target PUFAs leading to their fragmentation and the formation of covalent bonds to proteins. GSH-dependent glutathione peroxidase 4 (GPX4) reduces hydroperoxides to lipid alcohols (L-OH) limiting the formation of alkoxy radicals. *Gpx4* deficient mice are embryonic lethal, which points to the vital importance of GPX4. GPX4 activity depends on GSH, which synthesis requires cystine (Cys_2), the oxidized form of the amino acid cysteine. Cystine is imported into cells through cystine/glutamate antiporter. Depriving cells of cystine leads to drop in GSH level, degradation of ferritin, release of iron, and ferroptosis. Increased ROS and iron overload also promote ferroptosis. It can be prevented without restoring GSH level by lipophilic antioxidants such as vitamin E and iron chelators. Other forms of animal RCD such as necroptosis, pyroptosis, and NETosis are mostly involved in immunodefense and will be discussed in chapter IX.

In plants, PCD also plays an important role in development, stress response, and immunodefense. During development, some plant cells need to be removed; these cells die by RCD and their corpses are degraded by autolysis [229]. In response to stress, plants often undergo autophagy-mediated RCD [203].

At certain conditions, stressed cells do not die but instead enter a state of senescence, which is characterized by permanent cell cycle arrest and characteristic morphological, biochemical, and functional changes [230]. Senescence works in parallel with RCD to prevent replication of damaged cells and tumor development. Senescent cells develop distinctive metabolic features referred to as the senescence-associated secretory phenotype. They accumulate with aging and contribute to both regeneration and aging-associated disorders [231].

7.12. Regulation of Stress Response at the Level of Organism

Multicellular eukaryotes evolved into highly complex systems that consist of specialized tissues and cells. These cells preserve the ability to respond to stress. However cell-autonomous responses can also be regulated at the level of the organism. Communication between cells and tissues lets the organism produce an integrated stress response to the external and internal signals. In animals, this integration is controlled by the nervous and hormonal systems, whereas in plants, hormones implement such integration.

7.12.1. Non-autonomous Regulation of Stress Response in Animals

C. elegans is a useful organism to study how animals respond to stress at the organismal level because it is a very simple animal in which the function of every cell is known. Almost a third of the cells in *C. elegans* (302 out of 959) are neurons and many of them are involved in the coordination of stress response in this organism. Two *C. elegans* thermosensory AFD neurons sense the ambient temperature and pass the information to other cells through serotonergic signaling [232]. Excitation of AFD neurons leads to activation of HSF1 signaling and production of HSPs in somatic cells even in the absence of heat stress [233].

Several studies demonstrated that mitochondrial stress in *C. elegans* neural tissue leads to induction of a mitochondria-specific unfolded protein response in distal tissues [234]. This induction is mediated by a neural sub-circuit composed of three types of sensory neurons and an interneuron and involves neuropeptide FLP-2 and serotonin release [235, 236]. When stress is induced in the larva, increased stress resistance is maintained into adulthood, which suggests the involvement of epigenetic mechanisms.

In adult worms, germ cells divide while somatic cells are post-mitotic. Therefore, DNA damage response takes place only in the germ cells. Nevertheless, *C. elegans* responds to DNA damage by increased stress resistance in somatic cells too [237].

Perturbation of proteostasis in one tissue of *C. elegans* such as neurons or the intestine induces Hsp90 in other tissues through the transcription factor PQM-1 [238]. Loss of ER proteostasis in aging *C. elegans* can be reversed by expression of a constitutively active XBP1. Neuronally expressed XBP1 is sufficient to induce UPR and rescue homeostasis in distal, non-neuronal cells [239]. An analogous effect has been observed in the mouse; increased Xbp1s levels in Pomc neurons lead to activation of the Xbp1s in the liver [240]. Therefore the ability of tissue experiencing proteotoxic stress to induce molecular chaperones in distal tissues or "transcellular chaperone signaling" (TCS) may be common for all metazoans. Stressed cells can communicate to other cells through various signals from a change of in cell surface molecules and shape to the secretion of bioactive molecules [241].

In higher animals, neural and hormonal circuits increasingly control stress response, which includes responses to heat and cold. Sophisticated systems sense noxious stimuli and tissue damage through sensory nerve cells called nociceptors (from Latin nocere, harm) that are subjectively perceived by the brain as pain. Humans use membrane ion channels such as transient receptor potential (TRP) family of cation channels to sense temperature [242]. Thermo TRPs are expressed in sensory neurons and some other cells. Activation of cutaneous TRPs by very high or very low temperatures is perceived as heat or cold but also as pain. Sensations of heat and cold can also result from chemical activation of TRPV1 by capsaicin from chili pepper and TRPM8 by menthol, respectively. The exact mechanism of TRP activation by changes in temperature is not known. TRPV1 activation by high temperature may be mediated by conformational change leading to exposure of hydrophobic residues to solvent [243].

Then, information about body temperature is integrated in hypothalamus where some neurons are directly sensitive to heat or cold. These neurons also receive input from cutaneous and spinal thermoreceptors. When body temperature exceeds 37°C, the warmth-sensitive neurons are activated leading to sympathetic nervous system signaling that promotes dissipation of heat, such as vasodilatation, sweating, *etc*. If body temperature drops below 37°C, the cold-sensitive neurons increasingly discharge leading to signaling that promotes generation and conservation of heat [244].

Heat is generated by the adipose tissue and by muscles. Animals store energy in white adipose tissue (WAT), which has few mitochondria and brown adipose tissue (BAT) packed with mitochondria. BAT regulates animal energy expenditure [245]. In adult humans, most fat is represented by WAT. However, cold may induce cell progenitors inside the WAT to develop into BAT [246]. BAT thermogenesis during cold stress involves protein UCP1, which uncouples protons moving across the mitochondrial membrane from the synthesis of ATP and allows the energy to dissipate as heat [247]. Fatty acid oxidation is crucial for this process, which at the same time contributes to WAT oxidative stress and inflammation [248, 249]. BAT thermogenesis is regulated by thyroid hormones and sympathetic nervous system [244].

Muscles also can generate heat by shivering, when they do not generate motion but shake. Shivering is induced by a special motor center in the hypothalamus in response to cold. Muscles also generate heat by non-shivering thermogenesis through sarcoplasmic/endoplasmic reticulum Ca^{2+} ATPase (SERCA). In response to cold, regulators such as sarcolipin bind SERCA decreasing its affinity for calcium without affecting ATP hydrolysis, which switches SERCA from pumping Ca^{2+} to heat production [250]. In some conditions, this way of heat generation may play a major role in thermogenesis and BAT and muscle non-shivering thermogenesis can complement each other [251].

Prolonged exposure to heat can result in heat exhaustion characterized by excessive loss of fluids and electrolytes, dizziness, vomiting, and muscle cramps. If not treated, it may progress to heat stroke, when the organism cannot control its core temperature, which can lead to tissue damage and death.

In addition to a physiological response, heat and cold stress in animals lead to behaviors associated with seeking escape from too hot or too cold places. In some animals, regulation of core body temperature in response to the environment evolved as an additional adaptation to cold temperatures. For example, some mammals can hibernate during winter inhibiting their metabolism and decreasing their body temperature almost to the freezing point [252].

7.12.2. Stress Response in Plants

Plants use various signaling pathways to sense stress and to respond at the level of the organism. They include Ca^{2+} signaling, ROS, MAP kinase cascades, transcriptional factors such as Myb, and other means [253]. (p)ppGpp-guanosine tetra-and pentaphosphates that regulate the stringent response in bacteria have also been found in algae and higher plants where they participate in stress response [254]. Similar to animals, plants use hormones to regulate stress response [255]. Abscisic acid (ABA) is the main hormone that responds to cold, heat, drought, increased salinity, absence of light, and wounds [256]. Besides transcriptional regulation of stress response genes by ABA, posttranscriptional mechanisms such as regulation of mRNA decay, are also involved [257]. Interestingly, ABA is also found in mammals, specifically, it acts as a pro-survival factor in human and murine megakaryocytes [258].

7.13. Cross-Stress Resistance

Like bacteria, eukaryotic cells exposed to mild stress can acquire tolerance not only to stronger stress of the same kind but also to a different kind of stress [259]. For example, several *S. cerevisiae* strains have been isolated as thermotolerant and are found to be multiple stress-tolerant [260]. Increased stress resistance of these strains depends on the continuous expression of heat shock protein genes, increased trehalose levels, low ROS, and effective cell wall remodeling. Plants exposed to low levels of various stresses can acquire enhanced pest and disease resistance and abiotic stress tolerance, which is called defense priming [261]. Heat and cold stress-induced cross-tolerance to other abiotic stresses in plants is based on ROS signaling that activates multiple protective mechanisms [262]. In human cells, pre-stimulation with low dose UV improves UV-induced DNA damage repair through stimulation of the DDR network [263]. This phenomenon is related to the so-called hormetic effect, where a low dose of a harmful substance such as H_2O_2 has a positive effect on the organism, although high doses are toxic [264, 265].

Low-level stress in early life can improve stress resistance in adults [266]. The Caribbean fruit fly, *Anastrepha suspensa* exposed to anoxia at the pupal developmental stage has an increased stress resistance as an adult [267]. Mitochondrial stress in early life can lead to changes in DNA packaging, stress resistance, increased mitochondrial unfolded protein response, and increased life span [100, 268].

CONCLUDING REMARKS

Eukaryotic cells do not use their membranes to produce energy, and membrane stress is not as critical for them as for prokaryotes. Common sources of stress in eukaryotic cells are disturbances in protein production and folding, oxidative and DNA damage. The eukaryotic stress response involves changes in transcription and translation, protein modification and degradation, and epigenetic modifications. In multicellular organisms, hormones and the nervous system coordinate stress response of each cell.

CONSENT FOR PUBLICATION

Not applicable.

CONFLICT OF INTEREST

The author confirms that this chapter contents have no conflict of interest.

ACKNOWLEDGEMENTS

Declare none.

REFERENCES

[1] Szathmáry E. Toward major evolutionary transitions theory 2.0. Proc Natl Acad Sci USA 2015; 112(33): 10104-11.
[http://dx.doi.org/10.1073/pnas.1421398112] [PMID: 25838283]

[2] Castelle CJ, Banfield JF. Major New Microbial Groups Expand Diversity and Alter our Understanding of the Tree of Life. Cell 2018; 172(6): 1181-97.
[http://dx.doi.org/10.1016/j.cell.2018.02.016] [PMID: 29522741]

[3] Lyu Z, Whitman WB. Evolution of the archaeal and mammalian information processing systems: Towards an archaeal model for human disease. Cell Mol Life Sci 2017; 74(2): 183-212.
[http://dx.doi.org/10.1007/s00018-016-2286-y] [PMID: 27261368]

[4] Spang A, Eme L, Saw JH, *et al.* Asgard archaea are the closest prokaryotic relatives of eukaryotes. PLoS Genet 2018; 14(3)e1007080
[http://dx.doi.org/10.1371/journal.pgen.1007080] [PMID: 29596421]

[5] Rose RE, Pazos MA II, Curcio MJ, Fabris D. Global Epitranscriptomics Profiling of RNA Post-Transcriptional Modifications as an Effective Tool for Investigating the Epitranscriptomics of Stress Response. Mol Cell Proteomics 2016; 15(3): 932-44.
[http://dx.doi.org/10.1074/mcp.M115.054718] [PMID: 26733207]

[6] Nayak RR, Bernal WE, Lee JW, Kearns MJ, Cheung VG. Stress-induced changes in gene interactions in human cells. Nucleic Acids Res 2014; 42(3): 1757-71.
[http://dx.doi.org/10.1093/nar/gkt999] [PMID: 24170811]

[7] Galluzzi L, Yamazaki T, Kroemer G. Linking cellular stress responses to systemic homeostasis. Nat Rev Mol Cell Biol 2018; 19(11): 731-45.
[http://dx.doi.org/10.1038/s41580-018-0068-0] [PMID: 30305710]

[8] Pakos-Zebrucka K, Koryga I, Mnich K, Ljujic M, Samali A, Gorman AM. The integrated stress

response. EMBO Rep 2016; 17(10): 1374-95.
[http://dx.doi.org/10.15252/embr.201642195] [PMID: 27629041]

[9] Pavitt GD. Regulation of translation initiation factor eIF2B at the hub of the integrated stress response. Wiley Interdiscip Rev RNA 2018; 9(6)e1491
[http://dx.doi.org/10.1002/wrna.1491] [PMID: 29989343]

[10] Taniuchi S, Miyake M, Tsugawa K, Oyadomari M, Oyadomari S. Integrated stress response of vertebrates is regulated by four eIF2α kinases. Sci Rep 2016; 6: 32886.
[http://dx.doi.org/10.1038/srep32886] [PMID: 27633668]

[11] Starck SR, Tsai JC, Chen K, *et al.* Translation from the 5′ untranslated region shapes the integrated stress response. Science 2016; 351(6272)aad3867
[http://dx.doi.org/10.1126/science.aad3867] [PMID: 26823435]

[12] Mazor KM, Stipanuk MH. GCN2- and eIF2α-phosphorylation-independent, but ATF4-dependent, induction of CARE-containing genes in methionine-deficient cells. Amino Acids 2016; 48(12): 2831-42.
[http://dx.doi.org/10.1007/s00726-016-2318-9] [PMID: 27613409]

[13] Gallinetti J, Harputlugil E, Mitchell JR. Amino acid sensing in dietary-restriction-mediated longevity: Roles of signal-transducing kinases GCN2 and TOR. Biochem J 2013; 449(1): 1-10.
[http://dx.doi.org/10.1042/BJ20121098] [PMID: 23216249]

[14] Dalet A, Argüello RJ, Combes A, *et al.* Protein synthesis inhibition and GADD34 control IFN-β heterogeneous expression in response to dsRNA. EMBO J 2017; 36(6): 761-82.
[http://dx.doi.org/10.15252/embj.201695000] [PMID: 28100675]

[15] de la Parra C, Ernlund A, Alard A, Ruggles K, Ueberheide B, Schneider RJ. A widespread alternate form of cap-dependent mRNA translation initiation. Nat Commun 2018; 9(1): 3068.
[http://dx.doi.org/10.1038/s41467-018-05539-0] [PMID: 30076308]

[16] James CC, Smyth JW. Alternative mechanisms of translation initiation: An emerging dynamic regulator of the proteome in health and disease. Life Sci 2018; 212: 138-44.
[http://dx.doi.org/10.1016/j.lfs.2018.09.054] [PMID: 30290184]

[17] Leppek K, Das R, Barna M. Functional 5′ UTR mRNA structures in eukaryotic translation regulation and how to find them. Nat Rev Mol Cell Biol 2018; 19(3): 158-74.
[http://dx.doi.org/10.1038/nrm.2017.103] [PMID: 29165424]

[18] Zhou J, Wan J, Shu XE, *et al.* N^6-Methyladenosine Guides mRNA Alternative Translation during Integrated Stress Response. Mol Cell 2018; 69(4): 636-647.e7.
[http://dx.doi.org/10.1016/j.molcel.2018.01.019] [PMID: 29429926]

[19] Zhang X, Gao X, Coots RA, Conn CS, Liu B, Qian SB. Translational control of the cytosolic stress response by mitochondrial ribosomal protein L18. Nat Struct Mol Biol 2015; 22(5): 404-10.
[http://dx.doi.org/10.1038/nsmb.3010] [PMID: 25866880]

[20] Green KM, Glineburg MR, Kearse MG, *et al.* RAN translation at C9orf72-associated repeat expansions is selectively enhanced by the integrated stress response. Nat Commun 2017; 8(1): 2005.
[http://dx.doi.org/10.1038/s41467-017-02200-0] [PMID: 29222490]

[21] Hofmann S, Cherkasova V, Bankhead P, Bukau B, Stoecklin G. Translation suppression promotes stress granule formation and cell survival in response to cold shock. Mol Biol Cell 2012; 23(19): 3786-800.
[http://dx.doi.org/10.1091/mbc.e12-04-0296] [PMID: 22875991]

[22] Rodrigues LOCP, Graça RSF, Carneiro LAM. Integrated Stress Responses to Bacterial Pathogenesis Patterns. Front Immunol 2018; 9: 1306.
[http://dx.doi.org/10.3389/fimmu.2018.01306] [PMID: 29930559]

[23] van 't Wout EF, van Schadewijk A, van Boxtel R, *et al.* Virulence Factors of Pseudomonas aeruginosa Induce Both the Unfolded Protein and Integrated Stress Responses in Airway Epithelial Cells. PLoS

Pathog 2015; 11(6)e1004946
[http://dx.doi.org/10.1371/journal.ppat.1004946] [PMID: 26083346]

[24] Buchan JR, Parker R. Eukaryotic stress granules: The ins and outs of translation. Mol Cell 2009; 36(6): 932-41.
[http://dx.doi.org/10.1016/j.molcel.2009.11.020] [PMID: 20064460]

[25] Jayabalan AK, Sanchez A, Park RY, *et al.* NEDDylation promotes stress granule assembly. Nat Commun 2016; 7: 12125.
[http://dx.doi.org/10.1038/ncomms12125] [PMID: 27381497]

[26] Loll-Krippleber R, Brown GW. P-body proteins regulate transcriptional rewiring to promote DNA replication stress resistance. Nat Commun 2017; 8(1): 558.
[http://dx.doi.org/10.1038/s41467-017-00632-2] [PMID: 28916784]

[27] Arimoto-Matsuzaki K, Saito H, Takekawa M. TIA1 oxidation inhibits stress granule assembly and sensitizes cells to stress-induced apoptosis. Nat Commun 2016; 7: 10252.
[http://dx.doi.org/10.1038/ncomms10252] [PMID: 26738979]

[28] Gasch AP, Spellman PT, Kao CM, *et al.* Genomic expression programs in the response of yeast cells to environmental changes. Mol Biol Cell 2000; 11(12): 4241-57.
[http://dx.doi.org/10.1091/mbc.11.12.4241] [PMID: 11102521]

[29] Ho YH, Shishkova E, Hose J, Coon JJ, Gasch AP. Decoupling Yeast Cell Division and Stress Defense Implicates mRNA Repression in Translational Reallocation during Stress. Curr Biol 2018; 28(16): 2673-2680.e4.
[http://dx.doi.org/10.1016/j.cub.2018.06.044] [PMID: 30078561]

[30] Natkańska U, Skoneczna A, Sieńko M, Skoneczny M. The budding yeast orthologue of Parkinson's disease-associated DJ-1 is a multi-stress response protein protecting cells against toxic glycolytic products. Biochim Biophys Acta Mol Cell Res 2017; 1864(1): 39-50.
[http://dx.doi.org/10.1016/j.bbamcr.2016.10.016] [PMID: 27984092]

[31] Wanichthanarak K, Wongtosrad N, Petranovic D. Genome-wide expression analyses of the stationary phase model of ageing in yeast. Mech Ageing Dev 2015; 149: 65-74.
[http://dx.doi.org/10.1016/j.mad.2015.05.008] [PMID: 26079307]

[32] Crawford RA, Pavitt GD. Translational regulation in response to stress in Saccharomyces cerevisiae. Yeast 2019; 36(1): 5-21.
[http://dx.doi.org/10.1002/yea.3349] [PMID: 30019452]

[33] Granados AA, Pietsch JMJ, Cepeda-Humerez SA, Farquhar IL, Tkačik G, Swain PS. Distributed and dynamic intracellular organization of extracellular information. Proc Natl Acad Sci USA 2018; 115(23): 6088-93.
[http://dx.doi.org/10.1073/pnas.1716659115] [PMID: 29784812]

[34] Chatterjee M, Acar M. Heritable stress response dynamics revealed by single-cell genealogy. Sci Adv 2018; 4(4)e1701775
[http://dx.doi.org/10.1126/sciadv.1701775] [PMID: 29675464]

[35] Brion C, Pflieger D, Souali-Crespo S, Friedrich A, Schacherer J. Differences in environmental stress response among yeasts is consistent with species-specific lifestyles. Mol Biol Cell 2016; 27(10): 1694-705.
[http://dx.doi.org/10.1091/mbc.E15-12-0816] [PMID: 27009200]

[36] Shor E, Fox CA, Broach JR. The yeast environmental stress response regulates mutagenesis induced by proteotoxic stress. PLoS Genet 2013; 9(8)e1003680
[http://dx.doi.org/10.1371/journal.pgen.1003680] [PMID: 23935537]

[37] Murray JI, Whitfield ML, Trinklein ND, Myers RM, Brown PO, Botstein D. Diverse and specific gene expression responses to stresses in cultured human cells. Mol Biol Cell 2004; 15(5): 2361-74.
[http://dx.doi.org/10.1091/mbc.e03-11-0799] [PMID: 15004229]

[38] Stevens H, Williams AB, Michael WM. Cell-Type Specific Responses to DNA Replication Stress in Early C. elegans Embryos. PLoS One 2016; 11(10)e0164601 [http://dx.doi.org/10.1371/journal.pone.0164601] [PMID: 27727303]

[39] Vihervaara A, Duarte FM, Lis JT. Molecular mechanisms driving transcriptional stress responses. Nat Rev Genet 2018; 19(6): 385-97. [http://dx.doi.org/10.1038/s41576-018-0001-6] [PMID: 29556092]

[40] Xue Y, Acar M. Mechanisms for the epigenetic inheritance of stress response in single cells. Curr Genet 2018; 64(6): 1221-8. [http://dx.doi.org/10.1007/s00294-018-0849-1] [PMID: 29846762]

[41] Cramer T, Rosenberg T, Kisliouk T, Meiri N. PARP Inhibitor Affects Long-term Heat-stress Response *via* Changes in DNA Methylation. Neuroscience 2019; 399: 65-76. [http://dx.doi.org/10.1016/j.neuroscience.2018.12.018] [PMID: 30579833]

[42] Sabari BR, Zhang D, Allis CD, Zhao Y. Metabolic regulation of gene expression through histone acylations. Nat Rev Mol Cell Biol 2017; 18(2): 90-101. [http://dx.doi.org/10.1038/nrm.2016.140] [PMID: 27924077]

[43] Li X, Xiong X, Wang K, *et al.* Transcriptome-wide mapping reveals reversible and dynamic N(1)-methyladenosine methylome. Nat Chem Biol 2016; 12(5): 311-6. [http://dx.doi.org/10.1038/nchembio.2040] [PMID: 26863410]

[44] Morgan JT, Fink GR, Bartel DP. Excised linear introns regulate growth in yeast. Nature 2019; 565(7741): 606-11. [http://dx.doi.org/10.1038/s41586-018-0828-1] [PMID: 30651636]

[45] Parenteau J, Maignon L, Berthoumieux M, Catala M, Gagnon V, Abou Elela S. Introns are mediators of cell response to starvation. Nature 2019; 565(7741): 612-7. [http://dx.doi.org/10.1038/s41586-018-0859-7] [PMID: 30651641]

[46] Olejniczak M, Kotowska-Zimmer A, Krzyzosiak W. Stress-induced changes in miRNA biogenesis and functioning. Cell Mol Life Sci 2018; 75(2): 177-91. [http://dx.doi.org/10.1007/s00018-017-2591-0] [PMID: 28717872]

[47] Hwang J, Qi L. Quality Control in the Endoplasmic Reticulum: Crosstalk between ERAD and UPR pathways. Trends Biochem Sci 2018; 43(8): 593-605. [http://dx.doi.org/10.1016/j.tibs.2018.06.005] [PMID: 30056836]

[48] Schultz ML, Krus KL, Kaushik S, *et al.* Coordinate regulation of mutant NPC1 degradation by selective ER autophagy and MARCH6-dependent ERAD. Nat Commun 2018; 9(1): 3671. [http://dx.doi.org/10.1038/s41467-018-06115-2] [PMID: 30202070]

[49] Almanza A, Carlesso A, Chintha C, *et al.* Endoplasmic reticulum stress signalling - from basic mechanisms to clinical applications. FEBS J 2019; 286(2): 241-78. [http://dx.doi.org/10.1111/febs.14608] [PMID: 30027602]

[50] Frakes AE, Dillin A. The UPR^{ER}: Sensor and Coordinator of Organismal Homeostasis. Mol Cell 2017; 66(6): 761-71. [http://dx.doi.org/10.1016/j.molcel.2017.05.031] [PMID: 28622521]

[51] Gardner BM, Walter P. Unfolded proteins are Ire1-activating ligands that directly induce the unfolded protein response. Science 2011; 333(6051): 1891-4. [http://dx.doi.org/10.1126/science.1209126] [PMID: 21852455]

[52] Halbleib K, Pesek K, Covino R, *et al.* Activation of the Unfolded Protein Response by Lipid Bilayer Stress. Mol Cell 2017; 67(4): 673-684.e8. [http://dx.doi.org/10.1016/j.molcel.2017.06.012] [PMID: 28689662]

[53] Volmer R, van der Ploeg K, Ron D. Membrane lipid saturation activates endoplasmic reticulum unfolded protein response transducers through their transmembrane domains. Proc Natl Acad Sci USA

2013; 110(12): 4628-33.
[http://dx.doi.org/10.1073/pnas.1217611110] [PMID: 23487760]

[54] Covino R, Hummer G, Ernst R. Integrated Functions of Membrane Property Sensors and a Hidden Side of the Unfolded Protein Response. Mol Cell 2018; 71(3): 458-67.
[http://dx.doi.org/10.1016/j.molcel.2018.07.019] [PMID: 30075144]

[55] Tam AB, Roberts LS, Chandra V, *et al.* The UPR Activator ATF6 Responds to Proteotoxic and Lipotoxic Stress by Distinct Mechanisms. Dev Cell 2018; 46(3): 327-343.e7.
[http://dx.doi.org/10.1016/j.devcel.2018.04.023] [PMID: 30086303]

[56] Papaioannou A, Higa A, Jégou G, *et al.* Alterations of EDEM1 functions enhance ATF6 pro-survival signaling. FEBS J 2018; 285(22): 4146-64.
[http://dx.doi.org/10.1111/febs.14669] [PMID: 30281916]

[57] Tam AB, Koong AC, Niwa M. Ire1 has distinct catalytic mechanisms for XBP1/HAC1 splicing and RIDD. Cell Rep 2014; 9(3): 850-8.
[http://dx.doi.org/10.1016/j.celrep.2014.09.016] [PMID: 25437541]

[58] Walter F, O'Brien A, Concannon CG, Düssmann H, Prehn JHM. ER stress signaling has an activating transcription factor 6α (ATF6)-dependent "off-switch". J Biol Chem 2018; 293(47): 18270-84.
[http://dx.doi.org/10.1074/jbc.RA118.002121] [PMID: 30287689]

[59] Yang Y, Liu L, Naik I, Braunstein Z, Zhong J, Ren B. Transcription Factor C/EBP Homologous Protein in Health and Diseases. Front Immunol 2017; 8: 1612.
[http://dx.doi.org/10.3389/fimmu.2017.01612] [PMID: 29230213]

[60] Lin JH, Li H, Yasumura D, *et al.* IRE1 signaling affects cell fate during the unfolded protein response. Science 2007; 318(5852): 944-9.
[http://dx.doi.org/10.1126/science.1146361] [PMID: 17991856]

[61] Fink EE, Moparthy S, Bagati A, *et al.* XBP1-KLF9 Axis Acts as a Molecular Rheostat to Control the Transition from Adaptive to Cytotoxic Unfolded Protein Response. Cell Rep 2018; 25(1): 212-223.e4.
[http://dx.doi.org/10.1016/j.celrep.2018.09.013] [PMID: 30282030]

[62] Fregno I, Molinari M. Endoplasmic reticulum turnover: ER-phagy and other flavors in selective and non-selective ER clearance. F1000 Res 2018; 7: 454.
[http://dx.doi.org/10.12688/f1000research.13968.1] [PMID: 29744037]

[63] Brandt C, Nolte H, Henschke S, *et al.* Food Perception Primes Hepatic ER Homeostasis *via* Melanocortin-Dependent Control of mTOR Activation. Cell 2018; 175(5): 1321-1335.e20.
[http://dx.doi.org/10.1016/j.cell.2018.10.015] [PMID: 30445039]

[64] Higuchi-Sanabria R, Frankino PA, Paul JW III, Tronnes SU, Dillin A. A Futile Battle? Protein Quality Control and the Stress of Aging. Dev Cell 2018; 44(2): 139-63.
[http://dx.doi.org/10.1016/j.devcel.2017.12.020] [PMID: 29401418]

[65] Crookshank JA, Serrano D, Wang GS, *et al.* Changes in insulin, glucagon and ER stress precede immune activation in type 1 diabetes. J Endocrinol 2018; 239(2): 181-95.
[http://dx.doi.org/10.1530/JOE-18-0328] [PMID: 30139929]

[66] Hanaoka M, Ishikawa T, Ishiguro M, *et al.* Expression of ATF6 as a marker of pre-cancerous atypical change in ulcerative colitis-associated colorectal cancer: a potential role in the management of dysplasia. J Gastroenterol 2018; 53(5): 631-41.
[http://dx.doi.org/10.1007/s00535-017-1387-1] [PMID: 28884228]

[67] Valenzuela V, Jackson KL, Sardi SP, Hetz C. Gene Therapy Strategies to Restore ER Proteostasis in Disease. Molecular therapy: the journal of the American Society of Gene Therapy 2018; 26(6): 1404-3.
[http://dx.doi.org/10.1016/j.ymthe.2018.04.004]

[68] Sheng X, Nenseth HZ, Qu S, *et al.* IRE1α-XBP1s pathway promotes prostate cancer by activating c-MYC signaling. Nat Commun 2019; 10(1): 323.

[http://dx.doi.org/10.1038/s41467-018-08152-3] [PMID: 30679434]

[69] Chun Y, Kim J. Autophagy: An Essential Degradation Program for Cellular Homeostasis and Life. Cells 2018; 7(12)E278 [http://dx.doi.org/10.3390/cells7120278] [PMID: 30572663]

[70] Hansen M, Rubinsztein DC, Walker DW. Autophagy as a promoter of longevity: Insights from model organisms. Nat Rev Mol Cell Biol 2018; 19(9): 579-93. [http://dx.doi.org/10.1038/s41580-018-0033-y] [PMID: 30006559]

[71] Thorburn A. Autophagy and disease. J Biol Chem 2018; 293(15): 5425-30. [http://dx.doi.org/10.1074/jbc.R117.810739] [PMID: 29191833]

[72] Levine B, Klionsky DJ. Autophagy wins the 2016 Nobel Prize in Physiology or Medicine: Breakthroughs in baker's yeast fuel advances in biomedical research. Proc Natl Acad Sci USA 2017; 114(2): 201-5. [http://dx.doi.org/10.1073/pnas.1619876114] [PMID: 28039434]

[73] Corona Velazquez AF, Jackson WT. So Many Roads: The Multifaceted Regulation of Autophagy Induction. Mol Cell Biol 2018; 38(21): e00303-18. [http://dx.doi.org/10.1128/MCB.00303-18] [PMID: 30126896]

[74] Hurley JH, Young LN. Mechanisms of Autophagy Initiation. Annu Rev Biochem 2017; 86: 225-44. [http://dx.doi.org/10.1146/annurev-biochem-061516-044820] [PMID: 28301741]

[75] Harada K, Kotani T, Kirisako H, *et al.* Two distinct mechanisms target the autophagy-related E3 complex to the pre-autophagosomal structure. eLife 2019; 8e43088 [http://dx.doi.org/10.7554/eLife.43088] [PMID: 30810528]

[76] Nishimura T, Tamura N, Kono N, *et al.* Autophagosome formation is initiated at phosphatidylinositol synthase-enriched ER subdomains. EMBO J 2017; 36(12): 1719-35. [http://dx.doi.org/10.15252/embj.201695189] [PMID: 28495679]

[77] Hu X, Mullins RD. LC3 and STRAP regulate actin filament assembly by JMY during autophagosome formation. J Cell Biol 2019; 218(1): 251-66. [http://dx.doi.org/10.1083/jcb.201802157] [PMID: 30420355]

[78] Chowdhury S, Otomo C, Leitner A, *et al.* Insights into autophagosome biogenesis from structural and biochemical analyses of the ATG2A-WIPI4 complex. Proc Natl Acad Sci USA 2018; 115(42): E9792-801. [http://dx.doi.org/10.1073/pnas.1811874115] [PMID: 30185561]

[79] Kotani T, Kirisako H, Koizumi M, Ohsumi Y, Nakatogawa H. The Atg2-Atg18 complex tethers pre-autophagosomal membranes to the endoplasmic reticulum for autophagosome formation. Proc Natl Acad Sci USA 2018; 115(41): 10363-8. [http://dx.doi.org/10.1073/pnas.1806727115] [PMID: 30254161]

[80] Zhao YG, Zhang H. Formation and maturation of autophagosomes in higher eukaryotes: a social network. Curr Opin Cell Biol 2018; 53: 29-36. [http://dx.doi.org/10.1016/j.ceb.2018.04.003] [PMID: 29727743]

[81] Kaushik S, Cuervo AM. The coming of age of chaperone-mediated autophagy. Nat Rev Mol Cell Biol 2018; 19(6): 365-81. [http://dx.doi.org/10.1038/s41580-018-0001-6] [PMID: 29626215]

[82] Anding AL, Baehrecke EH. Cleaning House: Selective Autophagy of Organelles. Dev Cell 2017; 41(1): 10-22. [http://dx.doi.org/10.1016/j.devcel.2017.02.016] [PMID: 28399394]

[83] Palikaras K, Lionaki E, Tavernarakis N. Mechanisms of mitophagy in cellular homeostasis, physiology and pathology. Nat Cell Biol 2018; 20(9): 1013-22. [http://dx.doi.org/10.1038/s41556-018-0176-2] [PMID: 30154567]

[84] Wyant GA, Abu-Remaileh M, Frenkel EM, *et al.* NUFIP1 is a ribosome receptor for starvation-induced ribophagy. Science 2018; 360(6390): 751-8. [http://dx.doi.org/10.1126/science.aar2663] [PMID: 29700228]

[85] Iwama R, Ohsumi Y. Analysis of autophagy activated during changes in carbon source availability in yeast cells. J Biol Chem 2019; 294(14): 5590-603. [http://dx.doi.org/10.1074/jbc.RA118.005698] [PMID: 30755486]

[86] Doherty J, Baehrecke EH. Life, death and autophagy. Nat Cell Biol 2018; 20(10): 1110-7. [http://dx.doi.org/10.1038/s41556-018-0201-5] [PMID: 30224761]

[87] Gradzka S, Thomas OS, Kretz O, *et al.* Inhibitor of apoptosis proteins are required for effective fusion of autophagosomes with lysosomes. Cell Death Dis 2018; 9(5): 529. [http://dx.doi.org/10.1038/s41419-018-0508-y] [PMID: 29743550]

[88] Bohovych I, Khalimonchuk O. Sending Out an SOS: Mitochondria as a Signaling Hub. Front Cell Dev Biol 2016; 4: 109. [http://dx.doi.org/10.3389/fcell.2016.00109] [PMID: 27790613]

[89] Toyama EQ, Herzig S, Courchet J, *et al.* Metabolism. AMP-activated protein kinase mediates mitochondrial fission in response to energy stress. Science 2016; 351(6270): 275-81. [http://dx.doi.org/10.1126/science.aab4138] [PMID: 26816379]

[90] Ruan L, Zhou C, Jin E, *et al.* Cytosolic proteostasis through importing of misfolded proteins into mitochondria. Nature 2017; 543(7645): 443-6. [http://dx.doi.org/10.1038/nature21695] [PMID: 28241148]

[91] Munro D, Treberg JR. A radical shift in perspective: Mitochondria as regulators of reactive oxygen species. J Exp Biol 2017; 220(Pt 7): 1170-80. [http://dx.doi.org/10.1242/jeb.132142] [PMID: 28356365]

[92] Mailloux RJ. Mitochondrial Antioxidants and the Maintenance of Cellular Hydrogen Peroxide Levels. Oxid Med Cell Longev 2018; 20187857251 [http://dx.doi.org/10.1155/2018/7857251] [PMID: 30057684]

[93] Quirós PM, Mottis A, Auwerx J. Mitonuclear communication in homeostasis and stress. Nat Rev Mol Cell Biol 2016; 17(4): 213-26. [http://dx.doi.org/10.1038/nrm.2016.23] [PMID: 26956194]

[94] Hill S, Sataranatarajan K, Van Remmen H. Role of Signaling Molecules in Mitochondrial Stress Response. Front Genet 2018; 9: 225. [http://dx.doi.org/10.3389/fgene.2018.00225] [PMID: 30042784]

[95] Kim KH, Son JM, Benayoun BA, Lee C. The Mitochondrial-Encoded Peptide MOTS-c Translocates to the Nucleus to Regulate Nuclear Gene Expression in Response to Metabolic Stress. Cell Metab 2018; 28(3): 516-524.e7. [http://dx.doi.org/10.1016/j.cmet.2018.06.008] [PMID: 29983246]

[96] Moehle EA, Shen K, Dillin A. Mitochondrial Proteostasis in the Context of Cellular and Organismal Health and Aging. J Biol Chem 2018. [http://dx.doi.org/10.1074/jbc.TM117]

[97] Naresh NU, Haynes CM. Signaling and Regulation of the Mitochondrial Unfolded Protein Response. Cold Spring Harb Perspect Biol 2019; 11(6)a033944 [http://dx.doi.org/10.1101/cshperspect.a033944] [PMID: 30617048]

[98] Melber A, Haynes CM. UPR^{mt} regulation and output: a stress response mediated by mitochondrial-nuclear communication. Cell Res 2018; 28(3): 281-95. [http://dx.doi.org/10.1038/cr.2018.16] [PMID: 29424373]

[99] Münch C, Harper JW. Mitochondrial unfolded protein response controls matrix pre-RNA processing and translation. Nature 2016; 534(7609): 710-3.

[http://dx.doi.org/10.1038/nature18302] [PMID: 27350246]

[100] Tian Y, Garcia G, Bian Q, *et al.* Mitochondrial Stress Induces Chromatin Reorganization to Promote Longevity and UPR(mt). Cell 2016; 165(5): 1197-208. [http://dx.doi.org/10.1016/j.cell.2016.04.011] [PMID: 27133166]

[101] Civelek M, Mehrkens JF, Carstens NM, Fitzenberger E, Wenzel U. Inhibition of mitophagy decreases survival of Caenorhabditis elegans by increasing protein aggregation. Mol Cell Biochem 2019; 452(1-2): 123-31. [http://dx.doi.org/10.1007/s11010-018-3418-5] [PMID: 30094600]

[102] Rath E, Moschetta A, Haller D. Mitochondrial function - gatekeeper of intestinal epithelial cell homeostasis. Nature reviews Gastroenterology & hepatology 2018; 15(8): 497-516.

[103] Michelini LG, Benevento CE, Rossato FA, Siqueira-Santos ES, Castilho RF. Effects of partial inhibition of respiratory complex I on H2O 2 production by isolated brain mitochondria in different respiratory states. Neurochem Res 2014; 39(12): 2419-30. [http://dx.doi.org/10.1007/s11064-014-1446-4] [PMID: 25287903]

[104] Thangaraj A, Periyasamy P, Liao K, *et al.* HIV 1 TAT mediated microglial activation: role of mitochondrial dysfunction and defective mitophagy. Autophagy 2018; 14(9): 1596-619. [http://dx.doi.org/10.1080/15548627.2018.1476810] [PMID: 29966509]

[105] Kang D, Kirienko DR, Webster P, Fisher AL, Kirienko NV. Pyoverdine, a siderophore from Pseudomonas aeruginosa, translocates into C. elegans, removes iron, and activates a distinct host response. Virulence 2018; 9(1): 804-17. [http://dx.doi.org/10.1080/21505594.2018.1449508] [PMID: 29532717]

[106] Saxton RA, Sabatini DM. mTOR Signaling in Growth, Metabolism, and Disease. Cell 2017; 168(6): 960-76. [http://dx.doi.org/10.1016/j.cell.2017.02.004] [PMID: 28283069]

[107] Sancak Y, Peterson TR, Shaul YD, *et al.* The Rag GTPases bind raptor and mediate amino acid signaling to mTORC1. Science 2008; 320(5882): 1496-501. [http://dx.doi.org/10.1126/science.1157535] [PMID: 18497260]

[108] Kim J, Kim E. Rag GTPase in amino acid signaling. Amino Acids 2016; 48(4): 915-28. [http://dx.doi.org/10.1007/s00726-016-2171-x] [PMID: 26781224]

[109] Efeyan A, Zoncu R, Chang S, *et al.* Regulation of mTORC1 by the Rag GTPases is necessary for neonatal autophagy and survival. Nature 2013; 493(7434): 679-83. [http://dx.doi.org/10.1038/nature11745] [PMID: 23263183]

[110] Sabatini DM. Twenty-five years of mTOR: Uncovering the link from nutrients to growth. Proc Natl Acad Sci USA 2017; 114(45): 11818-25. [http://dx.doi.org/10.1073/pnas.1716173114] [PMID: 29078414]

[111] Gu X, Orozco JM, Saxton RA, *et al.* SAMTOR is an *S*-adenosylmethionine sensor for the mTORC1 pathway. Science 2017; 358(6364): 813-8. [http://dx.doi.org/10.1126/science.aao3265] [PMID: 29123071]

[112] Latimer MN, Freij KW, Cleveland BM, Biga PR. Physiological and Molecular Mechanisms of Methionine Restriction. Front Endocrinol (Lausanne) 2018; 9: 217. [http://dx.doi.org/10.3389/fendo.2018.00217] [PMID: 29780356]

[113] Castellano BM, Thelen AM, Moldavski O, *et al.* Lysosomal cholesterol activates mTORC1 *via* an SLC38A9-Niemann-Pick C1 signaling complex. Science 2017; 355(6331): 1306-11. [http://dx.doi.org/10.1126/science.aag1417] [PMID: 28336668]

[114] Efeyan A, Comb WC, Sabatini DM. Nutrient-sensing mechanisms and pathways. Nature 2015; 517(7534): 302-10. [http://dx.doi.org/10.1038/nature14190] [PMID: 25592535]

[115] Lu M, Li C. Nutrient sensing in pancreatic islets: lessons from congenital hyperinsulinism and monogenic diabetes. Ann N Y Acad Sci 2018; 1411(1): 65-82. [http://dx.doi.org/10.1111/nyas.13448] [PMID: 29044608]

[116] Stuhlmann T, Planells-Cases R, Jentsch TJ. LRRC8/VRAC anion channels enhance β-cell glucose sensing and insulin secretion. Nat Commun 2018; 9(1): 1974. [http://dx.doi.org/10.1038/s41467-018-04353-y] [PMID: 29773801]

[117] Ramírez S, Gómez-Valadés AG, Schneeberger M, *et al.* Mitochondrial Dynamics Mediated by Mitofusin 1 Is Required for POMC Neuron Glucose-Sensing and Insulin Release Control. Cell Metab 2017; 25(6): 1390-1399.e6. [http://dx.doi.org/10.1016/j.cmet.2017.05.010] [PMID: 28591639]

[118] Kroemer G, López-Otín C, Madeo F, de Cabo R. Carbotoxicity-Noxious Effects of Carbohydrates. Cell 2018; 175(3): 605-14. [http://dx.doi.org/10.1016/j.cell.2018.07.044] [PMID: 30340032]

[119] Garcia D, Shaw RJ. AMPK: Mechanisms of Cellular Energy Sensing and Restoration of Metabolic Balance. Mol Cell 2017; 66(6): 789-800. [http://dx.doi.org/10.1016/j.molcel.2017.05.032] [PMID: 28622524]

[120] Hardie DG. Keeping the home fires burning: AMP-activated protein kinase. J R Soc Interface 2018; 15(138)20170774 [http://dx.doi.org/10.1098/rsif.2017.0774] [PMID: 29343628]

[121] Chandrashekarappa DG, McCartney RR, O'Donnell AF, Schmidt MC. The β subunit of yeast AMP-activated protein kinase directs substrate specificity in response to alkaline stress. Cell Signal 2016; 28(12): 1881-93. [http://dx.doi.org/10.1016/j.cellsig.2016.08.016] [PMID: 27592031]

[122] Seebacher F. The evolution of metabolic regulation in animals. Comparative biochemistry and physiology Part B, Biochemistry & molecular biology 2018; 224: 195-203. [http://dx.doi.org/10.1016/j.cbpb.2017.11.002]

[123] Hewitt G, Korolchuk VI. Repair, Reuse, Recycle: The Expanding Role of Autophagy in Genome Maintenance. Trends Cell Biol 2017; 27(5): 340-51. [http://dx.doi.org/10.1016/j.tcb.2016.11.011] [PMID: 28011061]

[124] Webster CM, Pino EC, Carr CE, *et al.* Genome-wide RNAi Screen for Fat Regulatory Genes in C. elegans Identifies a Proteostasis-AMPK Axis Critical for Starvation Survival. Cell Rep 2017; 20(3): 627-40. [http://dx.doi.org/10.1016/j.celrep.2017.06.068] [PMID: 28723566]

[125] Honigberg SM. v Similar environments but diverse fates: Responses of budding yeast to nutrient deprivation. Microbial cell (Graz, Austria) 2016; 3(8): 302-28.

[126] Argüello-Miranda O, Liu Y, Wood NE, Kositangool P, Doncic A. Integration of Multiple Metabolic Signals Determines Cell Fate Prior to Commitment. Mol Cell 2018; 71(5): 733-744.e11. [http://dx.doi.org/10.1016/j.molcel.2018.07.041] [PMID: 30174289]

[127] Schwartz S. Cracking the epitranscriptome. RNA 2016; 22(2): 169-74. [http://dx.doi.org/10.1261/rna.054502.115] [PMID: 26787305]

[128] Morano KA, Grant CM, Moye-Rowley WS. The response to heat shock and oxidative stress in Saccharomyces cerevisiae. Genetics 2012; 190(4): 1157-95. [http://dx.doi.org/10.1534/genetics.111.128033] [PMID: 22209905]

[129] Nakazawa N, Yanata H, Ito N, Kaneta E, Takahashi K. Oxidative stress tolerance of a spore clone isolated from Shirakami kodama yeast depends on altered regulation of Msn2 leading to enhanced expression of ROS-degrading enzymes. J Gen Appl Microbiol 2018; 64(4): 149-57. [http://dx.doi.org/10.2323/jgam.2017.11.002] [PMID: 29607878]

[130] Marguerat S, Lawler K, Brazma A, Bähler J. Contributions of transcription and mRNA decay to gene expression dynamics of fission yeast in response to oxidative stress. RNA Biol 2014; 11(6): 702-14. [http://dx.doi.org/10.4161/rna.29196] [PMID: 25007214]

[131] Thompson DM, Lu C, Green PJ, Parker R. tRNA cleavage is a conserved response to oxidative stress in eukaryotes. RNA 2008; 14(10): 2095-103. [http://dx.doi.org/10.1261/rna.1232808] [PMID: 18719243]

[132] Silva GM, Finley D, Vogel C. K63 polyubiquitination is a new modulator of the oxidative stress response. Nat Struct Mol Biol 2015; 22(2): 116-23. [http://dx.doi.org/10.1038/nsmb.2955] [PMID: 25622294]

[133] Schosserer M, Minois N, Angerer TB, *et al.* Methylation of ribosomal RNA by NSUN5 is a conserved mechanism modulating organismal lifespan. Nat Commun 2015; 6: 6158. [http://dx.doi.org/10.1038/ncomms7158] [PMID: 25635753]

[134] Tonelli C, Chio IIC, Tuveson DA. Transcriptional Regulation by Nrf2. Antioxid Redox Signal 2018; 29(17): 1727-45. [http://dx.doi.org/10.1089/ars.2017.7342] [PMID: 28899199]

[135] Guo Y, Yu S, Zhang C, Kong AT. Epigenetic regulation of Keap1-Nrf2 signaling. Free radical biology & medicine 2015; 88(Pt B): 337-49. [http://dx.doi.org/10.1016/j.freeradbiomed.2015.06.013]

[136] Tebay LE, Robertson H, Durant ST, Vitale SR, Penning TM, Dinkova-Kostova AT, *et al.* Mechanisms of activation of the transcription factor Nrf2 by redox stressors, nutrient cues, and energy status and the pathways through which it attenuates degenerative disease. Free radical biology & medicine 2015; 88(Pt B): 108-46. [http://dx.doi.org/10.1016/j.freeradbiomed.2015.06.021]

[137] Kobayashi M, Li L, Iwamoto N, *et al.* The antioxidant defense system Keap1-Nrf2 comprises a multiple sensing mechanism for responding to a wide range of chemical compounds. Mol Cell Biol 2009; 29(2): 493-502. [http://dx.doi.org/10.1128/MCB.01080-08] [PMID: 19001094]

[138] Rada P, Rojo AI, Chowdhry S, McMahon M, Hayes JD, Cuadrado A. SCF/beta-TrCP promotes glycogen synthase kinase 3-dependent degradation of the Nrf2 transcription factor in a Keap1-independent manner. Mol Cell Biol 2011; 31(6): 1121-33. [http://dx.doi.org/10.1128/MCB.01204-10] [PMID: 21245377]

[139] Huang HC, Nguyen T, Pickett CB. Regulation of the antioxidant response element by protein kinase C-mediated phosphorylation of NF-E2-related factor 2. Proc Natl Acad Sci USA 2000; 97(23): 12475-80. [http://dx.doi.org/10.1073/pnas.220418997] [PMID: 11035812]

[140] Lushchak VI. Adaptive response to oxidative stress: Bacteria, fungi, plants and animals. Comparative biochemistry and physiology Toxicology & pharmacology. CBP 2011; 153(2): 175-90.

[141] Lacher SE, Lee JS, Wang X, Campbell MR, Bell DA, Slattery M. Beyond antioxidant genes in the ancient Nrf2 regulatory network. Free radical biology & medicine 2015; 88(Pt B): 452-65. [http://dx.doi.org/10.1016/j.freeradbiomed.2015.06.044]

[142] Pajares M, Cuadrado A, Rojo AI. Modulation of proteostasis by transcription factor NRF2 and impact in neurodegenerative diseases. Redox Biol 2017; 11: 543-53. [http://dx.doi.org/10.1016/j.redox.2017.01.006] [PMID: 28104575]

[143] Kobayashi EH, Suzuki T, Funayama R, *et al.* Nrf2 suppresses macrophage inflammatory response by blocking proinflammatory cytokine transcription. Nat Commun 2016; 7: 11624. [http://dx.doi.org/10.1038/ncomms11624] [PMID: 27211851]

[144] Skibinski G, Hwang V, Ando DM, *et al.* Nrf2 mitigates LRRK2- and α-synuclein-induced neurodegeneration by modulating proteostasis. Proc Natl Acad Sci USA 2017; 114(5): 1165-70.

[http://dx.doi.org/10.1073/pnas.1522872114] [PMID: 28028237]

[145] Tsakiri EN, Gumeni S, Iliaki KK, *et al.* Hyperactivation of Nrf2 increases stress tolerance at the cost of aging acceleration due to metabolic deregulation. Aging Cell 2019; 18(1)e12845 [PMID: 30537423]

[146] Taguchi K, Yamamoto M. The KEAP1-NRF2 System in Cancer. Front Oncol 2017; 7: 85. [http://dx.doi.org/10.3389/fonc.2017.00085] [PMID: 28523248]

[147] Huppke P, Weissbach S, Church JA, *et al.* Activating *de novo* mutations in NFE2L2 encoding NRF2 cause a multisystem disorder. Nat Commun 2017; 8(1): 818. [http://dx.doi.org/10.1038/s41467-017-00932-7] [PMID: 29018201]

[148] Wakabayashi N, Itoh K, Wakabayashi J, *et al.* Keap1-null mutation leads to postnatal lethality due to constitutive Nrf2 activation. Nat Genet 2003; 35(3): 238-45. [http://dx.doi.org/10.1038/ng1248] [PMID: 14517554]

[149] Mitchell S, Vargas J, Hoffmann A. Signaling *via* the NFκB system. Wiley Interdiscip Rev Syst Biol Med 2016; 8(3): 227-41. [http://dx.doi.org/10.1002/wsbm.1331] [PMID: 26990581]

[150] Sakauchi C, Wakatsuki H, Ichijo H, Hattori K. Pleiotropic properties of ASK1. Biochim Biophys Acta, Gen Subj 2017; 1861 (1 Pt A): 3030-8. [http://dx.doi.org/10.1016/j.bbagen.2016.09.028] [PMID: 27693599]

[151] Geleziunas R, Xu W, Takeda K, Ichijo H, Greene WC. HIV-1 Nef inhibits ASK1-dependent death signalling providing a potential mechanism for protecting the infected host cell. Nature 2001; 410(6830): 834-8. [http://dx.doi.org/10.1038/35071111] [PMID: 11298454]

[152] Rietman ML, Spijkerman AMW, Wong A, *et al.* Antioxidants linked with physical, cognitive and psychological frailty: Analysis of candidate biomarkers and markers derived from the MARK-AGE study. Mech Ageing Dev 2019; 177: 135-43. [http://dx.doi.org/10.1016/j.mad.2018.04.007] [PMID: 29719199]

[153] Poprac P, Jomova K, Simunkova M, Kollar V, Rhodes CJ, Valko M. Targeting Free Radicals in Oxidative Stress-Related Human Diseases. Trends Pharmacol Sci 2017; 38(7): 592-607. [http://dx.doi.org/10.1016/j.tips.2017.04.005] [PMID: 28551354]

[154] Gomez-Pastor R, Burchfiel ET, Thiele DJ. Regulation of heat shock transcription factors and their roles in physiology and disease. Nat Rev Mol Cell Biol 2018; 19(1): 4-19. [http://dx.doi.org/10.1038/nrm.2017.73] [PMID: 28852220]

[155] Prodromou C. Regulatory Mechanisms of Hsp90. Biochemistry & molecular biology journal 2017; 3(1): 2. [http://dx.doi.org/10.21767/2471-8084.100030]

[156] Krakowiak J, Zheng X, Patel N, *et al.* Hsf1 and Hsp70 constitute a two-component feedback loop that regulates the yeast heat shock response. eLife 2018; 7e31668 [http://dx.doi.org/10.7554/eLife.31668] [PMID: 29393852]

[157] Hentze N, Le Breton L, Wiesner J, Kempf G, Mayer MP. Molecular mechanism of thermosensory function of human heat shock transcription factor Hsf1. eLife 2016; 5e11576 [http://dx.doi.org/10.7554/eLife.11576] [PMID: 26785146]

[158] Vihervaara A, Mahat DB, Guertin MJ, *et al.* Transcriptional response to stress is pre-wired by promoter and enhancer architecture. Nat Commun 2017; 8(1): 255. [http://dx.doi.org/10.1038/s41467-017-00151-0] [PMID: 28811569]

[159] Solís EJ, Pandey JP, Zheng X, *et al.* Defining the Essential Function of Yeast Hsf1 Reveals a Compact Transcriptional Program for Maintaining Eukaryotic Proteostasis. Mol Cell 2016; 63(1): 60-71. [http://dx.doi.org/10.1016/j.molcel.2016.05.014] [PMID: 27320198]

[160] Li J, Labbadia J, Morimoto RI. Rethinking HSF1 in Stress, Development, and Organismal Health. Trends Cell Biol 2017; 27(12): 895-905.
[http://dx.doi.org/10.1016/j.tcb.2017.08.002] [PMID: 28890254]

[161] Barna J, Csermely P, Vellai T. Roles of heat shock factor 1 beyond the heat shock response. Cell Mol Life Sci 2018; 75(16): 2897-916.
[http://dx.doi.org/10.1007/s00018-018-2836-6] [PMID: 29774376]

[162] Jarolim S, Ayer A, Pillay B, *et al.* Saccharomyces cerevisiae genes involved in survival of heat shock. G3 (Bethesda) 2013; 3(12): 2321-33.
[http://dx.doi.org/10.1534/g3.113.007971] [PMID: 24142923]

[163] Castells-Roca L, García-Martínez J, Moreno J, Herrero E, Bellí G, Pérez-Ortín JE. Heat shock response in yeast involves changes in both transcription rates and mRNA stabilities. PLoS One 2011; 6(2)e17272
[http://dx.doi.org/10.1371/journal.pone.0017272] [PMID: 21364882]

[164] Petrova NV, Velichko AK, Razin SV, Kantidze OL. Early S-phase cell hypersensitivity to heat stress. Cell Cycle 2016; 15(3): 337-44.
[http://dx.doi.org/10.1080/15384101.2015.1127477] [PMID: 26689112]

[165] Weinhandl K, Winkler M, Glieder A, Camattari A. Carbon source dependent promoters in yeasts. Microb Cell Fact 2014; 13: 5.
[http://dx.doi.org/10.1186/1475-2859-13-5] [PMID: 24401081]

[166] Henderson CM, Block DE. Examining the role of membrane lipid composition in determining the ethanol tolerance of Saccharomyces cerevisiae. Appl Environ Microbiol 2014; 80(10): 2966-72.
[http://dx.doi.org/10.1128/AEM.04151-13] [PMID: 24610851]

[167] Hagiwara D, Sakamoto K, Abe K, Gomi K. Signaling pathways for stress responses and adaptation in Aspergillus species: stress biology in the post-genomic era. Biosci Biotechnol Biochem 2016; 80(9): 1667-80.
[http://dx.doi.org/10.1080/09168451.2016.1162085] [PMID: 27007956]

[168] Gasch AP, Yu FB, Hose J, *et al.* Single-cell RNA sequencing reveals intrinsic and extrinsic regulatory heterogeneity in yeast responding to stress. PLoS Biol 2017; 15(12)e2004050
[http://dx.doi.org/10.1371/journal.pbio.2004050] [PMID: 29240790]

[169] Solé C, Nadal-Ribelles M, de Nadal E, Posas F. A novel role for lncRNAs in cell cycle control during stress adaptation. Curr Genet 2015; 61(3): 299-308.
[http://dx.doi.org/10.1007/s00294-014-0453-y] [PMID: 25262381]

[170] Gunde-Cimerman N, Plemenitaš A, Oren A. Strategies of adaptation of microorganisms of the three domains of life to high salt concentrations. FEMS Microbiol Rev 2018; 42(3): 353-75.
[http://dx.doi.org/10.1093/femsre/fuy009] [PMID: 29529204]

[171] Jacobsen MD, Beynon RJ, Gethings LA, *et al.* Specificity of the osmotic stress response in *Candida albicans* highlighted by quantitative proteomics. Sci Rep 2018; 8(1): 14492.
[http://dx.doi.org/10.1038/s41598-018-32792-6] [PMID: 30262823]

[172] Weinisch L, Kühner S, Roth R, *et al.* Identification of osmoadaptive strategies in the halophile, heterotrophic ciliate Schmidingerothrix salinarum. PLoS Biol 2018; 16(1)e2003892
[http://dx.doi.org/10.1371/journal.pbio.2003892] [PMID: 29357351]

[173] Ene IV, Walker LA, Schiavone M, *et al.* Cell Wall Remodeling Enzymes Modulate Fungal Cell Wall Elasticity and Osmotic Stress Resistance. MBio 2015; 6(4)e00986
[http://dx.doi.org/10.1128/mBio.00986-15] [PMID: 26220968]

[174] Oparil S, Acelajado MC, Bakris GL, *et al.* Hypertension. Nat Rev Dis Primers 2018; 4: 18014.
[http://dx.doi.org/10.1038/nrdp.2018.14] [PMID: 29565029]

[175] Weinert TA, Hartwell LH. The RAD9 gene controls the cell cycle response to DNA damage in

Saccharomyces cerevisiae. Science 1988; 241(4863): 317-22. [http://dx.doi.org/10.1126/science.3291120] [PMID: 3291120]

[176] Blackford AN, Jackson SP. ATM, ATR, and DNA-PK: The Trinity at the Heart of the DNA Damage Response. Mol Cell 2017; 66(6): 801-17. [http://dx.doi.org/10.1016/j.molcel.2017.05.015] [PMID: 28622525]

[177] Imseng S, Aylett CH, Maier T. Architecture and activation of phosphatidylinositol 3-kinase related kinases. Curr Opin Struct Biol 2018; 49: 177-89. [http://dx.doi.org/10.1016/j.sbi.2018.03.010] [PMID: 29625383]

[178] Elledge SJ. The DNA Damage Response--Self-awareness for DNA: The 2015 Albert Lasker Basic Medical Research Award. JAMA 2015; 314(11): 1111-2. [http://dx.doi.org/10.1001/jama.2015.10387] [PMID: 26348099]

[179] Matsuoka S, Ballif BA, Smogorzewska A, *et al.* ATM and ATR substrate analysis reveals extensive protein networks responsive to DNA damage. Science 2007; 316(5828): 1160-6. [http://dx.doi.org/10.1126/science.1140321] [PMID: 17525332]

[180] Zou L, Elledge SJ. Sensing DNA damage through ATRIP recognition of RPA-ssDNA complexes. Science 2003; 300(5625): 1542-8. [http://dx.doi.org/10.1126/science.1083430] [PMID: 12791985]

[181] Kumar A, Mazzanti M, Mistrik M, *et al.* ATR mediates a checkpoint at the nuclear envelope in response to mechanical stress. Cell 2014; 158(3): 633-46. [http://dx.doi.org/10.1016/j.cell.2014.05.046] [PMID: 25083873]

[182] Barton LJ, Duan T, Ke W, *et al.* Nuclear lamina dysfunction triggers a germline stem cell checkpoint. Nat Commun 2018; 9(1): 3960. [http://dx.doi.org/10.1038/s41467-018-06277-z] [PMID: 30262885]

[183] Paull TT. Mechanisms of ATM Activation. Annu Rev Biochem 2015; 84: 711-38. [http://dx.doi.org/10.1146/annurev-biochem-060614-034335] [PMID: 25580527]

[184] Kaidi A, Jackson SP. KAT5 tyrosine phosphorylation couples chromatin sensing to ATM signalling. Nature 2013; 498(7452): 70-4. [http://dx.doi.org/10.1038/nature12201] [PMID: 23708966]

[185] Tresini M, Warmerdam DO, Kolovos P, *et al.* The core spliceosome as target and effector of non-canonical ATM signalling. Nature 2015; 523(7558): 53-8. [http://dx.doi.org/10.1038/nature14512] [PMID: 26106861]

[186] Ortega-Atienza S, Wong VC, DeLoughery Z, Luczak MW, Zhitkovich A. ATM and KAT5 safeguard replicating chromatin against formaldehyde damage. Nucleic Acids Res 2016; 44(1): 198-209. [http://dx.doi.org/10.1093/nar/gkv957] [PMID: 26420831]

[187] Guo Z, Kozlov S, Lavin MF, Person MD, Paull TT. ATM activation by oxidative stress. Science 2010; 330(6003): 517-21. [http://dx.doi.org/10.1126/science.1192912] [PMID: 20966255]

[188] Kozlov SV, Waardenberg AJ, Engholm-Keller K, Arthur JW, Graham ME, Lavin M. Reactive Oxygen Species (ROS)-Activated ATM-Dependent Phosphorylation of Cytoplasmic Substrates Identified by Large-Scale Phosphoproteomics Screen. Mol Cell Proteomics 2016; 15(3): 1032-47. [http://dx.doi.org/10.1074/mcp.M115.055723] [PMID: 26699800]

[189] Berger ND, Stanley FKT, Moore S, Goodarzi AA. ATM-dependent pathways of chromatin remodelling and oxidative DNA damage responses. Philos Trans R Soc Lond B Biol Sci 2017; 372(1731): 20160283.

[190] Eapen VV, Waterman DP, Bernard A, *et al.* A pathway of targeted autophagy is induced by DNA damage in budding yeast. Proc Natl Acad Sci USA 2017; 114(7): E1158-67. [http://dx.doi.org/10.1073/pnas.1614364114] [PMID: 28154131]

[191] Lane D, Levine A. p53 Research: the past thirty years and the next thirty years. Cold Spring Harb Perspect Biol 2010; 2(12)a000893
[http://dx.doi.org/10.1101/cshperspect.a000893] [PMID: 20463001]

[192] Hafner A, Bulyk ML, Jambhekar A, Lahav G. The multiple mechanisms that regulate p53 activity and cell fate. Nat Rev Mol Cell Biol 2019; 20(4): 199-210.
[http://dx.doi.org/10.1038/s41580-019-0110-x] [PMID: 30824861]

[193] Marnef A, Legube G. Organizing DNA repair in the nucleus: DSBs hit the road. Curr Opin Cell Biol 2017; 46: 1-8.
[http://dx.doi.org/10.1016/j.ceb.2016.12.003] [PMID: 28068556]

[194] Adar S, Hu J, Lieb JD, Sancar A. Genome-wide kinetics of DNA excision repair in relation to chromatin state and mutagenesis. Proc Natl Acad Sci USA 2016; 113(15): E2124-33.
[http://dx.doi.org/10.1073/pnas.1603388113] [PMID: 27036006]

[195] Wright RH, Lioutas A, Le Dily F, *et al.* ADP-ribose-derived nuclear ATP synthesis by NUDIX5 is required for chromatin remodeling. Science 2016; 352(6290): 1221-5.
[http://dx.doi.org/10.1126/science.aad9335] [PMID: 27257257]

[196] Williams AB, Schumacher B. DNA damage responses and stress resistance: Concepts from bacterial SOS to metazoan immunity. Mech Ageing Dev 2017; Pt A: 27-32.

[197] Corcoles-Saez I, Dong K, Johnson AL, *et al.* Essential Function of Mec1, the Budding Yeast ATM/ATR Checkpoint-Response Kinase, in Protein Homeostasis. Dev Cell 2018; 46(4): 495-503.e2.
[http://dx.doi.org/10.1016/j.devcel.2018.07.011] [PMID: 30130531]

[198] Pancholi NJ, Price AM, Weitzman MD. Take your PIKK: tumour viruses and DNA damage response pathways. Philos Trans R Soc Lond B Biol Sci 2017; 372(1732): 20160269.
[http://dx.doi.org/10.1098/rstb.2016.0269]

[199] Jeggo PA, Pearl LH, Carr AM. DNA repair, genome stability and cancer: a historical perspective. Nat Rev Cancer 2016; 16(1): 35-42.
[http://dx.doi.org/10.1038/nrc.2015.4] [PMID: 26667849]

[200] Pilié PG, Tang C, Mills GB, Yap TA. State-of-the-art strategies for targeting the DNA damage response in cancer. Nat Rev Clin Oncol 2019; 16(2): 81-104.
[http://dx.doi.org/10.1038/s41571-018-0114-z] [PMID: 30356138]

[201] Falcone C, Mazzoni C. External and internal triggers of cell death in yeast. Cell Mol Life Sci 2016; 73(11-12): 2237-50.
[http://dx.doi.org/10.1007/s00018-016-2197-y] [PMID: 27048816]

[202] Askew K, Li K, Olmos-Alonso A, *et al.* Coupled Proliferation and Apoptosis Maintain the Rapid Turnover of Microglia in the Adult Brain. Cell Rep 2017; 18(2): 391-405.
[http://dx.doi.org/10.1016/j.celrep.2016.12.041] [PMID: 28076784]

[203] Kabbage M, Kessens R, Bartholomay LC, Williams B. The Life and Death of a Plant Cell. Annu Rev Plant Biol 2017; 68: 375-404.
[http://dx.doi.org/10.1146/annurev-arplant-043015-111655] [PMID: 28125285]

[204] Weinlich R, Oberst A, Beere HM, Green DR. Necroptosis in development, inflammation and disease. Nat Rev Mol Cell Biol 2017; 18(2): 127-36.
[http://dx.doi.org/10.1038/nrm.2016.149] [PMID: 27999438]

[205] Nagata S. Apoptosis and Clearance of Apoptotic Cells. Annu Rev Immunol 2018; 36: 489-517.
[http://dx.doi.org/10.1146/annurev-immunol-042617-053010] [PMID: 29400998]

[206] Galluzzi L, Vitale I, Aaronson SA, *et al.* Molecular mechanisms of cell death: recommendations of the Nomenclature Committee on Cell Death 2018. Cell Death Differ 2018; 25(3): 486-541.
[http://dx.doi.org/10.1038/s41418-017-0012-4] [PMID: 29362479]

[207] Kerr JF, Wyllie AH, Currie AR. Apoptosis: a basic biological phenomenon with wide-ranging

implications in tissue kinetics. Br J Cancer 1972; 26(4): 239-57. [http://dx.doi.org/10.1038/bjc.1972.33] [PMID: 4561027]

[208] Yuan J, Shaham S, Ledoux S, Ellis HM, Horvitz HR. The C. elegans cell death gene ced-3 encodes a protein similar to mammalian interleukin-1 beta-converting enzyme. Cell 1993; 75(4): 641-52. [http://dx.doi.org/10.1016/0092-8674(93)90485-9] [PMID: 8242740]

[209] Aravind L, Dixit VM, Koonin EV. Apoptotic molecular machinery: vastly increased complexity in vertebrates revealed by genome comparisons. Science 2001; 291(5507): 1279-84. [http://dx.doi.org/10.1126/science.291.5507.1279] [PMID: 11181990]

[210] Zmasek CM, Godzik A. Evolution of the animal apoptosis network. Cold Spring Harb Perspect Biol 2013; 5(3)a008649 [http://dx.doi.org/10.1101/cshperspect.a008649] [PMID: 23457257]

[211] Saleh A, Srinivasula SM, Acharya S, Fishel R, Alnemri ES. Cytochrome c and dATP-mediated oligomerization of Apaf-1 is a prerequisite for procaspase-9 activation. J Biol Chem 1999; 274(25): 17941-5. [http://dx.doi.org/10.1074/jbc.274.25.17941] [PMID: 10364241]

[212] Dorstyn L, Akey CW, Kumar S. New insights into apoptosome structure and function. Cell Death Differ 2018; 25(7): 1194-208. [http://dx.doi.org/10.1038/s41418-017-0025-z] [PMID: 29765111]

[213] Kalkavan H, Green DR. MOMP, cell suicide as a BCL-2 family business. Cell Death Differ 2018; 25(1): 46-55. [http://dx.doi.org/10.1038/cdd.2017.179] [PMID: 29053143]

[214] Kale J, Osterlund EJ, Andrews DW. BCL-2 family proteins: changing partners in the dance towards death. Cell Death Differ 2018; 25(1): 65-80. [http://dx.doi.org/10.1038/cdd.2017.186] [PMID: 29149100]

[215] Bridgham JT, Wilder JA, Hollocher H, Johnson AL. All in the family: evolutionary and functional relationships among death receptors. Cell Death Differ 2003; 10(1): 19-25. [http://dx.doi.org/10.1038/sj.cdd.4401174] [PMID: 12655292]

[216] Scott FL, Stec B, Pop C, *et al.* The Fas-FADD death domain complex structure unravels signalling by receptor clustering. Nature 2009; 457(7232): 1019-22. [http://dx.doi.org/10.1038/nature07606] [PMID: 19118384]

[217] Galluzzi L, Bravo-San Pedro JM, Kepp O, Kroemer G. Regulated cell death and adaptive stress responses. Cell Mol Life Sci 2016; 73(11-12): 2405-10. [http://dx.doi.org/10.1007/s00018-016-2209-y] [PMID: 27048813]

[218] Kaiser WJ, Upton JW, Long AB, *et al.* RIP3 mediates the embryonic lethality of caspase-8-deficient mice. Nature 2011; 471(7338): 368-72. [http://dx.doi.org/10.1038/nature09857] [PMID: 21368762]

[219] Vanden Berghe T, Hassannia B, Vandenabeele P. An outline of necrosome triggers. Cell Mol Life Sci 2016; 73(11-12): 2137-52. [http://dx.doi.org/10.1007/s00018-016-2189-y] [PMID: 27052312]

[220] Elliott MR, Ravichandran KS. The Dynamics of Apoptotic Cell Clearance. Dev Cell 2016; 38(2): 147-60. [http://dx.doi.org/10.1016/j.devcel.2016.06.029] [PMID: 27459067]

[221] Barth ND, Marwick JA, Vendrell M, Rossi AG, Dransfield I. The “Phagocytic Synapse” and Clearance of Apoptotic Cells. Front Immunol 2017; 8: 1708. [http://dx.doi.org/10.3389/fimmu.2017.01708] [PMID: 29255465]

[222] Cullen SP, Henry CM, Kearney CJ, *et al.* Fas/CD95-induced chemokines can serve as “find-me” signals for apoptotic cells. Mol Cell 2013; 49(6): 1034-48. [http://dx.doi.org/10.1016/j.molcel.2013.01.025] [PMID: 23434371]

[223] Dixon SJ, Lemberg KM, Lamprecht MR, *et al.* Ferroptosis: an iron-dependent form of nonapoptotic cell death. Cell 2012; 149(5): 1060-72. [http://dx.doi.org/10.1016/j.cell.2012.03.042] [PMID: 22632970]

[224] Cao JY, Dixon SJ. Mechanisms of ferroptosis. Cell Mol Life Sci 2016; 73(11-12): 2195-209. [http://dx.doi.org/10.1007/s00018-016-2194-1] [PMID: 27048822]

[225] Latunde-Dada GO. Ferroptosis: Role of lipid peroxidation, iron and ferritinophagy. Biochimica et biophysica acta General subjects 2017; 1861(8): 1893-900.

[226] Conrad M, Kagan VE, Bayir H, *et al.* Regulation of lipid peroxidation and ferroptosis in diverse species. Genes Dev 2018; 32(9-10): 602-19. [http://dx.doi.org/10.1101/gad.314674.118] [PMID: 29802123]

[227] Hirschhorn T, Stockwell BR. The development of the concept of ferroptosis. Free radical biology & medicine 2019; 133: 130-43. [http://dx.doi.org/10.1016/j.freeradbiomed.2018.09.043]

[228] Yang WS, Kim KJ, Gaschler MM, Patel M, Shchepinov MS, Stockwell BR. Peroxidation of polyunsaturated fatty acids by lipoxygenases drives ferroptosis. Proc Natl Acad Sci USA 2016; 113(34): E4966-75. [http://dx.doi.org/10.1073/pnas.1603244113] [PMID: 27506793]

[229] Escamez S, Tuominen H. Contribution of cellular autolysis to tissular functions during plant development. Curr Opin Plant Biol 2017; 35: 124-30. [http://dx.doi.org/10.1016/j.pbi.2016.11.017] [PMID: 27936412]

[230] Hernandez-Segura A, Nehme J, Demaria M. Hallmarks of Cellular Senescence. Trends Cell Biol 2018; 28(6): 436-53. [http://dx.doi.org/10.1016/j.tcb.2018.02.001] [PMID: 29477613]

[231] de Magalhães JP, Passos JF. Stress, cell senescence and organismal ageing. Mech Ageing Dev 2018; 170: 2-9. [http://dx.doi.org/10.1016/j.mad.2017.07.001] [PMID: 28688962]

[232] Prahlad V, Cornelius T, Morimoto RI. Regulation of the cellular heat shock response in Caenorhabditis elegans by thermosensory neurons. Science 2008; 320(5877): 811-4. [http://dx.doi.org/10.1126/science.1156093] [PMID: 18467592]

[233] Tatum MC, Ooi FK, Chikka MR, *et al.* Neuronal serotonin release triggers the heat shock response in C. elegans in the absence of temperature increase. Curr Biol 2015; 25(2): 163-74. [http://dx.doi.org/10.1016/j.cub.2014.11.040] [PMID: 25557666]

[234] Durieux J, Wolff S, Dillin A. The cell-non-autonomous nature of electron transport chain-mediated longevity. Cell 2011; 144(1): 79-91. [http://dx.doi.org/10.1016/j.cell.2010.12.016] [PMID: 21215371]

[235] Berendzen KM, Durieux J, Shao LW, *et al.* Neuroendocrine Coordination of Mitochondrial Stress Signaling and Proteostasis. Cell 2016; 166(6): 1553-1563.e10. [http://dx.doi.org/10.1016/j.cell.2016.08.042] [PMID: 27610575]

[236] Shao LW, Niu R, Liu Y. Neuropeptide signals cell non-autonomous mitochondrial unfolded protein response. Cell Res 2016; 26(11): 1182-96. [http://dx.doi.org/10.1038/cr.2016.118] [PMID: 27767096]

[237] Ermolaeva MA, Segref A, Dakhovnik A, *et al.* DNA damage in germ cells induces an innate immune response that triggers systemic stress resistance. Nature 2013; 501(7467): 416-20. [http://dx.doi.org/10.1038/nature12452] [PMID: 23975097]

[238] O'Brien D, Jones LM, Good S, *et al.* A PQM-1-Mediated Response Triggers Transcellular Chaperone Signaling and Regulates Organismal Proteostasis. Cell Rep 2018; 23(13): 3905-19. [http://dx.doi.org/10.1016/j.celrep.2018.05.093] [PMID: 29949773]

[239] Taylor RC, Dillin A. XBP-1 is a cell-nonautonomous regulator of stress resistance and longevity. Cell 2013; 153(7): 1435-47.
[http://dx.doi.org/10.1016/j.cell.2013.05.042] [PMID: 23791175]

[240] Williams KW, Liu T, Kong X, *et al.* Xbp1s in Pomc neurons connects ER stress with energy balance and glucose homeostasis. Cell Metab 2014; 20(3): 471-82.
[http://dx.doi.org/10.1016/j.cmet.2014.06.002] [PMID: 25017942]

[241] Neves J, Demaria M, Campisi J, Jasper H. Of flies, mice, and men: evolutionarily conserved tissue damage responses and aging. Dev Cell 2015; 32(1): 9-18.
[http://dx.doi.org/10.1016/j.devcel.2014.11.028] [PMID: 25584795]

[242] Solinski HJ, Hoon MA. Cells and circuits for thermosensation in mammals. Neurosci Lett 2019; 690: 167-70.
[http://dx.doi.org/10.1016/j.neulet.2018.10.026] [PMID: 30355519]

[243] Sosa-Pagán JO, Iversen ES, Grandl J. TRPV1 temperature activation is specifically sensitive to strong decreases in amino acid hydrophobicity. Sci Rep 2017; 7(1): 549.
[http://dx.doi.org/10.1038/s41598-017-00636-4] [PMID: 28373693]

[244] Martínez-Sánchez N, Seoane-Collazo P, Contreras C, *et al.* Hypothalamic AMPK-ER Stress-JNK1 Axis Mediates the Central Actions of Thyroid Hormones on Energy Balance. Cell Metab 2017; 26(1): 212-229.e12.
[http://dx.doi.org/10.1016/j.cmet.2017.06.014] [PMID: 28683288]

[245] Müller MJ, Geisler C. From the past to future: from energy expenditure to energy intake to energy expenditure. Eur J Clin Nutr 2017; 71(3): 358-64.
[http://dx.doi.org/10.1038/ejcn.2016.231] [PMID: 27901032]

[246] Fukunaka A, Fukada T, Bhin J, *et al.* Zinc transporter ZIP13 suppresses beige adipocyte biogenesis and energy expenditure by regulating C/EBP-β expression. PLoS Genet 2017; 13(8)e1006950
[http://dx.doi.org/10.1371/journal.pgen.1006950] [PMID: 28854265]

[247] Nicholls DG. The hunt for the molecular mechanism of brown fat thermogenesis. Biochimie 2017; 134: 9-18.
[http://dx.doi.org/10.1016/j.biochi.2016.09.003] [PMID: 27621145]

[248] Guerra C, Koza RA, Walsh K, Kurtz DM, Wood PA, Kozak LP. Abnormal nonshivering thermogenesis in mice with inherited defects of fatty acid oxidation. J Clin Invest 1998; 102(9): 1724-31.
[http://dx.doi.org/10.1172/JCI4532] [PMID: 9802886]

[249] Lee J, Ellis JM, Wolfgang MJ. Adipose fatty acid oxidation is required for thermogenesis and potentiates oxidative stress-induced inflammation. Cell Rep 2015; 10(2): 266-79.
[http://dx.doi.org/10.1016/j.celrep.2014.12.023] [PMID: 25578732]

[250] Moraru A, Cakan-Akdogan G, Strassburger K, *et al.* THADA Regulates the Organismal Balance between Energy Storage and Heat Production. Dev Cell 2017; 41(1): 72-81.e6.
[http://dx.doi.org/10.1016/j.devcel.2017.03.016] [PMID: 28399403]

[251] Bal NC, Singh S, Reis FCG, *et al.* Both brown adipose tissue and skeletal muscle thermogenesis processes are activated during mild to severe cold adaptation in mice. J Biol Chem 2017; 292(40): 16616-25.
[http://dx.doi.org/10.1074/jbc.M117.790451] [PMID: 28794154]

[252] Cerri M. The Central Control of Energy Expenditure: Exploiting Torpor for Medical Applications. Annu Rev Physiol 2017; 79: 167-86.
[http://dx.doi.org/10.1146/annurev-physiol-022516-034133] [PMID: 27813827]

[253] Hilleary R, Gilroy S. Systemic signaling in response to wounding and pathogens. Curr Opin Plant Biol 2018; 43: 57-62.
[http://dx.doi.org/10.1016/j.pbi.2017.12.009] [PMID: 29351871]

[254] Boniecka J, Prusińska J, Dąbrowska GB, Goc A. Within and beyond the stringent response-RSH and (p)ppGpp in plants. Planta 2017; 246(5): 817-42.
[http://dx.doi.org/10.1007/s00425-017-2780-y] [PMID: 28948393]

[255] Verma V, Ravindran P, Kumar PP. Plant hormone-mediated regulation of stress responses. BMC Plant Biol 2016; 16: 86.
[http://dx.doi.org/10.1186/s12870-016-0771-y] [PMID: 27079791]

[256] Fujii H, Chinnusamy V, Rodrigues A, *et al. in vitro* reconstitution of an abscisic acid signalling pathway. Nature 2009; 462(7273): 660-4.
[http://dx.doi.org/10.1038/nature08599] [PMID: 19924127]

[257] Soma F, Mogami J, Yoshida T, *et al.* ABA-unresponsive SnRK2 protein kinases regulate mRNA decay under osmotic stress in plants. Nat Plants 2017; 3: 16204.
[http://dx.doi.org/10.1038/nplants.2016.204] [PMID: 28059081]

[258] Malara A, Fresia C, Di Buduo CA, *et al.* The Plant Hormone Abscisic Acid Is a Prosurvival Factor in Human and Murine Megakaryocytes. J Biol Chem 2017; 292(8): 3239-51.
[http://dx.doi.org/10.1074/jbc.M116.751693] [PMID: 28049729]

[259] Swiecilo A. Cross-stress resistance in Saccharomyces cerevisiae yeast--new insight into an old phenomenon. Cell stress & chaperones 2016; 6(1): 107.

[260] Kitichantaropas Y, Boonchird C, Sugiyama M, Kaneko Y, Harashima S, Auesukaree C. Cellular mechanisms contributing to multiple stress tolerance in Saccharomyces cerevisiae strains with potential use in high-temperature ethanol fermentation. AMB Express 2016; 6(1): 107.
[http://dx.doi.org/10.1186/s13568-016-0285-x] [PMID: 27826949]

[261] Martinez-Medina A, Flors V, Heil M, *et al.* Recognizing Plant Defense Priming. Trends Plant Sci 2016; 21(10): 818-22.
[http://dx.doi.org/10.1016/j.tplants.2016.07.009] [PMID: 27507609]

[262] Hossain MA, Li ZG, Hoque TS, Burritt DJ, Fujita M, Munné-Bosch S. Heat or cold priming-induced cross-tolerance to abiotic stresses in plants: key regulators and possible mechanisms. Protoplasma 2018; 255(1): 399-412.
[http://dx.doi.org/10.1007/s00709-017-1150-8] [PMID: 28776104]

[263] Drigeard Desgarnier MC, Rochette PJ. Enhancement of UVB-induced DNA damage repair after a chronic low-dose UVB pre-stimulation. DNA Repair (Amst) 2018; 63: 56-62.
[http://dx.doi.org/10.1016/j.dnarep.2018.01.008] [PMID: 29448173]

[264] Calabrese EJ. The Emergence of the Dose-Response Concept in Biology and Medicine. Int J Mol Sci 2016; 17(12)E2034
[http://dx.doi.org/10.3390/ijms17122034] [PMID: 27929392]

[265] Semchyshyn HM, Valishkevych BV. Hormetic Effect of H2O2 in Saccharomyces cerevisiae: Involvement of TOR and Glutathione Reductase. Dose Response 2016; 14(2)1559325816636130
[http://dx.doi.org/10.1177/1559325816636130] [PMID: 27099601]

[266] Gems D, Partridge L. Stress-response hormesis and aging: "that which does not kill us makes us stronger". Cell Metab 2008; 7(3): 200-3.
[http://dx.doi.org/10.1016/j.cmet.2008.01.001] [PMID: 18316025]

[267] Visser B, Williams CM, Hahn DA, Short CA, López-Martínez G. Hormetic benefits of prior anoxia exposure in buffering anoxia stress in a soil-pupating insect. J Exp Biol 2018; 221(Pt 6)jeb167825
[http://dx.doi.org/10.1242/jeb.167825] [PMID: 29367272]

[268] Merkwirth C, Jovaisaite V, Durieux J, *et al.* Two Conserved Histone Demethylases Regulate Mitochondrial Stress-Induced Longevity. Cell 2016; 165(5): 1209-23.
[http://dx.doi.org/10.1016/j.cell.2016.04.012] [PMID: 27133168]

CHAPTER 8

p53, Guardian of the Genome

Abstract: p53 is the most important tumor suppressor. It arrests cell cycle in response to stress, stimulates DNA repair, and induces senescence or apoptosis of cells that have unrepairable damage. p53 level and activity are kept low by the E3 ubiquitin ligase Mdm2, which ubiquitinates p53 targeting it for proteasomal degradation. Additional post-translational modifications of p53 such as phosphorylation, methylation, acetylation, and others, signal the cell's internal condition to p53 and modulate its activity.

Keywords: ATM, ATR, Apoptosis, Cell cycle, Cell cycle arrest, Cancer, DNA damage, Mdm2, p53, p63, p73, Tumor suppressor.

INTRODUCTION

Multicellularity is based on cooperation among cells for joint survival. The fate of individual cells needs to be subdued to common goals. Effective multicellularity requires control over proliferation, differentiation, resource allocation, division of labor, creation and maintenance of extracellular environment, and controlled cell death. The breakdown of the control mechanisms can lead to unrestricted proliferation, cancer, and death of the organism [1].

Multicellular organisms contain two types of cells; somatic cells, which form various tissues, and germ cells, which specialize in reproduction. Many tissues also contain stem cells, which have the capacity to differentiate into various cell types in order to regenerate the tissue. Stem cells proliferate and differentiate into specialized cells under strict control by morphogens, growth factors, and hormones.

Certain cells are required only temporarily, such as during development or immune response, and, after they serve their role, they need to be promptly eliminated by programmed cell death (PCD). Therefore, PCD is an indispensable part of multicellular life. In addition, damaged, infected, or malfunctioning cells also need to be removed by regulated cell death (RCD). Many stresses lead to DNA damage, which is very dangerous for cells. Genome alterations in somatic and stem cells can lead to cancer and tissue defects, whereas DNA defects in germ

Nadia Danilova

cells can be passed onto the next generation. Many thousands of DNA damage events take place in each human cell every day [2]. Most DNA damage is immediately repaired but some damage escapes DNA repair machinery and can lead to genome instability and mutations. Mutations accumulate even in stem cells, which do not divide frequently. For example, human adult stem cells accumulate mutations during a lifetime at a rate of approximately 40 new mutations per year [3]. Mutations in brain stem cells lead to the presence of islands of mutated neurons in adult brains, which may contribute to neurodegenerative diseases [4]. Somatic cells also show a steady increase in mutations with aging [5].

Cells with altered genomes need to be identified and immediately removed from the proliferating pool. It is achieved by induction of cell cycle arrest. At the next step, the fate of such cell is decided. First, repairs are induced. If the damage is unrepairable, the next choice is between permanent cell cycle arrest called senescence, and cell death.

These decisions are made by p53 transcriptional factor known as a "guardian of the genome" [6, 7]. Diverse stress signals such as DNA damage, telomere erosion, abnormal chromosome segregation, errors in ribosomal biogenesis, depletion of nucleotides, low glucose level, insufficient oxygenation, or oncogene activation are transmitted to p53 by a variety of mechanisms. p53 serves as a transcriptional "super hub" that integrates these signals and determines cell fate by inducing adjustment of homeostasis, cell repair, senescence, or cell death [8 - 10].

The crucial role of p53 is illustrated by the fact that more than half of human cancers have mutations in *TP53* gene, which encodes p53 while in many tumors with the wild-type *TP53*, various mechanisms repress p53 activity [11 - 13]. Cancer risk for individuals with germline mutations in *TP53* (Li-Fraumeni syndrome) is nearly 50% at the age of 31 and nearly 100% at the age of 70 [14]. Mice deficient for p53 also develop a variety of cancers at an early age [15].

In unstressed cells, p53 activity is kept low due to two inhibitors: Mdm2 and Mdm4 (also known as MdmX). Mdm2 is an E3 ubiquitin ligase, it ubiquitinates p53 targeting it for proteasomal degradation; in addition, both Mdm2 and Mdm4 inhibit p53 activity by direct binding [10, 16 - 18]. The decrease in Mdm2/4 activity leads to p53 upregulation.

8.1. Genes of p53 Family

In higher animals, there are three *TP53*-like genes — *Tp53*, *Tp63,* and *Tp73;* they encode multiple proteins that are generated by using alternative promoters and alternative splicing [19]. These proteins constitute p53 protein family. In humans, twelve p53 isoforms have been discovered that are the result of alternative promoters usage, alternative splicing of exon 2 and 9, and the use of different codons for translation initiation [20]. What is known as canonical p53 is p53alpha, the first identified and the most studied isoform.

Members of the p53 family can cooperate or compete, and regulate each other. p53, p63, and p73 can act in a partially redundant way, for example in the induction of cell cycle arrest, cell death, and metabolic regulation, but each protein has its own role [21]. In mammals, p63 is important for the development of epidermis and its derivatives, whereas p73 is indispensable for the development of the neural system [22 - 24].

Genes of p53/Mdm2 system and their regulation are very conserved for nearly 1 billion years from placozoa to humans [21, 25]. Almost all invertebrates have one gene for p53/p63/p73 protein and one Mdm gene, whereas teleost fish already have all three p53 family genes [26].

p53 family members have a similar structure with a transactivation domain (TAD) followed by a proline-rich region, DNA-binding domain (DBD), and oligomerization domain (OD) [19, 21] (Fig. **8A**). p53 TAD domain contains two sub-domains TAD1 and TAD2 that have different targets and act synergistically [10, 27]. TAD domains are natively unfolded and bind target DNA through the induced fit mechanism [28]. OD domain is responsible for p53 tetramerization. A nuclear export signal is also located in this domain; during p53 activation and tetramerization it is masked, leading to p53 retention in the nucleus. At C-terminus, p53 has an unstructured C-terminal basic domain (CTD). p53 slides along DNA *via* this domain to scan for p53 consensus recognition sequences [29].

A conserved feature of the p53 family is transcription from two promoters —one upstream from exon one, and one intronic that generates the N-terminally truncated isoforms [19]. Truncated isoforms oppose the activity of full-length proteins (Fig. **8B**). A major truncated p53 isoform is generated from an internal promoter in intron four, resulting in a p53 protein initiated at codon 133 (Δ133p53) [20]. Activated p53 stimulates expression of its inhibitor Δ133p53 [30] (Fig. **8B**).

8.2. p53 Activity

p53 DNA-binding motif, or so-called p53-response element (p53-RE), consists of the decamer RRRCWWGYYY, where R = purine, W = A or T, and Y = pyrimidine and, most commonly, a second related palindrome linked by a spacer region [31]. There are thousands of such sites in the human genome, which differ in their affinity for p53.

p53 family proteins bind DNA as tetramers in cooperation with general and specialized transcriptional co-regulators [10, 27, 28]. Tetramerization is necessary for binding and activation of the components of the core transcriptional machinery. Post-translation modifications regulate p53 activation and its affinity to interacting partners. C-terminal p53 domain also can modulate DNA binding. CTD domain stabilizes the p53-DNA bond, recruits various cofactors, and adopts different conformations depending on cellular context and cofactor availability. It has a higher affinity to ssDNA, supercoiled DNA, and to alternative DNA structures such as triplex DNA, DNA loops, cruciform structures, *etc.*; this can affect DNA binding through TAD domains [32].

p53 also directly interact with histone-modifying proteins such as Cbp and p300. In addition to chromatin opening by histone acetylation, these proteins stabilize p53 by interactions with the TAD domain and by induction of acetylation of p53 at C-terminus, which prevents p53 ubiquitination and degradation [27]. Mdm2 and Mdm4, which also bind p53 TAD, oppose Cbp/p300 activity.

p53 binding sites are common not only among promoters but also enhancers, which greatly expands the repertoire of p53-regulated genes [10, 33]. p53 is pre-bound to enhancers in genes involved in DNA damage response, which helps cells to mount a fast response [34]. Most p53-bound enhancers are within inaccessible chromatin that becomes accessible following DNA damage. This suggests that p53 may modulate chromatin accessibility.

p53 transcriptionally regulates hundreds of genes [35, 36]. They function in diverse processes including cell cycle arrest, apoptosis, senescence, DNA repair, repression of transposable elements, stem cells maintenance, development, glucose and lipid metabolism, mitochondrial health, autophagy, and immune response. Many genes are regulated by p53 indirectly; this includes most of those that are repressed in response to p53 activation [37].

p53 controls transcriptome also by changing RNA stability [38]. A study of the connection between p53 level and expression of its targets show that most of the difference in their mRNA levels arose post-transcriptionally [39].

p53 family proteins can regulate other proteins through direct interactions. p53 directly binds at least 80 proteins participating in signaling, cell cycle, transcription, DNA repair, regulation of reactive oxygen species, cytoskeleton, apoptosis, *etc.* [40, 41].

8.3. p53 Network Regulation

p53 network regulation is multileveled and complex. In its role as a super hub, p53 integrates diverse stress signals from cellular internal and external environment and generates a response that is customized to the type and level of the signals as well as to the cell type and status. Besides stress signaling, information about the levels of nutrients, ATP, growth factors and hormones also incorporates into the response. Multiple feedback loops and oscillations of the major components of p53 network keep response under control [39, 42]. Most regulation takes place post-translationally with the contribution of other types of regulation. Here we discuss examples of several modes of p53 regulation.

p53 promoter has binding sites for stress proteins and oncogenes including members of AP-1 family such as c-Fos and c-Jun, NFkB, and Myc/Max [43, 44]. p53 transcription fluctuates during the cell cycle, increasing ahead of S phase to ensure enough mRNA for a rapid response to DNA damage [45]. Stimulation of *tp53* transcription by oncogenic signaling probably serves a similar purpose, to have enough p53 to quickly prevent runaway proliferation [44]. p53 can regulate its own transcription in a positive feedback loop, which promotes a stronger and steeper response [46]. Cells use selective inhibition of p53 transcription when increased cell survival is necessary. For example, immunoglobulin class switch recombination and somatic hypermutation in B cells require DNA breaks, which would normally induce p53. To avoid p53 response, the oncogene Bcl6 binds *tp53* promoter and suppresses p53 expression in B cells [47].

p53's mRNA stability and translation are regulated by multiple mechanisms. For example, the p53 target Wig1 binds to 3'UTR of p53 mRNA and enhances its stability in a positive feedback loop [48]. DNA repair protein Ku inhibits p53 protein translation by binding to a stem-loop structure within p53 mRNA's 5' UTR. This binding is disrupted after Ku acetylation in response to genotoxic stress, which allows p53 translation [49]. Micro RNAs are also involved in the regulation of p53 translation. MiR-504 inhibits p53 translation through binding to 3' UTR of p53 mRNA [50]. MiR-504 resides within the fibroblast growth factor 13 (*FGF13*) gene. *FGF13* is inhibited by p53 under stress reducing miR-504 production and its inhibitory effects on p53 [51]. MiR-125b also inhibits p53 translation by binding to its 3'UTR [52]. The AU/U-rich element-binding protein HuR counterbalances this inhibition; it translocates to the nucleus during stress and binds at a site adjacent to the miR-125b site, causing restoration of p53 activity [53]. An additional mode of regulation of p53 translation is associated with the induction of p53 synthesis at the internal ribosome entry site (IRES), which is often taking place following DNA damage [54]. Such multiple inputs ensure a balanced regulation of p53 translation.

Most of p53 regulation takes place at the protein level. In the absence of stress, p53 half-life is only 6–20 min, which increases to hours in response to stress. Stress signals lead to extensive enzymatic posttranslational modifications of p53 that include ubiquitination, phosphorylation, acetylation, sumoylation, neddylation, methylation, and glycosylation [9]. Multiple enzymes participate in conveying stress signals to p53. The pattern of posttranslational modifications depends on the type of stress signal [55]. Modifications affect p53 conformation, activity, choice of binding partners, and determine the makeup of the p53-induced response [10, 56 - 58].

The core of p53 network is the p53/Mdm2 module. Ubiquitination of p53 by the E3 ubiquitin ligase Mdm2 leads to its proteasomal degradation and is the main regulator of p53 stability. Mdm2 is the most important p53 ubiquitin ligase, although other enzymes may also be involved [59]. Low levels of Mdm2 activity induce monoubiquitination and nuclear export of p53, whereas high levels promote p53 polyubiquitination and degradation. The ubiquitin-dependent protein degradation pathway can be reversed by the action of deubiquitinating enzymes such as Usp10, a cytoplasmic ubiquitin-specific protease. Translocation and stabilization of Usp10 are regulated by ATM-mediated phosphorylation of Usp10 [60]. Cyld also can deubiquitinate p53 in response to genotoxic stress [61].

Mdm2 can ubiquitinase itself for destruction; this ubiquitination can be antagonized by the deubiquitinase HAUSP [62]. Acetyltransferase p300-mediated acetylation of Mdm2 blocks self-ubiquitination, enabling HAUSP binding. Under genotoxic stress, the deacetylase sirtuin 1 (Sirt1) deacetylates Mdm2 promoting its self-ubiquitination [63].

As seen from the above examples, posttranscriptional modifications of p53 network members serve as a mechanism of p53 regulation. Altogether, p53 can be phosphorylated at approximately 24 sites by many kinases including ATR, ATM, p38, and checkpoint kinases [64 - 66]. ATM/p53 pathway is extremely sensitive to double-strand DNA damage. Targeting a single site in the human genome is sufficient to trigger a checkpoint response [67]. Pattern of p53 phosphorylation affects its stability, interaction with its regulators, and activity [55]. For example, phosphorylation at Ser15 blocks Mdm2 access to p53 preventing p53 ubiquitination and degradation. Phosphorylation of Mdm2 and Mdm4, thus modulating their activity, provides an additional way to regulate p53 network [68].

p53 acetylation is essential for its activation [56]. The viral protein HBZ suppresses p53 activity by inhibiting the acetyltransferases p300/CBP and HBO1 [69]. Several regulatory proteins bind to p53 in dependence on its acetylation status. For example, the oncoprotein SET inhibits p53 in unstressed cells but stress-induced acetylation of p53 C-terminal domain abolishes its binding to SET [70]. Acetylation of p53 at Lys382 seems to be the most important for its activation in response to genotoxic stress [71]. Acetylation of this site changes the kinetics of p53 binding to its response elements and increases p53 residence time on these elements. Increase in p53 concentration does not result in the activation of target genes unless acetylation also takes place.

More than 30 proteins can interact with p53, Mdm2, and Mdm4, acting as inhibitors or activators [41]. One of the most important inputs is through Mdm2 inhibition by Arf, an alternative reading frame product of the Ink4a/Arf locus, which is activated by excessive oncogenic signaling [72]. Perturbations of ribosome biogenesis lead to an imbalance of ribosomal proteins; some of them are able to bind Mdm2 and block its binding to p53, thus conveying ribosomal stress to p53 [73]. Small molecules also can regulate p53. During nutrient stress, C16-ceramide accumulates in cells and disrupts p53 binding to Mdm2, preventing p53 ubiquitination [74].

Members of p53 network often mutually regulate each other, which can be exemplified by p53/Mdm2. During DNA damage response, ATR and ATM phosphorylate Mdm2, leading to p53 activation. Then, the activated p53 transcriptionally upregulates Mdm2. Therefore increase in p53 activity upregulates Mdm2, which then lowers p53 resulting in decreased Mdm2; this leads to p53 and Mdm2 counterphase oscillation [68, 75].

P53 network regulation can also be achieved by sequestering p53 family proteins and their regulators into different cellular compartments; for example, during *Xenopus* oogenesis apoptosis takes place in somatic cells but not in oocytes where pro- and antiapoptotic molecules are separated [76].

The interplay between various p53 family isoforms also can affect the outcome of a stress response. Cellular pathways downstream from growth factor receptors negatively control p53 by inhibiting its phosphorylation, promoting translation of Mdm2, and enhancing Mdm2 stability and nuclear localization [77]. In contrast, truncated isoforms of the p53 family are positively regulated by growth factors; therefore, the final response is a sum of various signaling interactions [20] (Fig. **8B**).

p53 network receives inputs from other pathways, for example, p53 target cell cycle inhibitor p21 is also regulated by DNA damage response mediator Rad9 [78]. Some miRNAs induced during stress response can last for a long time providing a kind of memory. For example, p53-induced miRNAs miR-143-3p, miR-145-5p, miR-34a-5p, and miR-139-5p are long-lived miRNAs that regulate targets such as Mdm2 and p21. Such miRNAs induced during the first exposure to stress can persist for a long time and diminish response to the second exposure [79].

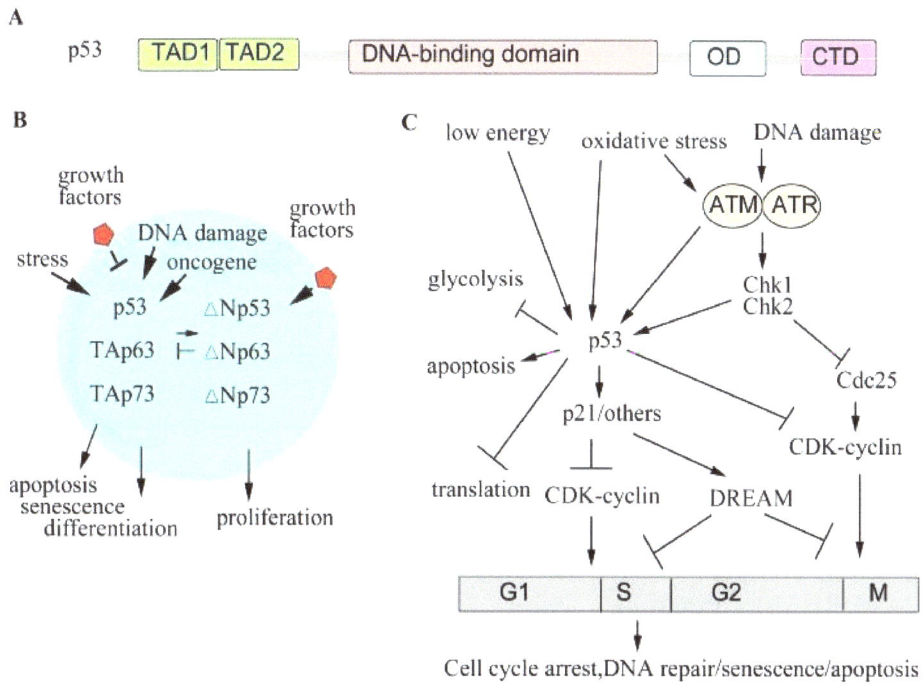

Fig. (8). p53 protein. **A**) p53 has two transactivation subdomains, DNA-binding domain, oligomerization domain, and C-terminal domain. **B**) Full length members of p53 family have similar partially overlapping activities, whereas truncated family members have opposing activities and often serve as their inhibitors. **C**) p53 is activated by many stress signals and induces different responses depending on stress intensity and cell state. Shown responses include cell cycle arrest, inhibition of glycolysis, inhibition of translation, and apoptosis.

When stress is over, p53 network needs to be down-regulated. Such down-regulation includes actions of deacetylases, phosphatases, and microRNAs [80, 81]. p53 negative regulator Wip1 phosphatase is induced by p53. It dephosphorylates Ser15 of p53. Dephosphorylation and deacetylation are necessary to make p53 available for interaction with the ubiquitin-ligase Mdm2 that targets p53 protein for degradation. Wip1 also dephosphorylates several p53-activating kinases, thus reducing p53 phosphorylation and activity. In addition, Wip1 activity promotes autophagy that degrades the proapoptotic p53 target Noxa [82]. These mechanisms bring p53 activity to its basal levels, and cell cycle can resume.

8.4. p53 Dictates Cell Fate

Activation of p53 can lead to different cell fates depending on stress intensity and cell state [83]. Major results of p53 activation include cell cycle arrest, activation of DNA repair, senescence, and cell death. Different pathways mediate these outcomes.

Cell cycle arrest is caused by p53-mediated transcriptional upregulation of cell cycle controllers such as cell-cycle inhibitor p21/CDKN1A, p27, and others [37, 84]. Cell cycle progression is regulated by MuvB complexes that switch between activator and repressor modes *via* the recruitment of different factors [85]. During cell cycle, cyclin-dependent kinases (CDKs) phosphorylate Rb-like proteins p107 and p130, which prevents them from joining MuvB. B-MYB and FOXM1 are recruited instead, turning MuvB complexes into activators. Low CDK activity during G0 phase leads to hypophosphorylation of p107 and p130, which let them join MuvB to form a repressor complex DREAM. DREAM suppresses the expression of ~250 genes that function in diverse processes associated with cell cycle, from DNA synthesis to mitosis. When p21 is induced by p53, it inhibits CDKs leading to a hypophosphorylated state of p107 and p130 and formation of the DREAM complex [37]. Other p53-induced cell cycle inhibitors may also promote DREAM formation.

p21 is regulated at the transcription level by p53 and other factors and, at the protein level, by proteasomal degradation. Several E3 ubiquitin ligases polyubiquitinate it, whereas deubiquitylase Usp11 reverses p21 polyubiquity-lation, preventing its degradation [86, 87]. Following cell cycle arrest, cells can recover, undergo senescence, or apoptosis with p21 playing an important role in cell fate choice [88, 89].

Induction of DNA repair is another common result of p53 activation [90]. p53 transcriptionally upregulates many genes involved in DNA repair [35]. Besides activation of DNA repair genes promoters, recent data show that p53 is bound to DNA damage response genes enhancers and poised for a fast activation during DNA damage [34].

p53 activation can lead to cell senescence, an irreversible cell cycle arrest that removes damaged cells from the proliferating pool [91]. Senescent cells have changes in chromatin organization and in metabolism that sustain p53 activation. It is thought that prolonged activation of p21 induces mitochondrial dysfunction and production of ROS, which maintains an ongoing DDR and promotes senescence [92]. A distinctive feature of senescent cells is secretion of cytokines and chemokines, which is known as senescence-associated secretory phenotype (SASP) [93]. Senescence protects the organism from cancer, stimulates regenerative responses but also contributes to inflammation and aging [94, 95].

Another cell fate induced by p53 is apoptosis and other regulated cell death pathways [96]. Intrinsic apoptosis is induced by mitochondrial outer membrane permeabilization by members of Bcl2 protein family, the pore-formers Bax and Bak [97, 98]. In the absence of stress, antiapoptotic Bcl2 proteins such as Bcl2, Bclw, and others bind and inactivate Bax/Bak. In response to stress, a fraction of induced p53 protein localizes to mitochondria where it binds antiapoptotic Bcl2 proteins releasing Bax/Bak. This shifts the equilibrium between pro- and anti-apoptotic Bcl2 proteins and promotes apoptosis. In addition, p53 can directly activate Bax [99]. Moreover, p53 transcriptionally upregulates proapoptotic Bcl2 genes including Bax and Bak and their activators Noxa and Puma.

P53 is also involved in other regulated cell death pathways such as ferroptosis, lipid peroxidation induced cell death. p53 can promote or inhibit ferroptosis by multiple mechanisms [100, 101].

The choice of cell fate is enabled by many interlinked feedback loops and depends on many factors including the cell type, the degree and nature of cellular damage, the intensity and duration of signaling to p53, the nature of transcriptional cofactors and epigenetic modulators involved, p53 oligomerization status, and the affinity of p53 for the promoters of specific genes. In human cells exposed to gamma irradiation, a minor DNA damage induces p53 activity in series of repeated pulses, which activates DNA repair and cell recovery, whereas extensive damage induces a sustained p53 signaling and leads to senescence [102]. In order to induce apoptosis, p53 level must reach a particular threshold, which depends on cell state. Moreover, this threshold increases with time due to the induction of antiapoptotic genes later in the response [103]. Differences in stress intensity and p53 post-translational modifications are associated with the activation of different negative and positive feedback loops that determine cell fate. In general, positive feedback loops such as those involving Pten and Wip1 are associated with cell fate stabilization and bistability, whereas negative feedback loops such as Mdm2 loop are associated with reversible fates [42, 104]. For example, p53 activates transcription of the pro-survival *lrp1* gene in response to both sub-lethal and lethal doses of stress, but Lrp1 protein is upregulated only in response to sub-lethal stress [105]. Lethal doses of stress upregulate p53 targets miRNAs miR-103 and mir-107; these miRNAs inhibit Lrp1 translation. This stress intensity-dependent miRNA feedback loop determines if the cell survives or dies. Positive p53/Wip1/miR-16 loop determines the choice between senescence and apoptosis with higher miR-16 levels favoring apoptosis [104].

The importance of cell state can be illustrated by the finding that inhibition of rRNA biosynthesis always induces cell cycle arrest, whereas apoptosis takes place only in cells with a high ribosome biogenesis rate [106]. In addition, there are variations in p53 signaling among individual cells, which allow some cells to escape p53 response [107].

8.5. Genome Protection by p53

Genome instability is the major enabler of cancer [108]. Recent studies increasingly emphasize the importance of p53 for genome stability. p53 protects the genome in multiple ways whereas its major inhibitors Mdm2 and Mdm4 induce genome instability [109].

First, p53 regulates DNA repair, which is important for genome stability. P53 upregulates DNA repair genes transcriptionally and directly participates in repair [34, 35, 90]. p53 represses aberrant homologous recombination, which requires the interaction of p53 with RPA protein, which binds single-stranded DNA, and recombinase RAD51. When cells with unrepairable DNA damage proliferate, that may result in mutations, chromosome breaks, and translocations [110].

Second, p53 regulates a large number of genes functioning in mitosis and mediates checkpoints controlling cell ploidy [37]. Aneuploidy is a common feature of cancer cells observed in 70% of human tumors [109, 111]. In addition, chromosomes in tumor cells often have translocations and deletions or amplification of various regions. Among causes of these defects is a loss of cell-cycle checkpoint control, which can lead to premature mitosis, and various errors in mitosis. During mitosis, replicated chromosomes need to be faithfully segregated into the daughter cells. Formation of the mitotic spindle is facilitated by centrosomes. Centrosome abnormalities such as centrosome amplification can cause chromosome transmission errors. Several mechanisms signal mitotic spindle dysfunction to p53. Centrosome overduplication activates PIDDosome complex, leading to a caspase-2-mediated Mdm2 cleavage and p53 stabilization [112]. The spindle assembly checkpoint (SAC) senses attachment of chromosomes to the spindle and inhibits cell division if some chromosomes are not attached. Defects in this checkpoint result in chromosome missegregation and p53 activation [113]. Loss of centrosomes, or over-extended mitosis leads to 53Bp1 and Usp28 mediated p53 activation and G1 arrest [106, 114]. Chromosome missegregation during anaphase also induces p53 through differential phosphorylation of histone H3.3 at Ser31 along the arms of lagging or misaligned chromosomes [115].

The Hippo pathway senses tetraploidization, inhibits proliferation, and activates p53 [116]. This pathway includes the kinases Lats1/2, whose activation leads to the inhibition of the oncogenic transcriptional factors Yap/Taz. In addition, Lats2 kinase binds and inhibits Mdm2 activating p53 [117]. Phosphorylation and activation of Lats2 in tetraploid cells may be caused by decreased Rho activity in these cells [116].

Chromosome ends are protected from DNA damage repair machinery by telomeres. Human telomeres are composed of an array of TTAGGG repeats bound by the shelterin complex. p53 prevents DNA damage in telomeres by binding to the non-canonical subtelomeric sites and enhancing processes involved in telomere maintenance and protection [118]. Telomere shortening leads to telomere crisis when telomeres cannot protect the chromosome's ends. It induces ATM/ATR-mediated response that leads to cell cycle arrest or even cell death. In cells that lost control over the cell cycle, telomere shortening promotes genome instability, fusion of telomeres, rearrangements, translocations, chromosome duplication and cancer [119]. Most human cancers escape telomere crisis by stimulating telomerase through activating mutations in the promoter of the telomerase subunit Tert, through amplification of *tert* gene, and through other mutations that increase telomerase activity. p53 inactivation further enables cancer progression [120].

Recent data show that p53 has a direct role in replication [121]. The core domain of p53 has 3'- 5'exonuclease activity [122], which is also a property of common DNA polymerases (POL); it serves to correct replication errors. Polymerases that function in translesion DNA synthesis are exonuclease-deficient and error-prone. p53 restricts the mutagenic impact of such polymerases and it's 3'- 5'exonuclease activity is necessary for this effect [121, 123]. p53 localizes at replication forks and enhances the processivity of DNA replication [124]. After replication arrest, p53 works with translesion nucleases and other factors to repair stalled or collapsed replication forks, which includes induction of recombination [125]. p53 also promotes restart of the replication forks by recruiting MRE11 DNA replication restart nuclease and suppressing RAD52 and POLtheta mutagenic pathways [126].

Regulation of gene expression involves epigenetic modifications of the chromatin such as methylation and acetylation of DNA and histones. This ensures the correct chromatin packing and provides controlled access of polymerases to DNA while silencing transposable elements [127]. p53 controls the quality of epigenome, removing cells that failed to reproduce correct methylation pattern [128].

p53 restricts retrotransposon activity in Drosophila and zebrafish models, and in human cells. In ovaries of p53-/- Drosophila female, retrotransposons are highly expressed [129]. p53 is activated during meiosis in Drosophila and mice due to double-strand breaks that are induced during meiotic recombination by the topoisomerase Spo11 [130]. Retrotransposon transcription is also increased in meiotic cells, and the increase is stimulated by double-strand breaks since p53/spo11 double mutants have no increase in transposon activity [129]. Therefore p53 collaborates with other mechanisms to silence retrotransposons movement activated during meiosis [129, 131, 132]. This lead to a hypothesis that cancers driven by p53 mutations could represent "transposopathy", a disease linked to movement and expansion of mobile elements [129, 132].

Similar to human p53, p53 family proteins in sea anemone, *C. elegans*, and Drosophila are responsible for DNA damage-induced apoptosis but their activity is restricted to germ cells. This finding suggests that preservation of the germline may be the ancestral function of these proteins. Drosophila p53 is induced during meiosis and kills cells with defective meiotic DNA repair [130]. *C. elegans* p53-like protein Cep-1 also maintains the fidelity of meiosis by killing germ cells with DNA damage through apoptotic mechanisms [133]. In mammals too, p53 family is involved in germ cells protection as well as implantation, spermatogenesis, embryonic development, and stem cell maintenance with different specialization of each family member [21, 134, 135]. At the pluripotent stage of the mouse embryo, approximately 35% of cells that are less fit than their neighbors are eliminated through p53-mTOR pathway [136]. Cell death is mediated by repression of mTOR activity in less-fit cells.

8.6. p53 as a Tumor Suppressor

The key role of p53 in tumor suppression is evident from its frequent mutations or inactivation in human cancers and tumor predisposition of p53 null animal models [11, 14, 15]. Elephants that have 100 times more cells than humans have no proportionally more tumors because their p53 is duplicated [137 - 139]. Many human tumors contain p53 proteins with missense mutations in their DBD that alter, reduce, or preclude p53 binding to DNA; some mutants have no 3'-5' exonuclease activity. These mutant proteins do not have tumor suppression activity; instead, some of them act as oncogenes [140, 141]. Animal models show the importance of p53 structural and post-transcriptional modification on tumorigenesis [142].

The exact mechanism of tumor suppression by p53 remains enigmatic. Initially, the tumor-suppressing activity of p53 was ascribed to its role in induction of cell cycle arrest and apoptosis. There is ample evidence that p53 induces cell cycle arrest and apoptosis in animal tumors. The ability of p53 to induce cell cycle arrest, senescence, and programmed cell death play a role in tumor suppression [143]. Tumors with wild-type p53 often show p53-mediated senescence response to chemotherapy [91]. However, a series of recent studies show that p53-mediated transcriptional programs that mediate acute DNA damage responses are distinct from tumor suppression programs [144, 145]. Cells from mice that completely lack p21, Puma, and Noxa, are deficient in their ability to undergo cell cycle arrest, apoptosis, or senescence. These mice, however, are not prone to tumorigenesis [146].

Tumorigenesis is a complex process with alterations in cell cycle regulation, in responses to growth factors and death signals, and in metabolism. Distinctive homeostatic features of cancer called "cancer hallmarks" have been identified from multiple studies [108]. They include genome instability, continuous proliferation, replicative immortality, absence of response to growth suppressors and cell death signals, aerobic glycolysis, induction of angiogenesis, invasiveness, metastasis, and inflammation. Although p53 is the main tumor suppressor, p63 and p73 also contribute to cancer prevention [147, 148]. Adding to this complexity, truncated p53 family members that have growth-promoting activities may act as oncogenes that oppose the tumor suppression imposed by the full-length isoforms [148 - 150].

To explain how p53 suppresses cancer, several hypotheses are suggested. They include maintenance of genomic stability, metabolism, mitochondrial health, non-apoptotic forms of cell death, stem cell control, cell migration, and regulation of communication between a tumor and its environment [12, 13, 151].

The discussed above genome protection by p53 may play a major role in tumor suppression. p53 guards cell cycle including mitosis, DNA replication, meiosis, telomere maintenance, and restricts transposon movement. A starting point for some tumors could be defects in cell cycle and leading to genome alterations in mitosis. For example, centrosome amplification is observed in premalignant conditions; it is increased in dysplasia and correlates with loss of p53 function [152].

Cancer-related metabolism has several characteristic features that support enhanced cell proliferation. They include pro-survival signaling through PI3/Akt/TOR pathway, changes in the metabolism of glucose and amino acids, usage of alternative nutrients, and interactions with the environment [153 - 155]. Cancer cells switch from aerobic respiration to glycolysis, which is known as the Warburg effect and also takes place in normal cells during fast growth [156, 157]. Glycolysis is a much less efficient way of ATP production than aerobic respiration (two ATP per glucose molecule instead of 32) but it is faster and, in addition, it generates intermediates necessary for building macromolecules. To compensate for the low efficiency of ATP production, cancer cells greatly upregulate transporters for glucose uptake [158]. The aggressiveness of tumors caused by p53 mutations in mesenchymal stem cells correlates with changes in the mitochondrial mass and with increased mitochondrial oxidation [159].

p53 family restrains tumor metabolism at multiple points. p53 suppresses glycolysis through inhibition of the glucose transporters Glu1/4 and several glycolytic enzymes and lowering export of the end product of glycolysis, lactate, from the cell [160, 161]. p53 also directly activates enzymes of the mitochondrial electron transport chain such as cytochrome c oxidase 2 (SCO2) [162]. These changes promote a shift from glycolysis to respiration and activation of ROS neutralizing mechanisms. Glycolysis can also be suppressed by TAp63 and TAp73, which effect is opposed by the activity of their truncated isoforms [149]. p53 family restricts the growth of cancer cells also by inhibition of lipids biosynthesis and promotion of their degradation [163, 164].

The activity of mTOR, a central homeostatic regulator of cell growth is increased in p53-deficient cancer cells [165]. mTOR is a downstream target of growth-promoting P13K/Akt pathway [166]. p53 inhibits mTOR by several mechanisms. It transactivates the phosphatase PTEN, which inhibits P13K/Akt/mTOR pathway and upregulates several mTOR inhibitors [77].

In cancer, autophagy can be tumor-suppressive or tumor-promoting depending on nutrient availability and microenvironment [167]. p53 activation is often associated with enhanced autophagy. Autophagic degradation of the inhibitory p53 isoform Delta133p53alpha promotes senescence [168]. On the other hand, autophagy represses p53 preventing tissue degeneration but promoting cancer spread [167, 169].

Tumors often develop from stem cells with genomic damage. Li-Fraumeni patients that lack p53 develop tumors mostly from stem cells in tissues with active cell renewal such as blood and epithelial lineages [170]. p53 suppresses tumorigenesis in stem cells by inducing their differentiation. In contrast to stem cells, somatic cells need to be reprogrammed into a pluripotent state to turn into cancer. p53 inactivation greatly increases the efficiency of somatic cell reprogramming into pluripotent stem cells (iPSCs) [171 - 173]. Reactivation of p53 at any point during this process prevents stem cell formation [174]. Thus p53 acts as a barrier to somatic cell reprogramming.

Recent studies show the importance of mitochondrial health for cell function including cancer suppression [175, 176]. p53 mutants that preserve the ability to regulate energy metabolism and to control ROS protect cells from tumorigenesis [177]. In response to mild mitochondrial damage, p53-inducible protein Mieap stimulates the formation of lysosome-like organelles that eliminate oxidized proteins. If the mitochondria cannot be repaired, Mieap induces vacuole-like structures to degrade them [178]. Mieap-deficient mice have a high frequency of cancer. p53 controls mitochondrial health also through ribonucleotide reductase (p53R2), which supplies deoxyribonucleotides for mitochondrial DNA synthesis [179] and through coordination of mitochondrial genome repair [180].

At a high level of ROS, p53 can promote an iron-depended cell death ferroptosis in cancer cells [100, 181]. However, activation of p53 target p21 leads to slower depletion of the antioxidant glutathione and reduced oxidation of lipids, which delays ferroptotic death [182].

p53 can inhibit cancer by suppressing invasion and metastasis of cancer cells. For example, p53 induces LIMA1 protein that inhibits cell movement by decreasing actin filament depolymerization [183]. A truncated isoform of p53 blocks this activity [184].

Tumors develop in interaction with their environment. Their secreted factors change neighboring cells such as fibroblasts and endothelial cells so that the resident cells start supporting tumor growth [185]. For example, stroma-associated pancreatic stellate cells support tumor growth through autophagic alanine secretion [186]. Cancer cells stimulate such autophagy and use the alanine secreted by stromal cells, as an alternative nutrient to fuel their tricarboxylic acid (TCA) cycle. In the stroma of pancreatic ductal adenocarcinoma, the glycan-binding protein galectin-1 is a major driver of cancer growth [187]. Cellular phenotypes and signaling of the tumor stroma resemble those in wound healing, suggesting that tumors act as wounds that do not heal [188].

Tumor growth over a certain size reduces oxygen availability to the cancer cells, which activates the hypoxia-inducible factor 1(HIF1). This stimulates the development of intra-tumor blood vessels and other growth-promoting adjustments [189]. p53 suppresses angiogenesis [190]. In contrast, p53 mutant proteins cooperate with HIF1 in promoting tumor growth [191].

Chronic inflammation predisposes to cancer [192]. Persistent NFkB signaling in damaged or infected tissues suppresses p53 activity. In turn, p53 inhibits inflammation. p53-expressing senescent cells secrete factors that skew macrophage polarization towards M1 phenotype that is capable of attacking senescent cells, whereas p53-deficient cells skew macrophages towards a growth-promoting M2 phenotype [193]. Altogether, p53 regulation of tumors' communications with their environment contributes to tumor suppression.

8.7. p53 Interaction with NFkB

p53 network functions in cells in parallel with another conserved stress response pathway mediated by a family of dimeric transcriptional factors NFkB. Briefly, in mammals, NFkB system includes five proteins p65/RelA, RelB, cRel, p50, and p52, which form homo- or heterodimeric transcriptional factors with different DNA-binding specificity [194]. In most cells, p50-RelA heterodimer is the most common NFkB factor. A family of IkB inhibitors controls the activity of the dimers and are themselves controlled by IKK kinases alpha and beta. IKK activation leads to phosphorylation of IkB inhibitors, which leads to their ubiquitination and proteasomal degradation. Freed NFkB dimers translocate to the nucleus and bind kB sites on DNA to induce expression of target genes. Many NFkB targets participate in immune response, others, such as acute phase proteins, are involved in stress response.

NFkB system is regulated through degradation of IkB inhibitors, processing of NFkB precursors, expression of NFkB monomer proteins, their posttranslational modifications (such as ubiquitination, phosphorylation, acetylation, and oxidation/reduction), and through several positive and negative feedback mechanisms [194 - 196]. This allows NFkB system to specifically respond to diverse stimuli in various cellular contexts. NFkB can be activated by various stress signals that range from bacterial and viral pathogens to endoplasmic reticulum stress, DNA damage, oxidative stress, and wounds. NFkB is a pro-survival factor and induces anti-apoptotic genes; in some settings, however, it may promote apoptosis as well. NFkB activation is usually transient, although in some cancers it is constitutively active [197, 198]. NFkB activation may also contribute to metastasis [199].

The canonical NFkB pathway involves a scaffold/adaptor protein IKKgamma also known as NEMO [194]. The pathway is activated by IKKa/b phosphorylation triggered by various signals, *e.g.* from receptors sensing infection. IKK phosphorylation depends on NEMO ubiquitination or NEMO's binding to K63-linked polyubiquitin chains. Such chains are a common feature of immune signaling. The activated complex binds and phosphorylates IkB proteins, leading to their degradation and release of NFkB transcriptional factors.

Non-canonical NFkB pathway functions during development and in inflammatory pathways. It is NEMO-independent but depends, instead, on NFkB-inducing kinase (NIK). NIK integrates signals from TNF family of receptors and activates IKKa kinase to phosphorylate p100, a precursor of NFkB protein p52, which normally is bound to IkB [200]. Upon phosphorylation, p100 is processed into p52, which then binds RelB and forms a functional NFkB dimer.

Besides activation through the upstream kinases, NFkB can be activated through alternative mechanisms. For example, IkBa protein is constitutively produced and destroyed, therefore any stress that leads to inhibition of protein synthesis would decrease IkBa level, which would free NFkB dimers [201, 202]. Genotoxic stress activates NFkB through sequential modification of NEMO by SUMO-1 and ubiquitin [194, 203].

Both innate and adaptive immune systems critically depend on NFkB [204]. NFkB targets include cytokines, chemokines, cyclooxygenase-2 (COX-2), adhesion molecules, inducible nitric oxide synthase (iNOS), antibacterial peptides, MHC molecules, and proteins involved in antigen presentation. In each type of specialized immune cells, NFkB system is uniquely adapted to guide cell differentiation and function. Malfunction of NFkB system can lead to disease. For example, human mutations that prevent IkBa phosphorylation on serine 32 or 36 necessary for degradation of this inhibitor lead to severe immunodeficiency [205]. Such patients have severe B cell deficiency and may have other defects including lack of lymph nodes, lymphocytosis, dysfunctional a/b T cells, and absence of g/d T cells. Some viruses can manipulate NFkB system to promote survival of infected cells or to prevent interferon response [206, 207].

Many stresses including those leading to DNA damage co-induce p53 and NFkB-dependent gene expression [208 - 210]. p53 and NFkB responses to stress are mostly antagonistic [211, 212]. For example, p53 activates pro-apoptotic Bcl2 family members while NFkB induces anti-apoptotic proteins including Bcl-xL, which protects the mitochondria from membrane permeabilization [213]. During B cell development, rearrangement of immunoglobulin genes requires the introduction of DNA double-strand breaks by Rag proteins. Upregulation of NFkB in these cells prevents activation of the standard p53-dependent DNA damage response [214]. Chronic inflammation with persistent NFkB signaling promotes tumorigenesis, whereas p53 inhibits both [192]. p53 and NFkB can directly compete for binding to some promoters and enhancers [215 - 217]. p53 and NFkB can also cooperate and regulate each other. IkBa is an NFkB inhibitor that can also form a complex with p53 and inactivate it, which places this protein at a crossroad of cellular response to stress [218]. Another example of p53 and NFkB cross-talk is IKKb activating phosphorylation of p53 at Ser392 to promote metabolic adaptation to glutamine deprivation [219].

CONCLUDING REMARKS

Multicellular organisms need to control the fate of each cell. P53 transcriptional factor performs this function in animals. P53 prevents the proliferation of defective cells. In its absence, such cells can proliferate and form tumors. Most human cancers have mutations in *TP53* gene or have p53 inactivated by other mechanisms. If something goes wrong within a cell, p53 inhibits cell cycle, promotes DNA repair, adjusts metabolism, and stimulates autophagy to restore homeostasis. If a defective cell cannot be repaired, p53 induces its death.

CONSENT FOR PUBLICATION

Not applicable.

CONFLICT OF INTEREST

The author confirms that this chapter contents have no conflict of interest.

ACKNOWLEDGEMENTS

Declare none.

REFERENCES

[1] Aktipis CA, Boddy AM, Jansen G, *et al.* Cancer across the tree of life: cooperation and cheating in multicellularity. Philos Trans R Soc Lond B Biol Sci 2015; 370(1673): 370. [http://dx.doi.org/10.1098/rstb.2014.0219] [PMID: 26056363]

[2] Friedberg EC, Aguilera A, Gellert M, *et al.* DNA repair: from molecular mechanism to human disease. DNA Repair (Amst) 2006; 5(8): 986-96. [http://dx.doi.org/10.1016/j.dnarep.2006.05.005] [PMID: 16955546]

[3] Blokzijl F, de Ligt J, Jager M, *et al.* Tissue-specific mutation accumulation in human adult stem cells during life. Nature 2016; 538(7624): 260-4. [http://dx.doi.org/10.1038/nature19768] [PMID: 27698416]

[4] Keogh MJ, Wei W, Aryaman J, *et al.* High prevalence of focal and multi-focal somatic genetic variants in the human brain. Nat Commun 2018; 9(1): 4257. [http://dx.doi.org/10.1038/s41467-018-06331-w] [PMID: 30323172]

[5] Lo Sardo V, Ferguson W, Erikson GA, Topol EJ, Baldwin KK, Torkamani A. Influence of donor age on induced pluripotent stem cells. Nat Biotechnol 2017; 35(1): 69-74. [http://dx.doi.org/10.1038/nbt.3749] [PMID: 27941802]

[6] Lane DP. Cancer. p53, guardian of the genome. Nature 1992; 358(6381): 15-6. [http://dx.doi.org/10.1038/358015a0] [PMID: 1614522]

[7] Levine AJ. Reviewing the future of the P53 field. Cell Death Differ 2018; 25(1): 1-2. [http://dx.doi.org/10.1038/cdd.2017.181] [PMID: 29227987]

[8] Lane D, Levine A. p53 Research: the past thirty years and the next thirty years. Cold Spring Harb Perspect Biol 2010; 2(12)a000893 [http://dx.doi.org/10.1101/cshperspect.a000893] [PMID: 20463001]

[9] Aylon Y, Oren M. The Paradox of p53: What, How, and Why? Cold Spring Harb Perspect Med 2016; 6(10)a026328 [http://dx.doi.org/10.1101/cshperspect.a026328] [PMID: 27413116]

[10] Sullivan KD, Galbraith MD, Andrysik Z, Espinosa JM. Mechanisms of transcriptional regulation by p53. Cell Death Differ 2018; 25(1): 133-43. [http://dx.doi.org/10.1038/cdd.2017.174] [PMID: 29125602]

[11] Hollstein M, Sidransky D, Vogelstein B, Harris CC. p53 mutations in human cancers. Science 1991; 253(5015): 49-53.
[http://dx.doi.org/10.1126/science.1905840] [PMID: 1905840]

[12] Kaiser AM, Attardi LD. Deconstructing networks of p53-mediated tumor suppression *in vivo*. Cell Death Differ 2018; 25(1): 93-103.
[http://dx.doi.org/10.1038/cdd.2017.171] [PMID: 29099489]

[13] Mello SS, Attardi LD. Deciphering p53 signaling in tumor suppression. Curr Opin Cell Biol 2018; 51: 65-72.
[http://dx.doi.org/10.1016/j.ceb.2017.11.005] [PMID: 29195118]

[14] Mai PL, Best AF, Peters JA, *et al.* Risks of first and subsequent cancers among TP53 mutation carriers in the National Cancer Institute Li-Fraumeni syndrome cohort. Cancer 2016; 122(23): 3673-81.
[http://dx.doi.org/10.1002/cncr.30248] [PMID: 27496084]

[15] Donehower LA, Harvey M, Slagle BL, *et al.* Mice deficient for p53 are developmentally normal but susceptible to spontaneous tumours. Nature 1992; 356(6366): 215-21.
[http://dx.doi.org/10.1038/356215a0] [PMID: 1552940]

[16] Haupt Y, Maya R, Kazaz A, Oren M. Mdm2 promotes the rapid degradation of p53. Nature 1997; 387(6630): 296-9.
[http://dx.doi.org/10.1038/387296a0] [PMID: 9153395]

[17] Kubbutat MH, Jones SN, Vousden KH. Regulation of p53 stability by Mdm2. Nature 1997; 387(6630): 299-303.
[http://dx.doi.org/10.1038/387299a0] [PMID: 9153396]

[18] Pant V, Lozano G. Limiting the power of p53 through the ubiquitin proteasome pathway. Genes Dev 2014; 28(16): 1739-51.
[http://dx.doi.org/10.1101/gad.247452.114] [PMID: 25128494]

[19] Murray-Zmijewski F, Lane DP, Bourdon JC. p53/p63/p73 isoforms: an orchestra of isoforms to harmonise cell differentiation and response to stress. Cell Death Differ 2006; 13(6): 962-72.
[http://dx.doi.org/10.1038/sj.cdd.4401914] [PMID: 16601753]

[20] Joruiz SM, Bourdon JC. p53 Isoforms: Key Regulators of the Cell Fate Decision. Cold Spring Harb Perspect Med 2016; 6(8)a026039
[http://dx.doi.org/10.1101/cshperspect.a026039] [PMID: 26801896]

[21] Chillemi G, Kehrloesser S, Bernassola F, *et al.* Structural Evolution and Dynamics of the p53 Proteins. Cold Spring Harb Perspect Med 2017; 7(4)a028308
[http://dx.doi.org/10.1101/cshperspect.a028308] [PMID: 27091942]

[22] Mills AA, Zheng B, Wang XJ, Vogel H, Roop DR, Bradley A. p63 is a p53 homologue required for limb and epidermal morphogenesis. Nature 1999; 398(6729): 708-13.
[http://dx.doi.org/10.1038/19531] [PMID: 10227293]

[23] Yang A, Walker N, Bronson R, *et al.* p73-deficient mice have neurological, pheromonal and inflammatory defects but lack spontaneous tumours. Nature 2000; 404(6773): 99-103.
[http://dx.doi.org/10.1038/35003607] [PMID: 10716451]

[24] Russo C, Osterburg C, Sirico A, *et al.* Protein aggregation of the p63 transcription factor underlies severe skin fragility in AEC syndrome. Proc Natl Acad Sci USA 2018; 115(5): E906-15.
[http://dx.doi.org/10.1073/pnas.1713773115] [PMID: 29339502]

[25] Åberg E, Saccoccia F, Grabherr M, Ore WYJ, Jemth P, Hultqvist G. Evolution of the p53-MDM2 pathway. BMC Evol Biol 2017; 17(1): 177.
[http://dx.doi.org/10.1186/s12862-017-1023-y] [PMID: 28774266]

[26] Tan BX, Liew HP, Chua JS, *et al.* Anatomy of Mdm2 and Mdm4 in evolution. J Mol Cell Biol 2017; 9(1): 3-15.

[http://dx.doi.org/10.1093/jmcb/mjx002] [PMID: 28077607]

[27] Raj N, Attardi LD. The Transactivation Domains of the p53 Protein. Cold Spring Harb Perspect Med 2017; 7(1)a026047
[http://dx.doi.org/10.1101/cshperspect.a026047] [PMID: 27864306]

[28] Krois AS, Dyson HJ, Wright PE. Long-range regulation of p53 DNA binding by its intrinsically disordered N-terminal transactivation domain. Proc Natl Acad Sci USA 2018; 115(48): E11302-10.
[http://dx.doi.org/10.1073/pnas.1814051115] [PMID: 30420502]

[29] Tafvizi A, Huang F, Fersht AR, Mirny LA, van Oijen AM. A single-molecule characterization of p53 search on DNA. Proc Natl Acad Sci USA 2011; 108(2): 563-8.
[http://dx.doi.org/10.1073/pnas.1016020107] [PMID: 21178072]

[30] Marcel V, Petit I, Murray-Zmijewski F, *et al.* Diverse p63 and p73 isoforms regulate Δ133p53 expression through modulation of the internal TP53 promoter activity. Cell Death Differ 2012; 19(5): 816-26.
[http://dx.doi.org/10.1038/cdd.2011.152] [PMID: 22075982]

[31] Hoh J, Jin S, Parrado T, Edington J, Levine AJ, Ott J. The p53MH algorithm and its application in detecting p53-responsive genes. Proc Natl Acad Sci USA 2002; 99(13): 8467-72.
[http://dx.doi.org/10.1073/pnas.132268899] [PMID: 12077306]

[32] Brázda V, Coufal J. Recognition of Local DNA Structures by p53 Protein. Int J Mol Sci 2017; 18(2)E375
[http://dx.doi.org/10.3390/ijms18020375] [PMID: 28208646]

[33] Su D, Wang X, Campbell MR, *et al.* Interactions of chromatin context, binding site sequence content, and sequence evolution in stress-induced p53 occupancy and transactivation. PLoS Genet 2015; 11(1)e1004885
[http://dx.doi.org/10.1371/journal.pgen.1004885] [PMID: 25569532]

[34] Younger ST, Rinn JL. p53 regulates enhancer accessibility and activity in response to DNA damage. Nucleic Acids Res 2017; 45(17): 9889-900.
[http://dx.doi.org/10.1093/nar/gkx577] [PMID: 28973438]

[35] Fischer M. Census and evaluation of p53 target genes. Oncogene 2017; 36(28): 3943-56.
[http://dx.doi.org/10.1038/onc.2016.502] [PMID: 28288132]

[36] Nguyen TT, Grimm SA, Bushel PR, *et al.* Revealing a human p53 universe. Nucleic Acids Res 2018; 46(16): 8153-67.
[http://dx.doi.org/10.1093/nar/gky720] [PMID: 30107566]

[37] Engeland K. Cell cycle arrest through indirect transcriptional repression by p53: I have a DREAM. Cell Death Differ 2018; 25(1): 114-32.
[http://dx.doi.org/10.1038/cdd.2017.172] [PMID: 29125603]

[38] Venkata Narayanan I, Paulsen MT, Bedi K, *et al.* Transcriptional and post-transcriptional regulation of the ionizing radiation response by ATM and p53. Sci Rep 2017; 7: 43598.
[http://dx.doi.org/10.1038/srep43598] [PMID: 28256581]

[39] Hafner A, Stewart-Ornstein J, Purvis JE, Forrester WC, Bulyk ML, Lahav G. p53 pulses lead to distinct patterns of gene expression albeit similar DNA-binding dynamics. Nat Struct Mol Biol 2017; 24(10): 840-7.
[http://dx.doi.org/10.1038/nsmb.3452] [PMID: 28825732]

[40] Braithwaite AW, Del Sal G, Lu X. Some p53-binding proteins that can function as arbiters of life and death. Cell Death Differ 2006; 13(6): 984-93.
[http://dx.doi.org/10.1038/sj.cdd.4401924] [PMID: 16575404]

[41] Inoue K, Fry EA, Frazier DP. Transcription factors that interact with p53 and Mdm2. Int J Cancer 2016; 138(7): 1577-85.
[http://dx.doi.org/10.1002/ijc.29663] [PMID: 26132471]

[42] Hat B, Kochańczyk M, Bogdał MN, Lipniacki T. Feedbacks, Bifurcations, and Cell Fate Decision-Making in the p53 System. PLOS Comput Biol 2016; 12(2)e1004787 [http://dx.doi.org/10.1371/journal.pcbi.1004787] [PMID: 26928575]

[43] Kirch HC, Flaswinkel S, Rumpf H, Brockmann D, Esche H. Expression of human p53 requires synergistic activation of transcription from the p53 promoter by AP-1, NF-kappaB and Myc/Max. Oncogene 1999; 18(17): 2728-38. [http://dx.doi.org/10.1038/sj.onc.1202626] [PMID: 10348347]

[44] Lee WP, Lan KH, Li CP, Chao Y, Lin HC, Lee SD. Akt phosphorylates myc-associated zinc finger protein (MAZ), releases P-MAZ from the p53 promoter, and activates p53 transcription. Cancer Lett 2016; 375(1): 9-19. [http://dx.doi.org/10.1016/j.canlet.2016.02.023] [PMID: 26902421]

[45] Reisman D, Takahashi P, Polson A, Boggs K. Transcriptional Regulation of the p53 Tumor Suppressor Gene in S-Phase of the Cell-Cycle and the Cellular Response to DNA Damage. Biochem Res Int 2012; 2012808934 [http://dx.doi.org/10.1155/2012/808934] [PMID: 22830025]

[46] Wang S, El-Deiry WS. p73 or p53 directly regulates human p53 transcription to maintain cell cycle checkpoints. Cancer Res 2006; 66(14): 6982-9. [http://dx.doi.org/10.1158/0008-5472.CAN-06-0511] [PMID: 16849542]

[47] Phan RT, Dalla-Favera R. The BCL6 proto-oncogene suppresses p53 expression in germinal-centre B cells. Nature 2004; 432(7017): 635-9. [http://dx.doi.org/10.1038/nature03147] [PMID: 15577913]

[48] Vilborg A, Wilhelm MT, Wiman KG. Regulation of tumor suppressor p53 at the RNA level. J Mol Med (Berl) 2010; 88(7): 645-52. [http://dx.doi.org/10.1007/s00109-010-0609-2] [PMID: 20306257]

[49] Lamaa A, Le Bras M, Skuli N, *et al.* A novel cytoprotective function for the DNA repair protein Ku in regulating p53 mRNA translation and function. EMBO Rep 2016; 17(4): 508-18. [http://dx.doi.org/10.15252/embr.201541181] [PMID: 26964895]

[50] Hu W, Chan CS, Wu R, *et al.* Negative regulation of tumor suppressor p53 by microRNA miR-504. Mol Cell 2010; 38(5): 689-99. [http://dx.doi.org/10.1016/j.molcel.2010.05.027] [PMID: 20542001]

[51] Bublik DR, Bursać S, Sheffer M, *et al.* Regulatory module involving FGF13, miR-504, and p53 regulates ribosomal biogenesis and supports cancer cell survival. Proc Natl Acad Sci USA 2017; 114(4): E496-505. [http://dx.doi.org/10.1073/pnas.1614876114] [PMID: 27994142]

[52] Le MT, Shyh-Chang N, Khaw SL, *et al.* Conserved regulation of p53 network dosage by microRNA-125b occurs through evolving miRNA-target gene pairs. PLoS Genet 2011; 7(9)e1002242 [http://dx.doi.org/10.1371/journal.pgen.1002242] [PMID: 21935352]

[53] Ahuja D, Goyal A, Ray PS. Interplay between RNA-binding protein HuR and microRNA-125b regulates p53 mRNA translation in response to genotoxic stress. RNA Biol 2016; 13(11): 1152-65. [http://dx.doi.org/10.1080/15476286.2016.1229734] [PMID: 27592685]

[54] Ji B, Harris BR, Liu Y, *et al.* Targeting IRES-Mediated p53 Synthesis for Cancer Diagnosis and Therapeutics. Int J Mol Sci 2017; 18(1)E93 [http://dx.doi.org/10.3390/ijms18010093] [PMID: 28054974]

[55] Murray-Zmijewski F, Slee EA, Lu X. A complex barcode underlies the heterogeneous response of p53 to stress. Nat Rev Mol Cell Biol 2008; 9(9): 702-12. [http://dx.doi.org/10.1038/nrm2451] [PMID: 18719709]

[56] Tang Y, Zhao W, Chen Y, Zhao Y, Gu W. Acetylation is indispensable for p53 activation. Cell 2008; 133(4): 612-26.

[http://dx.doi.org/10.1016/j.cell.2008.03.025] [PMID: 18485870]

[57] Tong Q, Mazur SJ, Rincon-Arano H, Rothbart SB, Kuznetsov DM, Cui G, *et al.* An acetyl-methyl switch drives a conformational change in p53. Structure (London, England : 1993) 2015; 23(2): 322-1. [http://dx.doi.org/10.1016/j.str.2014.12.010]

[58] Zhu J, Dou Z, Sammons MA, Levine AJ, Berger SL. Lysine methylation represses p53 activity in teratocarcinoma cancer cells. Proc Natl Acad Sci USA 2016; 113(35): 9822-7. [http://dx.doi.org/10.1073/pnas.1610387113] [PMID: 27535933]

[59] Sane S, Rezvani K. Essential Roles of E3 Ubiquitin Ligases in p53 Regulation. Int J Mol Sci 2017; 18(2)E442 [http://dx.doi.org/10.3390/ijms18020442] [PMID: 28218667]

[60] Yuan J, Luo K, Zhang L, Cheville JC, Lou Z. USP10 regulates p53 localization and stability by deubiquitinating p53. Cell 2010; 140(3): 384-96. [http://dx.doi.org/10.1016/j.cell.2009.12.032] [PMID: 20096447]

[61] Fernández-Majada V, Welz PS, Ermolaeva MA, *et al.* The tumour suppressor CYLD regulates the p53 DNA damage response. Nat Commun 2016; 7: 12508. [http://dx.doi.org/10.1038/ncomms12508] [PMID: 27561390]

[62] Tavana O, Gu W. Modulation of the p53/MDM2 interplay by HAUSP inhibitors. J Mol Cell Biol 2017; 9(1): 45-52. [http://dx.doi.org/10.1093/jmcb/mjw049] [PMID: 27927749]

[63] Nihira NT, Ogura K, Shimizu K, *et al.* Acetylation-dependent regulation of MDM2 E3 ligase activity dictates its oncogenic function. Sci Signal 2017; 10(466)eaai8026 [http://dx.doi.org/10.1126/scisignal.aai8026] [PMID: 28196907]

[64] Banin S, Moyal L, Shieh S, *et al.* Enhanced phosphorylation of p53 by ATM in response to DNA damage. Science 1998; 281(5383): 1674-7. [http://dx.doi.org/10.1126/science.281.5383.1674] [PMID: 9733514]

[65] Canman CE, Lim DS, Cimprich KA, *et al.* Activation of the ATM kinase by ionizing radiation and phosphorylation of p53. Science 1998; 281(5383): 1677-9. [http://dx.doi.org/10.1126/science.281.5383.1677] [PMID: 9733515]

[66] Yogosawa S, Yoshida K. Tumor suppressive role for kinases phosphorylating p53 in DNA damage-induced apoptosis. Cancer Sci 2018; 109(11): 3376-82. [http://dx.doi.org/10.1111/cas.13792] [PMID: 30191640]

[67] van den Berg J, G Manjón A, Kielbassa K, Feringa FM, Freire R, Medema RH. A limited number of double-strand DNA breaks is sufficient to delay cell cycle progression. Nucleic Acids Res 2018; 46(19): 10132-44. [http://dx.doi.org/10.1093/nar/gky786] [PMID: 30184135]

[68] Eliaš J. Positive effect of Mdm2 on p53 expression explains excitability of p53 in response to DNA damage. J Theor Biol 2017; 418: 94-104. [http://dx.doi.org/10.1016/j.jtbi.2017.01.038] [PMID: 28126523]

[69] Wright DG, Marchal C, Hoang K, *et al.* Human T-cell leukemia virus type-1-encoded protein HBZ represses p53 function by inhibiting the acetyltransferase activity of p300/CBP and HBO1. Oncotarget 2016; 7(2): 1687-706. [http://dx.doi.org/10.18632/oncotarget.6424] [PMID: 26625199]

[70] Wang D, Kon N, Lasso G, *et al.* Acetylation-regulated interaction between p53 and SET reveals a widespread regulatory mode. Nature 2016; 538(7623): 118-22. [http://dx.doi.org/10.1038/nature19759] [PMID: 27626385]

[71] Loffreda A, Jacchetti E, Antunes S, *et al.* Live-cell p53 single-molecule binding is modulated by C-terminal acetylation and correlates with transcriptional activity. Nat Commun 2017; 8(1): 313. [http://dx.doi.org/10.1038/s41467-017-00398-7] [PMID: 28827596]

[72] Weber JD, Taylor LJ, Roussel MF, Sherr CJ, Bar-Sagi D. Nucleolar Arf sequesters Mdm2 and activates p53. Nat Cell Biol 1999; 1(1): 20-6.
[http://dx.doi.org/10.1038/8991] [PMID: 10559859]

[73] Deisenroth C, Franklin DA, Zhang Y. The Evolution of the Ribosomal Protein-MDM2-p53 Pathway. Cold Spring Harb Perspect Med 2016; 6(12)a026138
[http://dx.doi.org/10.1101/cshperspect.a026138] [PMID: 27908926]

[74] Fekry B, Jeffries KA, Esmaeilniakooshkghazi A, *et al.* C_{16}-ceramide is a natural regulatory ligand of p53 in cellular stress response. Nat Commun 2018; 9(1): 4149.
[http://dx.doi.org/10.1038/s41467-018-06650-y] [PMID: 30297838]

[75] Lahav G, Rosenfeld N, Sigal A, *et al.* Dynamics of the p53-Mdm2 feedback loop in individual cells. Nat Genet 2004; 36(2): 147-50.
[http://dx.doi.org/10.1038/ng1293] [PMID: 14730303]

[76] Kloc M, Shirato Y, Bilinski S, Browder LW, Johnston J. Differential subcellular sequestration of proapoptotic and antiapoptotic proteins and colocalization of Bcl-x(L) with the germ plasm, in Xenopus laevis oocytes. Genesis 2007; 45(8): 523-31.
[http://dx.doi.org/10.1002/dvg.20322] [PMID: 17661400]

[77] Levine AJ, Harris CR, Puzio-Kuter AM. The interfaces between signal transduction pathways: IGF-1/mTor, p53 and the Parkinson Disease pathway. Oncotarget 2012; 3(11): 1301-7.
[http://dx.doi.org/10.18632/oncotarget.759] [PMID: 23211569]

[78] Lieberman HB, Panigrahi SK, Hopkins KM, Wang L, Broustas CG. p53 and RAD9, the DNA Damage Response, and Regulation of Transcription Networks. Radiat Res 2017; 187(4): 424-32.
[http://dx.doi.org/10.1667/RR003CC.1] [PMID: 28140789]

[79] Cabrita MA, Bose R, Vanzyl EJ, *et al.* The p53 protein induces stable miRNAs that have the potential to modify subsequent p53 responses. Gene 2017; 608: 86-94.
[http://dx.doi.org/10.1016/j.gene.2017.01.018] [PMID: 28119089]

[80] Jonak K, Kurpas M, Szoltysek K, Janus P, Abramowicz A, Puszynski K. A novel mathematical model of ATM/p53/NF- κB pathways points to the importance of the DDR switch-off mechanisms. BMC Syst Biol 2016; 10(1): 75.
[http://dx.doi.org/10.1186/s12918-016-0293-0] [PMID: 27526774]

[81] Lazo PA. Reverting p53 activation after recovery of cellular stress to resume with cell cycle progression. Cell Signal 2017; 33: 49-58.
[http://dx.doi.org/10.1016/j.cellsig.2017.02.005] [PMID: 28189587]

[82] Torii S, Yoshida T, Arakawa S, Honda S, Nakanishi A, Shimizu S. Identification of PPM1D as an essential Ulk1 phosphatase for genotoxic stress-induced autophagy. EMBO Rep 2016; 17(11): 1552-64.
[http://dx.doi.org/10.15252/embr.201642565] [PMID: 27670885]

[83] Hafner A, Bulyk ML, Jambhekar A, Lahav G. The multiple mechanisms that regulate p53 activity and cell fate. Nat Rev Mol Cell Biol 2019; 20(4): 199-210.
[http://dx.doi.org/10.1038/s41580-019-0110-x] [PMID: 30824861]

[84] Xiong Y, Hannon GJ, Zhang H, Casso D, Kobayashi R, Beach D. p21 is a universal inhibitor of cyclin kinases. Nature 1993; 366(6456): 701-4.
[http://dx.doi.org/10.1038/366701a0] [PMID: 8259214]

[85] Sadasivam S, Duan S, DeCaprio JA. The MuvB complex sequentially recruits B-Myb and FoxM1 to promote mitotic gene expression. Genes Dev 2012; 26(5): 474-89.
[http://dx.doi.org/10.1101/gad.181933.111] [PMID: 22391450]

[86] Biswas K, Sarkar S, Du K, Brautigan DL, Abbas T, Larner JM. The E3 Ligase CHIP Mediates p21 Degradation to Maintain Radioresistance. Mol Cancer Res 2017; 15(6): 651-9.
[http://dx.doi.org/10.1158/1541-7786.MCR-16-0466] [PMID: 28232384]

[87] Deng T, Yan G, Song X, *et al.* Deubiquitylation and stabilization of p21 by USP11 is critical for cell-cycle progression and DNA damage responses. Proc Natl Acad Sci USA 2018; 115(18): 4678-83. [http://dx.doi.org/10.1073/pnas.1714938115] [PMID: 29666278]

[88] Georgakilas AG, Martin OA, Bonner WM. p21: A Two-Faced Genome Guardian. Trends Mol Med 2017; 23(4): 310-9. [http://dx.doi.org/10.1016/j.molmed.2017.02.001] [PMID: 28279624]

[89] Kim YY, Jee HJ, Um JH, Kim YM, Bae SS, Yun J. Cooperation between p21 and Akt is required for p53-dependent cellular senescence. Aging Cell 2017; 16(5): 1094-103. [http://dx.doi.org/10.1111/acel.12639] [PMID: 28691365]

[90] Sengupta S, Harris CC. p53: traffic cop at the crossroads of DNA repair and recombination. Nat Rev Mol Cell Biol 2005; 6(1): 44-55. [http://dx.doi.org/10.1038/nrm1546] [PMID: 15688066]

[91] Tonnessen-Murray CA, Lozano G, Jackson JG. The Regulation of Cellular Functions by the p53 Protein: Cellular Senescence. Cold Spring Harb Perspect Med 2017; 7(2)a026112 [http://dx.doi.org/10.1101/cshperspect.a026112] [PMID: 27881444]

[92] Correia-Melo C, Marques FD, Anderson R, *et al.* Mitochondria are required for pro-ageing features of the senescent phenotype. EMBO J 2016; 35(7): 724-42. [http://dx.doi.org/10.15252/embj.201592862] [PMID: 26848154]

[93] Hernandez-Segura A, Nehme J, Demaria M. Hallmarks of Cellular Senescence. Trends Cell Biol 2018; 28(6): 436-53. [http://dx.doi.org/10.1016/j.tcb.2018.02.001] [PMID: 29477613]

[94] Ritschka B, Storer M, Mas A, *et al.* The senescence-associated secretory phenotype induces cellular plasticity and tissue regeneration. Genes Dev 2017; 31(2): 172-83. [http://dx.doi.org/10.1101/gad.290635.116] [PMID: 28143833]

[95] de Magalhães JP, Passos JF. Stress, cell senescence and organismal ageing. Mech Ageing Dev 2018; 170: 2-9. [http://dx.doi.org/10.1016/j.mad.2017.07.001] [PMID: 28688962]

[96] Aubrey BJ, Kelly GL, Janic A, Herold MJ, Strasser A. How does p53 induce apoptosis and how does this relate to p53-mediated tumour suppression? Cell Death Differ 2018; 25(1): 104-13. [http://dx.doi.org/10.1038/cdd.2017.169] [PMID: 29149101]

[97] Marchenko ND, Moll UM. Mitochondrial death functions of p53. Molecular & cellular oncology 2014; 1(2): e955995. [http://dx.doi.org/10.1080/23723548.2014.955995]

[98] Kale J, Osterlund EJ, Andrews DW. BCL-2 family proteins: changing partners in the dance towards death. Cell Death Differ 2018; 25(1): 65-80. [http://dx.doi.org/10.1038/cdd.2017.186] [PMID: 29149100]

[99] Chipuk JE, Kuwana T, Bouchier-Hayes L, *et al.* Direct activation of Bax by p53 mediates mitochondrial membrane permeabilization and apoptosis. Science 2004; 303(5660): 1010-4. [http://dx.doi.org/10.1126/science.1092734] [PMID: 14963330]

[100] Jiang L, Kon N, Li T, *et al.* Ferroptosis as a p53-mediated activity during tumour suppression. Nature 2015; 520(7545): 57-62. [http://dx.doi.org/10.1038/nature14344] [PMID: 25799988]

[101] Kang R, Kroemer G, Tang D. The tumor suppressor protein p53 and the ferroptosis network. Free radical biology & medicine 2018; 133: 162-8.

[102] Purvis JE, Karhohs KW, Mock C, Batchelor E, Loewer A, Lahav G. p53 dynamics control cell fate. Science 2012; 336(6087): 1440-4. [http://dx.doi.org/10.1126/science.1218351] [PMID: 22700930]

[103] Paek AL, Liu JC, Loewer A, Forrester WC, Lahav G. Cell-to-Cell Variation in p53 Dynamics Leads to Fractional Killing. Cell 2016; 165(3): 631-42.
[http://dx.doi.org/10.1016/j.cell.2016.03.025] [PMID: 27062928]

[104] Issler MVC, Mombach JCM. MicroRNA-16 feedback loop with p53 and Wip1 can regulate cell fate determination between apoptosis and senescence in DNA damage response. PLoS One 2017; 12(10)e0185794
[http://dx.doi.org/10.1371/journal.pone.0185794] [PMID: 28968438]

[105] Leslie PL, Franklin DA, Liu Y, Zhang Y. p53 Regulates the Expression of LRP1 and Apoptosis through a Stress Intensity-Dependent MicroRNA Feedback Loop. Cell Rep 2018; 24(6): 1484-95.
[http://dx.doi.org/10.1016/j.celrep.2018.07.010] [PMID: 30089260]

[106] Lambrus BG, Holland AJ. A New Mode of Mitotic Surveillance. Trends Cell Biol 2017; 27(5): 314-21.
[http://dx.doi.org/10.1016/j.tcb.2017.01.004] [PMID: 28188027]

[107] Reyes J, Chen JY, Stewart-Ornstein J, Karhohs KW, Mock CS, Lahav G. Fluctuations in p53 Signaling Allow Escape from Cell-Cycle Arrest. Mol Cell 2018; 71(4): 581-591.e5.
[http://dx.doi.org/10.1016/j.molcel.2018.06.031] [PMID: 30057196]

[108] Hanahan D, Weinberg RA. Hallmarks of cancer: the next generation. Cell 2011; 144(5): 646-74.
[http://dx.doi.org/10.1016/j.cell.2011.02.013] [PMID: 21376230]

[109] Eischen CM. Genome Stability Requires p53. Cold Spring Harb Perspect Med 2016; 6(6)a026096
[http://dx.doi.org/10.1101/cshperspect.a026096] [PMID: 27252396]

[110] Fitzgerald DM, Hastings PJ, Rosenberg SM. Stress-Induced Mutagenesis: Implications in Cancer and Drug Resistance. Annual review of cancer biology 2017; 1: 119-40.

[111] Levine MS, Holland AJ. The impact of mitotic errors on cell proliferation and tumorigenesis. Genes Dev 2018; 32(9-10): 620-38.
[http://dx.doi.org/10.1101/gad.314351.118] [PMID: 29802124]

[112] Fava LL, Schuler F, Sladky V, *et al.* The PIDDosome activates p53 in response to supernumerary centrosomes. Genes Dev 2017; 31(1): 34-45.
[http://dx.doi.org/10.1101/gad.289728.116] [PMID: 28130345]

[113] Thompson SL, Compton DA. Proliferation of aneuploid human cells is limited by a p53-dependent mechanism. J Cell Biol 2010; 188(3): 369-81.
[http://dx.doi.org/10.1083/jcb.200905057] [PMID: 20123995]

[114] Meitinger F, Anzola JV, Kaulich M, *et al.* 53BP1 and USP28 mediate p53 activation and G1 arrest after centrosome loss or extended mitotic duration. J Cell Biol 2016; 214(2): 155-66.
[http://dx.doi.org/10.1083/jcb.201604081] [PMID: 27432897]

[115] Hinchcliffe EH, Day CA, Karanjeet KB, *et al.* Chromosome missegregation during anaphase triggers p53 cell cycle arrest through histone H3.3 Ser31 phosphorylation. Nat Cell Biol 2016; 18(6): 668-75.
[http://dx.doi.org/10.1038/ncb3348] [PMID: 27136267]

[116] Ganem NJ, Cornils H, Chiu SY, *et al.* Cytokinesis failure triggers hippo tumor suppressor pathway activation. Cell 2014; 158(4): 833-48.
[http://dx.doi.org/10.1016/j.cell.2014.06.029] [PMID: 25126788]

[117] Aylon Y, Michael D, Shmueli A, Yabuta N, Nojima H, Oren M. A positive feedback loop between the p53 and Lats2 tumor suppressors prevents tetraploidization. Genes Dev 2006; 20(19): 2687-700.
[http://dx.doi.org/10.1101/gad.1447006] [PMID: 17015431]

[118] Tutton S, Azzam GA, Stong N, *et al.* Subtelomeric p53 binding prevents accumulation of DNA damage at human telomeres. EMBO J 2016; 35(2): 193-207.
[http://dx.doi.org/10.15252/embj.201490880] [PMID: 26658110]

[119] Maciejowski J, de Lange T. Telomeres in cancer: tumour suppression and genome instability. Nat Rev Mol Cell Biol 2017; 18(3): 175-86.
[http://dx.doi.org/10.1038/nrm.2016.171] [PMID: 28096526]

[120] Rodriguez-Brenes IA, Komarova NL, Wodarz D. The role of telomere shortening in carcinogenesis: A hybrid stochastic-deterministic approach. J Theor Biol 2019; 460: 144-52.
[http://dx.doi.org/10.1016/j.jtbi.2018.09.003] [PMID: 30315815]

[121] Gottifredi V, Wiesmüller L. The Tip of an Iceberg: Replication-Associated Functions of the Tumor Suppressor p53. Cancers (Basel) 2018; 10(8)E250
[http://dx.doi.org/10.3390/cancers10080250] [PMID: 30060597]

[122] Mummenbrauer T, Janus F, Müller B, Wiesmüller L, Deppert W, Grosse F. p53 Protein exhibits 3′-t--5′ exonuclease activity. Cell 1996; 85(7): 1089-99.
[http://dx.doi.org/10.1016/S0092-8674(00)81309-4] [PMID: 8674115]

[123] Avkin S, Sevilya Z, Toube L, *et al.* p53 and p21 regulate error-prone DNA repair to yield a lower mutation load. Mol Cell 2006; 22(3): 407-13.
[http://dx.doi.org/10.1016/j.molcel.2006.03.022] [PMID: 16678112]

[124] Klusmann I, Rodewald S, Müller L, *et al.* p53 Activity Results in DNA Replication Fork Processivity. Cell Rep 2016; 17(7): 1845-57.
[http://dx.doi.org/10.1016/j.celrep.2016.10.036] [PMID: 27829155]

[125] Hampp S, Kiessling T, Buechle K, *et al.* DNA damage tolerance pathway involving DNA polymerase ι and the tumor suppressor p53 regulates DNA replication fork progression. Proc Natl Acad Sci USA 2016; 113(30): E4311-9.
[http://dx.doi.org/10.1073/pnas.1605828113] [PMID: 27407148]

[126] Roy S, Tomaszowski KH, Luzwick JW, *et al.* p53 orchestrates DNA replication restart homeostasis by suppressing mutagenic RAD52 and POLθ pathways. eLife 2018; 7e31723
[http://dx.doi.org/10.7554/eLife.31723] [PMID: 29334356]

[127] Neri F, Rapelli S, Krepelova A, *et al.* Intragenic DNA methylation prevents spurious transcription initiation. Nature 2017; 543(7643): 72-7.
[http://dx.doi.org/10.1038/nature21373] [PMID: 28225755]

[128] Levine AJ. The p53 protein plays a central role in the mechanism of action of epigenetic drugs that alter the methylation of cytosine residues in DNA. Oncotarget 2017; 8(5): 7228-30.
[http://dx.doi.org/10.18632/oncotarget.14805] [PMID: 28129641]

[129] Wylie A, Jones AE, D'Brot A, *et al.* p53 genes function to restrain mobile elements. Genes Dev 2016; 30(1): 64-77.
[http://dx.doi.org/10.1101/gad.266098.115] [PMID: 26701264]

[130] Lu WJ, Chapo J, Roig I, Abrams JM. Meiotic recombination provokes functional activation of the p53 regulatory network. Science 2010; 328(5983): 1278-81.
[http://dx.doi.org/10.1126/science.1185640] [PMID: 20522776]

[131] Levine AJ, Ting DT, Greenbaum BD. P53 and the defenses against genome instability caused by transposons and repetitive elements. BioEssays 2016; 38(6): 508-13.
[http://dx.doi.org/10.1002/bies.201600031] [PMID: 27172878]

[132] Tiwari B, Jones AE, Abrams JM. Transposons, p53 and Genome Security. Trends Genet 2018; 34(11): 846-55.
[http://dx.doi.org/10.1016/j.tig.2018.08.003] [PMID: 30195581]

[133] Mateo AR, Kessler Z, Jolliffe AK, *et al.* The p53-like Protein CEP-1 Is Required for Meiotic Fidelity in C. elegans. Curr Biol 2016; 26(9): 1148-58.
[http://dx.doi.org/10.1016/j.cub.2016.03.036] [PMID: 27151662]

[134] Jain AK, Barton MC. p53: emerging roles in stem cells, development and beyond. Development 2018;

145(8)dev158360
[http://dx.doi.org/10.1242/dev.158360] [PMID: 29654218]

[135] Hu W, Feng Z, Teresky AK, Levine AJ. p53 regulates maternal reproduction through LIF. Nature 2007; 450(7170): 721-4.
[http://dx.doi.org/10.1038/nature05993] [PMID: 18046411]

[136] Bowling S, Di Gregorio A, Sancho M, *et al.* P53 and mTOR signalling determine fitness selection through cell competition during early mouse embryonic development. Nat Commun 2018; 9(1): 1763.
[http://dx.doi.org/10.1038/s41467-018-04167-y] [PMID: 29720666]

[137] Abegglen LM, Caulin AF, Chan A, *et al.* Potential Mechanisms for Cancer Resistance in Elephants and Comparative Cellular Response to DNA Damage in Humans. JAMA 2015; 314(17): 1850-60.
[http://dx.doi.org/10.1001/jama.2015.13134] [PMID: 26447779]

[138] Caulin AF, Maley CC. Peto's Paradox: evolution's prescription for cancer prevention. Trends in ecology & evolution 2011; 26(4): 175-82.
[http://dx.doi.org/10.1016/j.tree.2011.01.002]

[139] Haupt S, Haupt Y. P53 at the start of the 21st century: lessons from elephants. F1000 Res 2017; 6: 2041.
[http://dx.doi.org/10.12688/f1000research.12682.1] [PMID: 29250320]

[140] Pfister NT, Prives C. Transcriptional Regulation by Wild-Type and Cancer-Related Mutant Forms of p53. Cold Spring Harb Perspect Med 2017; 7(2)a026054
[http://dx.doi.org/10.1101/cshperspect.a026054] [PMID: 27836911]

[141] Kim MP, Lozano G. Mutant p53 partners in crime. Cell Death Differ 2018; 25(1): 161-8.
[http://dx.doi.org/10.1038/cdd.2017.185] [PMID: 29099488]

[142] Garcia PB, Attardi LD. Illuminating p53 function in cancer with genetically engineered mouse models. Semin Cell Dev Biol 2014; 27: 74-85.
[http://dx.doi.org/10.1016/j.semcdb.2013.12.014] [PMID: 24394915]

[143] Matt S, Hofmann TG. The DNA damage-induced cell death response: a roadmap to kill cancer cells. Cell Mol Life Sci 2016; 73(15): 2829-50.
[http://dx.doi.org/10.1007/s00018-016-2130-4] [PMID: 26791483]

[144] Christophorou MA, Ringshausen I, Finch AJ, Swigart LB, Evan GI. The pathological response to DNA damage does not contribute to p53-mediated tumour suppression. Nature 2006; 443(7108): 214-7.
[http://dx.doi.org/10.1038/nature05077] [PMID: 16957739]

[145] Brady CA, Jiang D, Mello SS, *et al.* Distinct p53 transcriptional programs dictate acute DNA-damage responses and tumor suppression. Cell 2011; 145(4): 571-83.
[http://dx.doi.org/10.1016/j.cell.2011.03.035] [PMID: 21565614]

[146] Valente LJ, Gray DH, Michalak EM, *et al.* p53 efficiently suppresses tumor development in the complete absence of its cell-cycle inhibitory and proapoptotic effectors p21, Puma, and Noxa. Cell Rep 2013; 3(5): 1339-45.
[http://dx.doi.org/10.1016/j.celrep.2013.04.012] [PMID: 23665218]

[147] Tomasini R, Tsuchihara K, Wilhelm M, *et al.* TAp73 knockout shows genomic instability with infertility and tumor suppressor functions. Genes Dev 2008; 22(19): 2677-91.
[http://dx.doi.org/10.1101/gad.1695308] [PMID: 18805989]

[148] Coates PJ, Nenutil R, Holcakova J, Nekulova M, Podhorec J, Svoboda M, *et al.* p63 isoforms in triple-negative breast cancer: DeltaNp63 associates with the basal phenotype whereas TAp63 associates with androgen receptor, lack of BRCA mutation, PTEN and improved survival. Virchows Archiv : an international journal of pathology 2018; 472(3): 351-9.

[149] Venkatanarayan A, Raulji P, Norton W, *et al.* IAPP-driven metabolic reprogramming induces regression of p53-deficient tumours *in vivo*. Nature 2015; 517(7536): 626-30.

[http://dx.doi.org/10.1038/nature13910] [PMID: 25409149]

[150] Gebel J, Luh LM, Coutandin D, *et al.* Mechanism of TAp73 inhibition by ΔNp63 and structural basis of p63/p73 hetero-tetramerization. Cell Death Differ 2016; 23(12): 1930-40. [http://dx.doi.org/10.1038/cdd.2016.83] [PMID: 27716744]

[151] Acosta J, Wang W, Feldser DM. Off and back-on again: a tumor suppressor's tale. Oncogene 2018; 37(23): 3058-69. [http://dx.doi.org/10.1038/s41388-018-0186-3] [PMID: 29540833]

[152] Lopes CAM, Mesquita M, Cunha AI, *et al.* Centrosome amplification arises before neoplasia and increases upon p53 loss in tumorigenesis. J Cell Biol 2018; 217(7): 2353-63. [http://dx.doi.org/10.1083/jcb.201711191] [PMID: 29739803]

[153] Pavlova NN, Thompson CB. The Emerging Hallmarks of Cancer Metabolism. Cell Metab 2016; 23(1): 27-47. [http://dx.doi.org/10.1016/j.cmet.2015.12.006] [PMID: 26771115]

[154] Vazquez A, Kamphorst JJ, Markert EK, Schug ZT, Tardito S, Gottlieb E. Cancer metabolism at a glance. J Cell Sci 2016; 129(18): 3367-73. [http://dx.doi.org/10.1242/jcs.181016] [PMID: 27635066]

[155] Shuvalov O, Petukhov A, Daks A, Fedorova O, Vasileva E, Barlev NA. One-carbon metabolism and nucleotide biosynthesis as attractive targets for anticancer therapy. Oncotarget 2017; 8(14): 23955-77. [http://dx.doi.org/10.18632/oncotarget.15053] [PMID: 28177894]

[156] Vander Heiden MG, Cantley LC, Thompson CB. Understanding the Warburg effect: the metabolic requirements of cell proliferation. Science 2009; 324(5930): 1029-33. [http://dx.doi.org/10.1126/science.1160809] [PMID: 19460998]

[157] Lunt SY, Vander Heiden MG. Aerobic glycolysis: meeting the metabolic requirements of cell proliferation. Annu Rev Cell Dev Biol 2011; 27: 441-64. [http://dx.doi.org/10.1146/annurev-cellbio-092910-154237] [PMID: 21985671]

[158] Levine AJ, Puzio-Kuter AM. The control of the metabolic switch in cancers by oncogenes and tumor suppressor genes. Science 2010; 330(6009): 1340-4. [http://dx.doi.org/10.1126/science.1193494] [PMID: 21127244]

[159] Lonetto G, Koifman G, Silberman A, Attery A, Solomon H, Levin-Zaidman S, *et al.* Mutant p53-dependent mitochondrial metabolic alterations in a mesenchymal stem cell-based model of progressive malignancy. Cell Death Differ 2018. [http://dx.doi.org/10.1038/s41418-018-0227-z] [PMID: 30413783]

[160] Bensaad K, Tsuruta A, Selak MA, *et al.* TIGAR, a p53-inducible regulator of glycolysis and apoptosis. Cell 2006; 126(1): 107-20. [http://dx.doi.org/10.1016/j.cell.2006.05.036] [PMID: 16839880]

[161] Labuschagne CF, Zani F, Vousden KH. Control of metabolism by p53 - Cancer and beyond. Biochim Biophys Acta Rev Cancer 2018; 1870(1): 32-42. [http://dx.doi.org/10.1016/j.bbcan.2018.06.001] [PMID: 29883595]

[162] Matoba S, Kang JG, Patino WD, *et al.* p53 regulates mitochondrial respiration. Science 2006; 312(5780): 1650-3. [http://dx.doi.org/10.1126/science.1126863] [PMID: 16728594]

[163] Su X, Gi YJ, Chakravarti D, *et al.* TAp63 is a master transcriptional regulator of lipid and glucose metabolism. Cell Metab 2012; 16(4): 511-25. [http://dx.doi.org/10.1016/j.cmet.2012.09.006] [PMID: 23040072]

[164] Moon SH, Huang CH, Houlihan SL, *et al.* p53 Represses the Mevalonate Pathway to Mediate Tumor Suppression. Cell 2019; 176(3): 564-580.e19. [http://dx.doi.org/10.1016/j.cell.2018.11.011] [PMID: 30580964]

[165] Akeno N, Miller AL, Ma X, Wikenheiser-Brokamp KA. p53 suppresses carcinoma progression by inhibiting mTOR pathway activation. Oncogene 2015; 34(5): 589-99. [http://dx.doi.org/10.1038/onc.2013.589] [PMID: 24469052]

[166] Saxton RA, Sabatini DM. mTOR Signaling in Growth, Metabolism, and Disease. Cell 2017; 168(6): 960-76. [http://dx.doi.org/10.1016/j.cell.2017.02.004] [PMID: 28283069]

[167] White E. Autophagy and p53. Cold Spring Harb Perspect Med 2016; 6(4)a026120 [http://dx.doi.org/10.1101/cshperspect.a026120] [PMID: 27037419]

[168] Horikawa I, Fujita K, Jenkins LM, *et al.* Autophagic degradation of the inhibitory p53 isoform Δ133p53α as a regulatory mechanism for p53-mediated senescence. Nat Commun 2014; 5: 4706. [http://dx.doi.org/10.1038/ncomms5706] [PMID: 25144556]

[169] Lee IH, Kawai Y, Fergusson MM, *et al.* Atg7 modulates p53 activity to regulate cell cycle and survival during metabolic stress. Science 2012; 336(6078): 225-8. [http://dx.doi.org/10.1126/science.1218395] [PMID: 22499945]

[170] Levine AJ, Puzio-Kuter AM, Chan CS, Hainaut P. The Role of the p53 Protein in Stem-Cell Biology and Epigenetic Regulation. Cold Spring Harb Perspect Med 2016; 6(9)a026153 [http://dx.doi.org/10.1101/cshperspect.a026153] [PMID: 27352800]

[171] Hong H, Takahashi K, Ichisaka T, *et al.* Suppression of induced pluripotent stem cell generation by the p53-p21 pathway. Nature 2009; 460(7259): 1132-5. [http://dx.doi.org/10.1038/nature08235] [PMID: 19668191]

[172] Kawamura T, Suzuki J, Wang YV, *et al.* Linking the p53 tumour suppressor pathway to somatic cell reprogramming. Nature 2009; 460(7259): 1140-4. [http://dx.doi.org/10.1038/nature08311] [PMID: 19668186]

[173] Utikal J, Polo JM, Stadtfeld M, *et al.* Immortalization eliminates a roadblock during cellular reprogramming into iPS cells. Nature 2009; 460(7259): 1145-8. [http://dx.doi.org/10.1038/nature08285] [PMID: 19668190]

[174] Yi L, Lu C, Hu W, Sun Y, Levine AJ. Multiple roles of p53-related pathways in somatic cell reprogramming and stem cell differentiation. Cancer Res 2012; 72(21): 5635-45. [http://dx.doi.org/10.1158/0008-5472.CAN-12-1451] [PMID: 22964580]

[175] Napoli M, Flores ER. The p53 family orchestrates the regulation of metabolism: physiological regulation and implications for cancer therapy. Br J Cancer 2017; 116(2): 149-55. [http://dx.doi.org/10.1038/bjc.2016.384] [PMID: 27884017]

[176] Rath E, Moschetta A, Haller D. Mitochondrial function - gatekeeper of intestinal epithelial cell homeostasis. Nature reviews Gastroenterology & hepatology 2018; 15(8): 497-516.

[177] Li T, Kon N, Jiang L, *et al.* Tumor suppression in the absence of p53-mediated cell-cycle arrest, apoptosis, and senescence. Cell 2012; 149(6): 1269-83. [http://dx.doi.org/10.1016/j.cell.2012.04.026] [PMID: 22682249]

[178] Nakamura Y, Arakawa H. Discovery of Mieap-regulated mitochondrial quality control as a new function of tumor suppressor p53. Cancer Sci 2017; 108(5): 809-17. [http://dx.doi.org/10.1111/cas.13208] [PMID: 28222492]

[179] Bourdon A, Minai L, Serre V, *et al.* Mutation of RRM2B, encoding p53-controlled ribonucleotide reductase (p53R2), causes severe mitochondrial DNA depletion. Nat Genet 2007; 39(6): 776-80. [http://dx.doi.org/10.1038/ng2040] [PMID: 17486094]

[180] Park JH, Zhuang J, Li J, Hwang PM. p53 as guardian of the mitochondrial genome. FEBS Lett 2016; 590(7): 924-34. [http://dx.doi.org/10.1002/1873-3468.12061] [PMID: 26780878]

[181] Dixon SJ, Lemberg KM, Lamprecht MR, *et al.* Ferroptosis: an iron-dependent form of nonapoptotic

cell death. Cell 2012; 149(5): 1060-72.
[http://dx.doi.org/10.1016/j.cell.2012.03.042] [PMID: 22632970]

[182] Tarangelo A, Magtanong L, Bieging-Rolett KT, *et al.* p53 Suppresses Metabolic Stress-Induced Ferroptosis in Cancer Cells. Cell Rep 2018; 22(3): 569-75.
[http://dx.doi.org/10.1016/j.celrep.2017.12.077] [PMID: 29346757]

[183] Ohashi T, Idogawa M, Sasaki Y, Tokino T. p53 mediates the suppression of cancer cell invasion by inducing LIMA1/EPLIN. Cancer Lett 2017; 390: 58-66.
[http://dx.doi.org/10.1016/j.canlet.2016.12.034] [PMID: 28093207]

[184] Gadea G, Arsic N, Fernandes K, *et al.* *TP53* drives invasion through expression of its Δ133p53β variant. eLife 2016; 5e14734
[http://dx.doi.org/10.7554/eLife.14734] [PMID: 27630122]

[185] LeBleu VS, Kalluri R. A peek into cancer-associated fibroblasts: origins, functions and translational impact. Dis Model Mech 2018; 11(4)dmm029447
[http://dx.doi.org/10.1242/dmm.029447] [PMID: 29686035]

[186] Sousa CM, Biancur DE, Wang X, *et al.* Pancreatic stellate cells support tumour metabolism through autophagic alanine secretion. Nature 2016; 536(7617): 479-83.
[http://dx.doi.org/10.1038/nature19084] [PMID: 27509858]

[187] Orozco CA, Martinez-Bosch N, Guerrero PE, *et al.* Targeting galectin-1 inhibits pancreatic cancer progression by modulating tumor-stroma crosstalk. Proc Natl Acad Sci USA 2018; 115(16): E3769-78.
[http://dx.doi.org/10.1073/pnas.1722434115] [PMID: 29615514]

[188] Foster DS, Jones RE, Ransom RC, Longaker MT, Norton JA. The evolving relationship of wound healing and tumor stroma. JCI Insight 2018; 3(18): 99911.
[http://dx.doi.org/10.1172/jci.insight.99911] [PMID: 30232274]

[189] Cavaco A, Rezaei M, Niland S, Eble JA. Collateral Damage Intended-Cancer-Associated Fibroblasts and Vasculature Are Potential Targets in Cancer Therapy. Int J Mol Sci 2017; 18(11)E2355
[http://dx.doi.org/10.3390/ijms18112355] [PMID: 29112161]

[190] Dameron KM, Volpert OV, Tainsky MA, Bouck N. Control of angiogenesis in fibroblasts by p53 regulation of thrombospondin-1. Science 1994; 265(5178): 1582-4.
[http://dx.doi.org/10.1126/science.7521539] [PMID: 7521539]

[191] Amelio I, Mancini M, Petrova V, *et al.* p53 mutants cooperate with HIF-1 in transcriptional regulation of extracellular matrix components to promote tumor progression. Proc Natl Acad Sci USA 2018; 115(46): E10869-78.
[http://dx.doi.org/10.1073/pnas.1808314115] [PMID: 30381462]

[192] Gudkov AV, Komarova EA. p53 and the Carcinogenicity of Chronic Inflammation. Cold Spring Harb Perspect Med 2016; 6(11)a026161
[http://dx.doi.org/10.1101/cshperspect.a026161] [PMID: 27549311]

[193] Lujambio A, Akkari L, Simon J, *et al.* Non-cell-autonomous tumor suppression by p53. Cell 2013; 153(2): 449-60.
[http://dx.doi.org/10.1016/j.cell.2013.03.020] [PMID: 23562644]

[194] Mitchell S, Vargas J, Hoffmann A. Signaling *via* the NFκB system. Wiley Interdiscip Rev Syst Biol Med 2016; 8(3): 227-41.
[http://dx.doi.org/10.1002/wsbm.1331] [PMID: 26990581]

[195] Häcker H, Karin M. Regulation and function of IKK and IKK-related kinases. Sci STKE 2006; 2006(357): re13.
[http://dx.doi.org/10.1126/stke.3572006re13] [PMID: 17047224]

[196] Göktuna SI, Diamanti MA, Chau TL. IKKs and tumor cell plasticity. FEBS J 2018; 285(12): 2161-81.
[http://dx.doi.org/10.1111/febs.14444] [PMID: 29582565]

[197] Perkins ND. The diverse and complex roles of NF-κB subunits in cancer. Nat Rev Cancer 2012; 12(2): 121-32.
[http://dx.doi.org/10.1038/nrc3204] [PMID: 22257950]

[198] Pires BRB, Silva RCMC, Ferreira GM, Abdelhay E. NF-kappaB: Two Sides of the Same Coin. Genes (Basel) 2018; 9(1)E24
[http://dx.doi.org/10.3390/genes9010024] [PMID: 29315242]

[199] Bakhoum SF, Ngo B, Laughney AM, *et al.* Chromosomal instability drives metastasis through a cytosolic DNA response. Nature 2018; 553(7689): 467-72.
[http://dx.doi.org/10.1038/nature25432] [PMID: 29342134]

[200] Sun SC. The non-canonical NF-κB pathway in immunity and inflammation. Nat Rev Immunol 2017; 17(9): 545-58.
[http://dx.doi.org/10.1038/nri.2017.52] [PMID: 28580957]

[201] Jiang HY, Wek SA, McGrath BC, *et al.* Phosphorylation of the alpha subunit of eukaryotic initiation factor 2 is required for activation of NF-kappaB in response to diverse cellular stresses. Mol Cell Biol 2003; 23(16): 5651-63.
[http://dx.doi.org/10.1128/MCB.23.16.5651-5663.2003] [PMID: 12897138]

[202] Janssens S, Pulendran B, Lambrecht BN. Emerging functions of the unfolded protein response in immunity. Nat Immunol 2014; 15(10): 910-9.
[http://dx.doi.org/10.1038/ni.2991] [PMID: 25232821]

[203] Huang TT, Wuerzberger-Davis SM, Wu ZH, Miyamoto S. Sequential modification of NEMO/IKKgamma by SUMO-1 and ubiquitin mediates NF-kappaB activation by genotoxic stress. Cell 2003; 115(5): 565-76.
[http://dx.doi.org/10.1016/S0092-8674(03)00895-X] [PMID: 14651848]

[204] Alexopoulou L, Holt AC, Medzhitov R, Flavell RA. Recognition of double-stranded RNA and activation of NF-kappaB by Toll-like receptor 3. Nature 2001; 413(6857): 732-8.
[http://dx.doi.org/10.1038/35099560] [PMID: 11607032]

[205] Boisson B, Puel A, Picard C, Casanova JL. Human IκBα Gain of Function: a Severe and Syndromic Immunodeficiency. J Clin Immunol 2017; 37(5): 397-412.
[http://dx.doi.org/10.1007/s10875-017-0400-z] [PMID: 28597146]

[206] Baratchian M, Davis CA, Shimizu A, *et al.* Distinct Activation Mechanisms of NF-κB Regulator Inhibitor of NF-κB Kinase (IKK) by Isoforms of the Cell Death Regulator Cellular FLICE-like Inhibitory Protein (cFLIP). J Biol Chem 2016; 291(14): 7608-20.
[http://dx.doi.org/10.1074/jbc.M116.718122] [PMID: 26865630]

[207] Chan B, Gonçalves Magalhães V, Lemmermann NAW, *et al.* The murine cytomegalovirus M35 protein antagonizes type I IFN induction downstream of pattern recognition receptors by targeting NF-κB mediated transcription. PLoS Pathog 2017; 13(5)e1006382
[http://dx.doi.org/10.1371/journal.ppat.1006382] [PMID: 28542326]

[208] Fang L, Choudhary S, Zhao Y, *et al.* ATM regulates NF-κB-dependent immediate-early genes *via* RelA Ser 276 phosphorylation coupled to CDK9 promoter recruitment. Nucleic Acids Res 2014; 42(13): 8416-32.
[http://dx.doi.org/10.1093/nar/gku529] [PMID: 24957606]

[209] Schäfer C, Göder A, Beyer M, *et al.* Class I histone deacetylases regulate p53/NF-κB crosstalk in cancer cells. Cell Signal 2017; 29: 218-25.
[http://dx.doi.org/10.1016/j.cellsig.2016.11.002] [PMID: 27838375]

[210] Dunphy G, Flannery SM, Almine JF, *et al.* Non-canonical Activation of the DNA Sensing Adaptor STING by ATM and IFI16 Mediates NF-κB Signaling after Nuclear DNA Damage. Mol Cell 2018; 71(5): 745-760.e5.
[http://dx.doi.org/10.1016/j.molcel.2018.07.034] [PMID: 30193098]

[211] Ak P, Levine AJ. p53 and NF-κB: different strategies for responding to stress lead to a functional antagonism. FASEB J 2010; 24(10): 3643-52.
[http://dx.doi.org/10.1096/fj.10-160549] [PMID: 20530750]

[212] Kawauchi K, Araki K, Tobiume K, Tanaka N. p53 regulates glucose metabolism through an IKK-N-kappaB pathway and inhibits cell transformation. Nat Cell Biol 2008; 10(5): 611-8.
[http://dx.doi.org/10.1038/ncb1724] [PMID: 18391940]

[213] Barroso-González J, Auclair S, Luan S, *et al.* PACS-2 mediates the ATM and NF-κB-dependent induction of anti-apoptotic Bcl-xL in response to DNA damage. Cell Death Differ 2016; 23(9): 1448-57.
[http://dx.doi.org/10.1038/cdd.2016.23] [PMID: 26943323]

[214] Bednarski JJ, Pandey R, Schulte E, *et al.* RAG-mediated DNA double-strand breaks activate a cell type-specific checkpoint to inhibit pre-B cell receptor signals. J Exp Med 2016; 213(2): 209-23.
[http://dx.doi.org/10.1084/jem.20151048] [PMID: 26834154]

[215] Dreyfus DH, Nagasawa M, Gelfand EW, Ghoda LY. Modulation of p53 activity by IkappaBalpha: evidence suggesting a common phylogeny between NF-kappaB and p53 transcription factors. BMC Immunol 2005; 6: 12.
[http://dx.doi.org/10.1186/1471-2172-6-12] [PMID: 15969767]

[216] Rahnamoun H, Lu H, Duttke SH, Benner C, Glass CK, Lauberth SM. Mutant p53 shapes the enhancer landscape of cancer cells in response to chronic immune signaling. Nat Commun 2017; 8(1): 754.
[http://dx.doi.org/10.1038/s41467-017-01117-y] [PMID: 28963538]

[217] Schumm K, Rocha S, Caamano J, Perkins ND. Regulation of p53 tumour suppressor target gene expression by the p52 NF-kappaB subunit. EMBO J 2006; 25(20): 4820-32.
[http://dx.doi.org/10.1038/sj.emboj.7601343] [PMID: 16990795]

[218] Morotti A, Crivellaro S, Panuzzo C, Carrà G, Guerrasio A, Saglio G. IκB-α: At the crossroad between oncogenic and tumor-suppressive signals. Oncol Lett 2017; 13(2): 531-4.
[http://dx.doi.org/10.3892/ol.2016.5465] [PMID: 28356925]

[219] Ishak Gabra MB, Yang Y, Lowman XH, Reid MA, Tran TQ, Kong M. IKKβ activates p53 to promote cancer cell adaptation to glutamine deprivation. Oncogenesis 2018; 7(11): 93.
[http://dx.doi.org/10.1038/s41389-018-0104-0] [PMID: 30478303]

CHAPTER 9

Autonomous Immunity

Abstract: Autonomous immunity is a set of immune mechanisms present in practically every cell of a multicellular organism. They include immune mechanisms based on nuclear acids, such as RNA interference (RNAi). Small RNAs generated from pathogen dsRNAs guide nucleases to the pathogenic nucleic acids. In addition, RNAi restricts the expression of transposable elements by establishing transcription-repressing chromatin over DNA regions that encode such elements. Other autonomous mechanisms prevent the entry of viruses into cells, detect and destroy non-self nucleic acids, restrict pathogen growth and replication, prevent pathogen release, and induce regulated cell death if the infection cannot be suppressed. Infected cells release interferons and other cytokines that alert neighboring cells and preventively induce defensive mechanisms in them. Pathogens express pathogen-associated molecular patterns (PAMPs), which are recognized by pattern-recognition receptors (PRRs). These receptors also recognize damaged "self" molecules expressing damage-associated molecular patterns. Infection can also be sensed indirectly through changes in the functioning of some key cellular molecules. Stress and infection can lead to the formation of inflammasomes triggering production and secretion of pro-inflammatory cytokines.

Keywords: Argonaute, Drosha, Dicer, Damage-associated molecular pattern, Danger theory, Effector immunity, Inflammasome, NLR, Pathogen-associated molecular patterns, Pattern recognition receptor, piRNAs, RNA interference, siRNAs, TLR.

INTRODUCTION

Multicellular organisms have multiple mechanisms to protect themselves from infection. Animals have specialized innate and adaptive immune systems that include immune cells and humoral factors such as complement. Non-immune cells too have an impressive arsenal of immunodefense mechanisms. These mechanisms are called intrinsic or autonomous since many of them can function independently of cells and factors of specialized immune systems. Autonomous mechanisms are also present in immune cells, where they are integrated into cells' function. Although being able to work independently, autonomous immune mechanisms typically communicate with the specialized immune systems through cytokines and other factors.

Nadia Danilova

9.1. RNA Interference

Gene silencing based on small complementary RNA or DNA guides is an ancient regulatory principle. In bacteria and archaea, it is carried out by CRISPR-Cas and Argonaute (Ago)-based immunodefense systems presented in chapter five. DNA viruses, common in prokaryotes, are the main targets of these systems. With the expansion of RNA viruses in eukaryotes, Argonaute-based RNA interference (RNAi) systems have evolved [1, 2]. RNAi uses small RNA guides to cleave or downregulate RNA targets. RNA guides are produced from double-stranded RNAs by RNase-III-like endonucleases. RNAi systems evolve differently in various species: *Caenorhabditis elegans* has 27 Ago proteins whereas humans have only eight [3]. The modern RNAi branches into several specialized systems that include microRNAs (miRNAs) functioning as gene regulators, small interfering RNAs (siRNAs) functioning in antiviral defense, and PIWI-interacting RNAs (piRNAs) functioning in protection against transposable elements.

Most animal mRNAs are regulated by miRNAs, 21–23 nucleotide RNAs that have sequence complementarity to the 3′ UTRs of mRNAs [4]. miRNAs are typically transcribed as long primary transcripts (pri-miRNAs) from miRNA genes. pri-miRNAs are processed in the nucleus by the RNase-III-like endonuclease Drosha into approximately 70 nucleotide pre-miRNAs with stem-loop structures. Then, pre-miRNAs are transported from the nucleus to the cytoplasm and further processed by the second RNase Dicer into short duplexes. The guide strand of an miRNA duplex is loaded onto the RNA-induced silencing complex (RISC), where it guides Ago nucleases to the complementary mRNA transcripts to arrest their translation and/or accelerate their degradation by uncapping and deadenylation. miRNA function is regulated by post-translational modification of Ago proteins, alternative ways of miRNAs production and maturation, sequence editing, changes in miRNAs transport, modulation of their binding to targets, and through other means [5, 6].

miRNAs are involved in stress and immune response in multiple ways. Stress changes miRNA biogenesis. It can lead to miRNA modifications that influence its activity or specificity [7, 8]. miRNAs are involved in the regulation of p53 and NFκB networks [9 - 11]. In response to DNA damage, p53 transcriptionally upregulates several miRNAs including miR-34 family [12]. miR-34a then regulates hundreds of genes that are involved in cell cycle arrest and apoptosis [13]. In addition, p53 interacts with Drosha complex and promotes post-transcriptional processing of several miRNAs with a growth-suppressing function such as miR-16-1 [14]. In contrast, p53 mutants interfere with Drosha complex assembly and miRNAs production. Correspondently, decreased miRNA levels are often observed in cancers with mutated p53.

miRNA biogenesis enzymes Drosha and Dicer are involved in DNA repair. They generate small RNAs at DNA double-strand break sites that are necessary for DNA repair [15].

In plants, miRNAs regulate the expression of immune disease resistance (R) genes. In the absence of infection, miRNAs suppress these genes [16]. In turn, overexpression of immune genes during infection suppresses miRNAs biosynthesis.

The small size of miRNAs means that they can migrate out of the cell into other cells and blood [17]. miRNAs from dying neurons are picked up by the astrocytes and interfere with their function [18]. Altered concentrations of some miRNAs in blood plasma may serve as a marker of Alzheimer's disease [19].

RNAi, as an antiviral mechanism, has been discovered in plants and it provides the main protection against viruses in many species including, besides plants, nematodes, and arthropods [20, 21]. siRNAs are generated from pathogen dsRNAs, followed by their amplification by host RNA-directed RNA polymerase [22]. siRNAs differ from miRNAs by higher specificity. They usually cleave their targets before translation.

RNAi is used in a new plant protection technique where dsRNA corresponding to a pathogen's protein is sprayed on the plant. This dsRNA is processed by the plant's enzymes into siRNAs, which target the corresponding pathogen mRNA. dsRNA is delivered to the distal tissue through the plant vascular system and is processed there [23]. In vertebrates, RNAi participation in antiviral defense is often masked by concurrent mechanisms such as interferon system [24]. In stem cells, where the interferon system is inactive, RNAi plays an important role in antiviral defense [25]. Cellular miRNAs also can contribute to antiviral immunity by repressing host factors essential for viral replication [26, 27]. Many viruses encode factors suppressing RNAi machinery [28].

RNase-III endonucleases including Drosha have an additional role in antiviral defense. They bind positive-strand RNA viruses at stem-loop sequences that correspond to the RNA-dependent RNA polymerase (RdRp). Such binding functions as a clamp that prevents RdRp translation leading to inhibition of virus replication [1].

Nuclear small RNAs have a function in genome regulation and preservation by establishing transcription-repressing chromatin at specific regions. In animals, PIWI-interacting RNAs (piRNAs) are the main nuclear small RNAs. They work in association with the PIWI subfamily of Argonaute proteins to suppress the expression of transposable elements, and guide genome rearrangement and epigenetic programming [29, 30]. piRNAs are typically 21–35 nucleotides in length and are transcribed from regions enriched in transposable elements called piRNA clusters [31 - 33]. RNA transcripts are cleaved, loaded on PIWI proteins, and trimmed by the 3' to 5' exonuclease Trimmer (PNLDC1 in mammals) [34]. piRNAs guide Ago-containing complexes to complementary RNAs to induce histone H3 lysine 9 di-and tri-methylation leading to heterochromatin formation [35, 36]. piRNAs are predominantly produced in gonads where they silence retrotransposons and other mobile elements in the germline [37].

If PIWI encounters in the cytoplasm a transcript of a transposon that escaped silencing, it cleaves the transcripts and generates additional piRNAs. This is called the Ping-Pong amplification loop [38]. This loop serves to specifically amplify piRNAs targeting active transposons.

PIWI pathway can influence many cellular processes including germ cell specification, stem cell maintenance, metabolism, and neuronal functions [30, 39 - 41]. Stress-induced deficiency of PIWI pathway in *Caenorhabditis elegans* can result in temporal sterility [42]. Similarly, in mice, deficiency of the exonuclease PNLDC1 that trims piRNAs leads to male sterility [43]. Aberrant expression of the PIWI pathway proteins is found in many types of cancer [44]. Herpesvirus integrated into genome changes piRNA processing and PIWI protein function in human embryos, which may have long-term health consequences [45].

An alternative way to defend the genome against mobile genetic elements is found in ciliates [46]. Ciliates have a transcriptionally active macronucleus and a germline micronucleus, which is transcriptionally silent during vegetative growth. The macronucleus develops during sexual conjugation from a copy of micronucleus through the elimination of repetitive DNA sequences by PIWI proteins guided by small RNAs.

9.2. Pathogen-associated Molecular Patterns and Pattern Recognition Receptors

Pathogens have macromolecules and structural motifs absent from their eukaryotic hosts such as peptidoglycan cell wall of Gram-positive bacteria. In 1989, Charles Janeway hypothesized that immune cells recognize such pathogen-associated molecular patterns (PAMPs) using pattern recognition receptors (PRRs) [47]. He predicted that PRRs might recognize three categories of patterns: "microbial non-self", "induced or aberrant self", and "missing self".

Drosophila Toll receptor was the first discovered PRR. It controls secretion of the antifungal peptide drosomycin [48]. Among 10 human Toll-like receptors (TLRs),

TLR3, 7, 8, and 9 are expressed in endosomes and others are found on the cell surface [49, 50]. TLRs recognize a variety of PAMPs. TLR3 recognizes dsRNA, TLR7 and TLR8 — viral ssRNA, TLR9 — unmethylated CpG DNA, and TLR5 — flagellin [51, 52]. TLR2 recognizes fungal chitin more than five oligomers in size [53]. Recognition of complex glycolipids and proteins by TLR4 and TLR2 often requires co-receptors, accessory molecules, and heterodimerization of TLR2 with TLR1 or TLR6 [49]. TLR10 inhibits other TLRs [54].

Another example of PRRs is C-type lectin receptors (CLRs) that bind carbohydrate ligands. Some CLRs are secreted proteins; others are expressed on the cell surface [55]. CLRs are important for the detection and clearance of bacterial and fungal infections as well as for many physiological functions.

Retinoic acid-inducible gene-I (RIG-I)-like receptors (RLRs) family are cytosolic RNA helicases that sense viral RNAs [56]. RLRs can also recognize cellular RNAs misprocessed during infection [57].

Nucleotide-binding domain (NBD) and leucine-rich repeat (LRR)-containing receptors (NLRs) are found both in animals and in plants [58, 59]. They are activated by a wide variety of microbial and "induced or aberrant self" ligands.

PRR signaling often depends on the formation of higher-order oligomeric complexes, signalosomes. Components of a signalosome interact through homologous domains such as Toll/IL1 receptor (TIR) motif [60]. Activated TLRs, except TLR3, facilitate the formation of a signalosome that is based on MyD88 adaptor (myddosome) [61]. Many CLRs have cytoplasmic immunoreceptors tyrosine-based activation motif (ITAM) or inhibitory ITIM motif. Upon ligand recognition, ITAM motif is phosphorylated and recruits Syk family of tyrosine kinases that phosphorylates multiple targets [62]. A member of NLR family NOD2 recruits its downstream kinase RIP2 through its DD/CARD domain, which leads to the formation of long cylindrical filaments necessary for NFκB induction [63]. Many signaling cascades are formed on mitochondrial membranes [64].

Bacterial pathogens usually carry multiple PAMPs including DNA, carbohydrates, lipids, and protein-based motifs. In contrast, recognition of viruses is mostly based on specific features of their RNAs and DNAs. Some PAMPs are produced by alive pathogens but absent in dead ones. An example is the small compound cyclic di-adenosine-monophosphate produced by bacteria, which can directly bind and activate STING immune receptor [65].

Although a broader repertoire of PRRs and a higher level of their expression are found in specialized immune cells, non-immune cells also express PRRs [66]. Receptor repertoire is highly variable between non-immune cells reflecting diverse requirements of different cell types for self-defense. For example, epithelial cells have a higher probability to encounter a pathogen and typically express more PRRs [67, 68].

Cell response to a pathogen depends on the integration of information from many immune receptors. It was noted that non-pathogenic microbes typically activated a single or few receptor families, whereas pathogens activated receptors from multiple families [69]. Various pathogens activate different combinations of PRRs, which let the cell generate a response tailored to a particular pathogen [70].

PRRs are involved in various processes besides immunity. For example, TLR signaling is involved in chronic pain [71]. Excessive signaling from PRRs can lead to various disorders such as chronic inflammation, hypertension, and cancer [72].

9.3. Danger Theory

Numerous commensal bacteria that reside in human gut and skin express the same PAMPs as pathogenic bacteria but do not elicit an immune response. To explain this puzzle, a seminal hypothesis was developed by Polly Matzinger. She suggested that tissue damage is even more important for the activation of the immune system than the presence of foreign entity [73, 74]. This idea also aligns with Janeway prediction that some PRR receptors might recognize "induced or aberrant self" [47].

Tissue damage is associated with the release of diverse molecules able to stimulate an immune response. They became known as Damage Associated Molecular Patterns (DAMPs). Examples include ATP, HSPs, uric acid, DNA and formyl peptides from mitochondrial, nuclear DNA, non-coding RNAs, histones, calreticulin, heme, and cholesterol [75 - 77]. Modified endogenous molecules such as oxidized membrane lipids, misfolded proteins that expose their hydrophobic regions, and glycated proteins are also potent DAMPs. Some DAMPs are directly recognized by PRRs. In addition, damaged tissues produce alarmins such as HMGB1, S100 proteins, IL1α, and IL33 that stimulate immune response [78].

A vital insight of the danger theory is that the immune system can sense not only a pathogen itself but the damage/changes caused by it. The danger theory stimulated other theories that further exploited this idea. "Guard theory" was suggested to explain the detection of pathogens by plant R-proteins; it hypothesized that these proteins recognize alterations in plant cells caused by effectors produced by a pathogen [79]. This mechanism is also called effector immunity since it senses changes in a limited number of host proteins caused by bacterial effectors [80].

"Patterns of pathogenesis" theory builds on the same idea for an explanation of recognition of infection by animal PRRs such as NLRs. It also points to the importance of other signals associated with pathogen growth, secretion, and anomalous localization of biomolecules associated with infection [81]. For example, the presence of any DNA in the cytosol, normally devoid of these molecules, induces an immune response.

Even substantial deviation from homeostasis induced by pathogenic organisms can trigger immune response [82]. Since the subject of recognition in effector-triggered immunity is not the pathogen but its effect, it complements pattern-recognition-based immunity, which recognizes pathogen molecules. The two types of immunity can interact at the signaling level and can synergize or antagonize each other [83, 84].

The length of exposure to the immune challenge may also be important. Some data suggest that the immune system detects sudden changes in the organism but is tolerant to slow changes or to chronic exposure to an antigen. This idea forms the basis of the discontinuity theory of immunity [85].

Changes in cellular homeostasis also can trigger an immune response, which in *C. elegans* has been termed "surveillance immunity" [82, 86]. This term can be broadened to include various mechanisms, such as effector immunity, that monitor the integrity of the cellular environment. Surveillance immunity allows cells to detect infection by pathogens for which there is no recognizable PAMPs as well as by pathogens that produce factors blocking PRRs.

Typically, infection is associated with activation of several signaling cascades. Some provide information about the pathogens, others signal change/damage of the intracellular environment. Experimental data support the idea that pathogen recognized as both foreign and harmful would elicit an enhanced immune response [87]. Surveillance immunity and PRR-triggered immunity complement each other and interact at the signaling level [83, 84]. They can converge on activating the same responses. For example, both PRR-based and effector immunity pathways activate NADPH oxidase [88].

9.4. Interferons

Recognition of PAMPs, DAMPs, or "altered self" leads to activation of effectors that eliminate the pathogen or induce cell death. In addition, infected cells produce cytokines that activate immune mechanisms in neighboring cells and recruit specialized immune cells to the site of infection. Among cytokines produced by infected animal cells interferons (IFNs) play an especially important role in autonomous immunity.

IFNs are subdivided into three major types depending on the receptors they use for signaling [89]. Type I IFNs act by inducing dimerization of IFNAR1 and IFNAR2 receptor chains [90]. IFN gamma belongs to type II; it binds a different receptor composed from IFNGR1 and IFNGR2 subunits [91]. Type III IFNs signal through the high-affinity receptor subunit IFNLR1 and the low-affinity subunit IL10R2, which is also used by other cytokines [92, 93].

IFN type I and III are induced when cellular PRRs such as TLRs bind their ligands, which leads to activation of IFN regulatory factors (IRF) [89]. There are 9 IRFs in mammals, which play different roles in the immune response. IRF3 and IRF7 are the most common downstream targets of PRR signaling. IRFs are present in the cytoplasm in an inactive monomeric form. Immune signaling leads to their post-translational modification, dimerization, nuclear translocation, and induction of IFN transcription. IFNs are translated even during general translational arrest induced *e.g.* in the course of viral infection by PKR kinase. Selective translation of IFNs is achieved by induction by IRF3 of a cofactor of phosphatase C GADD34 that decreases eIF2α phosphorylation to permit pulses of IFN synthesis [94].

Viruses are major activators of IFNs although they are also induced by some bacterial infections [95]. IFN gamma is activated by cytokines and is produced mostly in immune cells [96]. Type I IFNs can be induced in virtually all cells, although dendritic cells typically produce higher amounts. Type III IFNs are produced mostly by epithelial cells [93, 97].

When released from infected cells, IFNs prepare neighboring cells to fight infection by increasing the expression of immune sensors and effectors. IFNs also coordinate the response of innate and adaptive immune cells.

Downstream from IFN receptors, signals are transmitted by members of the Janus kinase (JAK) family that phosphorylate STAT family members [98]. IFN type I and III induce the formation of a complex, which contains Stat1, Stat2, and Irf9. Type II IFN signaling leads to the formation of Stat1 homodimer [96]. Various non-canonical pathways have also been reported [99]. IFN signaling is regulated by modulation of IFN surface receptor levels, receptor phosphorylation, acetylation, ubiquitination, and expression of various factors affecting the transmission of a signal from the IFN receptor.

IFN signaling induces expression of interferon-stimulated genes (ISGs), which inhibit virus propagation by various mechanisms [100 - 103]. ISG effectors act at viral entry, inhibit their transcription, translation, replication, assembly of new virions, and their exit from the cells. In addition, ISGs include immune sensors and positive and negative regulators of the IFN pathway. Multiple points of attack by ISGs make it harder for a virus to escape host defense or evolve new escape mechanisms.

IFN effects range from pathogen growth inhibition to cytostatic and pro-apoptotic. Despite the similar mode of signaling, the spectrum of target genes differs among various IFNs and cell types. For example, type III IFNs are tailored to function in epithelial cells where type I can also be induced. The sets of ISGs induced by type I and type III are very similar but the kinetics of their expression differ with an earlier and shorter response to type I IFNs and a delayed and prolonged response to type III [104].

Human and mouse cells treated with IFN show features of "trained immunity". In several generations of these cells, ISGs are induced more rapidly and strongly than in non-treated cells. This epigenetic transcriptional memory is achieved through histone modification chromatin marks that let RNA polymerase II to remain associated with the ISGs promoters [105] and by a metabolic shift to glycolysis [106].

The importance of IFN pathways in antimicrobial defense is underscored by the multiplicity of mechanisms that pathogens evolved to inhibit IFNs [107 - 109]. In the absence of infection, IFNs are expressed at a very low level providing cells with the basal level of ISGs, which sets a threshold for an immune response to infection [110]. IFNs are powerful antimicrobial agents but can also cause tissue damage and autoimmune conditions during prolonged or deregulated production [111, 112].

9.5. Inflammasome

Inflammasomes are protein complexes that assemble in response to infection, tissue damage, or disturbances in cellular homeostasis. They activate proinflammatory caspases, which then cleave cytokine precursors to generate active cytokines such as IL1β and IL18 [113, 114]. The composition of inflammasomes varies but typically includes a PRR, such as a member of NLRs family, the adaptor protein ASC, and pro-caspase 1. The most common inflammasome-forming PRRs include pyrin, NLRP3, NLR1, NLRC4, and AIM2. Inflammasome formation starts with priming, a signal that leads to an increased PRR and pro-cytokine expression. Priming can be induced through PRR-mediated signaling or in response to cytokines. For example, NLRC4 inflammasome can be primed by TNF [115]. Second, activation signal, leads to a PRR conformational change that starts inflammasome assembly, which contains 10-12 NLR molecules [116].

NLRP3 is the most studied inflammasome and has an especially wide specificity range [117]. It is induced by various ligands from pathogens, by diverse DAMPs, and by environmental stressors. Several particulate substances are known NLRP3 stimulants; they include urate crystals, cholesterol crystals, and silica.

NLRP3 molecule is composed of a pyrin domain (PYD), a nucleotide-binding domain, and leucine-rich repeat (LRR) domain. In the absence of triggers, NLRP3 is present in the cytoplasm in a metastable state that involves inter-domain molecular contacts, which are stabilized in complex with Hsp90 and its co-chaperones [118, 119]. Pyrin domain of NLRP3 is sheltered in this complex. Other PRRs forming inflammasomes are also held inactive in complexes with various factors [120].

NLRP3 activation is associated with a conformational change, exposure of the PYD domain, and oligomerization. Activated NLRP3 binds the pyrin domain of ASC; this induces ASC polymerization into micro-sized specks [121]. ASC complex recruits pro-caspase-1 through its caspase recruitment domain (CARD), which leads to autoproteolytic caspase-1 activation. Caspase-1 then cleaves various substrates including pro-IL1β and gasdermin D leading to IL1β release and/or pyroptosis. ASC specks can be released from cells and internalized by surrounding cells [122].

The exact mechanism of NLR-based inflammasome activation is not known but several models have been proposed. One model suggests that animal NLRs function similarly to plant NLRs, as postulated in the "guard theory" — they sense not the pathogen itself but cellular changes induced by it [79]. Bacterial entry into host cells depends on the actin cytoskeleton, which, in turn, depends on Rho family GTPases. Accordingly, many bacteria produce effectors that modify and block these GTPases [123]. This makes Rho family GTPases a candidate for recognition by PRRs. Two NLR sensors Nod1/2 are thought to recognize the overactivated state of Rho GTPases [124]. Pyrin inflammasome is activated in response to bacterial effectors that modify the switch region in RhoA GTPase by glycosylation, deamidation, and other modifications [125, 126]. There is no direct interaction between Rho GTPases and PRRs. Colchicine, which disrupts microtubule assembly, inhibits pyrin activation suggesting microtubule role in this process [120]. An alternative way of pyrin inflammasome activation involves metabolites of bile acids modified by gut microbiota. Only bacteria-modified metabolites but not the principal bile acids activate pyrin inflammasome [127].

Drop in K^+ ions concentration is suggested as a common mechanism of NLRP3 inflammasome activation behind diverse triggers [128]. In macrophages, the K^+ efflux channel triggering NLRP3 inflammasome activation is identified as TWIK2 channel [129]. A *Bacillus cereus* toxin forms pores in the cell membrane, which also causes efflux of K^+ and NLRP3 inflammasome activation [130]. RNA viruses that induce lytic cell death also activate inflammasome through K^+ efflux mechanism [131]. Phospholipase C-mediated Ca^{2+} mobilization is involved in inflammasome activation by hepatitis C virus core protein [132].

NEK7 kinase is believed to act downstream of K^+ efflux to regulate NLRP3 oligomerization and activation [133]. NEK7 is involved in microtubule nucleation, the activity of the centrosome, mitotic spindle formation, and cytokinesis. NLRP3 activation is restricted by NEK7 to the interphase of the cell cycle [134]. Small molecule imiquimod that inhibits mitochondrial respiratory complex also activates NLRP3 through NEK7 [135]. Therefore, NEK7 may play a broader role in NLRP3 inflammasome regulation [136, 137].

Recently, a new likely NLRP3 ligand has been identified. The trans-Golgi tubular network is the place where proteins are sorted into vesicles for exocytosis, secretion, or lysosomal degradation. Phosphatidylinositol-4-phosphate (PtdIns4P) is involved in these processes. Various stressors lead to disassembly of the trans-Golgi network, thus releasing PtdIns4P. NLRP3 then binds to PtdIns4P, which triggers inflammasome formation [138].

Inflammasome activation needs to be tightly regulated to avoid uncontrolled inflammation. NLRP3 inflammasome can be inhibited at various steps [137]. NLRP3 expression is controlled by miRNAs. Post-translational modification such as ubiquitination, phosphorylation, and thiol-nitrosylation can decrease NLRP3 activity, prevent oligomerization, and promote degradation. Sequestration of NLRP3, ASC, or caspase-1, clearance of damaged mitochondria, and induction of autophagy also negatively regulates inflammasomes [137]. The ketone metabolite β-hydroxybutyrate induced by starvation and high-intensity exercise inhibits inflammasome by blocking K^+ efflux [139]. NLRP3-related NFκB signaling inhibition also suppresses inflammation [140].

Unlike NLRP3, ligands for NLRC4 and AIM2 inflammasomes are known. AIM2 inflammasome is activated by foreign DNA and altered or mislocalized “self” DNA [141 - 144]. In NLRC4 inflammasome, NAIP receptors function as sensors. NAIP5 recognizes bacterial flagellin, whereas NAIP2 is specific to the needle protein Cpr1 from the bacterial secretion system type three [115, 145, 146]. NAIPs binding to their ligands leads to their association with NLRC4 inducing its conformational change and activation. Activated NLRC4 catalyzes the activation of inactive NLRC4 molecules by prion-like polymerization leading to the formation of wheel-like structures [116, 147]. A unique aspect of NLRC4 is its ability to interact directly with pro-caspase-1 through their CARD domains, although the presence of the adaptor ASC enhances activation [115]. NLRC4 can be additionally regulated by phosphorylation by several kinases [115].

NLRP1B inflammasome activation is associated with N-terminal proteasomal degradation of NLRP1B molecule [148, 149]. Enzymes produced by bacterial pathogens can mediate such degradation.

Besides caspase-1-based canonical inflammasomes, human caspase-4, -5 and murine caspase-11 form non-canonical inflammasomes. These caspases directly sense the cytosolic presence of lipopolysaccharide (LPS) of Gram-negative bacteria [150, 151]. Caspase-11 can also bind oxidized phospholipids (oxPAPC) [152]. It binds oxPAPC through its catalytic domain and LPS through CARD domains, which leads to distinct inflammasome-dependent activities. Both LPS and oxPAPC induce caspase-11 dependent secretion of IL1β but only LPS induces pyroptosis.

Expression of caspases varies between tissues. In human intestinal epithelial cells caspase-4, not caspase-1, is important for IL18 production [153]. Accordingly, intestinal pathogens produce caspase-4 inhibitors in order to suppress the immune response of epithelial cells. Microbiota-associated metabolites taurine, spermine, and histamine modulate NLRP6 inflammasome activation, IL18 production, and IL18-mediated secretion of antibacterial peptides [154]. Taurine is inflammasome activator, whereas spermine and histamine are inhibitors.

Besides interleukins, inflammasome activation leads to the secretion of other proinflammatory compounds such as high mobility box 1 (HMGB1) [155]. Inflammasome can induce the inflammatory form of regulated cell death called pyroptosis, which is also accompanied by cytokine release [156]. Cytokines that are produced in excess can have detrimental side effects [157]. Such excess may lead to inflammation in the absence of infection as in several autoimmune conditions called inflammasomopathies [158]. They include Familial Mediterranean Fever caused by mutations in pyrin and cryopyrin-associated periodic syndrome caused by mutations in NLRP3 [117]. Inflammasome-mediated IL1β secretion mediates low-grade chronic inflammation associated with aging [159].

9.6. Major Sensors of Autonomous Immune System

9.6.1. Sensing Foreign RNA

Foreign RNA is sensed by multiple PRRs. An important sign of viral infection is the presence of dsRNA, which is produced by most viruses at some point during their replication cycle [160, 161]. dsRNA is sensed by protein kinase R, one of the four mammalian kinases that phosphorylate eIF2α leading to inhibition of protein synthesis during integrated stress response [162]. PKR binding of viral dsRNA or "self" RNA with structural defects such as bulges and internal loops leads to its homodimerization, autophosphorylation, and a conformational change that exposes the binding site for its substrate eIF2α [163]. PKR activation is enhanced by IFNs. Since PKR activity slows down virus replication, viruses developed measures to inhibit it; for example, human cytomegalovirus evades PKR by producing a factor that inhibits eIF2α binding to PKR [164].

Another important class of dsRNA sensors is oligoadenylate synthetases (OASs) that produce 2',5'-oligoadenylate oligomers from ATP. OASs enzymes are activated by cytosolic dsRNA of more than 17 base pairs in length [165]. 3'-end polyuridine ssRNA sequence is necessary for full activation of human OAS1 [166]. The structural characteristics of RNA are also important for OAS activation. Both PKR and OAS bind human non-coding RNA nc886 that can adopt two conformations. The conformer that strongly activates OAS1 has a unique apical stem-loop structure [167]. Mechanisms that switch the inactive nc886 conformer into the active one, though unknown, could be involved in recognition of "self" from "non-self" RNAs. OAS enzymes are also induced by IFNs among other interferon-stimulated genes.

2',5'-oligoadenylate oligomers produced by OASs enzymes bind ribonuclease L (RNase L) leading to its dimerization and activation [168]. RNase L then digests all RNAs, cellular and viral, which arrests protein synthesis and virus replication. RNase L has a preference for targets representing tRNAs and Y RNAs [169]. The importance of OAS/RNase L pathway in immunodefense is illustrated by multiple strategies that viruses use to suppress it, including direct inhibition of enzymes, sequestration of dsRNA, production of 2',5'-oligoadenylate analogs and 2',5'-oligoadenylate phosphodiesterases [170, 171].

Viral and bacterial RNAs have several structural features absent from animal RNAs. Viruses have blunt-ended dsRNA at the 5'end. Other differences from host RNAs include 5'-terminal structures. Eukaryotic mRNA has methylguanosine (m7)G cap linked to the triphosphorylated terminal residue. In addition, terminal nucleotides are methylated at their 2′O positions; a cap with one 2'O methyl group is designated as cap1, a cap with two 2'O methyl groups as cap2, and non-methylated cap as cap0. Many viral RNAs also have a triphosphorylated 5'end but it is typically under-methylated. Moreover, some viral genomes have unmodified 5'-triphosphates (PPP-RNA) or diphosphates (PP-RNA) at the 5' end. These features of viral RNAs are recognized in the cytoplasm of infected cells by RLR family receptors RIG-I, MDA5, and LGP2 [172]. Each RLR has its own preferences, for example, 5' end structure is important for RIG-1 activation, whereas the long length of viral RNA is more important for MDA5 [173 - 176]. Various 5′ modifications affect the binding specificity of these sensors. For example, cap N1-2'O-methylation prevents RIG-I activation [177]. RLRs can also recognize cellular RNAs misprocessed during stress or infection [57].

RIG-1 and MDA5 are multidomain proteins that, in the absence of a target, acquire an autoinhibited form with head to tail contact. After target recognition, they undergo ATP-dependent conformational changes to form filamentous oligomers on their substrates. These oligomers are recognized at the mitochondrial membrane by MAVS (mitochondrial antiviral signaling) protein. It leads to MAVS polymerization into prion-like aggregates and recruitment and phosphorylation of IRF3/7, which leads to IFN induction [178 - 181].

Recent data show that immune response to "self" RNAs with double-strand regions is actively suppressed. Such regions are present in RNA polymerase II (pol II)-transcribed mRNAs from transposable elements that form duplex RNA structures. IFN-induced ADAR (Adenosine deaminase, RNA specific) mutator proteins edit adenosine by deamination to inosine in such dsRNA [182 - 184]. It makes duplex RNA unstable preventing activation of dsRNA sensors. Mutations in *ADAR* genes and in *IFIH1* that encodes MDA5 lead to Aicardi-Goutières syndrome (AGS), an autoinflammatory condition associated with spontaneous IFN production and neurodegeneration. AGS mutants of MDA5 have reduced sensitivity to structural irregularities of duplex RNA structures derived from "self" RNAs [185].

Editing of "self" RNAs has its downside. ADAR1 activity in tumors limits sensing of innate dsRNA ligands and increases survival of the tumor cells [186, 187]. Some viruses take advantage of ADAR1 to make their RNAs edited, and elude innate immunity [184].

Presence of foreign RNA in cells can also be detected in endosomes by TLRs. TLR3 recognizes dsRNA, which leads to IFN pathway activation [188, 189]. TLR3 can also be activated by single strand RNA that has stem structures with bulge/loops [190]. TLR10 binds dsRNA in endosomes competing with TLR3 and inhibiting interferon induction [191]. TLR3 is important for the protection of brain cells against herpes virus infection [192]. Human endosomal TLR7 and

TLR8 bind ssRNA of microbial origin that contains successive uridines [193 - 195].

Due to the importance of viral RNA structure for recognition by host immune factors, viruses produce proteins that shield its sensitive features such as dsRNA blunt ends. Moreover, viruses can borrow caps from hosts, and methylate their terminal nucleotides at their 2'O positions [107, 196].

9.6.2. Sensing Foreign DNA

Presence of foreign DNA in cells is sensed by several mechanisms. The cytoplasm of a healthy cell does not contain DNA. Therefore its presence in the cytoplasm is a danger signal, as it points to viral or bacterial infection or damage to nuclear DNA or mitochondria. Cyclic guanosine monophosphate (GMP)-adenosine monophosphate (AMP) synthase (cGAS) is a cytosolic DNA sensor. In the presence of DNA, it produces the second messenger 2’3’-cyclic GMP-AMP (cGAMP) [197 - 199]. The dynamics of cGAS dimerization associated with its activation depend on DNA concentration and length [200]. Human cGAS evolved to recognize longer DNAs, which increased their specificity compared to other mammals [200]. cGAS can be activated not only by microbial DNA but also by misplaced self DNA such as DNA fragments leaking into the cytoplasm during DNA damage [201]. Mitochondrial DNA released into the cytoplasm due to mitochondrial damage during infection or stress also activates cGAS [202].

cGAMP produced by activated cGAS binds the adaptor protein STING (stimulator of IFN genes) which is located on the membranes of the endoplasmic reticulum (ER) [203]. Activated STING moves from ER to Golgi where it binds TBK1 kinase; this complex phosphorylates IRF3 leading to IFN production, activation of NFκB, inflammation, or cell death [199, 201, 204, 205]. STING is regulated by phosphorylation, palmitoylation, sumoylation, ubiquitination, and other mechanisms [178]. An alternate isoform of STING that lacks transmembrane domain sequesters cGAMP and inhibits immune response [206]. In virus-infected cells, cGAMP can be packaged into the viral particles that deliver cGAMP to target cells where it can initiate an immune response [207].

STING can be activated independently of cGAS by Gram-positive bacteria. They produce the metabolites c-di-GMP and c-di-AMP, which directly activate STING [208]. Only live bacteria produce these metabolites, therefore the immune response is adjusted in dependence on bacteria viability. Live bacteria induce inflammasome activation and IFN production, whereas non-replicating bacteria prime inflammasomes but do not activate them and induce a lower level of IFNs [209]. Structurally similar synthetic compounds including cyclic dinucleotides mimicking cGAMP and derivatives of amidobenzimidazole are studied as stimulators of immunity for treatment of cancer, infectious diseases, and as vaccine adjuvants [210].

STING interacts with other dsDNA sensors such as IFI16 [211, 212]. STING can also suppress viral infection by inhibiting translation during infection by RNA viruses [213]. This effect requires RLR signaling to STING. These data suggest that STING acts as a signaling hub that tailors the cellular response to the type of viral infection.

Interferon-inducible protein 16 (IFI16) is the cytoplasmic and nuclear dsDNA sensor. It forms filamentous oligomers when activated by dsDNA [214]. In cytoplasm, IFI16 recognizes incomplete HIV reverse transcripts generated during abortive HIV infection and is thought to play a major role in the pyroptotic death of resting CD4+ T cells [215]. In nucleus, it recognizes unchromatinized dsDNA that forms atypical structures such as supercoil and quadruplex [214]. IFI16 forms filaments on DNA of cells infected with herpes simplex virus 1 (HSV-1). These filaments recruit other antiviral restriction factors, which prevent recruitment of RNA polymerase II to viral genes and their transcription [216].

Other ALR family members are also important for antiviral defense. IFIX acts as a nuclear antiviral factor that suppresses immediate-early and early viral gene expression [217]. AIM2 recognizes cytosolic dsDNA and forms caspase-1-activating inflammasomes [141 - 143]. IFI16 has an alternative isoform that lacks the pyrin domain [218]. This isoform interacts with AIM2 and prevents inflammasome formation. It also sequesters the cytoplasmic dsDNA and makes it unavailable for AIM2 sensing.

Immune response to cytosolic DNA can also be activated by its conversion into RNA by Pol III, which then leads to RIG-1 activation [219, 220]. Multiple DNA-binding factors such as the DNA repair proteins Ku70, Ku80, DNA-PK, Rad50, MRE11, helicase DDX41, and others also can induce an immune response by binding to cytosolic DNA [221, 222].

In endosomes, foreign DNA is recognized by TLR9. TLR9 binds unmethylated bacterial DNA, which activates MyD88/IRF7/NFκB/IFN pathway [223]. Mitochondrial DNA resembles bacterial genomes in circular structure and non-methylated CpG islands and can stimulate PRRs such as TLR9 or cGAS [76]. White blood cells eject the mitochondrial DNA as web filament structures in response to oligodeoxynucleotides of class C [224]. This initiates a rapid INF response in the neighboring cells.

DNA and RNA sensing pathways coordinate their activity through direct interaction, cytokine signaling, and other mechanisms [95, 225, 226]. Overactivation of RNA and DNA sensors can lead to inflammatory conditions and cancer. For example, cytoplasmic self-DNA recognized by RNA polymerase III/ RIG-1 pathway results in upregulation of APOBEC3A and can lead to chromosomal DNA mutations [227]. Viruses have evolved diverse strategies to prevent their recognition by targeting nucleic acid sensors, adaptor proteins, and transcription factors involved in IFN activation [228, 229].

9.6.3. Nucleotide-binding Domain (NBD) and Leucine-rich Repeat (LRR)-containing Receptors (NLRs)

NLR receptors are described as a tightly folded "hair trigger" device into which unlimited number of various detection platforms can be integrated [58]. NLRs are activated by a wide variety of microbial antigen, DAMPs, and changes that are induced in host cells by pathogens [230]. Humans have 23 NLR genes; sea urchin has 203, whereas in wheat this family expanded to 402 and in apple —to 737 members [58]. 23 human NLRs are divided into four families: A, B, C, and P. Several NLR members form inflammasomes.

In plants, the most dominant plant resistance (R) genes responsible for the effector -triggered immunity belong to NLR family [58, 80, 231]. NLR receptors recognize changes in "self" proteins that are targeted by bacterial effectors; these proteins are called "sensors". In this way, a limited number of "sensors" can recognize a large number of bacterial effectors. In some cases, NLRs can be associated with decoy proteins that mimic targets of pathogen effectors [232]. In the absence of infection, the expression of R genes is suppressed by microRNAs [16].

In animals, initial studies of NLR receptors NOD1 and NOD2 suggested that these receptors recognize peptidoglycans from gram-negative and gram-positive bacteria [233, 234]. Later studies suggest that NOD1 senses activation of small Rho GTPases by bacterial effector proteins [124]. Most other NLR receptors may use a similar principle and recognize changes in various host molecules induced by a pathogen instead of the pathogen per se. NLRP3 discussed in the Inflammasome section may represent one of such receptors. NLR inflammasomes connect stress responses with immunity since not only do they respond to infection but also convert various stress signals into stimulation of immune response.

Several NLRs act as negative regulators of immune response initiated by other PRR families. NLRC3 sequesters stimulator of interferon genes (STING) preventing IFN response [235]. Viral DNA binding to NLRC3 releases STING and permits IFN response. NLRP11 inhibits TLR signaling by targeting the adaptor protein TRAF6 for degradation through its ubiquitination [236]. The general effect of NLRP11 expression is inhibition of NF-κB and IFNs [237].

NLRs are also involved in the regulation of metabolism, cell differentiation, and embryonic development [84]. For example, in germ cells, NLRP14 targets TBK1 kinase for ubiquitination and degradation to negatively regulate the cytosolic nucleic acid sensing [238]. This prevents immune response to the paternal DNA during fertilization.

9.6.4. Uncommon PRRs

During infection, upregulation of multiple classes of sensors has been observed [239]. Many pathogens have the means to avoid recognition by PRRs, therefore, some pathways may be more effective than others in recognition of a specific pathogen. Some PRRs are adapted to function in specific cell types. For example, CLR lectin RegIIIa is expressed in the epithelial cells of the small intestine. It combines the functions of a sensor and effector. It binds the membrane phospholipids of Gram-positive bacteria and forms a hexameric membrane-permeabilizing pore killing the bacteria [240]. Another example is a receptor selectively expressed in the colonic epithelia — Lypd8. It is secreted into the gut lumen where it binds flagellated bacteria, suppressing their mobility [241].

Disturbance of cell membranes during viral entry can be sensed independently of known viral sensors. Mucosal surfaces sense viruses through a mechanism that involves viral O-linked glycans and results in induction of CXCR3 chemokines [242]. This attracts neutrophils and stimulates production of mucins to restore the damaged mucus layer. Cells can sense fusion of their membrane with a viral envelope as a danger signal through the non-canonical STING pathway [243, 244].

Complement also plays a role in antiviral defense. C3b complement component covalently attaches to any moiety including viral capsids. The virus carries attached C3b into the cell where it triggers mitochondrial MAVS-dependent signaling. In addition, viruses with attached C3b undergo rapid proteasomal degradation, which prevents their replication [245]. Enteroviruses avoid this system using a complement-degrading protease.

In human and other animals, airways bitter and sweet taste receptors can work as sensors for bacteria that use quorum sensing for communication. Bitter taste receptors from T2R family on the epithelial cells layering airways recognize bacterial quorum sensing molecules and trigger NO (a bactericide) production

[246 - 248]. This has health implications since, for example, people with a non-functional T2R38 have a higher risk of developing chronic rhinosinusitis [246].

9.7. Activation of Immune Mechanisms by Genetic Defects

Malfunction of basic cellular processes such as DNA replication, ribosome biogenesis, translation, respiration, or protein turnover can lead to activation of the autonomous immune mechanisms in the absence of tissue damage, toxins, or pathogens. In *C. elegans*, disruption of essential cellular processes by RNAi leads to expression of immune effectors [249]. Disruption of ribosome biogenesis leads to accumulation of unprocessed pre-rRNAs and DNA damage, which activates sensors such as MDA5 and TLR3, and results in IFN and NF-κB induction in zebrafish and human cells [250]. Defects in genes involved in the metabolism of endogenous nucleic acids lead to a buildup of abnormal products that activate immune mechanisms [251]. For example, non-spliced mRNA serves as a ligand for TLR3 [252]. Deficiency of DNase II in macrophages leads to accumulation of non-digested DNA that is sensed by PRRs and leads to IFN activation [253]. Defects in DNA methylation change the expression of endogenous viruses and triggers IFN response [254]. Malfunction of mitochondria can also induce an immune response [76]. These data suggest that many human disorders caused by mutations in genes unrelated to the immune system can be accompanied by activation of the autonomous immune mechanisms, which can lead to inflammation and worsen the outcome of the primary disease.

9.8. Immune Effectors

Immune effectors interfere with various stages of pathogen's life cycle. Hundreds of immune effectors are induced by interferons among interferon-stimulated genes (ISGs), whereas many others are interferon-independent. Subsets of ISGs can be induced both by canonical IFN-JAK-STAT pathway and by other mechanisms such as the non-canonical IFN signaling, TNF/NFκB, nucleotide synthesis inhibitors, and retinoic acid [99, 100, 102, 255]. Examples of immune effectors with known mechanism of action are discussed below.

Among the most important antiviral effectors are mutator proteins. Cytidine deaminases of APOBEC3 family include seven proteins (3A-3H). They deaminate cytidine into uridine leading to G to A hypermutations [256]. These enzymes preferentially bind ssDNA [257, 258]. Several APOBEC3 family members can incorporate themselves into virions and, during reverse transcription, edit deoxycytidine residues into deoxyuridine on the nascent DNA strand. Resulting hypermutation makes the provirus defective and stops viral propagation. APOBEC3 also restricts the mobility of retroelements [259 - 261]. Human HIV virus produces the accessory factor Vif, which protects viruses from APOBEC3-dependent restriction by blocking its incorporation into virions and by increasing its degradation through the ubiquitin pathway [262 - 264]. The interface of APOBEC3G and Vif contact is a subject of long adaptive evolution in primates [265]. HIV-1 protease also protects the virus by cleaving the highly restrictive APOBEC3G isoform.

Since APOBEC3 proteins are mutators, they can mutate host genes too, specifically, during DNA replication when ssDNA is exposed at the replication fork [266]. Increased APOBEC3 expression is found in human cancer where it may be the major cause of mutagenesis of cancer genomes, including deactivating mutations of p53 [267, 268]. Genome instability, increased mitogenic signaling, and replication stress promote APOBEC3 mutagenesis [269].

Viperin is one of the most highly induced ISGs. It belongs to the family of radical SAM (S-adenosylmethionine) enzymes. Sam enzyme cleaves a C-S bond in SAM to generate a 3-amino-3-carboxypropyl radical that forms a bond with the enzyme's [4Fe-4S] cluster [270]. This intermediate converts to an organic radical in the presence of a substrate. Viperin restricts many DNA and RNA viruses and several mechanisms of action have been suggested [271, 272]. Recent data show that viperin converts cytidine triphosphate (CTP) into 3'-deoxy-3',4'-didehy-ro-CTP (ddhCTP) [273]. ddhCTP acts as a chain terminator for RNA-dependent RNA polymerases that replicate viral RNA genomes into mRNAs. ddhCTP directly inhibits replication of Zika virus and may be a unifying factor behind antiviral activities of viperin.

Several antiviral factors decrease the levels of cellular metabolites necessary for pathogen replication [255]. SAMHD1 is a dGTP-stimulated triphosphohydrolase that breaks dNTPs into corresponding deoxynucleosides and inorganic triphosphate. SAMHD1 inhibits reverse transcription of retroviruses by depleting the cellular pool of dNTPs [274]. SAMHD1 is regulated by phosphorylation in dependence on cell cycle and in some settings can inhibit immune response [275]. Another factor, indoleamine 2,3-dioxygenase 1 (IDO1) can inhibit intracellular pathogens by tryptophan depletion [276].

Myxovirus resistance (Mx) proteins are also induced by IFNs. They belong to the dynamin superfamily of large GTPases that catalyze fusion and fission of biological membranes [277, 278]. Human MxA has broad antiviral activity against RNA and some DNA viruses but not against herpes simplex virus 1 or HIV-1; those are restricted by MxB [279, 280]. MxB restricts herpes viruses by blocking the uncoating of incoming viral DNA from the viral capsid. It requires the formation of MxB dimers or oligomers [281].

Interferon-induced tetratricopeptide repeats (IFITs) proteins are also highly expressed ISGs [103]. They are both sensors of foreign RNA and effectors. IFITs contain multiple copies of tetratricopeptide repeats and are able to interact with various proteins and RNAs, which constitute the basis of their immune activity. IFITs inhibit translation of mRNAs with "non-self" cap structures [282]. They bind 5'-ends of viral RNA as heterodimers and prevent recruitment of cap-binding eIF4F and ribosome assembly [283]. IFITs can also recognize bacterial RNAs that have no caps.

Interferon induced transmembrane proteins (IFITMs) block infection of many viruses. S-palmitoylation and association with membranes are required for their antiviral activity [284, 285]. IFITM3 is located in endocytic vesicles that fuse with the envelope of an incoming virus and facilitates the delivery of the virus to lysosomes [286]. IFITM3 may also interfere with the budding of new enveloped viruses from the host cell. IFITM3 incorporates into HIV-1 virions and inhibits the fusion of virus progeny [287]. It is constitutively highly expressed in human pulmonary endothelial cells and is important for intrinsic immunity against influenza viruses [288].

NCOA7 effector inhibits viruses that enter the cell by endocytosis [289]. It inhibits fusion of the viral and endosomal membranes. NCOA7 also increases vacuolar H+-ATPase (V-ATPase) activity, which promotes cytoplasmic vesicles acidification, increased activity of lysosomal proteases, and degradation of endocytosed viruses.

TRIM5 is a PRR that recognizes retroviral capsids, and it is also an effector. It assembles into a hexagonal lattice on capsids, causing their destabilization and disassembly [290, 291]. TRIM5 also has E3 ubiquitin ligase activity associated with its RING domain. Hexagonal assembly induces TRIM5 polyubiquitination, which leads to the induction of an antiviral state and promotes proteasomal destruction of the virus and TRIM5 itself [292].

CCCH-type zinc finger protein ZAP binds RNAs from some viruses such as HIV-1, Ebola virus, and others and recruits cellular degradation machinery to degrade them [293, 294]. ZAP also inhibits retrotransposition [295]. ZAP recognize RNAs enriched in CG nucleotides in comparison to host RNAs [296].

Tetherin/BST2 is also an IFN-induced antiviral factor that prevents the release of budding enveloped virions from cells. It is a membrane protein that holds virions to the cell membrane and each other [255, 297]. High mannose glycosylation is required for tetherin activity [298].

Similar to the superinfection exclusion state induced in bacteria infected by some viruses, viral infection of animal cells can also prevent infection by other viruses [299]. Viruses can also restrict infection by bacterial pathogens. Cytomegalovirus infected mice have an improved response to bacterial infection [300]. Similarly, latent herpes virus infection increases immunity against bacterial pathogens such as *Listeria monocytogenes* and *Yersinia pestis* [301]. Latency-induced protection depends on upregulation of the basal level of immune system activity. Therefore, a lifelong persistence of herpes viruses common in humans may provide some benefits.

Many pathogens enter cells in vacuoles either as a result of phagocytosis or by stimulating their own uptake. Vacuoles shelter pathogens from cytosolic PRRs. Many factors secreted by pathogens aim to preserve the integrity of the pathogen-containing vacuoles (PCVs) and to prevent their fusion with lysosomes. Some immune effectors target intracellular pathogens that reside in PCVs. Among them are Mx proteins and various interferon-inducible GTPases including the immunity-related GTPase (IRG) and guanylate-binding protein (GBP) families [302 - 305].

Carbohydrate-binding protein galectin-3 detects proteins associated with PCVs that are products of bacterial secretion systems and directs GTPases to the PCV [306]. The main effect of these effectors is damage to PCV membranes resulting in pathogen release into the cytosol where they activate an immune response. Interferon-inducible GTPases can also directly target membranes of bacteria and protozoa, as well as replication complexes of RNA viruses [307]. In addition, GBPs have been reported to contribute to other antimicrobial mechanisms such as the formation of inflammasomes [304].

Activation of various mechanisms that deplete pathogens of Mn, Fe, and Zn cations is also an efficient way to restrict pathogen growth. For example, host cell uses Nramp1 transporter to deplete phagosomes of divalent cations needed by *Salmonella typhimurium* to survive inside a modified phagosome in macrophages. To counteract this mechanism, the bacterium produces PspA, which increases divalent metal transport [308].

Some effectors can function extracellularly. Interferon-induced plasminogen activator inhibitor 1 (PAI-1) inhibits the cleavage of influenza virus glycoprotein preventing virion maturation [309]. Polymorphism in the human gene encoding this inhibitor affects sensitivity to influenza.

These examples show how immune effectors restrict infection at multiple stages and by diverse mechanisms. For each pathogen, a specific combination of effectors is especially effective [100, 310]. This may be due to differences in physiology between pathogens and to the type of factors they use to inactivate immune effectors. Many factors may modulate the efficiency of pathogen restriction. Different immune mechanisms cross talk and can potentiate or inhibit each other.

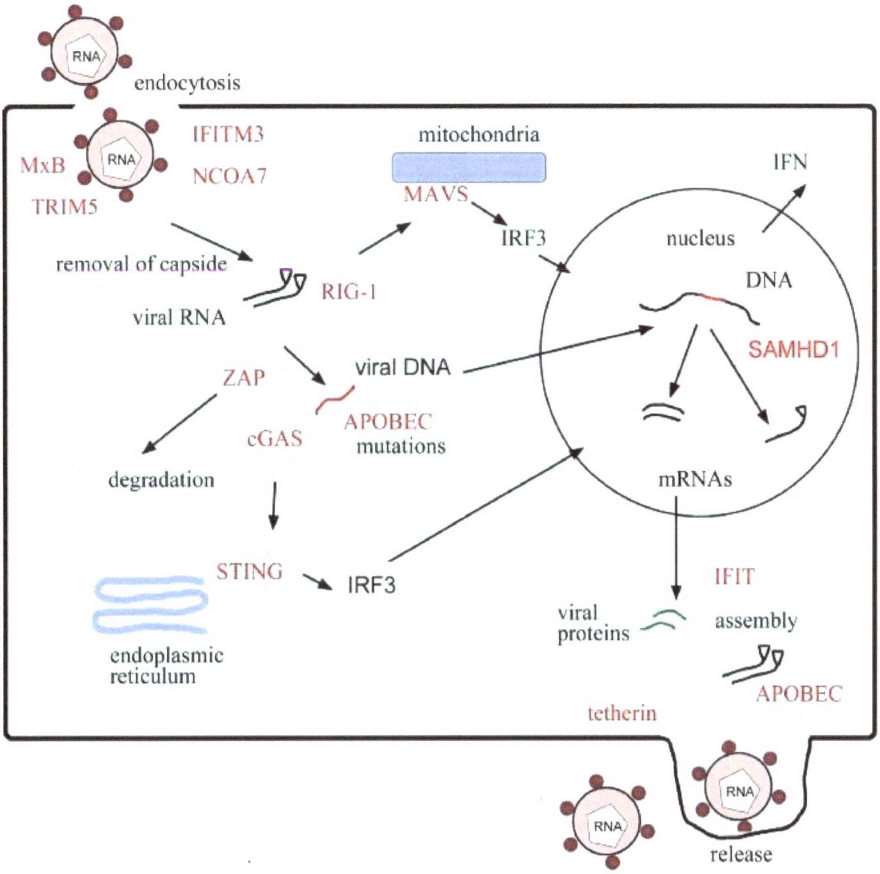

Fig. (9). Diverse immune mechanisms restrict a virus such as HIV at various stages of its life cycle. Viral fusion with cellular membranes can be inhibited by IFITM3 and NCOA7. MxB inhibits viral uncoating. TRIM5 assembles on viral capsids causing their disassembly and degradation. RIG-1 recognizes viral genomic RNA and signals through MAVS to induce IFNs. ZAP also recognizes viral RNA and targets it for degradation. During reverse transcription of viral RNA into DNA, APOBEC introduces mutations that render the viral progeny defective. Viral DNA activates cGAS/STING pathway that also leads to IFN induction. At the next step, viral DNA integrates into the host DNA. Its transcription results in new viral genomic RNA and mRNAs. This step can be inhibited by SAMHD1. mRNAs are translated into viral proteins. IFIT proteins inhibit the translation of viral mRNAs. Next, viral proteins and RNAs are assembled to form new virions. APOBEC incorporates into the new virions and continues its mutagenic activity in the newly-infected cells. Tetherin inhibits the release of new virions.

9.9. Toxic Compounds as Immune Effectors

Similar to prokaryotes eukaryotes use toxic compounds for their protection. Toxic compounds designed to kill pathogens include ROS, RNS, antibiotics, antimicrobial peptides, and plant toxins. ROS is a universal weapon used both by plants and animals [311, 312]. In mammalian cells, major ROS producers are the NOX family of NADPH oxidases [313]. They reduce molecular oxygen into superoxide, which then generates various ROS. Seven human NOX enzymes differ in their expression patterns and activation mechanisms. NOX2 is a key enzyme responsible for the respiratory burst in macrophages and other phagocytes. NOX1 is mostly expressed in colon epithelial cells [314]. Another ROS producer in intestinal cells is the spermine oxidase that converts spermine into spermidine with H_2O_2 release [315]. NO production in intestinal cells by the inducible NO synthase also contributes to immunodefense. The mucosal surfaces of airways are protected by dual oxidases generating H_2O_2 and by lactoperoxidase [316]. Increased airways ROS interferes with the correct splicing of viral transcripts. Host-generated H_2O_2 can subvert pathogen signaling in some bacteria adapted to low oxygen environments [317]. H_2O_2 induces oxidation of phosphotyrosine by bacterial peroxidases, which inactivates the enzymes involved in polysaccharide biosynthesis and metabolism and weakens bacteria.

Many cells of multicellular organisms produce antimicrobial peptides (AMPs). A single cell in the human gut can produce more than 20 AMPs with different activities in response to PRRs activation [318]. This mechanism is especially important in the protection of fetuses, whose immunity is immature. AMPs are produced by amnion and fetal cells [319, 320].

9.10. Regulated Cell Death as Immunodefense

Regulated cell death (RCD) is the last resort for an infected cell. Depending on the type of infection and cellular conditions, death can proceed through apoptosis or through the regulated necrosis pathway, which is subdivided into several subtypes including necroptosis, pyroptosis, ferroptosis, *etc*. [321]. During apoptosis, the cellular content remains enclosed by membranes that prevent its leakage outside the cell. Apoptotic bodies are efficiently removed by neighboring cells and macrophages in a process of efferocytosis without induction of immune response [322]. The mechanism of apoptosis is discussed in chapter 7. Many viral infections lead to apoptosis through the intrinsic pathway or through the extrinsic cell death receptor-mediated pathway [323, 324]. Bacterial infections also can lead to apoptosis. Intrinsic apoptosis is the main mechanism that restricts *Clostridium difficile* infection of intestinal epithelial cells [325]. Pathogenic *Escherichia coli* produces factors suppressing apoptosis [326].

Contrary to apoptotic cell death, which does not activate immune mechanisms, necrosis leads to permeabilization of cellular membranes and release of cellular content. It includes many DAMPs that attract immune cells and activate inflammation [327 - 330]. The main mechanism of necroptosis involves activation of the receptor-interacting protein kinase RIPK3 by three different adaptors, RIPK1, TRIF, and DAI [328, 331]. RIPK1 is the main pro-necrotic kinase that responds to several death receptors, IFNγ, TCR, and genotoxic stress by forming an amyloid-like activating complex with RIPK3, the so-called canonical necrosome. TRIF and DAI respond correspondently to TLR3/4 and DNA viruses and form non-canonical necrosomes. TNF also can activate RIPK3, if caspase 8 that promotes the apoptotic pathway is inhibited [332]. Upon activation, RIPK3 recruits and phosphorylates the necroptosis executioner MLKL. MLKL then undergoes oligomerization to form a second amyloid-like structure and translocates to the intracellular and plasma membranes where it induces membrane rupture [333]. MLKL can also be activated independently of RIPK3 through IFNγ/STAT1 pathway [334]. The importance of necroptosis in immunodefense is illustrated by *E. coli*'s production of a protease that cleaves proteins participating in necroptosis [335].

Pyroptosis is another proinflammatory RCD. It is induced by cleavage of gasdermin family of protein by activated inflammatory caspases [336 - 338]. Uncleaved gasdermin D (GSDMD) is autoinhibited. Cleavage by caspases releases its N-terminal domain (GSDMD-NT), which has a pore-forming ability. It binds to the plasma membrane and oligomerizes to form pores [339, 340]. This leads to osmotic lysis and release of pro-inflammatory cytokines such as IL1β and IL18, alarmins, and DAMPs. GSDMD-NT binds to phosphatidylinositol phosphates and phosphatidylserine, which are present only on the cell membrane inner leaflet, therefore GSDMD-NT does not damage neighboring cells. It can, however, bind cardiolipin on bacterial membranes and kills bacteria *in vitro*. During pyroptosis, bacteria become trapped within the cellular debris, which justifies the name pyroptosis-induced intracellular traps (PITs) [156]. PITs then are cleared by efferocytosis. Gasdermin-NT domains can also permeabilize mitochondrial membrane releasing cytochrome c and activating the apoptosome [341].

NETosis is an RCD first discovered in neutrophils that is associated with the formation of neutrophil extracellular traps (NETs) —strands of decondensed DNA in complex with proteins forming large web-like structures that trap microorganisms [342, 343]. NETosis starts with calcium release, which stimulates peptidyl arginine deiminase 4 (PAD4) migration from the cytoplasm into the nucleus where it catalyzes the conversion of arginine into citrulline. Citrullination converts positively charged arginine into a neutral residue. It decreases the positive charge of histones, lessening their affinity to negatively charged DNA and compromising the stability of nucleosomes. Citrullination and proteolysis of histones lead to chromatin decondensation. Chromatin swelling causes rupture of the nuclear envelope and plasma membrane and drives NET release [344]. NETs are decorated with antimicrobial proteins, including neutrophil elastase, myeloperoxidase, and cathepsin G that help to kill captured microorganisms. NETs also stimulate inflammation and IFN production.

NETosis is induced by microbes but also by other stimuli such as cytokines and antigen-antibody complexes through different pathways [343]. NETosis can be induced by changes in blood plasma redox balance, which is supported by albumin's thiols. Oxidation of albumin disrupts the plasma redox balance and causes NETosis [345]. Silica and nanoparticles also can induce NETosis [346]. In addition to the conventional NETosis, a pathway has been described that restricts NET formation to mitochondrial DNA, which allows the neutrophils to preserve membrane integrity after NET release and survive [347]. Lymphocytes are also capable of ejecting mitochondrial DNA webs in response to stimulation of pattern recognition receptors [224]. *Staphylococcus aureus* escape phagocytic clearance by producing nucleases and adenosine synthase A [348]. Adenosine synthase A convert adenosine in NETs into deoxyadenosine (dAdo), which migrates into phagocytes and kills them. Plant root cells are also able to release extracellular DNA traps; some plant pathogens escape them by producing DNases [349]. NETs contribute to immunopathology by inducing the formation of autoantibodies and serving as scaffolds for thrombosis [350]. Choice of a specific death pathway by an infected cell is determined by its cellular conditions and by pathogen-produced factors that interfere with particular RCDs.

CONCLUDING REMARKS

Most cells of multicellular organisms have autonomous innate immune mechanisms. Some of them are based on RNA interference; others - on PRRs that recognize pathogen-associated molecular patterns. Immunity based on PRRs detects pathogens directly. An alternative indirect way of sensing infection is by detecting modification of cellular molecules or functions caused by the pathogen, which is the principle of the effector-triggered immunity. The number of key proteins and pathways targeted by pathogens is limited; therefore their surveillance is an economic (degenerate) way of detecting a wide variety of pathogens using a small number of sensors. Anomalous localization of biomolecules, for example, DNA in the cytoplasm, also activates immune response as does change in cell homeostasis. Various types of immune surveillance complement each other providing a multi-layered coordinated defense system.

CONSENT FOR PUBLICATION

Not applicable.

CONFLICT OF INTEREST

The author confirms that this chapter contents have no conflict of interest.

ACKNOWLEDGEMENTS

Declare none.

REFERENCES

[1] Aguado LC, Schmid S, May J, *et al.* RNase III nucleases from diverse kingdoms serve as antiviral effectors. Nature 2017; 547(7661): 114-7. [http://dx.doi.org/10.1038/nature22990] [PMID: 28658212]

[2] Koonin EV. Evolution of RNA- and DNA-guided antivirus defense systems in prokaryotes and eukaryotes: common ancestry *vs* convergence. Biol Direct 2017; 12(1): 5. [http://dx.doi.org/10.1186/s13062-017-0177-2] [PMID: 28187792]

[3] Youngman EM, Claycomb JM. From early lessons to new frontiers: the worm as a treasure trove of small RNA biology. Front Genet 2014; 5: 416. [http://dx.doi.org/10.3389/fgene.2014.00416] [PMID: 25505902]

[4] Treiber T, Treiber N, Meister G. Regulation of microRNA biogenesis and its crosstalk with other cellular pathways. Nat Rev Mol Cell Biol 2019; 20(1): 5-20. [http://dx.doi.org/10.1038/s41580-018-0059-1] [PMID: 30228348]

[5] Golden RJ, Chen B, Li T, *et al.* An Argonaute phosphorylation cycle promotes microRNA-mediated silencing. Nature 2017; 542(7640): 197-202. [http://dx.doi.org/10.1038/nature21025] [PMID: 28114302]

[6] Gebert LFR, MacRae IJ. Regulation of microRNA function in animals. Nat Rev Mol Cell Biol 2019; 20(1): 21-37. [http://dx.doi.org/10.1038/s41580-018-0045-7] [PMID: 30108335]

[7] Kucherenko MM, Shcherbata HR. Stress-dependent miR-980 regulation of Rbfox1/A2bp1 promotes ribonucleoprotein granule formation and cell survival. Nat Commun 2018; 9(1): 312. [http://dx.doi.org/10.1038/s41467-017-02757-w] [PMID: 29358748]

[8] Olejniczak M, Kotowska-Zimmer A, Krzyzosiak W. Stress-induced changes in miRNA biogenesis and functioning. Cell Mol Life Sci 2018; 75(2): 177-91. [http://dx.doi.org/10.1007/s00018-017-2591-0] [PMID: 28717872]

[9] Zhu S, Pan W, Song X, *et al.* The microRNA miR-23b suppresses IL-17-associated autoimmune inflammation by targeting TAB2, TAB3 and IKK-α. Nat Med 2012; 18(7): 1077-86. [http://dx.doi.org/10.1038/nm.2815] [PMID: 22660635]

[10] Cabrita MA, Bose R, Vanzyl EJ, *et al.* The p53 protein induces stable miRNAs that have the potential to modify subsequent p53 responses. Gene 2017; 608: 86-94. [http://dx.doi.org/10.1016/j.gene.2017.01.018] [PMID: 28119089]

[11] Liu J, Zhang C, Zhao Y, Feng Z. MicroRNA Control of p53. J Cell Biochem 2017; 118(1): 7-14. [http://dx.doi.org/10.1002/jcb.25609] [PMID: 27216701]

[12] He L, He X, Lim LP, *et al.* A microRNA component of the p53 tumour suppressor network. Nature 2007; 447(7148): 1130-4.

[http://dx.doi.org/10.1038/nature05939] [PMID: 17554337]

[13] Chou CH, Chang NW, Shrestha S, *et al.* miRTarBase 2016: updates to the experimentally validated miRNA-target interactions database. Nucleic Acids Res 2016; 44(D1): D239-47. [http://dx.doi.org/10.1093/nar/gkv1258] [PMID: 26590260]

[14] Suzuki HI, Yamagata K, Sugimoto K, Iwamoto T, Kato S, Miyazono K. Modulation of microRNA processing by p53. Nature 2009; 460(7254): 529-33. [http://dx.doi.org/10.1038/nature08199] [PMID: 19626115]

[15] Lu WT, Hawley BR, Skalka GL, *et al.* Drosha drives the formation of DNA:RNA hybrids around DNA break sites to facilitate DNA repair. Nat Commun 2018; 9(1): 532. [http://dx.doi.org/10.1038/s41467-018-02893-x] [PMID: 29416038]

[16] Cai Q, Liang C, Wang S, *et al.* The disease resistance protein SNC1 represses the biogenesis of microRNAs and phased siRNAs. Nat Commun 2018; 9(1): 5080. [http://dx.doi.org/10.1038/s41467-018-07516-z] [PMID: 30498229]

[17] Ressel S, Rosca A, Gordon K, Buck AH. Extracellular RNA in viral-host interactions: Thinking outside the cell. Wiley Interdiscip Rev RNA 2019; 10(4)e1535 [http://dx.doi.org/10.1002/wrna.1535] [PMID: 30963709]

[18] Hoye ML, Regan MR, Jensen LA, *et al.* Motor neuron-derived microRNAs cause astrocyte dysfunction in amyotrophic lateral sclerosis. Brain 2018; 141(9): 2561-75. [http://dx.doi.org/10.1093/brain/awy182] [PMID: 30007309]

[19] Fransquet PD, Ryan J. Micro RNA as a potential blood-based epigenetic biomarker for Alzheimer's disease. Clin Biochem 2018; 58: 5-14. [http://dx.doi.org/10.1016/j.clinbiochem.2018.05.020] [PMID: 29885309]

[20] Hamilton AJ, Baulcombe DC. A species of small antisense RNA in posttranscriptional gene silencing in plants. Science 1999; 286(5441): 950-2. [http://dx.doi.org/10.1126/science.286.5441.950] [PMID: 10542148]

[21] Waterhouse PM, Wang MB, Lough T. Gene silencing as an adaptive defence against viruses. Nature 2001; 411(6839): 834-42. [http://dx.doi.org/10.1038/35081168] [PMID: 11459066]

[22] Wang XB, Wu Q, Ito T, *et al.* RNAi-mediated viral immunity requires amplification of virus-derived siRNAs in Arabidopsis thaliana. Proc Natl Acad Sci USA 2010; 107(1): 484-9. [http://dx.doi.org/10.1073/pnas.0904086107] [PMID: 19966292]

[23] Koch A, Biedenkopf D, Furch A, *et al.* An RNAi-Based Control of Fusarium graminearum Infections Through Spraying of Long dsRNAs Involves a Plant Passage and Is Controlled by the Fungal Silencing Machinery. PLoS Pathog 2016; 12(10)e1005901 [http://dx.doi.org/10.1371/journal.ppat.1005901] [PMID: 27737019]

[24] Maillard PV, van der Veen AG, Poirier EZ, Reis ESC. Slicing and dicing viruses: antiviral RNA interference in mammals. Embo j 2019; 38(8) [http://dx.doi.org/10.15252/embj.2018100941]

[25] Wu X, Kwong AC, Rice CM. Antiviral resistance of stem cells. Curr Opin Immunol 2019; 56: 50-9. [http://dx.doi.org/10.1016/j.coi.2018.10.004] [PMID: 30352329]

[26] Triboulet R, Mari B, Lin YL, *et al.* Suppression of microRNA-silencing pathway by HIV-1 during virus replication. Science 2007; 315(5818): 1579-82. [http://dx.doi.org/10.1126/science.1136319] [PMID: 17322031]

[27] Li Q, Lowey B, Sodroski C, *et al.* Cellular microRNA networks regulate host dependency of hepatitis C virus infection. Nat Commun 2017; 8(1): 1789. [http://dx.doi.org/10.1038/s41467-017-01954-x] [PMID: 29176620]

[28] Fareh M, van Lopik J, Katechis I, *et al.* Viral suppressors of RNAi employ a rapid screening mode to

discriminate viral RNA from cellular small RNA. Nucleic Acids Res 2018; 46(6): 3187-97. [http://dx.doi.org/10.1093/nar/gkx1316] [PMID: 29325071]

[29] Ross RJ, Weiner MM, Lin H. PIWI proteins and PIWI-interacting RNAs in the soma. Nature 2014; 505(7483): 353-9. [http://dx.doi.org/10.1038/nature12987] [PMID: 24429634]

[30] Ozata DM, Gainetdinov I, Zoch A, O'Carroll D, Zamore PD. PIWI-interacting RNAs: small RNAs with big functions. Nat Rev Genet 2019; 20(2): 89-108. [http://dx.doi.org/10.1038/s41576-018-0073-3] [PMID: 30446728]

[31] Andersen PR, Tirian L, Vunjak M, Brennecke J. A heterochromatin-dependent transcription machinery drives piRNA expression. Nature 2017; 549(7670): 54-9. [http://dx.doi.org/10.1038/nature23482] [PMID: 28847004]

[32] Chang TH, Mattei E, Gainetdinov I, Colpan C, Weng Z, Zamore PD. Maelstrom Represses Canonical Polymerase II Transcription within Bi-directional piRNA Clusters in Drosophila melanogaster. Mol Cell 2019; 73(2): 291-303.e6. [http://dx.doi.org/10.1016/j.molcel.2018.10.038] [PMID: 30527661]

[33] Gainetdinov I, Colpan C, Arif A, Cecchini K, Zamore PD. A Single Mechanism of Biogenesis, Initiated and Directed by PIWI Proteins, Explains piRNA Production in Most Animals. Mol Cell 2018; 71(5): 775-790.e5. [http://dx.doi.org/10.1016/j.molcel.2018.08.007] [PMID: 30193099]

[34] Nishida KM, Sakakibara K, Iwasaki YW, *et al.* Hierarchical roles of mitochondrial Papi and Zucchini in Bombyx germline piRNA biogenesis. Nature 2018; 555(7695): 260-4. [http://dx.doi.org/10.1038/nature25788] [PMID: 29489748]

[35] Jih G, Iglesias N, Currie MA, *et al.* Unique roles for histone H3K9me states in RNAi and heritable silencing of transcription. Nature 2017; 547(7664): 463-7. [http://dx.doi.org/10.1038/nature23267] [PMID: 28682306]

[36] Zhang D, Tu S, Stubna M, *et al.* The piRNA targeting rules and the resistance to piRNA silencing in endogenous genes. Science 2018; 359(6375): 587-92. [http://dx.doi.org/10.1126/science.aao2840] [PMID: 29420292]

[37] Vagin VV, Sigova A, Li C, Seitz H, Gvozdev V, Zamore PD. A distinct small RNA pathway silences selfish genetic elements in the germline. Science 2006; 313(5785): 320-4. [http://dx.doi.org/10.1126/science.1129333] [PMID: 16809489]

[38] Czech B, Hannon GJ. One Loop to Rule Them All: The Ping-Pong Cycle and piRNA-Guided Silencing. Trends Biochem Sci 2016; 41(4): 324-37. [http://dx.doi.org/10.1016/j.tibs.2015.12.008] [PMID: 26810602]

[39] Dufourt J, Bontonou G, Chartier A, *et al.* piRNAs and Aubergine cooperate with Wispy poly(A) polymerase to stabilize mRNAs in the germ plasm. Nat Commun 2017; 8(1): 1305. [http://dx.doi.org/10.1038/s41467-017-01431-5] [PMID: 29101389]

[40] Czech B, Munafò M, Ciabrelli F, *et al.* piRNA-Guided Genome Defense: From Biogenesis to Silencing. Annu Rev Genet 2018; 52: 131-57. [http://dx.doi.org/10.1146/annurev-genet-120417-031441] [PMID: 30476449]

[41] Rojas-Ríos P, Simonelig M. piRNAs and PIWI proteins: regulators of gene expression in development and stem cells. Development 2018; 145(17)dev161786 [http://dx.doi.org/10.1242/dev.161786] [PMID: 30194260]

[42] Heestand B, Simon M, Frenk S, Titov D, Ahmed S. Transgenerational Sterility of Piwi Mutants Represents a Dynamic Form of Adult Reproductive Diapause. Cell Rep 2018; 23(1): 156-71. [http://dx.doi.org/10.1016/j.celrep.2018.03.015] [PMID: 29617657]

[43] Nishimura T, Nagamori I, Nakatani T, *et al.* PNLDC1, mouse pre-piRNA Trimmer, is required for meiotic and post-meiotic male germ cell development. EMBO Rep 2018; 19(3)e44957

[http://dx.doi.org/10.15252/embr.201744957] [PMID: 29444933]

[44] Litwin M, Szczepańska-Buda A, Piotrowska A, Dzięgiel P, Witkiewicz W. The meaning of PIWI proteins in cancer development. Oncol Lett 2017; 13(5): 3354-62.
[http://dx.doi.org/10.3892/ol.2017.5932] [PMID: 28529570]

[45] Liu S, Huang S, Chen F, *et al.* Genomic Analyses from Non-invasive Prenatal Testing Reveal Genetic Associations, Patterns of Viral Infections, and Chinese Population History. Cell 2018; 175(2): 347-359.e14.
[http://dx.doi.org/10.1016/j.cell.2018.08.016] [PMID: 30290141]

[46] Swart EC, Nowacki M. The eukaryotic way to defend and edit genomes by sRNA-targeted DNA deletion. Ann N Y Acad Sci 2015; 1341: 106-14.
[http://dx.doi.org/10.1111/nyas.12636] [PMID: 25581723]

[47] Janeway CA Jr. Approaching the asymptote? Evolution and revolution in immunology. Cold Spring Harb Symp Quant Biol 1989; 54(Pt 1): 1-13.
[http://dx.doi.org/10.1101/SQB.1989.054.01.003] [PMID: 2700931]

[48] Lemaitre B, Nicolas E, Michaut L, Reichhart JM, Hoffmann JA. The dorsoventral regulatory gene cassette spätzle/Toll/cactus controls the potent antifungal response in Drosophila adults. Cell 1996; 86(6): 973-83.
[http://dx.doi.org/10.1016/S0092-8674(00)80172-5] [PMID: 8808632]

[49] Takeda K, Akira S. Toll-like receptors. Current protocols in immunology 2015; 109: 14.2.1-0.
[http://dx.doi.org/10.1002/0471142735.im1412s109]

[50] Dowling JK, Mansell A. Toll-like receptors: the swiss army knife of immunity and vaccine development. Clinical & translational immunology 2016; 5(5): e85.
[http://dx.doi.org/10.1038/cti.2016.22]

[51] Medzhitov R, Preston-Hurlburt P, Janeway CA Jr. A human homologue of the Drosophila Toll protein signals activation of adaptive immunity. Nature 1997; 388(6640): 394-7.
[http://dx.doi.org/10.1038/41131] [PMID: 9237759]

[52] Yang RB, Mark MR, Gray A, *et al.* Toll-like receptor-2 mediates lipopolysaccharide-induced cellular signalling. Nature 1998; 395(6699): 284-8.
[http://dx.doi.org/10.1038/26239] [PMID: 9751057]

[53] Fuchs K, Cardona Gloria Y, Wolz OO, *et al.* The fungal ligand chitin directly binds TLR2 and triggers inflammation dependent on oligomer size. EMBO Rep 2018; 19(12)e46065
[http://dx.doi.org/10.15252/embr.201846065] [PMID: 30337494]

[54] Hess NJ, Jiang S, Li X, Guan Y, Tapping RI. TLR10 Is a B Cell Intrinsic Suppressor of Adaptive Immune Responses. J Immunol 2017; 198(2): 699-707.
[http://dx.doi.org/10.4049/jimmunol.1601335] [PMID: 27956526]

[55] Brown GD, Willment JA, Whitehead L. C-type lectins in immunity and homeostasis. Nat Rev Immunol 2018; 18(6): 374-89.
[http://dx.doi.org/10.1038/s41577-018-0004-8] [PMID: 29581532]

[56] Yoneyama M, Onomoto K, Jogi M, Akaboshi T, Fujita T. Viral RNA detection by RIG-I-like receptors. Curr Opin Immunol 2015; 32: 48-53.
[http://dx.doi.org/10.1016/j.coi.2014.12.012] [PMID: 25594890]

[57] Zhao Y, Ye X, Dunker W, Song Y, Karijolich J. RIG-I like receptor sensing of host RNAs facilitates the cell-intrinsic immune response to KSHV infection. Nat Commun 2018; 9(1): 4841.
[http://dx.doi.org/10.1038/s41467-018-07314-7] [PMID: 30451863]

[58] Jones JD, Vance RE, Dangl JL. Intracellular innate immune surveillance devices in plants and animals. Science 2016; 354(6316)aaf6395
[http://dx.doi.org/10.1126/science.aaf6395] [PMID: 27934708]

[59] Borrelli GM, Mazzucotelli E, Marone D, *et al.* Regulation and Evolution of NLR Genes: A Close Interconnection for Plant Immunity. Int J Mol Sci 2018; 19(6)E1662 [http://dx.doi.org/10.3390/ijms19061662] [PMID: 29867062]

[60] Nanson JD, Kobe B, Ve T. Death, TIR, and RHIM: Self-assembling domains involved in innate immunity and cell-death signaling. J Leukoc Biol 2019; 105(2): 363-75. [http://dx.doi.org/10.1002/JLB.MR0318-123R] [PMID: 30517972]

[61] Balka KR, De Nardo D. Understanding early TLR signaling through the Myddosome. J Leukoc Biol 2019; 105(2): 339-51. [http://dx.doi.org/10.1002/JLB.MR0318-096R] [PMID: 30256449]

[62] Barrow AD, Trowsdale J. You say ITAM and I say ITIM, let's call the whole thing off: the ambiguity of immunoreceptor signalling. Eur J Immunol 2006; 36(7): 1646-53. [http://dx.doi.org/10.1002/eji.200636195] [PMID: 16783855]

[63] Pellegrini E, Desfosses A, Wallmann A, *et al.* RIP2 filament formation is required for NOD2 dependent NF-κB signalling. Nat Commun 2018; 9(1): 4043. [http://dx.doi.org/10.1038/s41467-018-06451-3] [PMID: 30279485]

[64] Banoth B, Cassel SL. Mitochondria in innate immune signaling. Transl Res 2018; 202: 52-68. [http://dx.doi.org/10.1016/j.trsl.2018.07.014] [PMID: 30165038]

[65] Blander JM, Barbet G. Exploiting vita-PAMPs in vaccines. Curr Opin Pharmacol 2018; 41: 128-36. [http://dx.doi.org/10.1016/j.coph.2018.05.012] [PMID: 29890457]

[66] Campbell RA, Schwertz H, Hottz ED, *et al.* Human megakaryocytes possess intrinsic antiviral immunity through regulated induction of IFITM3. Blood 2019; 133(19): 2013-26. [http://dx.doi.org/10.1182/blood-2018-09-873984] [PMID: 30723081]

[67] McClure R, Massari P. TLR-Dependent Human Mucosal Epithelial Cell Responses to Microbial Pathogens. Front Immunol 2014; 5: 386. [http://dx.doi.org/10.3389/fimmu.2014.00386] [PMID: 25161655]

[68] Günther J, Seyfert HM. The first line of defence: insights into mechanisms and relevance of phagocytosis in epithelial cells. Semin Immunopathol 2018; 40(6): 555-65. [http://dx.doi.org/10.1007/s00281-018-0701-1] [PMID: 30182191]

[69] Franz KM, Kagan JC. Innate Immune Receptors as Competitive Determinants of Cell Fate. Mol Cell 2017; 66(6): 750-60. [http://dx.doi.org/10.1016/j.molcel.2017.05.009] [PMID: 28622520]

[70] Netea MG, Joosten LA, Latz E, *et al.* Trained immunity: A program of innate immune memory in health and disease. Science 2016; 352(6284)aaf1098 [http://dx.doi.org/10.1126/science.aaf1098] [PMID: 27102489]

[71] Zhang ZJ, Guo JS, Li SS, *et al.* TLR8 and its endogenous ligand miR-21 contribute to neuropathic pain in murine DRG. J Exp Med 2018; 215(12): 3019-37. [http://dx.doi.org/10.1084/jem.20180800] [PMID: 30455267]

[72] McAlpine W, Sun L, Wang KW, *et al.* Excessive endosomal TLR signaling causes inflammatory disease in mice with defective SMCR8-WDR41-C9ORF72 complex function. Proc Natl Acad Sci USA 2018; 115(49): E11523-31. [http://dx.doi.org/10.1073/pnas.1814753115] [PMID: 30442666]

[73] Matzinger P. Tolerance, danger, and the extended family. Annu Rev Immunol 1994; 12: 991-1045. [http://dx.doi.org/10.1146/annurev.iy.12.040194.005015] [PMID: 8011301]

[74] Matzinger P. The danger model: A renewed sense of self. Science 2002; 296(5566): 301-5. [http://dx.doi.org/10.1126/science.1071059] [PMID: 11951032]

[75] Schaefer L. Complexity of danger: the diverse nature of damage-associated molecular patterns. J Biol Chem 2014; 289(51): 35237-45.

[http://dx.doi.org/10.1074/jbc.R114.619304] [PMID: 25391648]

[76] Boyapati RK, Tamborska A, Dorward DA, Ho GT. Advances in the understanding of mitochondrial DNA as a pathogenic factor in inflammatory diseases. F1000 Res 2017; 6: 169. [http://dx.doi.org/10.12688/f1000research.10397.1] [PMID: 28299196]

[77] Glück S, Guey B, Gulen MF, *et al.* Innate immune sensing of cytosolic chromatin fragments through cGAS promotes senescence. Nat Cell Biol 2017; 19(9): 1061-70. [http://dx.doi.org/10.1038/ncb3586] [PMID: 28759028]

[78] Bertheloot D, Latz E. HMGB1, IL-1alpha, IL-33 and S100 proteins: dual-function alarmins. Cellular & molecular immunology 2017; 14(1): 43-64.

[79] Dangl JL, Jones JD. Plant pathogens and integrated defence responses to infection. Nature 2001; 411(6839): 826-33. [http://dx.doi.org/10.1038/35081161] [PMID: 11459065]

[80] Khan M, Subramaniam R, Desveaux D. Of guards, decoys, baits and traps: pathogen perception in plants by type III effector sensors. Curr Opin Microbiol 2016; 29: 49-55. [http://dx.doi.org/10.1016/j.mib.2015.10.006] [PMID: 26599514]

[81] Vance RE, Isberg RR, Portnoy DA. Patterns of pathogenesis: discrimination of pathogenic and nonpathogenic microbes by the innate immune system. Cell Host Microbe 2009; 6(1): 10-21. [http://dx.doi.org/10.1016/j.chom.2009.06.007] [PMID: 19616762]

[82] Colaço HG, Moita LF. Initiation of innate immune responses by surveillance of homeostasis perturbations. FEBS J 2016; 283(13): 2448-57. [http://dx.doi.org/10.1111/febs.13730] [PMID: 27037950]

[83] Hatsugai N, Igarashi D, Mase K, *et al.* A plant effector-triggered immunity signaling sector is inhibited by pattern-triggered immunity. EMBO J 2017; 36(18): 2758-69. [http://dx.doi.org/10.15252/embj.201796529] [PMID: 28811287]

[84] Fekete T, Bencze D, Szabo A, *et al.* Regulatory NLRs Control the RLR-Mediated Type I Interferon and Inflammatory Responses in Human Dendritic Cells. Front Immunol 2018; 9: 2314. [http://dx.doi.org/10.3389/fimmu.2018.02314] [PMID: 30344524]

[85] Pradeu T, Vivier E. The discontinuity theory of immunity. Sci Immunol 2016; 1(1)AAG0479 [http://dx.doi.org/10.1126/sciimmunol.aag0479] [PMID: 28239677]

[86] Pukkila-Worley R. Surveillance Immunity: An Emerging Paradigm of Innate Defense Activation in Caenorhabditis elegans. PLoS Pathog 2016; 12(9)e1005795 [http://dx.doi.org/10.1371/journal.ppat.1005795] [PMID: 27631629]

[87] Gutschow MV, Mason JC, Lane KM, *et al.* Combinatorial processing of bacterial and host-derived innate immune stimuli at the single-cell level. Mol Biol Cell 2019; 30(2): 282-92. [http://dx.doi.org/10.1091/mbc.E18-07-0423] [PMID: 30462580]

[88] Kadota Y, Liebrand TWH, Goto Y, *et al.* Quantitative phosphoproteomic analysis reveals common regulatory mechanisms between effector- and PAMP-triggered immunity in plants. New Phytol 2019; 221(4): 2160-75. [http://dx.doi.org/10.1111/nph.15523] [PMID: 30300945]

[89] Negishi H, Taniguchi T, Yanai H. The Interferon (IFN) Class of Cytokines and the IFN Regulatory Factor (IRF) Transcription Factor Family. Cold Spring Harb Perspect Biol 2018; 10(11)a028423 [http://dx.doi.org/10.1101/cshperspect.a028423] [PMID: 28963109]

[90] Schreiber G. The molecular basis for differential type I interferon signaling. J Biol Chem 2017; 292(18): 7285-94. [http://dx.doi.org/10.1074/jbc.R116.774562] [PMID: 28289098]

[91] Fenimore J, A Young H. Regulation of IFN-γ Expression. Adv Exp Med Biol 2016; 941: 1-19. [http://dx.doi.org/10.1007/978-94-024-0921-5_1] [PMID: 27734406]

[92] Wack A, Terczyńska-Dyla E, Hartmann R. Guarding the frontiers: the biology of type III interferons. Nat Immunol 2015; 16(8): 802-9.
[http://dx.doi.org/10.1038/ni.3212] [PMID: 26194286]

[93] Wells AI, Coyne CB. Type III Interferons in Antiviral Defenses at Barrier Surfaces. Trends Immunol 2018; 39(10): 848-58.
[http://dx.doi.org/10.1016/j.it.2018.08.008] [PMID: 30219309]

[94] Dalet A, Argüello RJ, Combes A, *et al.* Protein synthesis inhibition and GADD34 control IFN-β heterogeneous expression in response to dsRNA. EMBO J 2017; 36(6): 761-82.
[http://dx.doi.org/10.15252/embj.201695000] [PMID: 28100675]

[95] Cheng Y, Schorey JS. *Mycobacterium tuberculosis*-induced IFN-β production requires cytosolic DNA and RNA sensing pathways. J Exp Med 2018; 215(11): 2919-35.
[http://dx.doi.org/10.1084/jem.20180508] [PMID: 30337468]

[96] Ivashkiv LB. IFNγ: signalling, epigenetics and roles in immunity, metabolism, disease and cancer immunotherapy. Nat Rev Immunol 2018; 18(9): 545-58.
[http://dx.doi.org/10.1038/s41577-018-0029-z] [PMID: 29921905]

[97] Kotenko SV, Durbin JE. Contribution of type III interferons to antiviral immunity: location, location, location. J Biol Chem 2017; 292(18): 7295-303.
[http://dx.doi.org/10.1074/jbc.R117.777102] [PMID: 28289095]

[98] Majoros A, Platanitis E, Kernbauer-Hölzl E, Rosebrock F, Müller M, Decker T. Canonical and Non-Canonical Aspects of JAK-STAT Signaling: Lessons from Interferons for Cytokine Responses. Front Immunol 2017; 8: 29.
[http://dx.doi.org/10.3389/fimmu.2017.00029] [PMID: 28184222]

[99] Wang W, Xu L, Su J, Peppelenbosch MP, Pan Q. Transcriptional Regulation of Antiviral Interferon-Stimulated Genes. Trends Microbiol 2017; 25(7): 573-84.
[http://dx.doi.org/10.1016/j.tim.2017.01.001] [PMID: 28139375]

[100] Schoggins JW, Wilson SJ, Panis M, *et al.* A diverse range of gene products are effectors of the type I interferon antiviral response. Nature 2011; 472(7344): 481-5.
[http://dx.doi.org/10.1038/nature09907] [PMID: 21478870]

[101] Schneider WM, Chevillotte MD, Rice CM. Interferon-stimulated genes: a complex web of host defenses. Annu Rev Immunol 2014; 32: 513-45.
[http://dx.doi.org/10.1146/annurev-immunol-032713-120231] [PMID: 24555472]

[102] Green R, Ireton RC, Gale M Jr. Interferon-stimulated genes: New platforms and computational approaches. Mammalian genome : Official journal of the International Mammalian Genome Society 2018; 29(7-8): 593-602.
[http://dx.doi.org/10.1007/s00335-018-9755-6]

[103] Schoggins JW. Recent advances in antiviral interferon-stimulated gene biology. F1000 Res 2018; 7: 309.
[http://dx.doi.org/10.12688/f1000research.12450.1] [PMID: 29568506]

[104] Pervolaraki K, Rastgou Talemi S, Albrecht D, *et al.* Differential induction of interferon stimulated genes between type I and type III interferons is independent of interferon receptor abundance. PLoS Pathog 2018; 14(11)e1007420
[http://dx.doi.org/10.1371/journal.ppat.1007420] [PMID: 30485383]

[105] Kamada R, Yang W, Zhang Y, *et al.* Interferon stimulation creates chromatin marks and establishes transcriptional memory. Proc Natl Acad Sci USA 2018; 115(39): E9162-71.
[http://dx.doi.org/10.1073/pnas.1720930115] [PMID: 30201712]

[106] Sohrabi Y, Godfrey R, Findeisen HM. Altered Cellular Metabolism Drives Trained Immunity. Trends Endocrinol Metab 2018; 29(9): 602-5.
[http://dx.doi.org/10.1016/j.tem.2018.03.012] [PMID: 29627292]

[107] Schulz KS, Mossman KL. Viral Evasion Strategies in Type I IFN Signaling - A Summary of Recent Developments. Front Immunol 2016; 7: 498. [http://dx.doi.org/10.3389/fimmu.2016.00498] [PMID: 27891131]

[108] García-Sastre A. Ten Strategies of Interferon Evasion by Viruses. Cell Host Microbe 2017; 22(2): 176-84. [http://dx.doi.org/10.1016/j.chom.2017.07.012] [PMID: 28799903]

[109] Freundt EC, Drappier M, Michiels T. Innate Immune Detection of Cardioviruses and Viral Disruption of Interferon Signaling. Front Microbiol 2018; 9: 2448. [http://dx.doi.org/10.3389/fmicb.2018.02448] [PMID: 30369921]

[110] Gough DJ, Messina NL, Clarke CJ, Johnstone RW, Levy DE. Constitutive type I interferon modulates homeostatic balance through tonic signaling. Immunity 2012; 36(2): 166-74. [http://dx.doi.org/10.1016/j.immuni.2012.01.011] [PMID: 22365663]

[111] Lombardi A, Tsomos E, Hammerstad SS, Tomer Y. Interferon alpha: The key trigger of type 1 diabetes. J Autoimmun 2018; 94: 7-15. [http://dx.doi.org/10.1016/j.jaut.2018.08.003] [PMID: 30115527]

[112] Taft J, Bogunovic D. The Goldilocks Zone of Type I IFNs: Lessons from Human Genetics. J Immunol 2018; 201(12): 3479-85. [http://dx.doi.org/10.4049/jimmunol.1800764] [PMID: 30530500]

[113] Martinon F, Burns K, Tschopp J. The inflammasome: a molecular platform triggering activation of inflammatory caspases and processing of proIL-beta. Mol Cell 2002; 10(2): 417-26. [http://dx.doi.org/10.1016/S1097-2765(02)00599-3] [PMID: 12191486]

[114] de Zoete MR, Palm NW, Zhu S, Flavell RA. Inflammasomes. Cold Spring Harb Perspect Biol 2014; 6(12)a016287 [http://dx.doi.org/10.1101/cshperspect.a016287] [PMID: 25324215]

[115] Duncan JA, Canna SW. The NLRC4 Inflammasome. Immunol Rev 2018; 281(1): 115-23. [http://dx.doi.org/10.1111/imr.12607] [PMID: 29247997]

[116] Hu Z, Zhou Q, Zhang C, *et al.* Structural and biochemical basis for induced self-propagation of NLRC4. Science 2015; 350(6259): 399-404. [http://dx.doi.org/10.1126/science.aac5489] [PMID: 26449475]

[117] de Torre-Minguela C, Mesa Del Castillo P, Pelegrín P. The NLRP3 and Pyrin Inflammasomes: Implications in the Pathophysiology of Autoinflammatory Diseases. Front Immunol 2017; 8: 43. [http://dx.doi.org/10.3389/fimmu.2017.00043] [PMID: 28191008]

[118] Mayor A, Martinon F, De Smedt T, Pétrilli V, Tschopp J. A crucial function of SGT1 and HSP90 in inflammasome activity links mammalian and plant innate immune responses. Nat Immunol 2007; 8(5): 497-503. [http://dx.doi.org/10.1038/ni1459] [PMID: 17435760]

[119] Siligardi G, Zhang M, Prodromou C. The Stoichiometric Interaction of the Hsp90-Sgt1-Rar1 Complex by CD and SRCD Spectroscopy. Front Mol Biosci 2018; 4: 95. [http://dx.doi.org/10.3389/fmolb.2017.00095] [PMID: 29387685]

[120] Gao W, Yang J, Liu W, Wang Y, Shao F. Site-specific phosphorylation and microtubule dynamics control Pyrin inflammasome activation. Proc Natl Acad Sci USA 2016; 113(33): E4857-66. [http://dx.doi.org/10.1073/pnas.1601700113] [PMID: 27482109]

[121] Gambin Y, Giles N, O'Carroll A, Polinkovsky M, Hunter D, Sierecki E. Single-Molecule Fluorescence Reveals the Oligomerization and Folding Steps Driving the Prion-like Behavior of ASC. J Mol Biol 2018; 430(4): 491-508. [http://dx.doi.org/10.1016/j.jmb.2017.12.013] [PMID: 29288634]

[122] Franklin BS, Latz E, Schmidt FI. The intra- and extracellular functions of ASC specks. Immunol Rev

2018; 281(1): 74-87.
[http://dx.doi.org/10.1111/imr.12611] [PMID: 29247990]

[123] Aktories K. Bacterial protein toxins that modify host regulatory GTPases. Nat Rev Microbiol 2011; 9(7): 487-98.
[http://dx.doi.org/10.1038/nrmicro2592] [PMID: 21677684]

[124] Keestra AM, Winter MG, Auburger JJ, *et al.* Manipulation of small Rho GTPases is a pathogen-induced process detected by NOD1. Nature 2013; 496(7444): 233-7.
[http://dx.doi.org/10.1038/nature12025] [PMID: 23542589]

[125] Xu H, Yang J, Gao W, *et al.* Innate immune sensing of bacterial modifications of Rho GTPases by the Pyrin inflammasome. Nature 2014; 513(7517): 237-41.
[http://dx.doi.org/10.1038/nature13449] [PMID: 24919149]

[126] Heilig R, Broz P. Function and mechanism of the pyrin inflammasome. Eur J Immunol 2018; 48(2): 230-8.
[http://dx.doi.org/10.1002/eji.201746947] [PMID: 29148036]

[127] Alimov I, Menon S, Cochran N, *et al.* Bile acid analogues are activators of pyrin inflammasome. J Biol Chem 2019; 294(10): 3359-66.
[http://dx.doi.org/10.1074/jbc.RA118.005103] [PMID: 30647128]

[128] Muñoz-Planillo R, Kuffa P, Martínez-Colón G, Smith BL, Rajendiran TM, Núñez G. K^+ efflux is the common trigger of NLRP3 inflammasome activation by bacterial toxins and particulate matter. Immunity 2013; 38(6): 1142-53.
[http://dx.doi.org/10.1016/j.immuni.2013.05.016] [PMID: 23809161]

[129] Di A, Xiong S, Ye Z, *et al.* The TWIK2 Potassium Efflux Channel in Macrophages Mediates NLRP3 Inflammasome-Induced Inflammation. Immunity 2018; 49(1): 56-65.e4.
[http://dx.doi.org/10.1016/j.immuni.2018.04.032] [PMID: 29958799]

[130] Mathur A, Feng S, Hayward JA, *et al.* A multicomponent toxin from Bacillus cereus incites inflammation and shapes host outcome *via* the NLRP3 inflammasome. Nat Microbiol 2019; 4(2): 362-74.
[http://dx.doi.org/10.1038/s41564-018-0318-0] [PMID: 30531979]

[131] da Costa LS, Outlioua A, Anginot A, Akarid K, Arnoult D. RNA viruses promote activation of the NLRP3 inflammasome through cytopathogenic effect-induced potassium efflux. Cell Death Dis 2019; 10(5): 346.
[http://dx.doi.org/10.1038/s41419-019-1579-0] [PMID: 31024004]

[132] Negash AA, Olson RM, Griffin S, Gale M Jr. Modulation of calcium signaling pathway by hepatitis C virus core protein stimulates NLRP3 inflammasome activation. PLoS Pathog 2019; 15(2)e1007593
[http://dx.doi.org/10.1371/journal.ppat.1007593] [PMID: 30811485]

[133] He Y, Zeng MY, Yang D, Motro B, Núñez G. NEK7 is an essential mediator of NLRP3 activation downstream of potassium efflux. Nature 2016; 530(7590): 354-7.
[http://dx.doi.org/10.1038/nature16959] [PMID: 26814970]

[134] Shi H, Wang Y, Li X, *et al.* NLRP3 activation and mitosis are mutually exclusive events coordinated by NEK7, a new inflammasome component. Nat Immunol 2016; 17(3): 250-8.
[http://dx.doi.org/10.1038/ni.3333] [PMID: 26642356]

[135] Groß CJ, Mishra R, Schneider KS, *et al.* K^+ Efflux-Independent NLRP3 Inflammasome Activation by Small Molecules Targeting Mitochondria. Immunity 2016; 45(4): 761-73.
[http://dx.doi.org/10.1016/j.immuni.2016.08.010] [PMID: 27692612]

[136] Liston A, Masters SL. Homeostasis-altering molecular processes as mechanisms of inflammasome activation. Nat Rev Immunol 2017; 17(3): 208-14.
[http://dx.doi.org/10.1038/nri.2016.151] [PMID: 28163301]

[137] Poudel B, Gurung P. An update on cell intrinsic negative regulators of the NLRP3 inflammasome. J

Leukoc Biol 2018.
[http://dx.doi.org/10.1002/JLB.3MIR0917-350R] [PMID: 29377242]

[138] Chen J, Chen ZJ. PtdIns4P on dispersed trans-Golgi network mediates NLRP3 inflammasome activation. Nature 2018; 564(7734): 71-6.
[http://dx.doi.org/10.1038/s41586-018-0761-3] [PMID: 30487600]

[139] Youm YH, Nguyen KY, Grant RW, *et al.* The ketone metabolite β-hydroxybutyrate blocks NLRP3 inflammasome-mediated inflammatory disease. Nat Med 2015; 21(3): 263-9.
[http://dx.doi.org/10.1038/nm.3804] [PMID: 25686106]

[140] Afonina IS, Zhong Z, Karin M, Beyaert R. Limiting inflammation-the negative regulation of NF-κB and the NLRP3 inflammasome. Nat Immunol 2017; 18(8): 861-9.
[http://dx.doi.org/10.1038/ni.3772] [PMID: 28722711]

[141] Hornung V, Ablasser A, Charrel-Dennis M, *et al.* AIM2 recognizes cytosolic dsDNA and forms a caspase-1-activating inflammasome with ASC. Nature 2009; 458(7237): 514-8.
[http://dx.doi.org/10.1038/nature07725] [PMID: 19158675]

[142] Fernandes-Alnemri T, Yu JW, Datta P, Wu J, Alnemri ES. AIM2 activates the inflammasome and cell death in response to cytoplasmic DNA. Nature 2009; 458(7237): 509-13.
[http://dx.doi.org/10.1038/nature07710] [PMID: 19158676]

[143] Lugrin J, Martinon F. The AIM2 inflammasome: Sensor of pathogens and cellular perturbations. Immunol Rev 2018; 281(1): 99-114.
[http://dx.doi.org/10.1111/imr.12618] [PMID: 29247998]

[144] Matyszewski M, Morrone SR, Sohn J. Digital signaling network drives the assembly of the AIM2-ASC inflammasome. Proc Natl Acad Sci USA 2018; 115(9): E1963-72.
[http://dx.doi.org/10.1073/pnas.1712860115] [PMID: 29440442]

[145] Zhao Y, Yang J, Shi J, *et al.* The NLRC4 inflammasome receptors for bacterial flagellin and type III secretion apparatus. Nature 2011; 477(7366): 596-600.
[http://dx.doi.org/10.1038/nature10510] [PMID: 21918512]

[146] Tenthorey JL, Haloupek N, López-Blanco JR, *et al.* The structural basis of flagellin detection by NAIP5: A strategy to limit pathogen immune evasion. Science 2017; 358(6365): 888-93.
[http://dx.doi.org/10.1126/science.aao1140] [PMID: 29146805]

[147] Hu Z, Yan C, Liu P, *et al.* Crystal structure of NLRC4 reveals its autoinhibition mechanism. Science 2013; 341(6142): 172-5.
[http://dx.doi.org/10.1126/science.1236381] [PMID: 23765277]

[148] Chui AJ, Okondo MC, Rao SD, *et al.* N-terminal degradation activates the NLRP1B inflammasome. Science 2019; 364(6435): 82-5.
[http://dx.doi.org/10.1126/science.aau1208] [PMID: 30872531]

[149] Sandstrom A, Mitchell PS, Goers L, Mu EW, Lesser CF, Vance RE. Functional degradation: A mechanism of NLRP1 inflammasome activation by diverse pathogen enzymes. Science 2019; 364(6435)eaau1330
[http://dx.doi.org/10.1126/science.aau1330] [PMID: 30872533]

[150] Shi J, Zhao Y, Wang Y, *et al.* Inflammatory caspases are innate immune receptors for intracellular LPS. Nature 2014; 514(7521): 187-92.
[http://dx.doi.org/10.1038/nature13683] [PMID: 25119034]

[151] Rathinam VAK, Zhao Y, Shao F. Innate immunity to intracellular LPS. Nat Immunol 2019; 20(5): 527-33.
[http://dx.doi.org/10.1038/s41590-019-0368-3] [PMID: 30962589]

[152] Zanoni I, Tan Y, Di Gioia M, *et al.* An endogenous caspase-11 ligand elicits interleukin-1 release from living dendritic cells. Science 2016; 352(6290): 1232-6.
[http://dx.doi.org/10.1126/science.aaf3036] [PMID: 27103670]

[153] Pallett MA, Crepin VF, Serafini N, *et al.* Bacterial virulence factor inhibits caspase-4/11 activation in intestinal epithelial cells. Mucosal Immunol 2017; 10(3): 602-12. [http://dx.doi.org/10.1038/mi.2016.77] [PMID: 27624779]

[154] Levy M, Thaiss CA, Zeevi D, *et al.* Microbiota-Modulated Metabolites Shape the Intestinal Microenvironment by Regulating NLRP6 Inflammasome Signaling. Cell 2015; 163(6): 1428-43. [http://dx.doi.org/10.1016/j.cell.2015.10.048] [PMID: 26638072]

[155] Jun HK, Jung YJ, Choi BK. Inflammasome activators induce fibronectin expression and release in macrophages. Cellular microbiology 2017; 19(5): cmi.12695. [http://dx.doi.org/10.1111/cmi.12695]

[156] Jorgensen I, Zhang Y, Krantz BA, Miao EA. Pyroptosis triggers pore-induced intracellular traps (PITs) that capture bacteria and lead to their clearance by efferocytosis. J Exp Med 2016; 213(10): 2113-28. [http://dx.doi.org/10.1084/jem.20151613] [PMID: 27573815]

[157] Mishra BB, Lovewell RR, Olive AJ, *et al.* Nitric oxide prevents a pathogen-permissive granulocytic inflammation during tuberculosis. Nat Microbiol 2017; 2: 17072. [http://dx.doi.org/10.1038/nmicrobiol.2017.72] [PMID: 28504669]

[158] Harapas CR, Steiner A, Davidson S, Masters SL. An Update on Autoinflammatory Diseases: Inflammasomopathies. Curr Rheumatol Rep 2018; 20(7): 40. [http://dx.doi.org/10.1007/s11926-018-0750-4] [PMID: 29846819]

[159] Furman D, Chang J, Lartigue L, *et al.* Expression of specific inflammasome gene modules stratifies older individuals into two extreme clinical and immunological states. Nat Med 2017; 23(2): 174-84. [http://dx.doi.org/10.1038/nm.4267] [PMID: 28092664]

[160] Weber F, Wagner V, Rasmussen SB, Hartmann R, Paludan SR. Double-stranded RNA is produced by positive-strand RNA viruses and DNA viruses but not in detectable amounts by negative-strand RNA viruses. J Virol 2006; 80(10): 5059-64. [http://dx.doi.org/10.1128/JVI.80.10.5059-5064.2006] [PMID: 16641297]

[161] McFadden MJ, Gokhale NS, Horner SM. Protect this house: cytosolic sensing of viruses. Curr Opin Virol 2017; 22: 36-43. [http://dx.doi.org/10.1016/j.coviro.2016.11.012] [PMID: 27951430]

[162] Pakos-Zebrucka K, Koryga I, Mnich K, Ljujic M, Samali A, Gorman AM. The integrated stress response. EMBO Rep 2016; 17(10): 1374-95. [http://dx.doi.org/10.15252/embr.201642195] [PMID: 27629041]

[163] Hull CM, Bevilacqua PC. Discriminating Self and Non-Self by RNA: Roles for RNA Structure, Misfolding, and Modification in Regulating the Innate Immune Sensor PKR. Acc Chem Res 2016; 49(6): 1242-9. [http://dx.doi.org/10.1021/acs.accounts.6b00151] [PMID: 27269119]

[164] Vincent HA, Ziehr B, Moorman NJ. Mechanism of Protein Kinase R Inhibition by Human Cytomegalovirus pTRS1. J Virol 2017; 91(5): e01574-16. [http://dx.doi.org/10.1128/JVI.01574-16] [PMID: 27974558]

[165] Donovan J, Dufner M, Korennykh A. Structural basis for cytosolic double-stranded RNA surveillance by human oligoadenylate synthetase 1. Proc Natl Acad Sci USA 2013; 110(5): 1652-7. [http://dx.doi.org/10.1073/pnas.1218528110] [PMID: 23319625]

[166] Vachon VK, Calderon BM, Conn GL. A novel RNA molecular signature for activation of 2′-5′ oligoadenylate synthetase-1. Nucleic Acids Res 2015; 43(1): 544-52. [http://dx.doi.org/10.1093/nar/gku1289] [PMID: 25477390]

[167] Calderon BM, Conn GL. A human cellular noncoding RNA activates the antiviral protein 2′-5--oligoadenylate synthetase 1. J Biol Chem 2018; 293(41): 16115-24. [http://dx.doi.org/10.1074/jbc.RA118.004747] [PMID: 30126839]

[168] Li Y, Banerjee S, Wang Y, *et al.* Activation of RNase L is dependent on OAS3 expression during infection with diverse human viruses. Proc Natl Acad Sci USA 2016; 113(8): 2241-6. [http://dx.doi.org/10.1073/pnas.1519657113] [PMID: 26858407]

[169] Donovan J, Rath S, Kolet-Mandrikov D, Korennykh A. Rapid RNase L-driven arrest of protein synthesis in the dsRNA response without degradation of translation machinery. RNA 2017; 23(11): 1660-71. [http://dx.doi.org/10.1261/rna.062000.117] [PMID: 28808124]

[170] Ullmer W, Semler BL. Diverse Strategies Used by Picornaviruses to Escape Host RNA Decay Pathways. Viruses 2016; 8(12)E335 [http://dx.doi.org/10.3390/v8120335] [PMID: 27999393]

[171] Drappier M, Jha BK, Stone S, *et al.* A novel mechanism of RNase L inhibition: Theiler's virus L* protein prevents 2-5A from binding to RNase L. PLoS Pathog 2018; 14(4)e1006989 [http://dx.doi.org/10.1371/journal.ppat.1006989] [PMID: 29652922]

[172] Leung DW, Amarasinghe GK. When your cap matters: structural insights into self *vs* non-self recognition of 5′ RNA by immunomodulatory host proteins. Curr Opin Struct Biol 2016; 36: 133-41. [http://dx.doi.org/10.1016/j.sbi.2016.02.001] [PMID: 26916433]

[173] Goubau D, Schlee M, Deddouche S, *et al.* Antiviral immunity *via* RIG-I-mediated recognition of RNA bearing 5′-diphosphates. Nature 2014; 514(7522): 372-5. [http://dx.doi.org/10.1038/nature13590] [PMID: 25119032]

[174] Devarkar SC, Wang C, Miller MT, *et al.* Structural basis for m7G recognition and 2′-O-methyl discrimination in capped RNAs by the innate immune receptor RIG-I. Proc Natl Acad Sci USA 2016; 113(3): 596-601. [http://dx.doi.org/10.1073/pnas.1515152113] [PMID: 26733676]

[175] Yoneyama M, Jogi M, Onomoto K. Regulation of antiviral innate immune signaling by stress-induced RNA granules. J Biochem 2016; 159(3): 279-86. [http://dx.doi.org/10.1093/jb/mvv122] [PMID: 26748340]

[176] Dias Junior AG, Sampaio NG, Rehwinkel J. A Balancing Act: MDA5 in Antiviral Immunity and Autoinflammation. Trends Microbiol 2019; 27(1): 75-85. [http://dx.doi.org/10.1016/j.tim.2018.08.007] [PMID: 30201512]

[177] Schuberth-Wagner C, Ludwig J, Bruder AK, *et al.* A Conserved Histidine in the RNA Sensor RIG-I Controls Immune Tolerance to N1-2'O-Methylated Self RNA. Immunity 2015; 43(1): 41-51. [http://dx.doi.org/10.1016/j.immuni.2015.06.015] [PMID: 26187414]

[178] Liu S, Cai X, Wu J, *et al.* Phosphorylation of innate immune adaptor proteins MAVS, STING, and TRIF induces IRF3 activation. Science 2015; 347(6227)aaa2630 [http://dx.doi.org/10.1126/science.aaa2630] [PMID: 25636800]

[179] Fang R, Jiang Q, Zhou X, *et al.* MAVS activates TBK1 and IKKε through TRAFs in NEMO dependent and independent manner. PLoS Pathog 2017; 13(11)e1006720 [http://dx.doi.org/10.1371/journal.ppat.1006720] [PMID: 29125880]

[180] Liu B, Zhang M, Chu H, *et al.* The ubiquitin E3 ligase TRIM31 promotes aggregation and activation of the signaling adaptor MAVS through Lys63-linked polyubiquitination. Nat Immunol 2017; 18(2): 214-24. [http://dx.doi.org/10.1038/ni.3641] [PMID: 27992402]

[181] Jiang QX. Structural variability in the RLR-MAVS pathway and sensitive detection of viral RNAs. Medicinal chemistry (Shariqah (United Arab Emirates)) 2018. [http://dx.doi.org/10.2174/1573406415666181219101613]

[182] Mannion NM, Greenwood SM, Young R, *et al.* The RNA-editing enzyme ADAR1 controls innate immune responses to RNA. Cell Rep 2014; 9(4): 1482-94. [http://dx.doi.org/10.1016/j.celrep.2014.10.041] [PMID: 25456137]

[183] Chung H, Calis JJA, Wu X, *et al.* Human ADAR1 Prevents Endogenous RNA from Triggering Translational Shutdown. Cell 2018; 172(4): 811-824.e14.
[http://dx.doi.org/10.1016/j.cell.2017.12.038] [PMID: 29395325]

[184] Pfaller CK, Donohue RC, Nersisyan S, Brodsky L, Cattaneo R. Extensive editing of cellular and viral double-stranded RNA structures accounts for innate immunity suppression and the proviral activity of ADAR1p150. PLoS Biol 2018; 16(11)e2006577
[http://dx.doi.org/10.1371/journal.pbio.2006577] [PMID: 30496178]

[185] Ahmad S, Mu X, Yang F, *et al.* Breaching Self-Tolerance to Alu Duplex RNA Underlies MDA5-Mediated Inflammation. Cell 2018; 172(4): 797-810.e13.
[http://dx.doi.org/10.1016/j.cell.2017.12.016] [PMID: 29395326]

[186] Gannon HS, Zou T, Kiessling MK, *et al.* Identification of ADAR1 adenosine deaminase dependency in a subset of cancer cells. Nat Commun 2018; 9(1): 5450.
[http://dx.doi.org/10.1038/s41467-018-07824-4] [PMID: 30575730]

[187] Ishizuka JJ, Manguso RT, Cheruiyot CK, *et al.* Loss of ADAR1 in tumours overcomes resistance to immune checkpoint blockade. Nature 2019; 565(7737): 43-8.
[http://dx.doi.org/10.1038/s41586-018-0768-9] [PMID: 30559380]

[188] Alexopoulou L, Holt AC, Medzhitov R, Flavell RA. Recognition of double-stranded RNA and activation of NF-kappaB by Toll-like receptor 3. Nature 2001; 413(6857): 732-8.
[http://dx.doi.org/10.1038/35099560] [PMID: 11607032]

[189] Chen L, Feng Z, Yue H, *et al.* Exosomes derived from HIV-1-infected cells promote growth and progression of cancer *via* HIV TAR RNA. Nat Commun 2018; 9(1): 4585.
[http://dx.doi.org/10.1038/s41467-018-07006-2] [PMID: 30389917]

[190] Tatematsu M, Nishikawa F, Seya T, Matsumoto M. Toll-like receptor 3 recognizes incomplete stem structures in single-stranded viral RNA. Nat Commun 2013; 4: 1833.
[http://dx.doi.org/10.1038/ncomms2857] [PMID: 23673618]

[191] Lee SM, Yip TF, Yan S, *et al.* Recognition of Double-Stranded RNA and Regulation of Interferon Pathway by Toll-Like Receptor 10. Front Immunol 2018; 9: 516.
[http://dx.doi.org/10.3389/fimmu.2018.00516] [PMID: 29616030]

[192] Sato R, Kato A, Chimura T, *et al.* Combating herpesvirus encephalitis by potentiating a TLR3-mTORC2 axis. Nat Immunol 2018; 19(10): 1071-82.
[http://dx.doi.org/10.1038/s41590-018-0203-2] [PMID: 30201994]

[193] Diebold SS, Kaisho T, Hemmi H, Akira S, Reis e Sousa C. Innate antiviral responses by means of TLR7-mediated recognition of single-stranded RNA. Science 2004; 303(5663): 1529-31.
[http://dx.doi.org/10.1126/science.1093616] [PMID: 14976261]

[194] Heil F, Hemmi H, Hochrein H, *et al.* Species-specific recognition of single-stranded RNA *via* toll-like receptor 7 and 8. Science 2004; 303(5663): 1526-9.
[http://dx.doi.org/10.1126/science.1093620] [PMID: 14976262]

[195] Zhang Z, Ohto U, Shibata T, *et al.* Structural Analyses of Toll-like Receptor 7 Reveal Detailed RNA Sequence Specificity and Recognition Mechanism of Agonistic Ligands. Cell Rep 2018; 25(12): 3371-3381.e5.
[http://dx.doi.org/10.1016/j.celrep.2018.11.081] [PMID: 30566863]

[196] Krug RM. Viral proteins that bind double-stranded RNA: countermeasures against host antiviral responses 2014; 34(6): 464-8.
[http://dx.doi.org/10.1089/jir.2014.0005]

[197] Ablasser A, Goldeck M, Cavlar T, *et al.* cGAS produces a 2′-5′-linked cyclic dinucleotide second messenger that activates STING. Nature 2013; 498(7454): 380-4.
[http://dx.doi.org/10.1038/nature12306] [PMID: 23722158]

[198] Sun L, Wu J, Du F, Chen X, Chen ZJ. Cyclic GMP-AMP synthase is a cytosolic DNA sensor that activates the type I interferon pathway. Science 2013; 339(6121): 786-91. [http://dx.doi.org/10.1126/science.1232458] [PMID: 23258413]

[199] Wu J, Sun L, Chen X, *et al.* Cyclic GMP-AMP is an endogenous second messenger in innate immune signaling by cytosolic DNA. Science 2013; 339(6121): 826-30. [http://dx.doi.org/10.1126/science.1229963] [PMID: 23258412]

[200] Zhou W, Whiteley AT, de Oliveira Mann CC, *et al.* Structure of the Human cGAS-DNA Complex Reveals Enhanced Control of Immune Surveillance. Cell 2018; 174(2): 300-311.e11. [http://dx.doi.org/10.1016/j.cell.2018.06.026] [PMID: 30007416]

[201] Li T, Chen ZJ. The cGAS-cGAMP-STING pathway connects DNA damage to inflammation, senescence, and cancer. J Exp Med 2018; 215(5): 1287-99. [http://dx.doi.org/10.1084/jem.20180139] [PMID: 29622565]

[202] Sun B, Sundström KB, Chew JJ, *et al.* Dengue virus activates cGAS through the release of mitochondrial DNA. Sci Rep 2017; 7(1): 3594. [http://dx.doi.org/10.1038/s41598-017-03932-1] [PMID: 28620207]

[203] Ishikawa H, Barber GN. STING is an endoplasmic reticulum adaptor that facilitates innate immune signalling. Nature 2008; 455(7213): 674-8. [http://dx.doi.org/10.1038/nature07317] [PMID: 18724357]

[204] Larkin B, Ilyukha V, Sorokin M, Buzdin A, Vannier E, Poltorak A. Cutting Edge: Activation of STING in T Cells Induces Type I IFN Responses and Cell Death. J Immunol 2017; 199(2): 397-402. [http://dx.doi.org/10.4049/jimmunol.1601999] [PMID: 28615418]

[205] Ogawa E, Mukai K, Saito K, Arai H, Taguchi T. The binding of TBK1 to STING requires exocytic membrane traffic from the ER. Biochem Biophys Res Commun 2018; 503(1): 138-45. [http://dx.doi.org/10.1016/j.bbrc.2018.05.199] [PMID: 29870684]

[206] Wang PH, Fung SY, Gao WW, *et al.* A novel transcript isoform of STING that sequesters cGAMP and dominantly inhibits innate nucleic acid sensing. Nucleic Acids Res 2018; 46(8): 4054-71. [http://dx.doi.org/10.1093/nar/gky186] [PMID: 29547894]

[207] Gentili M, Kowal J, Tkach M, *et al.* Transmission of innate immune signaling by packaging of cGAMP in viral particles. Science 2015; 349(6253): 1232-6. [http://dx.doi.org/10.1126/science.aab3628] [PMID: 26229115]

[208] Moretti J, Roy S, Bozec D, *et al.* STING Senses Microbial viability to Orchestrate Stress-Mediated Autophagy of the Endoplasmic Reticulum. Cell 2017; 171(4): 809-823.e13. [http://dx.doi.org/10.1016/j.cell.2017.09.034] [PMID: 29056340]

[209] Webster SJ, Brode S, Ellis L, *et al.* Detection of a microbial metabolite by STING regulates inflammasome activation in response to Chlamydia trachomatis infection. PLoS Pathog 2017; 13(6)e1006383 [http://dx.doi.org/10.1371/journal.ppat.1006383] [PMID: 28570638]

[210] Ramanjulu JM, Pesiridis GS, Yang J, *et al.* Design of amidobenzimidazole STING receptor agonists with systemic activity. Nature 2018; 564(7736): 439-43. [http://dx.doi.org/10.1038/s41586-018-0705-y] [PMID: 30405246]

[211] Almine JF, O'Hare CA, Dunphy G, *et al.* IFI16 and cGAS cooperate in the activation of STING during DNA sensing in human keratinocytes. Nat Commun 2017; 8: 14392. [http://dx.doi.org/10.1038/ncomms14392] [PMID: 28194029]

[212] Jønsson KL, Laustsen A, Krapp C, *et al.* IFI16 is required for DNA sensing in human macrophages by promoting production and function of cGAMP. Nat Commun 2017; 8: 14391. [http://dx.doi.org/10.1038/ncomms14391] [PMID: 28186168]

[213] Franz KM, Neidermyer WJ, Tan YJ, Whelan SPJ, Kagan JC. STING-dependent translation inhibition

restricts RNA virus replication. Proc Natl Acad Sci USA 2018; 115(9): E2058-67. [http://dx.doi.org/10.1073/pnas.1716937115] [PMID: 29440426]

[214] Stratmann SA, Morrone SR, van Oijen AM, Sohn J. The innate immune sensor IFI16 recognizes foreign DNA in the nucleus by scanning along the duplex. eLife 2015; 4e11721 [http://dx.doi.org/10.7554/eLife.11721] [PMID: 26673078]

[215] Monroe KM, Yang Z, Johnson JR, *et al.* IFI16 DNA sensor is required for death of lymphoid CD4 T cells abortively infected with HIV. Science 2014; 343(6169): 428-32. [http://dx.doi.org/10.1126/science.1243640] [PMID: 24356113]

[216] Merkl PE, Knipe DM. Role for a Filamentous Nuclear Assembly of IFI16, DNA, and Host Factors in Restriction of Herpesviral Infection. MBio 2019; 10(1): e02621-18. [http://dx.doi.org/10.1128/mBio.02621-18] [PMID: 30670617]

[217] Crow MS, Cristea IM. Human antiviral protein IFIX suppresses viral gene expression during HSV-1 infection and is counteracted by virus-induced proteasomal degradation. Molecular & cellular proteomics. MCP 2017; 16(4) (Suppl. 1): S200-s14. [PMID: 28077445]

[218] Wang PH, Ye ZW, Deng JJ, *et al.* Inhibition of AIM2 inflammasome activation by a novel transcript isoform of IFI16. EMBO Rep 2018; 19(10)e45737 [http://dx.doi.org/10.15252/embr.201845737] [PMID: 30104205]

[219] Chiu YH, Macmillan JB, Chen ZJ. RNA polymerase III detects cytosolic DNA and induces type I interferons through the RIG-I pathway. Cell 2009; 138(3): 576-91. [http://dx.doi.org/10.1016/j.cell.2009.06.015] [PMID: 19631370]

[220] Carter-Timofte ME, Hansen AF, Christiansen M, Paludan SR, Mogensen TH. Mutations in RNA Polymerase III genes and defective DNA sensing in adults with varicella-zoster virus CNS infection. Genes Immun 2019; 20(3): 214-23. [http://dx.doi.org/10.1038/s41435-018-0027-y] [PMID: 29728610]

[221] Roth S, Rottach A, Lotz-Havla AS, *et al.* Rad50-CARD9 interactions link cytosolic DNA sensing to IL-1β production. Nat Immunol 2014; 15(6): 538-45. [http://dx.doi.org/10.1038/ni.2888] [PMID: 24777530]

[222] Scutts SR, Ember SW, Ren H, *et al.* DNA-PK Is Targeted by Multiple Vaccinia Virus Proteins to Inhibit DNA Sensing. Cell Rep 2018; 25(7): 1953-1965.e4. [http://dx.doi.org/10.1016/j.celrep.2018.10.034] [PMID: 30428360]

[223] Hemmi H, Takeuchi O, Kawai T, *et al.* A Toll-like receptor recognizes bacterial DNA. Nature 2000; 408(6813): 740-5. [http://dx.doi.org/10.1038/35047123] [PMID: 11130078]

[224] Ingelsson B, Söderberg D, Strid T, *et al.* Lymphocytes eject interferogenic mitochondrial DNA webs in response to CpG and non-CpG oligodeoxynucleotides of class C. Proc Natl Acad Sci USA 2018; 115(3): E478-87. [http://dx.doi.org/10.1073/pnas.1711950115] [PMID: 29295921]

[225] Liu Y, Lin R, Olagnier D. RIGulation of STING expression: at the crossroads of viral RNA and DNA sensing pathways. Inflamm Cell Signal 2017; 4(1)e1491 [PMID: 28191486]

[226] Zevini A, Olagnier D, Hiscott J. Crosstalk between Cytoplasmic RIG-I and STING Sensing Pathways. Trends Immunol 2017; 38(3): 194-205. [http://dx.doi.org/10.1016/j.it.2016.12.004] [PMID: 28073693]

[227] Suspène R, Mussil B, Laude H, *et al.* Self-cytoplasmic DNA upregulates the mutator enzyme APOBEC3A leading to chromosomal DNA damage. Nucleic Acids Res 2017; 45(6): 3231-41. [http://dx.doi.org/10.1093/nar/gkx001] [PMID: 28100701]

[228] Chan YK, Gack MU. Viral evasion of intracellular DNA and RNA sensing. Nat Rev Microbiol 2016;

14(6): 360-73.
[http://dx.doi.org/10.1038/nrmicro.2016.45] [PMID: 27174148]

[229] Christensen MH, Paludan SR. Viral evasion of DNA-stimulated innate immune responses. Cellular & molecular immunology 2017; 14(1): 4-13.

[230] Motta V, Soares F, Sun T, Philpott DJ. NOD-like receptors: versatile cytosolic sentinels. Physiol Rev 2015; 95(1): 149-78.
[http://dx.doi.org/10.1152/physrev.00009.2014] [PMID: 25540141]

[231] Wang D, Sha Y, Hu J, Yang T, Piao X, Zhang X. Genetic signatures of plant resistance genes with known function within and between species. Genetica 2018; 146(6): 517-28.
[http://dx.doi.org/10.1007/s10709-018-0044-9] [PMID: 30315424]

[232] Moffett P. Using Decoys to Detect Pathogens: An Integrated Approach. Trends Plant Sci 2016; 21(5): 369-70.
[http://dx.doi.org/10.1016/j.tplants.2016.04.002] [PMID: 27079492]

[233] Girardin SE, Boneca IG, Carneiro LA, *et al.* Nod1 detects a unique muropeptide from gram-negative bacterial peptidoglycan. Science 2003; 300(5625): 1584-7.
[http://dx.doi.org/10.1126/science.1084677] [PMID: 12791997]

[234] Inohara N, Ogura Y, Fontalba A, *et al.* Host recognition of bacterial muramyl dipeptide mediated through NOD2. Implications for Crohn's disease. J Biol Chem 2003; 278(8): 5509-12.
[http://dx.doi.org/10.1074/jbc.C200673200] [PMID: 12514169]

[235] Li X, Deng M, Petrucelli AS, *et al.* Viral DNA Binding to NLRC3, an Inhibitory Nucleic Acid Sensor, Unleashes STING, a Cyclic Dinucleotide Receptor that Activates Type I Interferon. Immunity 2019; 50(3): 591-599.e6.
[http://dx.doi.org/10.1016/j.immuni.2019.02.009] [PMID: 30893587]

[236] Wu C, Su Z, Lin M, *et al.* NLRP11 attenuates Toll-like receptor signalling by targeting TRAF6 for degradation *via* the ubiquitin ligase RNF19A. Nat Commun 2017; 8(1): 1977.
[http://dx.doi.org/10.1038/s41467-017-02073-3] [PMID: 29215004]

[237] Ellwanger K, Becker E, Kienes I, *et al.* The NLR family pyrin domain-containing 11 protein contributes to the regulation of inflammatory signaling. J Biol Chem 2018; 293(8): 2701-10.
[http://dx.doi.org/10.1074/jbc.RA117.000152] [PMID: 29301940]

[238] Abe T, Lee A, Sitharam R, Kesner J, Rabadan R, Shapira SD. Germ-Cell-Specific Inflammasome Component NLRP14 Negatively Regulates Cytosolic Nucleic Acid Sensing to Promote Fertilization. Immunity 2017; 46(4): 621-34.
[http://dx.doi.org/10.1016/j.immuni.2017.03.020] [PMID: 28423339]

[239] Martínez I, Oliveros JC, Cuesta I, *et al.* Apoptosis, Toll-like, RIG-I-like and NOD-like Receptors Are Pathways Jointly Induced by Diverse Respiratory Bacterial and Viral Pathogens. Front Microbiol 2017; 8: 276.
[http://dx.doi.org/10.3389/fmicb.2017.00276] [PMID: 28298903]

[240] Mukherjee S, Zheng H, Derebe MG, *et al.* Antibacterial membrane attack by a pore-forming intestinal C-type lectin. Nature 2014; 505(7481): 103-7.
[http://dx.doi.org/10.1038/nature12729] [PMID: 24256734]

[241] Okumura R, Kurakawa T, Nakano T, *et al.* Lypd8 promotes the segregation of flagellated microbiota and colonic epithelia. Nature 2016; 532(7597): 117-21.
[http://dx.doi.org/10.1038/nature17406] [PMID: 27027293]

[242] Iversen MB, Reinert LS, Thomsen MK, *et al.* An innate antiviral pathway acting before interferons at epithelial surfaces. Nat Immunol 2016; 17(2): 150-8.
[http://dx.doi.org/10.1038/ni.3319] [PMID: 26595890]

[243] Holm CK, Jensen SB, Jakobsen MR, *et al.* Virus-cell fusion as a trigger of innate immunity dependent on the adaptor STING. Nat Immunol 2012; 13(8): 737-43.

[http://dx.doi.org/10.1038/ni.2350] [PMID: 22706339]

[244] Holm CK, Rahbek SH, Gad HH, *et al.* Influenza A virus targets a cGAS-independent STING pathway that controls enveloped RNA viruses. Nat Commun 2016; 7: 10680. [http://dx.doi.org/10.1038/ncomms10680] [PMID: 26893169]

[245] Tam JC, Bidgood SR, McEwan WA, James LC. Intracellular sensing of complement C3 activates cell autonomous immunity. Science 2014; 345(6201)1256070 [http://dx.doi.org/10.1126/science.1256070] [PMID: 25190799]

[246] Douglas JE, Cohen NA. Taste Receptors Mediate Sinonasal Immunity and Respiratory Disease. Int J Mol Sci 2017; 18(2)E437 [http://dx.doi.org/10.3390/ijms18020437] [PMID: 28218655]

[247] Hariri BM, McMahon DB, Chen B, *et al.* Flavones modulate respiratory epithelial innate immunity: Anti-inflammatory effects and activation of the T2R14 receptor. J Biol Chem 2017; 292(20): 8484-97. [http://dx.doi.org/10.1074/jbc.M116.771949] [PMID: 28373278]

[248] Yan CH, Hahn S, McMahon D, Bonislawski D, Kennedy DW, Adappa ND, *et al.* Nitric oxide production is stimulated by bitter taste receptors ubiquitously expressed in the sinonasal cavity. American journal of rhinology & allergy 2017; 31(2): 85-92. [http://dx.doi.org/10.2500/ajra.2017.31.4424]

[249] Melo JA, Ruvkun G. Inactivation of conserved C. elegans genes engages pathogen- and xenobiotic-associated defenses. Cell 2012; 149(2): 452-66. [http://dx.doi.org/10.1016/j.cell.2012.02.050] [PMID: 22500807]

[250] Danilova N, Wilkes M, Bibikova E, Youn MY, Sakamoto KM, Lin S. Innate immune system activation in zebrafish and cellular models of Diamond Blackfan Anemia. Sci Rep 2018; 8(1): 5165. [http://dx.doi.org/10.1038/s41598-018-23561-6] [PMID: 29581525]

[251] Ahlers LR, Goodman AG. Nucleic acid sensing and innate immunity: signaling pathways controlling viral pathogenesis and autoimmunity. Curr Clin Microbiol Rep 2016; 3(3): 132-41. [http://dx.doi.org/10.1007/s40588-016-0043-5] [PMID: 27857881]

[252] Karikó K, Ni H, Capodici J, Lamphier M, Weissman D. mRNA is an endogenous ligand for Toll-like receptor 3. J Biol Chem 2004; 279(13): 12542-50. [http://dx.doi.org/10.1074/jbc.M310175200] [PMID: 14729660]

[253] Okabe Y, Kawane K, Nagata S. IFN regulatory factor (IRF) 3/7-dependent and -independent gene induction by mammalian DNA that escapes degradation. Eur J Immunol 2008; 38(11): 3150-8. [http://dx.doi.org/10.1002/eji.200838559] [PMID: 18991290]

[254] Chiappinelli KB, Strissel PL, Desrichard A, *et al.* Inhibiting DNA Methylation Causes an Interferon Response in Cancer *via* dsRNA Including Endogenous Retroviruses. Cell 2015; 162(5): 974-86. [http://dx.doi.org/10.1016/j.cell.2015.07.011] [PMID: 26317466]

[255] Chemudupati M, Kenney AD, Bonifati S, *et al.* From APOBEC to ZAP: Diverse mechanisms used by cellular restriction factors to inhibit virus infections. Biochim Biophys Acta Mol Cell Res 2019; 1866(3): 382-94. [http://dx.doi.org/10.1016/j.bbamcr.2018.09.012] [PMID: 30290238]

[256] Sheehy AM, Gaddis NC, Choi JD, Malim MH. Isolation of a human gene that inhibits HIV-1 infection and is suppressed by the viral Vif protein. Nature 2002; 418(6898): 646-50. [http://dx.doi.org/10.1038/nature00939] [PMID: 12167863]

[257] Yu X, Yu Y, Liu B, *et al.* Induction of APOBEC3G ubiquitination and degradation by an HIV-1 Vif-Cul5-SCF complex. Science 2003; 302(5647): 1056-60. [http://dx.doi.org/10.1126/science.1089591] [PMID: 14564014]

[258] Shi K, Carpenter MA, Banerjee S, *et al.* Structural basis for targeted DNA cytosine deamination and mutagenesis by APOBEC3A and APOBEC3B. Nat Struct Mol Biol 2017; 24(2): 131-9. [http://dx.doi.org/10.1038/nsmb.3344] [PMID: 27991903]

[259] Esnault C, Heidmann O, Delebecque F, *et al.* APOBEC3G cytidine deaminase inhibits retrotransposition of endogenous retroviruses. Nature 2005; 433(7024): 430-3. [http://dx.doi.org/10.1038/nature03238] [PMID: 15674295]

[260] Richardson SR, Narvaiza I, Planegger RA, Weitzman MD, Moran JV. APOBEC3A deaminates transiently exposed single-strand DNA during LINE-1 retrotransposition. eLife 2014; 3e02008 [http://dx.doi.org/10.7554/eLife.02008] [PMID: 24843014]

[261] Renner TM, Bélanger K, Goodwin LR, Campbell M, Langlois MA. Characterization of molecular attributes that influence LINE-1 restriction by all seven human APOBEC3 proteins. Virology 2018; 520: 127-36. [http://dx.doi.org/10.1016/j.virol.2018.05.015] [PMID: 29860216]

[262] Sheehy AM, Gaddis NC, Malim MH. The antiretroviral enzyme APOBEC3G is degraded by the proteasome in response to HIV-1 Vif. Nat Med 2003; 9(11): 1404-7. [http://dx.doi.org/10.1038/nm945] [PMID: 14528300]

[263] Jäger S, Kim DY, Hultquist JF, *et al.* Vif hijacks CBF-β to degrade APOBEC3G and promote HIV-1 infection. Nature 2011; 481(7381): 371-5. [http://dx.doi.org/10.1038/nature10693] [PMID: 22190037]

[264] Binning JM, Smith AM, Hultquist JF, *et al.* Fab-based inhibitors reveal ubiquitin independent functions for HIV Vif neutralization of APOBEC3 restriction factors. PLoS Pathog 2018; 14(1)e1006830 [http://dx.doi.org/10.1371/journal.ppat.1006830] [PMID: 29304101]

[265] Compton AA, Emerman M. Convergence and divergence in the evolution of the APOBEC3G-Vif interaction reveal ancient origins of simian immunodeficiency viruses. PLoS Pathog 2013; 9(1)e1003135 [http://dx.doi.org/10.1371/journal.ppat.1003135] [PMID: 23359341]

[266] Mertz TM, Harcy V, Roberts SA. Risks at the DNA Replication Fork: Effects upon Carcinogenesis and Tumor Heterogeneity. Genes (Basel) 2017; 8(1)E46 [http://dx.doi.org/10.3390/genes8010046] [PMID: 28117753]

[267] Burns MB, Lackey L, Carpenter MA, *et al.* APOBEC3B is an enzymatic source of mutation in breast cancer. Nature 2013; 494(7437): 366-70. [http://dx.doi.org/10.1038/nature11881] [PMID: 23389445]

[268] Middlebrooks CD, Banday AR, Matsuda K, *et al.* Association of germline variants in the APOBEC3 region with cancer risk and enrichment with APOBEC-signature mutations in tumors. Nat Genet 2016; 48(11): 1330-8. [http://dx.doi.org/10.1038/ng.3670] [PMID: 27643540]

[269] Chou WC, Chen WT, Hsiung CN, *et al.* B-Myb Induces APOBEC3B Expression Leading to Somatic Mutation in Multiple Cancers. Sci Rep 2017; 7: 44089. [http://dx.doi.org/10.1038/srep44089] [PMID: 28276478]

[270] Dong M, Kathiresan V, Fenwick MK, *et al.* Organometallic and radical intermediates reveal mechanism of diphthamide biosynthesis. Science 2018; 359(6381): 1247-50. [http://dx.doi.org/10.1126/science.aao6595] [PMID: 29590073]

[271] Honarmand Ebrahimi K, Carr SB, McCullagh J, *et al.* The radical-SAM enzyme Viperin catalyzes reductive addition of a 5′-deoxyadenosyl radical to UDP-glucose *in vitro*. FEBS Lett 2017; 591(16): 2394-405. [http://dx.doi.org/10.1002/1873-3468.12769] [PMID: 28752893]

[272] Chakravarti A, Selvadurai K, Shahoei R, *et al.* Reconstitution and substrate specificity for isopentenyl pyrophosphate of the antiviral radical SAM enzyme viperin. J Biol Chem 2018; 293(36): 14122-33. [http://dx.doi.org/10.1074/jbc.RA118.003998] [PMID: 30030381]

[273] Gizzi AS, Grove TL, Arnold JJ, *et al.* A naturally occurring antiviral ribonucleotide encoded by the

human genome. Nature 2018; 558(7711): 610-4. [http://dx.doi.org/10.1038/s41586-018-0238-4] [PMID: 29925952]

[274] Goldstone DC, Ennis-Adeniran V, Hedden JJ, *et al.* HIV-1 restriction factor SAMHD1 is a deoxynucleoside triphosphate triphosphohydrolase. Nature 2011; 480(7377): 379-82. [http://dx.doi.org/10.1038/nature10623] [PMID: 22056990]

[275] Colomer-Lluch M, Ruiz A, Moris A, Prado JG. Restriction Factors: From Intrinsic Viral Restriction to Shaping Cellular Immunity Against HIV-1. Front Immunol 2018; 9: 2876. [http://dx.doi.org/10.3389/fimmu.2018.02876] [PMID: 30574147]

[276] Rabbani MAG, Barik S. 5-Hydroxytryptophan, a major product of tryptophan degradation, is essential for optimal replication of human parainfluenza virus. Virology 2017; 503: 46-51. [http://dx.doi.org/10.1016/j.virol.2017.01.007] [PMID: 28113063]

[277] Haller O, Staeheli P, Schwemmle M, Kochs G. Mx GTPases: dynamin-like antiviral machines of innate immunity. Trends Microbiol 2015; 23(3): 154-63. [http://dx.doi.org/10.1016/j.tim.2014.12.003] [PMID: 25572883]

[278] Coers J, Brown HM, Hwang S, Taylor GA. Partners in anti crime: how interferon-inducible GTPases and autophagy proteins team up in cell-intrinsic host defense. Curr Opin Immunol 2018; 54: 93-101. [http://dx.doi.org/10.1016/j.coi.2018.06.008] [PMID: 29986303]

[279] Graf L, Dick A, Sendker F, *et al.* Effects of allelic variations in the human myxovirus resistance protein A on its antiviral activity. J Biol Chem 2018; 293(9): 3056-72. [http://dx.doi.org/10.1074/jbc.M117.812784] [PMID: 29330299]

[280] Crameri M, Bauer M, Caduff N, *et al.* MxB is an interferon-induced restriction factor of human herpesviruses. Nat Commun 2018; 9(1): 1980. [http://dx.doi.org/10.1038/s41467-018-04379-2] [PMID: 29773792]

[281] Staeheli P, Haller O. Human MX2/MxB: a Potent Interferon-Induced Postentry Inhibitor of Herpesviruses and HIV-1. J Virol 2018; 92(24)e00709-18 [http://dx.doi.org/10.1128/JVI.00709-18] [PMID: 30258007]

[282] Abbas YM, Pichlmair A, Górna MW, Superti-Furga G, Nagar B. Structural basis for viral 5′-PP--RNA recognition by human IFIT proteins. Nature 2013; 494(7435): 60-4. [http://dx.doi.org/10.1038/nature11783] [PMID: 23334420]

[283] Johnson B, VanBlargan LA, Xu W, *et al.* Human IFIT3 Modulates IFIT1 RNA Binding Specificity and Protein Stability. Immunity 2018; 48(3): 487-499.e5. [http://dx.doi.org/10.1016/j.immuni.2018.01.014] [PMID: 29525521]

[284] Yount JS, Moltedo B, Yang YY, *et al.* Palmitoylome profiling reveals S-palmitoylation-dependent antiviral activity of IFITM3. Nat Chem Biol 2010; 6(8): 610-4. [http://dx.doi.org/10.1038/nchembio.405] [PMID: 20601941]

[285] Smith SE, Busse DC, Binter S, *et al.* Interferon-Induced Transmembrane Protein 1 Restricts Replication of Viruses That Enter Cells *via* the Plasma Membrane. J Virol 2019; 93(6)e02003-18 [http://dx.doi.org/10.1128/JVI.02003-18] [PMID: 30567988]

[286] Spence JS, He R, Hoffmann HH, *et al.* IFITM3 directly engages and shuttles incoming virus particles to lysosomes. Nat Chem Biol 2019; 15(3): 259-68. [http://dx.doi.org/10.1038/s41589-018-0213-2] [PMID: 30643282]

[287] Compton AA, Roy N, Porrot F, *et al.* Natural mutations in IFITM3 modulate post-translational regulation and toggle antiviral specificity. EMBO Rep 2016; 17(11): 1657-71. [http://dx.doi.org/10.15252/embr.201642771] [PMID: 27601221]

[288] Sun X, Zeng H, Kumar A, Belser JA, Maines TR, Tumpey TM. Constitutively Expressed IFITM3 Protein in Human Endothelial Cells Poses an Early Infection Block to Human Influenza Viruses. J Virol 2016; 90(24): 11157-67. [http://dx.doi.org/10.1128/JVI.01254-16] [PMID: 27707929]

[289] Doyle T, Moncorgé O, Bonaventure B, *et al.* The interferon-inducible isoform of NCOA7 inhibits endosome-mediated viral entry. Nat Microbiol 2018; 3(12): 1369-76. [http://dx.doi.org/10.1038/s41564-018-0273-9] [PMID: 30478388]

[290] Quinn CM, Wang M, Fritz MP, *et al.* Dynamic regulation of HIV-1 capsid interaction with the restriction factor TRIM5α identified by magic-angle spinning NMR and molecular dynamics simulations. Proc Natl Acad Sci USA 2018; 115(45): 11519-24. [http://dx.doi.org/10.1073/pnas.1800796115] [PMID: 30333189]

[291] Wagner JM, Christensen DE, Bhattacharya A, *et al.* General Model for Retroviral Capsid Pattern Recognition by TRIM5 Proteins. J Virol 2018; 92(4): e01563-17. [PMID: 29187540]

[292] Fletcher AJ, Vaysburd M, Maslen S, *et al.* Trivalent RING Assembly on Retroviral Capsids Activates TRIM5 Ubiquitination and Innate Immune Signaling. Cell Host Microbe 2018; 24(6): 761-775.e6. [http://dx.doi.org/10.1016/j.chom.2018.10.007] [PMID: 30503508]

[293] Gao G, Guo X, Goff SP. Inhibition of retroviral RNA production by ZAP, a CCCH-type zinc finger protein. Science 2002; 297(5587): 1703-6. [http://dx.doi.org/10.1126/science.1074276] [PMID: 12215647]

[294] Li MM, Lau Z, Cheung P, *et al.* TRIM25 Enhances the Antiviral Action of Zinc-Finger Antiviral Protein (ZAP). PLoS Pathog 2017; 13(1)e1006145 [http://dx.doi.org/10.1371/journal.ppat.1006145] [PMID: 28060952]

[295] Goodier JL, Pereira GC, Cheung LE, Rose RJ, Kazazian HH Jr. The Broad-Spectrum Antiviral Protein ZAP Restricts Human Retrotransposition. PLoS Genet 2015; 11(5)e1005252 [http://dx.doi.org/10.1371/journal.pgen.1005252] [PMID: 26001115]

[296] Takata MA, Gonçalves-Carneiro D, Zang TM, *et al.* CG dinucleotide suppression enables antiviral defence targeting non-self RNA. Nature 2017; 550(7674): 124-7. [http://dx.doi.org/10.1038/nature24039] [PMID: 28953888]

[297] Neil SJ, Zang T, Bieniasz PD. Tetherin inhibits retrovirus release and is antagonized by HIV-1 Vpu. Nature 2008; 451(7177): 425-30. [http://dx.doi.org/10.1038/nature06553] [PMID: 18200009]

[298] Waheed AA, Gitzen A, Swiderski M, Freed EO. High-Mannose But Not Complex-Type Glycosylation of Tetherin Is Required for Restriction of HIV-1 Release. Viruses 2018; 10(1)E26 [http://dx.doi.org/10.3390/v10010026] [PMID: 29303997]

[299] Laliberte JP, Moss B. A novel mode of poxvirus superinfection exclusion that prevents fusion of the lipid bilayers of viral and cellular membranes. J Virol 2014; 88(17): 9751-68. [http://dx.doi.org/10.1128/JVI.00816-14] [PMID: 24920806]

[300] Smithey MJ, Venturi V, Davenport MP, *et al.* Lifelong CMV infection improves immune defense in old mice by broadening the mobilized TCR repertoire against third-party infection. Proc Natl Acad Sci USA 2018; 115(29): E6817-25. [http://dx.doi.org/10.1073/pnas.1719451115] [PMID: 29967140]

[301] Barton ES, White DW, Cathelyn JS, *et al.* Herpesvirus latency confers symbiotic protection from bacterial infection. Nature 2007; 447(7142): 326-9. [http://dx.doi.org/10.1038/nature05762] [PMID: 17507983]

[302] Kim BH, Shenoy AR, Kumar P, Das R, Tiwari S, MacMicking JD. A family of IFN-γ-inducible 65-kD GTPases protects against bacterial infection. Science 2011; 332(6030): 717-21. [http://dx.doi.org/10.1126/science.1201711] [PMID: 21551061]

[303] Pilla-Moffett D, Barber MF, Taylor GA, Coers J. Interferon-Inducible GTPases in Host Resistance, Inflammation and Disease. J Mol Biol 2016; 428(17): 3495-513. [http://dx.doi.org/10.1016/j.jmb.2016.04.032] [PMID: 27181197]

[304] Santos JC, Broz P. Sensing of invading pathogens by GBPs: At the crossroads between cell-autonomous and innate immunity. J Leukoc Biol 2018; 104(4): 729-35. [http://dx.doi.org/10.1002/JLB.4MR0118-038R] [PMID: 30020539]

[305] Tretina K, Park ES, Maminska A, MacMicking JD. Interferon-induced guanylate-binding proteins: Guardians of host defense in health and disease. J Exp Med 2019; 216(3): 482-500. [http://dx.doi.org/10.1084/jem.20182031] [PMID: 30755454]

[306] Feeley EM, Pilla-Moffett DM, Zwack EE, *et al.* Galectin-3 directs antimicrobial guanylate binding proteins to vacuoles furnished with bacterial secretion systems. Proc Natl Acad Sci USA 2017; 114(9): E1698-706. [http://dx.doi.org/10.1073/pnas.1615771114] [PMID: 28193861]

[307] Ngo CC, Man SM. Mechanisms and functions of guanylate-binding proteins and related interferon-inducible GTPases: Roles in intracellular lysis of pathogens. Cellular microbiology 2017; 19(12): cmi.12791.

[308] Karlinsey JE, Maguire ME, Becker LA, Crouch ML, Fang FC. The phage shock protein PspA facilitates divalent metal transport and is required for virulence of Salmonella enterica sv. Typhimurium. Mol Microbiol 2010; 78(3): 669-85. [http://dx.doi.org/10.1111/j.1365-2958.2010.07357.x] [PMID: 20807201]

[309] Dittmann M, Hoffmann HH, Scull MA, *et al.* A serpin shapes the extracellular environment to prevent influenza A virus maturation. Cell 2015; 160(4): 631-43. [http://dx.doi.org/10.1016/j.cell.2015.01.040] [PMID: 25679759]

[310] OhAinle M, Helms L, Vermeire J, *et al.* A virus-packageable CRISPR screen identifies host factors mediating interferon inhibition of HIV. eLife 2018; 7e39823 [http://dx.doi.org/10.7554/eLife.39823] [PMID: 30520725]

[311] Kadota Y, Shirasu K, Zipfel C. Regulation of the NADPH Oxidase RBOHD During Plant Immunity Plant & cell physiology 2015; 56(8): 1472-80. [http://dx.doi.org/10.1093/pcp/pcv063]

[312] Gaudet RG, Bradfield CJ, MacMicking JD. Evolution of Cell-Autonomous Effector Mechanisms in Macrophages *versus* Non-Immune Cells. Microbiology spectrum 2016; 4(6): MCHD-0050. [http://dx.doi.org/10.1128/microbiolspec.MCHD-0050-2016]

[313] Panday A, Sahoo MK, Osorio D, Batra S. NADPH oxidases: an overview from structure to innate immunity-associated pathologies. Cellular & molecular immunology 2015; 12(1): 5-23. [http://dx.doi.org/10.1038/cmi.2014.89]

[314] den Hartog G, Chattopadhyay R, Ablack A, *et al.* Regulation of Rac1 and Reactive Oxygen Species Production in Response to Infection of Gastrointestinal Epithelia. PLoS Pathog 2016; 12(1)e1005382 [http://dx.doi.org/10.1371/journal.ppat.1005382] [PMID: 26761793]

[315] Gobert AP, Wilson KT. Polyamine- and NADPH-dependent generation of ROS during Helicobacter pylori infection: A blessing in disguise. Free radical biology & medicine 2017; 105: 16-27.

[316] Strengert M, Jennings R, Davanture S, Hayes P, Gabriel G, Knaus UG. Mucosal reactive oxygen species are required for antiviral response: role of Duox in influenza a virus infection. Antioxid Redox Signal 2014; 20(17): 2695-709. [http://dx.doi.org/10.1089/ars.2013.5353] [PMID: 24128054]

[317] Alvarez LA, Kovačič L, Rodríguez J, *et al.* NADPH oxidase-derived H2O2 subverts pathogen signaling by oxidative phosphotyrosine conversion to PB-DOPA. Proc Natl Acad Sci USA 2016; 113(37): 10406-11. [http://dx.doi.org/10.1073/pnas.1605443113] [PMID: 27562167]

[318] Johnston DG, Corr SC. Toll-Like Receptor Signalling and the Control of Intestinal Barrier Function. Methods Mol Biol 2016; 1390: 287-300. [http://dx.doi.org/10.1007/978-1-4939-3335-8_18] [PMID: 26803636]

[319] Buhimschi IA, Jabr M, Buhimschi CS, Petkova AP, Weiner CP, Saed GM. The novel antimicrobial peptide beta3-defensin is produced by the amnion: a possible role of the fetal membranes in innate immunity of the amniotic cavity. Am J Obstet Gynecol 2004; 191(5): 1678-87. [http://dx.doi.org/10.1016/j.ajog.2004.03.081] [PMID: 15547542]

[320] Kai-Larsen Y, Gudmundsson GH, Agerberth B. A review of the innate immune defence of the human foetus and newborn, with the emphasis on antimicrobial peptides. Acta paediatrica (Oslo, Norway : 1992) 2014; 103(10): 1000-8. [http://dx.doi.org/10.1111/apa.12700]

[321] Jorgensen I, Rayamajhi M, Miao EA. Programmed cell death as a defence against infection. Nat Rev Immunol 2017; 17(3): 151-64. [http://dx.doi.org/10.1038/nri.2016.147] [PMID: 28138137]

[322] Nagata S. Apoptosis and Clearance of Apoptotic Cells. Annu Rev Immunol 2018; 36: 489-517. [http://dx.doi.org/10.1146/annurev-immunol-042617-053010] [PMID: 29400998]

[323] Kvansakul M, Caria S, Hinds MG. The Bcl-2 Family in Host-Virus Interactions. Viruses 2017; 9(10)E290 [http://dx.doi.org/10.3390/v9100290] [PMID: 28984827]

[324] Zhou X, Jiang W, Liu Z, Liu S, Liang X. Virus Infection and Death Receptor-Mediated Apoptosis. Viruses 2017; 9(11)E316 [http://dx.doi.org/10.3390/v9110316] [PMID: 29077026]

[325] Saavedra PHV, Huang L, Ghazavi F, *et al.* Apoptosis of intestinal epithelial cells restricts Clostridium difficile infection in a model of pseudomembranous colitis. Nat Commun 2018; 9(1): 4846. [http://dx.doi.org/10.1038/s41467-018-07386-5] [PMID: 30451870]

[326] Pollock GL, Oates CVL, Giogha C, *et al.* Distinct roles of the anti-apoptotic effectors NleB and NleF from enteropathogenic *Escherichia coli*. Infect Immun 2017; 85(4): e01071-16. [http://dx.doi.org/10.1128/IAI.01071-16] [PMID: 28138023]

[327] Holler N, Zaru R, Micheau O, *et al.* Fas triggers an alternative, caspase-8-independent cell death pathway using the kinase RIP as effector molecule. Nat Immunol 2000; 1(6): 489-95. [http://dx.doi.org/10.1038/82732] [PMID: 11101870]

[328] Grootjans S, Vanden Berghe T, Vandenabeele P. Initiation and execution mechanisms of necroptosis: an overview. Cell Death Differ 2017; 24(7): 1184-95. [http://dx.doi.org/10.1038/cdd.2017.65] [PMID: 28498367]

[329] Weinlich R, Oberst A, Beere HM, Green DR. Necroptosis in development, inflammation and disease. Nat Rev Mol Cell Biol 2017; 18(2): 127-36. [http://dx.doi.org/10.1038/nrm.2016.149] [PMID: 27999438]

[330] Nailwal H, Chan FK. Necroptosis in anti-viral inflammation. Cell Death Differ 2019; 26(1): 4-13. [http://dx.doi.org/10.1038/s41418-018-0172-x] [PMID: 30050058]

[331] Vanden Berghe T, Hassannia B, Vandenabeele P. An outline of necrosome triggers. Cell Mol Life Sci 2016; 73(11-12): 2137-52. [http://dx.doi.org/10.1007/s00018-016-2189-y] [PMID: 27052312]

[332] Kaiser WJ, Upton JW, Long AB, *et al.* RIP3 mediates the embryonic lethality of caspase-8-deficient mice. Nature 2011; 471(7338): 368-72. [http://dx.doi.org/10.1038/nature09857] [PMID: 21368762]

[333] Johnston A, Wang Z. Necroptosis: MLKL Polymerization. J Nat Sci 2018; 4(7)e513 [PMID: 30294675]

[334] Günther C, He GW, Kremer AE, *et al.* The pseudokinase MLKL mediates programmed hepatocellular necrosis independently of RIPK3 during hepatitis. J Clin Invest 2016; 126(11): 4346-60. [http://dx.doi.org/10.1172/JCI87545] [PMID: 27756058]

[335] Pearson JS, Giogha C, Mühlen S, *et al.* EspL is a bacterial cysteine protease effector that cleaves RHIM proteins to block necroptosis and inflammation. Nat Microbiol 2017; 2: 16258. [http://dx.doi.org/10.1038/nmicrobiol.2016.258] [PMID: 28085133]

[336] Shi J, Zhao Y, Wang K, *et al.* Cleavage of GSDMD by inflammatory caspases determines pyroptotic cell death. Nature 2015; 526(7575): 660-5. [http://dx.doi.org/10.1038/nature15514] [PMID: 26375003]

[337] Man SM, Karki R, Kanneganti TD. Molecular mechanisms and functions of pyroptosis, inflammatory caspases and inflammasomes in infectious diseases. Immunol Rev 2017; 277(1): 61-75. [http://dx.doi.org/10.1111/imr.12534] [PMID: 28462526]

[338] Frank D, Vince JE. Pyroptosis *versus* necroptosis: similarities, differences, and crosstalk. Cell Death Differ 2019; 26(1): 99-114. [http://dx.doi.org/10.1038/s41418-018-0212-6] [PMID: 30341423]

[339] Ding J, Wang K, Liu W, *et al.* Pore-forming activity and structural autoinhibition of the gasdermin family. Nature 2016; 535(7610): 111-6. [http://dx.doi.org/10.1038/nature18590] [PMID: 27281216]

[340] Liu X, Zhang Z, Ruan J, *et al.* Inflammasome-activated gasdermin D causes pyroptosis by forming membrane pores. Nature 2016; 535(7610): 153-8. [http://dx.doi.org/10.1038/nature18629] [PMID: 27383986]

[341] Rogers C, Erkes DA, Nardone A, Aplin AE, Fernandes-Alnemri T, Alnemri ES. Gasdermin pores permeabilize mitochondria to augment caspase-3 activation during apoptosis and inflammasome activation. Nat Commun 2019; 10(1): 1689. [http://dx.doi.org/10.1038/s41467-019-09397-2] [PMID: 30976076]

[342] Brinkmann V, Reichard U, Goosmann C, *et al.* Neutrophil extracellular traps kill bacteria. Science 2004; 303(5663): 1532-5. [http://dx.doi.org/10.1126/science.1092385] [PMID: 15001782]

[343] de Bont CM, Koopman WJH, Boelens WC, Pruijn GJM. Stimulus-dependent chromatin dynamics, citrullination, calcium signalling and ROS production during NET formation. Biochim Biophys Acta Mol Cell Res 2018; 1865 (11 Pt A): 1621-9. [http://dx.doi.org/10.1016/j.bbamcr.2018.08.014] [PMID: 30327203]

[344] Neubert E, Meyer D, Rocca F, *et al.* Chromatin swelling drives neutrophil extracellular trap release. Nat Commun 2018; 9(1): 3767. [http://dx.doi.org/10.1038/s41467-018-06263-5] [PMID: 30218080]

[345] Inoue M, Nakashima R, Enomoto M, *et al.* Plasma redox imbalance caused by albumin oxidation promotes lung-predominant NETosis and pulmonary cancer metastasis. Nat Commun 2018; 9(1): 5116. [http://dx.doi.org/10.1038/s41467-018-07550-x] [PMID: 30504805]

[346] Rizzi M, Carniato F, Tonello S, *et al.* Charged molecular silica trigger *in vitro* NETosis in human granulocytes *via* both oxidative and autophagic pathways. Eur Rev Med Pharmacol Sci 2018; 22(20): 7058-68. [PMID: 30402874]

[347] Yousefi S, Stojkov D, Germic N, *et al.* Untangling "NETosis" from NETs. Eur J Immunol 2019; 49(2): 221-7. [http://dx.doi.org/10.1002/cji.201747053] [PMID: 30629284]

[348] Winstel V, Missiakas D, Schneewind O. *Staphylococcus aureus* targets the purine salvage pathway to kill phagocytes. Proc Natl Acad Sci USA 2018; 115(26): 6846-51. [http://dx.doi.org/10.1073/pnas.1805622115] [PMID: 29891696]

[349] Tran TM, MacIntyre A, Hawes M, Allen C. Escaping Underground Nets: Extracellular DNases Degrade Plant Extracellular Traps and Contribute to Virulence of the Plant Pathogenic Bacterium

Ralstonia solanacearum. PLoS Pathog 2016; 12(6)e1005686
[http://dx.doi.org/10.1371/journal.ppat.1005686] [PMID: 27336156]

[350] Sørensen OE, Borregaard N. Neutrophil extracellular traps - the dark side of neutrophils. J Clin Invest 2016; 126(5): 1612-20.
[http://dx.doi.org/10.1172/JCI84538] [PMID: 27135878]

CHAPTER 10

Stress Response Meets Autonomous Immunity

Abstract: Stress response contributes to autonomous immune responses. Many stresses including infection induce integrated stress response (ISR). ISR is mediated by a set of kinases PKR, PERK, HRI, and GCN2. Viral dsRNA activates PKR. Viral proteins are produced in the endoplasmic reticulum and may cause ER stress, which activates PERK. Infection also can cause iron deficiency sensed by HRI kinase and amino acid shortage sensed by GCN2. The arrest of translation caused by ISR inhibits viral replication and activates NFkB, a major regulator of immunity. Therefore, ISR acts as an immunodefense mechanism. Other stress responses that contribute to immune defense include unfolded protein response (UPR). UPR activates degradation of ER-associated mRNAs and proteins including viral ones. DNA damage response leads to NFkB activation, and so does oxidative stress through various mechanisms. The major stress response factor p53 has anti-viral activity. Autophagy activated by many stresses also plays a role in immunodefense by degrading intracellular pathogens.

Keywords: DNA damage response, ERAD, Homeostasis, Immune response, Integrated stress response, NLRP3 inflammasome, Nrf2, Oxidative stress, RIDD, Unfolded protein response, Xenophagy.

INTRODUCTION

Individual cells of multicellular organisms are equipped with both stress response mechanisms and autonomous immune mechanisms. Most cells express pattern-recognition receptors (PRRs) that recognize pathogen-associated molecular patterns (PAMPs) [1]. Immunity based on PRRs provides direct information about the presence of a pathogen. An alternative way to sense infection is by detecting the modification of cellular molecules or functions caused by a pathogen including changes in homeostasis [2 - 4]. This type of immunity includes effector immunity and surveillance immunity. It allows cells to sense pathogens for which there are no recognizable PAMPs as well as pathogens that produce factors blocking PRRs.

Infection is facilitated when cells experience external and internal stress, such as starvation, cold, irradiation, and others. In turn, disruption of homeostasis caused by pathogens often induces a stress response. For example, hepatitis C virus induces oxidative stress and endoplasmic reticulum (ER) stress [5]. When cells

Nadia Danilova

face both stress and infection, they need to develop an integrated response through the communication of stress response mechanisms with autonomous immune mechanisms. Here, we discuss how these mechanisms interact.

10.1. Homeostatic Perturbations and Immune Response

In simple animals such as the worm *Caenorhabditis elegans*, immune response overlaps with stress response. Mitochondrial stress in *C. elegans* leads to the secretion of lysozymes and antibacterial peptides [6]. Mitochondrial surveillance mechanisms, unfolded protein response (UPRmt), mitophagy, and ROS production contribute to *C. elegans* immune resistance [7]. *C. elegans* lacking the regulator of mitochondrial unfolded protein response ATFS-1 are susceptible to the pathogen *Pseudomonas aeruginosa,* whereas upregulation of this transcriptional factor improves immune resistance. Chaperone HSP-60, which is induced by mitochondrial stress, potentiates antibacterial immunity through stimulation of p38 MAP kinase pathway [8]. Therefore, surveillance of homeostasis in *C. elegans* is also responsible in turn, infection in *C. elegans* upregulates the phylogenetically conserved Ethanol and Stress Response Element (ESRE) network, which responds to a variety of abiotic stresses [9]. This underscores the close evolutionary connection between stress and immune response [4].

Induction of immune response as a reaction to perturbation of homeostasis (surveillance immunity) may be an ancient mechanism that in complex multicellular organisms became masked and overridden by specialized immune mechanisms [10].

Pathogens can disrupt homeostasis by factors that subvert metabolism to benefit pathogen propagation, by factors that inhibit immune responses, and by competing for cellular resources and depleting the nutrients. Increased reactive oxygen species (ROS), mitochondrial stress, an overload of ER with viral proteins, cell stress due to enhanced metabolic demand, and DNA damage often result from infection. These changes in homeostasis induce a stress response and can directly trigger inflammation in the absence of PAMPs or DAMPs [11 - 13].

NLRP3 inflammasome is activated by changes in K+ and Cl- levels [14, 15]. In this way, NLRP3 inflammasome can indirectly sense RNA viruses that induce lytic cell death and bacteria producing pore-forming toxins. In both cases, it leads to K+ efflux and inflammasome activation [16, 17].

High salt stress also leads to inflammasome activation. Macrophages monitor tissue osmolarity and activate NLRP3 and NLRC4 inflammasomes in response to high salt, which leads to the inflammatory response [18]. Decrease in salt consumption in humans leads to a lower level of pro-inflammatory cytokines and increased level of anti-inflammatory IL-10 [19].

High glucose and other metabolic disturbances also activate inflammasomes [20, 21]. Therefore, inflammasomes convert various stress signals into stimulation of immune response.

In contrast, starvation stress can suppress inflammasome activation. Starvation is associated with production of the ketone metabolite β-hydroxybutyrate, which inhibits inflammasome by blocking K^+ efflux [22]. Low energy level also can suppress inflammation through AMPK, an energy sensor [23]. AMPK inhibits NLRP3 inflammasome activation and NFκB induction [24, 25].

10.2. Integrated Stress Response and Immune Response

Many stresses induce the integrated stress response (ISR) leading to global inhibition of protein translation [26]. Infections also can induce ISR. Viral dsRNA directly activates ISR kinase PKR. Viral proteins are produced and folded in the ER and can overload the system, leading to endoplasmic reticulum stress and activation of another ISR kinase – PERK [5, 27]. Other ISR kinases also can sense alterations introduced by pathogens in the infected host. HRI kinase senses iron deficiency caused by *P. aeruginosa's* infection in airway epithelial cells [28]. GCN2 kinase senses amino acid shortage caused by bacterial infection [29].

During ISR, regular translation is suppressed and transcripts involved in stress response are mostly translated from the alternative start codons upstream from the conventional start codon AUG [30]. 5-UTRs of many mRNAs also contain upstream open reading frames (uORFs) that encode small peptides. These peptides may control translation of the main ORF and serve as MHC class I ligands that are recognized by the adaptive immune system [31]. The choice between conventional and alternative start codons is controlled by helicases and mRNA structure and is influenced by stress [32].

Inhibition of global translation associated with ISR response is sufficient to directly activate NFκB. It reduces the production of inhibitory IκBs and disrupts the IκB/NFκB equilibrium, which leads to an increased level of free NFκB proteins [33].

ISR inhibition of protein synthesis creates a problem for the production of proteins such as interferons that are crucial for the prevention of virus spreading. To allow IFN synthesis during ISR, cells use a neat regulatory loop. ISR can be terminated by GADD34, which dephosphorylates eIRF2α. GADD34 expression depends on the IFN regulator IRF3, which permits pulses of IFN synthesis in ISR-arrested cells [34].

The arrest of protein translation caused by ISR can inhibit viral replication. Stalled translation pre-initiation complexes that form during ISR are sequestrated into stress granules (SGs) [35, 36]. In this way, viral mRNAs can be silenced in SGs with their following destruction in P bodies (PBs) [37, 38]. Proteins encoded by interferon-stimulated genes (ISGs) such as OASL1 are also found in SG where they bind viral RNAs and promote IFN response [39]. Many viruses actively suppress the formation of SGs or manipulate SGs and PBs to enhance their replication [40, 41].

A short-lived ISR has a pro-survival effect and helps to restore homeostasis and fight infection. However, a prolonged ISR is a cell death inducer and can worsen the outcome of infection. For example, *Listeria monocytogenes* causes ISR activation that is detrimental for the resolution of infection in mice [42]. The shutdown of protein translation during sepsis contributes to kidney damage [43].

10.3. Unfolded Protein Response and Autonomous Immunity

Most of the secreted and membrane proteins are synthesized, modified, and folded in the endoplasmic reticulum. Increase in unfolded and misfolded proteins in ER leads to unfolded protein response (UPR) mediated by three signaling branches: activation of PERK kinase, which inhibits cap-dependent translation through phosphorylation of eIF2α; processing of the transcription factor ATF6 precursor into an active transcription factor; and generation of active XBP1 factor by IRE1 kinase [44]. ATF6 and XBP1 induce expression of genes responsible for the restoration of ER homeostasis through the increase of ER capacity and through stimulation of degradation of unfolded proteins and ER-associated mRNAs (ERAD). Many stresses such as ROS, hypoxia, temperature, infections, and others can interfere with protein folding. Genetic mutations also can cause protein misfolding. This makes ER a hub that integrates various stress signals.

UPR is connected to the autonomous immune defense mechanisms at multiple levels. The immune response often involves production and secretion of various immune mediators and effectors and therefore is critically dependent on the efficiency of ER protein production. Accordingly, UPR has a role in almost every immune mechanism. In several types of specialized immune cells, some UPR sensors are constitutively active. In dendritic cells, IRE1 and its target XBP1 are active in the absence of ER stress and are necessary for the development of these cells and for their function in antigen presentation [45, 46]. This pathway is also necessary for differentiation of B cells into plasma cells, maturation of esoinophils, and macrophage response to TLR signaling [47, 48].

UPR is a pro-survival mechanism; it serves to restore homeostasis, increase ER protein-producing capacity, and improve protein folding. If homeostasis is not restored, UPR switches to a pro-apoptotic mode. A long lasting ER stress leads to degradation of ER-associated mRNAs by IRE1 by so-called regulated IRE1-dependent decay (RIDD). This generates RNA fragments that can be recognized by cytoplasmic RNA sensors such as RIG-I and lead to IFN induction [49, 50]. IFN can also be induced through JNK kinase, which is IRE1 target [51]. TLR signaling has variable effects on UPR depending on the cell status; from prevention of IRE1 inactivating dephosphorylation, which would augment UPR, to suppression of ER-induced apoptosis [47].

UPR is associated with inflammatory phenotype mediated by several mechanisms [52]. The decrease in global translation reduces the production of inhibitory IκBs, which increases free NFκB [33]. In addition, IRE1 can trigger NFκB activation through TRAF2/Nod1/2 pathway [53], whereas ATF6 can activate NFκB through the AKT pathway [54]. Activated IRE1 induces thioredoxin-interacting protein (TXNIP) [11]. TXNIP then activates NLRP3 inflammasome leading to IL1β production [55]. Several ATF6-related factors are activated by proteolysis during UPR, leading to induction of the acute phase proteins [56]. These proteins also potentiate inflammation. Reciprocally, pro-inflammatory stimuli such as cytokines and signaling through TLR receptors can trigger ER stress and lead to induction of UPR [47].

UPR is important in antiviral immunity. During infection with enveloped viruses, the envelope proteins are produced and glycosylated in ER. Viral proteins can induce UPR because of their difference from cellular proteins but also because large number of viral proteins that are produced and transported through ER overloads it [5, 27]. UPR leads to increased degradation of viral mRNA through RIDD mechanism as well as to increased ERAD, which would degrade viral proteins [57, 58]. Therefore UPR works as an antiviral defense mechanism.

Viruses developed multiple mechanisms to circumvent UPR. Some viruses actively suppress UPR; others manipulate it to their advantage. Some aspects of UPR, such as ER expansion and upregulation of chaperones may increase viral replication. Viruses exploit components of the ER including endocytic, secretory, and autophagy pathways [59]. Viruses can redirect ERAD mechanism to degrade the host molecules involved in antiviral defense [57].

UPR is also involved in the immune response to bacterial infections. Virulence factors produced by *P. aeruginosa* induce both UPR and ISR in the host cells

[28]. Cyclic-di-adenosine monophosphate produced by Gram-positive bacteria triggers ER expansion, chaperone gene upregulation, and ER autophagy [60]. Similar to viral infection, induction of UPR can facilitate or obstruct immune response to bacterial pathogens. For example, ER stress caused by lipoprotein accumulation suppresses immunity against bacterial pathogens [61].

Prolong UPR is associated with activation of proapoptotic pathways and suppression of NFκB. ATF4, which is selectively transcribed under PERK-induced inhibition of protein synthesis, controls the expression of the gene encoding the pro-apoptotic CHOP. CHOP expression promotes immunosuppressive activities of myeloid [62] and adaptive immune cells [63].

UPR connections to the immune system are conserved in evolution. In plants, IRE1 is activated in response to pathogens and is required for immunodefense [54].

10.4. DDR and Autonomous Immunity

DNA damage response (DDR) is an orchestrated response to DNA damage that leads to cell cycle arrest and activation of DNA repair but, if the damage cannot be repaired, can also induce senescence and apoptosis [64]. DDR allows cells to distinguish between chromosomal DNA damage and foreign DNA. ATM kinase phosphorylates the histone H2AX, which marks the DNA ends and serves to discriminate "self" and "non-self" genomes [65]. DDR can restrict viral DNA replication. The cellular DNA repair complex composed of Mre11, Rad50, and Nbs1 (MRN) forms at the site of DNA double-strand break, which recruits and activates ATM [66]. MRN can bind adenovirus DNA and stitch viral genomes into concatemers too large to be packaged into virions [67]. To prevent concatemerization, the virus produces factors that degrade members of the MRN complex [68].

p53 is an important mediator of DDR [69]. It contributes to immune surveillance by inducing cell death of infected cells, modulating metabolism, stimulating IFNs, and other means [70].

DDR has many other links to immunity. DDR stimulates NFκB activity [71, 72]. During genotoxic stress, NFκB subunit NEMO migrates to the nucleus and forms a signalosome that also includes ATM kinase, poly(ADP-rybosyl) transferase PARP1, and the SUMO E3 ligase PIASy [73, 74]. After NEMO becomes SUMOylated by PIASy and phosphorylated by ATM, it is ubiquitinated and exported from the nucleus. ATM is also exported into the cytoplasm and together with NEMO participates in activation of the NFκB upstream kinase TAK1 leading to IKK activation. Activated NFκB induces expression of targets typical for the NEMO-dependent conventional pathway. NFκB targets such as BCL-xL promote cell survival in response to DNA damage [75]. Among NFκB targets are several inhibitors that prevent overactivation of NFκB in response to DNA damage. They include IκB inhibitors. Also, among NFκB targets are proteases that selectively remove SUMO residues from NEMO, terminating DDR-dependent NFκB signaling. In cells with severe DNA damage, PARP1 is cleaved by caspases, which also terminates NFκB signaling and permits initiation of apoptosis.

Radiotherapy and chemotherapy drugs that cause DNA double-strand breaks induce interferons [76]. The mechanism of IFN induction is connected to DNA fragments that are generated during DNA double-strand break repair. Bloom syndrome (BLM) helicase and exonuclease EXO1 play a major role in the generation of these fragments. Some of them can migrate into the cytosol where they are recognized by the STING-dependent DNA sensors. To prevent immune activation, DNA fragments are destroyed by 3'-5' exonuclease TREX1 located on the outer side of the nuclear membrane [77, 78]. Deficiency of this enzyme, malfunction of various DNA repair proteins, or excessive DNA damage can cause accumulation of DNA fragments in the cytoplasm and stimulation of NFκB and IFN pathways [78 - 80]. Similarly, unrepaired DNA lesions in cells from ataxia telangiectasia patients and *Atm* (-/-) mice induce IFNs through leakage of DNA fragments into the cytoplasm [81].

DDR can signal directly to the immune system by inducing ligands for receptors expressed on natural killer and cytotoxic T cells, both of which are capable of killing stressed cells [82]. Immunostimulatory effect of transient DDR on the organism is largely beneficial, whereas persistent DDR-dependent immune stimulation may be harmful [83]. Persistent DDR induces inflammation and senescence [84] and contributes to aging and cancer [85, 86].

Reciprocally, chronic inflammatory diseases associated with increased ROS/RNS production can lead to DNA damage [87]. Fast division of activated immune cells also can be associated with increased DNA damage [88]. In plants, DNA damage is observed after NLR receptors' activation [89].

Pathogens have a complex relationship with DDR. Many induce DDR because their replication intermediates resemble DNA damage, or they produce genotoxic factors such as endonucleases, or they induce ROS, or their oncoproteins force the cells into S phase and cause replicative stress [68, 90].

DDR induced by pathogens can lead to p53 activation and apoptosis through the mitochondrial pathway contributing to immune defense. Correspondently, many viruses actively suppress DDR. As mentioned above, adenoviruses degrade MRN complex [67]. The parvovirus MVM controls DDR by disabling the ATR kinase pathway [91]. However, some viruses need DDR for an efficient infection [68, 90]. They efficiently manipulate and exploit components of DDR. For example, the polyomavirus JC depends on the stimulation of the ATM-dependent NFκB activation [92]. Herpes simplex virus 1 (HSV1) uses ATR/CHK1 pathway for its replication [93]. DDR inhibits ribosome biogenesis through nucleolar Treache protein; some viruses can mimic this effect through factors that bind Treache and shut down ribosome biogenesis in the absence of DDR [94].

10.5. p53 and Autonomous Immunity

The master regulator of stress response p53 strongly influences immunity. p53-deficient mice have a higher load of viral infection, whereas "super p53" mice that have an additional copy of p53 gene show an enhanced resistance to viruses [95]. In fact, p53 was discovered as a cellular target silenced by proteins from several viruses [70, 96]. These proteins inactivate p53 by variable mechanisms. For example, human papilloma viruses produce E6 protein that binds cellular ubiquitin ligase E6AP; this complex then recruits and degrades p53 [97]. Several adenoviral proteins suppress p53 [98]. E1B-55k together with E4-orf6 forms a complex with p53 that drives p53 ubiquitination and proteasomal degradation. E1A cooperates with the cellular protein FUBP1 to downregulate p53-regulated promoters. E4-ORF3 suppresses activation of p53-regulated genes by induction of the heterochromatin at their promoters [99].

Bacterial pathogens also suppress p53 [100]. *Helicobacter pylori* produce CagA, which enhances p53 degradation and CagE, which promotes expression of the dominant negative forms of p53, Δ133p53, and Δ160p53. *Chlamydia trachomatis* stimulates proteasomal degradation of p53 [101].

An obvious mechanism by which p53 can interfere with pathogens is by inducing apoptosis of infected cells. However, recent studies suggest that p53 has many more ways to influence immune response. p53 can directly restrict the growth of some pathogens such as SV40 virus. Its large T-antigen (T-ag) is known to sequester and inactivate p53. Recent data show that p53 binds to T-ag promoter preventing its transcriptional activation by Sp1 and halts viral propagation [102].

A key role of p53 in controlling transposable elements (TEs) has been discussed in chapter 8. p53 silences TE by several mechanisms including regulation of PIWI pathway, DNA methylation, and chromatin modifications [103, 104]. A major p53 activator is DNA damage, which can also activate the movement of transposable elements [104]. Preventive activation of PIWI pathway by p53 then would increase the efficiency of TE silencing.

p53 promoter contains a functional IFN-stimulated response element, and IFNα/β signaling leads to increased p53 transcription [105]. On the other hand, p53 was found to directly interact with IRF9, the central component of IFN signaling pathway, enhancing IFN response [106]. p53 can also regulate some IFN targets. One example is cytidine deaminase APOBEC3 (A3) family that is an IFNs target important in defense against retroviral infection. p53 regulates A3 genes under stress conditions and immune challenges [107]. It stimulates the expression of most A3 genes except A3B, which is inhibited. A3B is highly expressed in many cancers in an NFκB-dependent manner [108]. Notably, p53 mutants promote A3B expression [107]. p53 can also enhance induction of A3 genes by IFNs. TLR3 receptor that signals through IFNs is also upregulated by p53 [109]. In addition, p53 prevents IFN overactivation by upregulating expression of ISG15 [110]. This ubiquitin-like modifier is necessary for the accumulation of USP18, which is an inhibitor of IFN pathway [111].

Whereas p53 interactions with IFN pathways are mostly cooperative, they are much more complex with NFκB-dependent pathways. Although many stressors, such as DNA damage, activate both p53 and NFκB networks and at some settings, NFκB can promote apoptosis as does p53, these networks differ fundamentally. p53 is a tumor suppressor, and NFκB is an oncogene. p53 responds mostly to internal stresses such as genotoxic stress. Cell cycle arrest and elimination of stressed cells from the proliferating pool by induction of senescence or apoptosis are major strategies employed by p53. In addition, p53 shifts metabolism from glycolysis to oxidative phosphorylation, which also limits cells' proliferation capability. In contrast, the main task of NFκB is a response to external stressors such as wound and infection, which often requires increased cell proliferation during regeneration and immune response. Accordingly, these networks often respond in opposing ways to the same signaling [112, 113]. For example, PI3K/Akt signaling inhibits p53 but activates NFκB. Oncogene Arf activates p53 but inhibits NFκB [114]. In addition, p53 and NFκB compete for the same histone acetyltrasferase enzymes and the stronger signal wins. Therefore, usually when one pathway is up, the other is down.

p53 has been recently shown to suppress autoimmune and inflammatory conditions in both mice and humans [115 - 117]. p53-deficient mice have higher NFκB activity and inflammation [118]. One example of p53 and NFκB cross-talk is the regulation of inflammation by a p53 target miR-23b. miR-23b suppresses IL17-associated inflammation by targeting NFκB upstream regulators TAB2, TAB3, and IKKα; in turn, IL17 downregulates miR-23b in inflammatory lesions [119].

Another example of p53 and NFκB pathways interaction is the ubiquitin-specific peptidase 48 (USP48). This peptidase is activated by CK2 kinase phosphorylation in response to cytokine signaling; it associates with RelA and trims K48 ubiquitin chains associated with RelA thus increasing RelA stability [120]. At the same time, USP48 binds Mdm2 and stabilizes it, which leads to increased Mdm2-mediated ubiquitination and degradation of p53 [121]. Many points of p53/NFκB interaction allow cells to integrate multiple inputs from the cell interior and exterior and to make an 'informed' choice between these pathways.

An additional function of p53 in immunity has been discovered recently. p53 helps to safely dispose of apoptotic bodies. A p53 target DD1α promotes intermolecular contact at junctions of apoptotic cells and macrophages, which stimulates efferocytosis [122].

10.6. Oxidative Stress and Autonomous Immunity

Cellular stress of diverse origin is often accompanied by dysregulation of redox balance and oxidative stress. Furthermore, many cell types can generate ROS and RNS to kill intracellular pathogens, which also can lead to oxidative stress. Autonomous immunity and oxidative stress response interact at several levels (Fig. **10**).

Early phases of oxidative stress activate NFκB through diverse mechanisms, whereas prolonged stress can suppress it and promote death pathways instead [123]. One example of NFκB activation by oxidative stress involves oxidation of cysteines in NFκB p50 subunit or in IKK complex [124]. In macrophages, ROS produced by mitochondria induces the formation of disulfide bonds in NEMO, leading to a direct IKK complex activation.

ATM kinase responsible for DNA damage response is also a ROS sensor. Oxidized ATM forms an activated disulfide-cross-linked dimer [125]. ATM then activates NFκB pathway through RelA phosphorylation [126]. ROS, in particular, the superoxide anion, are involved in neutrophils migration and extracellular traps formation [127 - 129]. Oxidative stress can lead to inflammation and tissue damage. Defects in ROS generation also can have negative effects. For example, a decreased ROS production caused by mutations in a subunit of the phagocyte NADPH oxidase (NOX2) leads to susceptibility to autoimmune diseases, possibly due to defective clearance of dying cells [130]. In the intestine, overproduction of ROS is associated with ileitis and ulcerative colitis, whereas impaired ROS generation is linked to Crohn's disease and pancolitis [131].

Mitochondria are both the main source of ROS and a hub for autonomous defense pathways, which include the mitochondrial signaling molecule MAVS [132]. In dendritic cells, mtROS modulates the immune response in dependence on which type of viral sensing pathway is stimulated. It lowers TLR9-dependent signaling and IFN production *via* blocking phosphorylation of the IFN-inducer IRF7. In contrast, elevated mtROS augments RIG-I-stimulated expression of IFNs, increases MAVS expression, and stimulates phosphorylation of IRF3 [133]. Moreover, ROS can directly induce MAVS oligomerization and IFN production in the absence of infection [134].

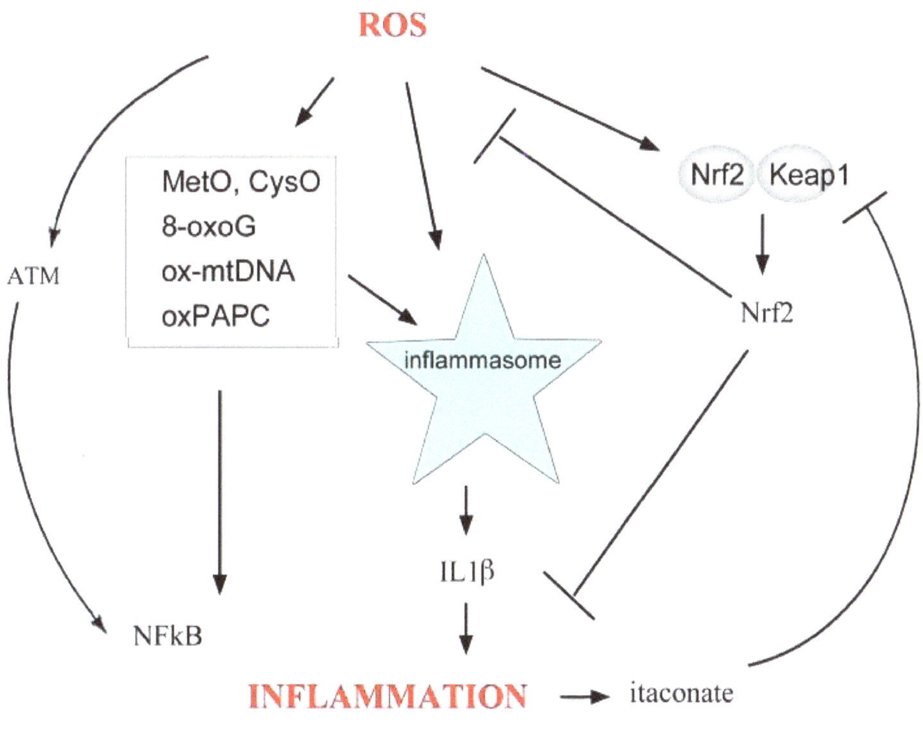

Fig. (10). ROS induces both inflammation and antioxidative mechanisms that inhibit inflammation. ATM senses ROS, which leads to NFkB activation. ROS oxidizes macromolecules making them recognizable by immune receptors, which triggers NFkB activation and, inflammation. The anti-oxidative factor Nrf2 is inhibited by Keap1. ROS oxidizes Keap1, releasing Nrf2. Nrf2 stimulates antioxidant genes thus attenuating oxidative stress. It also directly suppresses expression of IL1b and inflammation. An additional way of shutting the inflammation off is through production of itaconate, which alkylates Keap's cysteine residues leading to Nrf2 release.

Reports on effects of ROS and oxidative stress on inflammasome priming and activation are ambiguous. Mitochondrial stress can promote inflammasome activation [135, 136]. In response to ROS, mitochondrial lipid cardiolipin is translocated to the outer mitochondrial membrane and promotes inflammasome assembly [137]. At the same time, some reports demonstrate that ROS and RNS can suppress inflammasome activation [138, 139], whereas IL1β secretion in gout disease is redox-independent [140].

Oxidized endogenous molecules can modulate the immune response in dependence on cellular content. Oxidized phospholipids derived from 1-palmitoyl-2-arachidonyl-sn-glycero-3-phosphorylcholine (oxPAPC) are recognized by caspase-11, which also directly binds LPS. In dendritic cells, oxPAPC promotes inflammasome formation and IL1β release [141]. However, in macrophages, oxPAPC competes with activation of caspase-11 by LPS, suppresses pyroptosis, and inhibits IL1β release [142].

One of the main oxidative damages is the conversion of DNA guanine into 7,8-dihydro-8-oxoguanine (8-oxoG). It is repaired by 8-oxoguanine DNA glycosylase-1 (OGG1) [143]. OGG1 forms a complex with the excised 8-oxoG, inducing its nuclear export. The cytoplasmic OGG1/8-oxoG then activates NFκB. OGG1 enzyme is also involved in transcriptional regulation of inflammation. Many redox-responsive genes have guanine-rich clusters in their promoters that, when oxidized, bind OGG1 leading to changes in gene expression [144]. Specifically, OGG1 bound to 8-oxoG upstream from NFκB binding motifs serves as an epigenetic mark that increases NFκB DNA occupancy, which leads to an increased expression of pro-inflammatory genes. A synthetic inhibitor of OGG1 that prevents OGG1 binding decreases NFκB DNA occupancy and suppresses inflammation [145].

Nrf2 pathway plays a central role in oxidative stress response by regulating its antioxidative targets through AREs (antioxidant response elements) [146]. In the cytoplasm, Keap1 targets Nrf2 to ubiquitination and degradation. Keap1 also binds and promotes degradation of IKKβ, a component of the IKK kinase complex that activates NFκB [147]. Keap1, therefore, restricts inflammatory signaling. In addition, Nrf2 binds to the proximal regions of promoters of genes that encode proinflammatory cytokines such as IL6 and IL1β and inhibits their expression [148]. Moreover, Nrf2 is part of a negative feedback loop that regulates inflammation. Activation of inflammation is associated with the conversion of a component of tricarboxylic acid cycle aconitate into itaconate [149]. Itaconate alkylates cysteine residues on Keap1, releasing Nrf2 anti-inflammatory activity [150]. This limits inflammatory genes expression. In addition, itaconate inhibits IFN type I production, whereas interferons upregulate itaconate production [150]. Itaconate/Nrf2 signaling results in a decrease in STING mRNA stability and therefore reduces the IFN production [151]. This may lead to increased susceptibility to infection with DNA viruses.

Accordingly, the inflammation is prolonged in Nrf2-deficient mice [152].

Systemic activation of Nrf2 suppresses the multi-organ inflammation in autoimmune scurfy mutant mice [153]. Anti-inflammatory effect of the ketogenic diet is also associated with increased expression of Nrf2 [154]. Hyperactivation of Nrf2 increases stress tolerance but at the cost of metabolic dysregulation and aging acceleration [155].

In turn, activation of immune response modulates the oxidative stress response. For example, activation of pattern-recognition receptors such as TLR2, in monocytes increases their resistance to oxidative stress [156]. This effect is mediated by increased levels of Nrf2 and its targets: glutathione, glutathione-related enzymes, Mn superoxide dismutase, thioredoxin and thioredoxin-related antioxidant enzymes such as thioredoxin reductase and peroxidase. Therefore cells can increase their antioxidant potential during pathogen attack.

Oxidative stress and Nrf2 status influence the outcome of many infections. To mitigate oxidative stress imposed by the host, pathogens use various regulators, detoxification enzymes, and subversion mechanisms [157 - 159]. Nrf2 is a frequent target of pathogens. Respiratory syncytial virus (RSV) downregulates the expression of antioxidant enzymes by inducing Nrf2 deacetylation and proteasomal degradation [160]. However, some viruses benefit from induction of cytoprotective Nrf2 targets [161].

10.7. Autophagy and Immunodefense

Autophagy is an evolutionarily conserved process of delivery and degradation of cytoplasmic proteins and organelles in lysosomes [162]. Autophagy is induced by stress such as starvation, energy shortage, and damage to macromolecules. A class of selective autophagy called xenophagy is involved in defense against intracellular pathogens [163, 164]. It is important for protection from bacterial infections [165, 166] but also plays a role in defense against viruses [167] and eukaryotic pathogens such as malaria parasites and *Trypanosoma cruzi* [168, 169].

Intracellular pathogens are targeted for xenophagy through various pathways. *Salmonella typhimurium* is the most studied pathogen controlled by xenophagy. After infection of epithelial cells, the bacteria reside in Salmonella-containing vacuoles. They damage the vacuolar membranes leading to exposure of membranes' glycans. The glycans are recognized by galectin-8, which is used by the host cells to survey the integrity of endosomal and lysosomal membranes [170]. Galectin-8-bound vacuoles containing Salmonella recruit the autophagy receptor NDP52 and the whole vesicle is targeted for autophagy.

A central protein in autophagosome biogenesis is LC3, which is conjugated with phosphatidylethanolamine in membranes by Atg proteins [171]. LC3 and its homologues can be deposited on pathogen-containing vacuoles, causing recruitment of interferon-inducible GTPases and disruption of vacuoles [172].

When pathogens escape from vacuoles or phagosomes into the cytosol, they can be recognized by PRRs, ubiquitinated, and captured by autophagic mechanisms [164]. Ubiquitin chains on bacteria are recognized by autophagy receptors p62, NDP52, and optineurin. These receptors commonly have an LC3-interacting region and can simultaneously bind ubiquitinated bacteria and LC3 on autophagosomal membranes. In this way, autophagy receptors target bacteria to autophagosomes. Linear, K63, and K48 ubiquitin chains are recognized by different autophagy receptors.

Multiple E3 ubiquitin ligases are implicated in the ubiquitination-mediated autophagy. For example, LRSAM1 ligase is involved in xenophagy of *S. typhimurium* [173]. LRSAM1 is a multifunctional protein, which recognizes misfolded proteins and plays a role in the sorting of internalized cell-surface receptors [174]. LRSAM1 contains a leucine-rich repeat (LRR) domain similar to those used by many PRRs, which possibly can be used for pathogen recognition. Self-association and auto ubiquitination of LRSAM1 has been observed after contact with bacteria [173, 175]. Another E3 ubiquitin ligase involved in xenophagy is Parkin, which is also responsible for mitophagy. Parkin plays a role in ubiquitin-mediated autophagy of *S. typhimurium* and *Mycobacterium tuberculosis* [176]. ARIH1 and HOIP E3 ubiquitin ligases also participate in ubiquitination of *S. typhimurium* [177]. HOIP is part of the LUBAC complex, which synthesizes linear (Met-1-linked) polyubiquitin chains on bacterial surfaces already coated with ubiquitin [178]. LUBAC-synthesized polyubiquitin recruits autophagy receptor optineurin for xenophagy. Polyubiquitin coat also recruits NEMO leading to NFκB activation [178, 179]. This results in the activation of pro-inflammatory mechanisms and restriction of bacterial proliferation.

Autophagy is induced during infection by many viruses but its impact on the outcome of infection varies between viruses and cell types [180]. Selective autophagy is induced by hepatitis C virus RNA replication in the infected cells with the entailing accumulation of LC3-positive ubiquitinated aggregates near the replication complex. However, with the progression of infection, the virus impairs the autophagy [181]. Zika virus induces Sting/NFκB-dependent autophagy in Drosophila brain, which restricts the infection [167], whereas in most mammalian models autophagy provides no clear benefits [182]. Many E3 ubiquitin ligases from TRIM family are involved in virus-induced autophagy; some of them, such as TRIM5a, can directly recognize their targets [183].

Xenophagy can also be induced through PRR interaction with components of autophagic machinery [164]. For example, Nod2 recognizes bacteria at the site of their entry and recruits ATG16L1 to the plasma membrane to initiate the formation of autophagosome around the bacteria [184]. The membrane complement regulatory protein CD46 acts as a receptor for several viruses and bacteria. In *Neisseria gonorrhoeae*, its engagement induces autophagic response through activation of VPS34/Beclin1 [185, 186].

In addition to xenophagy, autophagy contributes to other aspects of the immune response. In LC3-asssociated phagocytosis (LAP), LC3 family proteins are conjugated with phagosome membranes, which promotes phagolysosomes maturation [187]. LAP facilitates the immunologically silent clearance of phagocyted cargo and defects in autophagy are associated with autoimmunity.

Recent data show a connection between autophagy and the antiviral cGAS/STING pathway. DNA presence in cytosol induces cGAMP production by cGAS, which activates STING pathway. Activated STING translocates to the ER compartment adjacent to Golgi. This region then serves as a site of LC3 conjugation with membrane phosphatidylethanolamine [188]. Interestingly, STING from the sea anemone induces autophagy but not interferons, suggesting that autophagy is the primordial function of this pathway.

Autophagy also collaborates with the complement system. The cytoplasmic complement factor C3 interacts with ATG16L1 to promote cell-protecting autophagy during stress [185].

In addition, autophagy regulates IFN activation, inflammation, and pyroptosis through selective autophagy of components of the corresponding pathways [189]. Autophagy removes cellular components such as aggregated proteins and damaged mitochondria that are able to induce inflammation. Mutations in the autophagy gene ATG16L are associated with increased risk for Crohn disease; they prevent autophagic removal of overactivated IRE1α complexes during ER stress [190]. In general, autophagy restricts excessive inflammation [189].

Another way how autophagy contributes to immunodefense has been found recently. When bacterial infection impairs secretion of lysozyme *via* the Golgi apparatus in Paneth cells, they switch to secretory autophagy to deliver this antimicrobial agent [191].

Autophagy is involved in the differentiation of immune cells [192]. It is also involved in induction and execution of adaptive immune responses including processing and presentation of antigens from pathogens on MHC II class molecules to T lymphocytes as well as in the maturation of T and B cells [193].

Many pathogens developed the means to inhibit xenophagy, for example by inhibiting the expression of Agt proteins or by producing factors that degrade these proteins [164]. For example, RavZ from *Legionella pneumophila* inhibits host autophagy through irreversible Atg8/LC3 de-conjugation from bacteria [163, 194]. *S. typhimurium* effector SopA is an E3 ligase that blocks the activity of the host E3 ubiquitin ligase TRIM56 by occluding the surface interacting with the E2-conjugating enzyme [195]. Moreover, SopA ubiquitinates TRIM56 and targets it for proteasomal degradation. Pathogens can also manipulate host autophagy to their advantage [196]. Some intracellular pathogens such as *Anaplasma phagocytophilum* or *Yersinia pseudotuberculosis* stimulate autophagy and reside in autophagosomes preventing their maturation and fusion with lysosomes [164]. Some viruses also learned to hijack autophagy and use it in their life cycle [197].

10.8. Balancing Stress and Immune Responses

A cell exposed to both stress and infection needs to generate an integrated response. Integration of stress and immune signaling takes place at the levels of ligands, signal transducers, and effector mechanisms. Relatively few signaling pathways such as MAPK3K/ERK1/2/p38/JNK, TAK1/NFκB, and IRF3/IRF7/IFN collect signals from multiple PRRs receptors [198]. Signaling sharing between different pathways enables the integration of signals from diverse stimuli, signal amplification or inhibition, and prioritization of response. Non-infectious microorganisms and tissue injury typically activate a small number of receptors whereas highly virulent pathogens activate diverse receptors from distinct families that use dissimilar signaling pathways [199]. In general, pathogens that are recognized as both foreign and harmful induce stronger immune response [200].

Stress can stimulate immunity. In animals, p53 is the hub of stress response pathways. It suppresses infections through promoting cell death of infected cells, stimulation of IFNs, and other mechanisms [70]. Many pathogens induce stress in host cells. For example, hepatitis virus C induces oxidative stress, UPR, ISR, and autophagy [5]. Stress response stimulates immune mechanisms and helps control infection at an early stage. In addition, many stress signals converge on the activation of inflammasomes [201]. DNA damage, ISR, UPR, and oxidative stress can all promote NFκB and IFN activation and potentiate immunity.

However, the effect of stress on autonomous immunity strongly depends on its type, duration, intensity, and pathogen' reactions. Short-term stress often stimulates immunity, whereas prolong, non-compensated stress can suppress immunity [202]. Oxidative stress response, DDR, UPR, ISR, and other stress responses stimulate immunodefense during short and mild stress but promote cell death instead during prolong and strong stress. HSP70 is a negative regulator of

NLRP3 inflammasome activation [203]. Increased temperature suppresses immune genes in honeybee [204]. In plants, heat, and drought typically enhance plant susceptibility to pathogens [205]. Nutritional phosphate stress also suppresses immunity in *Arabidopsis thaliana* response, which shows plants prioritize nutrition over defense [206].

Some damage associated with infection is collateral damage caused by overly active stress responses. For example, activation of unfolded protein response contributes to Zika virus-induced brain damage [207]. Correspondently, the outcome of infection depends not only on the immune response but also on the ability of the host to limit and repair the damage caused by infection. This ability is known as disease tolerance [208]. Prevention of overactivation of stress response and inflammation increases host tolerance [209]. Efficient repair also improves tolerance. For example, methionine sulfoxide reductase, which repairs proteins damaged by ROS during immune response in plants is required for plant resistance to infection [210]. Metabolic adaptations also increase stress tolerance; induction of the iron-sequestering ferritin, which counters the iron-driven oxidative inhibition of liver gluconeogenesis, increases tolerance to sepsis caused by infection [211]. In some cases, infection can increase stress tolerance; for example, infection in plants can increase their salt tolerance [212].

CONCLUDING REMARKS

In cells that simultaneously experience stress and a pathogen attack, stress response modulates the autonomous immune response. At a low level of stress, practically all stress responses are pro-survival. However, strong and prolonged stress promotes cell death. Accordingly, the effect of stress on immune response depends on the stress' duration and intensity. Diverse stressors including hypo- or hyperosmolarity, irradiation, increased ROS, ion imbalance, metabolic disturbances, and others activate inflammasomes and stimulate immunity. DNA damage response is associated with activation of IFN and NFκB pathways. ER stress contributes to immunodefense through degradation of viral RNA and proteins and induction of inflammation. p53, the main hub of the cellular stress response, stimulates immunodefense by inducing apoptosis in infected cells, stimulating IFN, and other mechanisms. Intense and prolonged stress of any kind can lead to self-damage and compromised the immune response. How cells allocate resources between stress and immune responses depends on many factors including cell type, the level of nutrients and energy, the type, intensity, and duration of stress, and the type of infection.

CONSENT FOR PUBLICATION

Not applicable.

CONFLICT OF INTEREST

The author confirms that this chapter contents have no conflict of interest.

ACKNOWLEDGEMENTS

Declared none.

REFERENCES

[1] Janeway CA Jr, Medzhitov R. Innate immune recognition. Annu Rev Immunol 2002; 20: 197-216. [http://dx.doi.org/10.1146/annurev.immunol.20.083001.084359] [PMID: 11861602]

[2] Stuart LM, Paquette N, Boyer L. Effector-triggered *versus* pattern-triggered immunity: how animals sense pathogens. Nat Rev Immunol 2013; 13(3): 199-206. [http://dx.doi.org/10.1038/nri3398] [PMID: 23411798]

[3] Jones JD, Vance RE, Dangl JL. Intracellular innate immune surveillance devices in plants and animals. Science 2016; 354(6316)aaf6395 [http://dx.doi.org/10.1126/science.aaf6395] [PMID: 27934708]

[4] Pukkila-Worley R. Surveillance Immunity: An Emerging Paradigm of Innate Defense Activation in Caenorhabditis elegans. PLoS Pathog 2016; 12(9)e1005795 [http://dx.doi.org/10.1371/journal.ppat.1005795] [PMID: 27631629]

[5] Ríos-Ocampo WA, Navas MC, Faber KN, Daemen T, Moshage H. The cellular stress response in hepatitis C virus infection: A balancing act to promote viral persistence and host cell survival. Virus Res 2019; 263: 1-8. [http://dx.doi.org/10.1016/j.virusres.2018.12.013] [PMID: 30599163]

[6] Pellegrino MW, Nargund AM, Kirienko NV, Gillis R, Fiorese CJ, Haynes CM. Mitochondrial UPR-regulated innate immunity provides resistance to pathogen infection. Nature 2014; 516(7531): 414-7. [http://dx.doi.org/10.1038/nature13818] [PMID: 25274306]

[7] Kwon S, Kim EJE, Lee SV. Mitochondria-mediated defense mechanisms against pathogens in Caenorhabditis elegans. BMB Rep 2018; 51(6): 274-9. [http://dx.doi.org/10.5483/BMBRep.2018.51.6.111] [PMID: 29764564]

[8] Jeong DE, Lee D, Hwang SY, *et al.* Mitochondrial chaperone HSP-60 regulates anti-bacterial immunity *via* p38 MAP kinase signaling. EMBO J 2017; 36(8): 1046-65. [http://dx.doi.org/10.15252/embj.201694781] [PMID: 28283579]

[9] Tjahjono E, Kirienko NV. A conserved mitochondrial surveillance pathway is required for defense against Pseudomonas aeruginosa. PLoS Genet 2017; 13(6)e1006876 [http://dx.doi.org/10.1371/journal.pgen.1006876] [PMID: 28662060]

[10] Colaço HG, Moita LF. Initiation of innate immune responses by surveillance of homeostasis perturbations. FEBS J 2016; 283(13): 2448-57. [http://dx.doi.org/10.1111/febs.13730] [PMID: 27037950]

[11] Lerner AG, Upton JP, Praveen PV, *et al.* IRE1α induces thioredoxin-interacting protein to activate the NLRP3 inflammasome and promote programmed cell death under irremediable ER stress. Cell Metab 2012; 16(2): 250-64. [http://dx.doi.org/10.1016/j.cmet.2012.07.007] [PMID: 22883233]

[12] Groß CJ, Mishra R, Schneider KS, *et al.* K^+ Efflux-Independent NLRP3 Inflammasome Activation by Small Molecules Targeting Mitochondria. Immunity 2016; 45(4): 761-73. [http://dx.doi.org/10.1016/j.immuni.2016.08.010] [PMID: 27692612]

[13] Liston A, Masters SL. Homeostasis-altering molecular processes as mechanisms of inflammasome activation. Nat Rev Immunol 2017; 17(3): 208-14. [http://dx.doi.org/10.1038/nri.2016.151] [PMID: 28163301]

[14] Di A, Xiong S, Ye Z, *et al.* The TWIK2 Potassium Efflux Channel in Macrophages Mediates NLRP3 Inflammasome-Induced Inflammation. Immunity 2018; 49(1): 56-65.e4. [http://dx.doi.org/10.1016/j.immuni.2018.04.032] [PMID: 29958799]

[15] Green JP, Yu S, Martín-Sánchez F, *et al.* Chloride regulates dynamic NLRP3-dependent ASC oligomerization and inflammasome priming. Proc Natl Acad Sci USA 2018; 115(40): E9371-80. [http://dx.doi.org/10.1073/pnas.1812744115] [PMID: 30232264]

[16] da Costa LS, Outlioua A, Anginot A, Akarid K, Arnoult D. RNA viruses promote activation of the NLRP3 inflammasome through cytopathogenic effect-induced potassium efflux. Cell Death Dis 2019; 10(5): 346. [http://dx.doi.org/10.1038/s41419-019-1579-0] [PMID: 31024004]

[17] Mathur A, Feng S, Hayward JA, *et al.* A multicomponent toxin from Bacillus cereus incites inflammation and shapes host outcome *via* the NLRP3 inflammasome. Nat Microbiol 2019; 4(2): 362-74. [http://dx.doi.org/10.1038/s41564-018-0318-0] [PMID: 30531979]

[18] Ip WK, Medzhitov R. Macrophages monitor tissue osmolarity and induce inflammatory response through NLRP3 and NLRC4 inflammasome activation. Nat Commun 2015; 6: 6931. [http://dx.doi.org/10.1038/ncomms7931] [PMID: 25959047]

[19] Yi B, Titze J, Rykova M, *et al.* Effects of dietary salt levels on monocytic cells and immune responses in healthy human subjects: a longitudinal study. Transl Res 2015; 166(1): 103-10. [http://dx.doi.org/10.1016/j.trsl.2014.11.007] [PMID: 25497276]

[20] Schroder K, Zhou R, Tschopp J. The NLRP3 inflammasome: a sensor for metabolic danger? Science 2010; 327(5963): 296-300. [http://dx.doi.org/10.1126/science.1184003] [PMID: 20075245]

[21] Christ A, Günther P, Lauterbach MAR, *et al.* Western Diet Triggers NLRP3-Dependent Innate Immune Reprogramming. Cell 2018; 172(1-2): 162-175.e14. [http://dx.doi.org/10.1016/j.cell.2017.12.013] [PMID: 29328911]

[22] Youm YH, Nguyen KY, Grant RW, *et al.* The ketone metabolite β-hydroxybutyrate blocks NLRP3 inflammasome-mediated inflammatory disease. Nat Med 2015; 21(3): 263-9. [http://dx.doi.org/10.1038/nm.3804] [PMID: 25686106]

[23] Hardie DG. Keeping the home fires burning: AMP-activated protein kinase. J R Soc Interface 2018; 15(138)20170774 [http://dx.doi.org/10.1098/rsif.2017.0774] [PMID: 29343628]

[24] Cordero MD, Williams MR, Ryffel B. AMP-Activated Protein Kinase Regulation of the NLRP3 Inflammasome during Aging. Trends Endocrinol Metab 2018; 29(1): 8-17. [http://dx.doi.org/10.1016/j.tem.2017.10.009] [PMID: 29150317]

[25] Zhao P, Wong KI, Sun X, *et al.* TBK1 at the Crossroads of Inflammation and Energy Homeostasis in Adipose Tissue. Cell 2018; 172(4): 731-743.e12. [http://dx.doi.org/10.1016/j.cell.2018.01.007] [PMID: 29425491]

[26] Pakos-Zebrucka K, Koryga I, Mnich K, Ljujic M, Samali A, Gorman AM. The integrated stress response. EMBO Rep 2016; 17(10): 1374-95. [http://dx.doi.org/10.15252/embr.201642195] [PMID: 27629041]

[27] Siddiquey MNA, Zhang H, Nguyen CC, Domma AJ, Kamil JP. The Human Cytomegalovirus Endoplasmic Reticulum-Resident Glycoprotein UL148 Activates the Unfolded Protein Response. J Virol 2018; 92(20): e00896-18. [http://dx.doi.org/10.1128/JVI.00896-18] [PMID: 30045994]

[28] van 't Wout EF, van Schadewijk A, van Boxtel R, *et al.* Virulence Factors of Pseudomonas aeruginosa Induce Both the Unfolded Protein and Integrated Stress Responses in Airway Epithelial Cells. PLoS Pathog 2015; 11(6)e1004946 [http://dx.doi.org/10.1371/journal.ppat.1004946] [PMID: 26083346]

[29] Rodrigues LOCP, Graça RSF, Carneiro LAM. Integrated Stress Responses to Bacterial Pathogenesis Patterns. Front Immunol 2018; 9: 1306. [http://dx.doi.org/10.3389/fimmu.2018.01306] [PMID: 29930559]

[30] Young SK, Wek RC. Upstream Open Reading Frames Differentially Regulate Gene-specific Translation in the Integrated Stress Response. J Biol Chem 2016; 291(33): 16927-35. [http://dx.doi.org/10.1074/jbc.R116.733899] [PMID: 27358398]

[31] Starck SR, Tsai JC, Chen K, *et al.* Translation from the 5′ untranslated region shapes the integrated stress response. Science 2016; 351(6272)aad3867 [http://dx.doi.org/10.1126/science.aad3867] [PMID: 26823435]

[32] Guenther UP, Weinberg DE, Zubradt MM, *et al.* The helicase Ded1p controls use of near-cognate translation initiation codons in 5′ UTRs. Nature 2018; 559(7712): 130-4. [http://dx.doi.org/10.1038/s41586-018-0258-0] [PMID: 29950728]

[33] Mitchell S, Vargas J, Hoffmann A. Signaling *via* the NFκB system. Wiley Interdiscip Rev Syst Biol Med 2016; 8(3): 227-41. [http://dx.doi.org/10.1002/wsbm.1331] [PMID: 26990581]

[34] Dalet A, Argüello RJ, Combes A, *et al.* Protein synthesis inhibition and GADD34 control IFN-β heterogeneous expression in response to dsRNA. EMBO J 2017; 36(6): 761-82. [http://dx.doi.org/10.15252/embj.201695000] [PMID: 28100675]

[35] Buchan JR, Parker R. Eukaryotic stress granules: the ins and outs of translation. Mol Cell 2009; 36(6): 932-41. [http://dx.doi.org/10.1016/j.molcel.2009.11.020] [PMID: 20064460]

[36] Moon SL, Morisaki T, Khong A, Lyon K, Parker R, Stasevich TJ. Multicolour single-molecule tracking of mRNA interactions with RNP granules. Nat Cell Biol 2019; 21(2): 162-8. [http://dx.doi.org/10.1038/s41556-018-0263-4] [PMID: 30664789]

[37] Nikolic J, Civas A, Lama Z, Lagaudrière-Gesbert C, Blondel D. Rabies Virus Infection Induces the Formation of Stress Granules Closely Connected to the Viral Factories. PLoS Pathog 2016; 12(10)e1005942 [http://dx.doi.org/10.1371/journal.ppat.1005942] [PMID: 27749929]

[38] McCormick C, Khaperskyy DA. Translation inhibition and stress granules in the antiviral immune response. Nat Rev Immunol 2017; 17(10): 647-60. [http://dx.doi.org/10.1038/nri.2017.63] [PMID: 28669985]

[39] Kang JS, Hwang YS, Kim LK, *et al.* OASL1 Traps Viral RNAs in Stress Granules to Promote Antiviral Responses. Mol Cells 2018; 41(3): 214-23. [PMID: 29463066]

[40] Poblete-Durán N, Prades-Pérez Y, Vera-Otarola J, Soto-Rifo R, Valiente-Echeverría F. Who Regulates Whom? An Overview of RNA Granules and Viral Infections. Viruses 2016; 8(7)E180 [http://dx.doi.org/10.3390/v8070180] [PMID: 27367717]

[41] Sharma NR, Majerciak V, Kruhlak MJ, Zheng ZM. KSHV inhibits stress granule formation by viral ORF57 blocking PKR activation. PLoS Pathog 2017; 13(10)e1006677 [http://dx.doi.org/10.1371/journal.ppat.1006677] [PMID: 29084250]

[42] Valderrama C, Clark A, Urano F, Unanue ER, Carrero JA. Listeria monocytogenes induces an interferon-enhanced activation of the integrated stress response that is detrimental for resolution of infection in mice. Eur J Immunol 2017; 47(5): 830-40. [http://dx.doi.org/10.1002/eji.201646856] [PMID: 28267207]

[43] Hato T, Maier B, Syed F, *et al.* Bacterial sepsis triggers an antiviral response that causes translation shutdown. J Clin Invest 2019; 129(1): 296-309. [http://dx.doi.org/10.1172/JCI123284] [PMID: 30507610]

[44] Karagöz GE, Acosta-Alvear D, Walter P. The Unfolded Protein Response: Detecting and Responding to Fluctuations in the Protein-Folding Capacity of the Endoplasmic Reticulum. Cold Spring Harb Perspect Biol 2019; 11(9)a033886 [http://dx.doi.org/10.1101/cshperspect.a033886] [PMID: 30670466]

[45] Osorio F, Tavernier SJ, Hoffmann E, *et al.* The unfolded-protein-response sensor IRE-1α regulates the function of CD8α+ dendritic cells. Nat Immunol 2014; 15(3): 248-57. [http://dx.doi.org/10.1038/ni.2808] [PMID: 24441789]

[46] Tavernier SJ, Osorio F, Vandersarren L, *et al.* Regulated IRE1-dependent mRNA decay sets the threshold for dendritic cell survival. Nat Cell Biol 2017; 19(6): 698-710. [http://dx.doi.org/10.1038/ncb3518] [PMID: 28459443]

[47] Grootjans J, Kaser A, Kaufman RJ, Blumberg RS. The unfolded protein response in immunity and inflammation. Nat Rev Immunol 2016; 16(8): 469-84. [http://dx.doi.org/10.1038/nri.2016.62] [PMID: 27346803]

[48] Tavernier SJ, Lambrecht BN, Janssens S. The Unfolded Protein Response in the Immune Cell Development: Putting the Caretaker in the Driving Seat. Curr Top Microbiol Immunol 2018; 414: 45-72. [http://dx.doi.org/10.1007/82_2017_1] [PMID: 28702709]

[49] Eckard SC, Rice GI, Fabre A, *et al.* The SKIV2L RNA exosome limits activation of the RIG-I-like receptors. Nat Immunol 2014; 15(9): 839-45. [http://dx.doi.org/10.1038/ni.2948] [PMID: 25064072]

[50] Soonthornvacharin S, Rodriguez-Frandsen A, Zhou Y, *et al.* Systems-based analysis of RIG-I-dependent signalling identifies KHSRP as an inhibitor of RIG-I receptor activation. Nat Microbiol 2017; 2: 17022. [http://dx.doi.org/10.1038/nmicrobiol.2017.22] [PMID: 28248290]

[51] Urano F, Wang X, Bertolotti A, *et al.* Coupling of stress in the ER to activation of JNK protein kinases by transmembrane protein kinase IRE1. Science 2000; 287(5453): 664-6. [http://dx.doi.org/10.1126/science.287.5453.664] [PMID: 10650002]

[52] Reverendo M, Mendes A, Argüello RJ, Gatti E, Pierre P. At the crossway of ER-stress and proinflammatory responses. FEBS J 2019; 286(2): 297-310. [http://dx.doi.org/10.1111/febs.14391] [PMID: 29360216]

[53] Keestra-Gounder AM, Byndloss MX, Seyffert N, *et al.* NOD1 and NOD2 signalling links ER stress with inflammation. Nature 2016; 532(7599): 394-7. [http://dx.doi.org/10.1038/nature17631] [PMID: 27007849]

[54] Janssens S, Pulendran B, Lambrecht BN. Emerging functions of the unfolded protein response in immunity. Nat Immunol 2014; 15(10): 910-9. [http://dx.doi.org/10.1038/ni.2991] [PMID: 25232821]

[55] Zhou R, Tardivel A, Thorens B, Choi I, Tschopp J. Thioredoxin-interacting protein links oxidative stress to inflammasome activation. Nat Immunol 2010; 11(2): 136-40. [http://dx.doi.org/10.1038/ni.1831] [PMID: 20023662]

[56] Zhang K, Shen X, Wu J, *et al.* Endoplasmic reticulum stress activates cleavage of CREBH to induce a systemic inflammatory response. Cell 2006; 124(3): 587-99. [http://dx.doi.org/10.1016/j.cell.2005.11.040] [PMID: 16469704]

[57] Frabutt DA, Zheng YH. Arms Race between Enveloped Viruses and the Host ERAD Machinery. Viruses 2016; 8(9)E255 [http://dx.doi.org/10.3390/v8090255] [PMID: 27657106]

[58] Cervantes-Ortiz SL, Zamorano Cuervo N, Grandvaux N. Respiratory Syncytial Virus and Cellular Stress Responses: Impact on Replication and Physiopathology. Viruses 2016; 8(5)E124 [http://dx.doi.org/10.3390/v8050124] [PMID: 27187445]

[59] Robinson M, Schor S, Barouch-Bentov R, Einav S. Viral journeys on the intracellular highways. Cell Mol Life Sci 2018; 75(20): 3693-714. [http://dx.doi.org/10.1007/s00018-018-2882-0] [PMID: 30043139]

[60] Moretti J, Roy S, Bozec D, *et al.* STING Senses Microbial viability to Orchestrate Stress-Mediated Autophagy of the Endoplasmic Reticulum. Cell 2017; 171(4): 809-823.e13. [http://dx.doi.org/10.1016/j.cell.2017.09.034] [PMID: 29056340]

[61] Singh J, Aballay A. Endoplasmic Reticulum Stress Caused by Lipoprotein Accumulation Suppresses Immunity against Bacterial Pathogens and Contributes to Immunosenescence. MBio 2017; 8(3): e00778-17. [http://dx.doi.org/10.1128/mBio.00778-17] [PMID: 28559483]

[62] Thevenot PT, Sierra RA, Raber PL, *et al.* The stress-response sensor chop regulates the function and accumulation of myeloid-derived suppressor cells in tumors. Immunity 2014; 41(3): 389-401. [http://dx.doi.org/10.1016/j.immuni.2014.08.015] [PMID: 25238096]

[63] Cao Y, Trillo-Tinoco J, Sierra RA, *et al.* ER stress-induced mediator C/EBP homologous protein thwarts effector T cell activity in tumors through T-bet repression. Nat Commun 2019; 10(1): 1280. [http://dx.doi.org/10.1038/s41467-019-09263-1] [PMID: 30894532]

[64] Elledge SJ. The DNA Damage Response-Self-awareness for DNA: The 2015 Albert Lasker Basic Medical Research Award. JAMA 2015; 314(11): 1111-2. [http://dx.doi.org/10.1001/jama.2015.10387] [PMID: 26348099]

[65] Shah GA, O'Shea CC. Viral and Cellular Genomes Activate Distinct DNA Damage Responses. Cell 2015; 162(5): 987-1002. [http://dx.doi.org/10.1016/j.cell.2015.07.058] [PMID: 26317467]

[66] Deshpande RA, Lee JH, Paull TT. Rad50 ATPase activity is regulated by DNA ends and requires coordination of both active sites. Nucleic Acids Res 2017; 45(9): 5255-68. [http://dx.doi.org/10.1093/nar/gkx173] [PMID: 28369545]

[67] Stracker TH, Carson CT, Weitzman MD. Adenovirus oncoproteins inactivate the Mre11-Rad50-NBS1 DNA repair complex. Nature 2002; 418(6895): 348-52. [http://dx.doi.org/10.1038/nature00863] [PMID: 12124628]

[68] Weitzman MD, Fradet-Turcotte A. Virus DNA Replication and the Host DNA Damage Response. Annu Rev Virol 2018; 5(1): 141-64. [http://dx.doi.org/10.1146/annurev-virology-092917-043534] [PMID: 29996066]

[69] Lane D, Levine A. p53 Research: the past thirty years and the next thirty years. Cold Spring Harb Perspect Biol 2010; 2(12)a000893 [http://dx.doi.org/10.1101/cshperspect.a000893] [PMID: 20463001]

[70] Miciak J, Bunz F. Long story short: p53 mediates innate immunity. Biochim Biophys Acta 2016; 1865(2): 220-7. [PMID: 26951863]

[71] Huang TT, Wuerzberger-Davis SM, Wu ZH, Miyamoto S. Sequential modification of NEMO/IKKgamma by SUMO-1 and ubiquitin mediates NF-kappaB activation by genotoxic stress. Cell 2003; 115(5): 565-76. [http://dx.doi.org/10.1016/S0092-8674(03)00895-X] [PMID: 14651848]

[72] McCool KW, Miyamoto S. DNA damage-dependent NF-κB activation: NEMO turns nuclear signaling inside out. Immunol Rev 2012; 246(1): 311-26. [http://dx.doi.org/10.1111/j.1600-065X.2012.01101.x] [PMID: 22435563]

[73] Stilmann M, Hinz M, Arslan SC, Zimmer A, Schreiber V, Scheidereit C. A nuclear poly(ADP-ribose--dependent signalosome confers DNA damage-induced IkappaB kinase activation. Mol Cell 2009; 36(3): 365-78.
[http://dx.doi.org/10.1016/j.molcel.2009.09.032] [PMID: 19917246]

[74] Kaur K, Park H, Pandey N, Azuma Y, De Guzman RN. Identification of a new small ubiquitin-like modifier (SUMO)-interacting motif in the E3 ligase PIASy. J Biol Chem 2017; 292(24): 10230-8.
[http://dx.doi.org/10.1074/jbc.M117.789982] [PMID: 28455449]

[75] Barroso-González J, Auclair S, Luan S, *et al.* PACS-2 mediates the ATM and NF-κB-dependent induction of anti-apoptotic Bcl-xL in response to DNA damage. Cell Death Differ 2016; 23(9): 1448-57.
[http://dx.doi.org/10.1038/cdd.2016.23] [PMID: 26943323]

[76] Brzostek-Racine S, Gordon C, Van Scoy S, Reich NC. The DNA damage response induces IFN. J Immunol 2011; 187(10): 5336-45.
[http://dx.doi.org/10.4049/jimmunol.1100040] [PMID: 22013119]

[77] Stetson DB, Ko JS, Heidmann T, Medzhitov R. Trex1 prevents cell-intrinsic initiation of autoimmunity. Cell 2008; 134(4): 587-98.
[http://dx.doi.org/10.1016/j.cell.2008.06.032] [PMID: 18724932]

[78] Erdal E, Haider S, Rehwinkel J, Harris AL, McHugh PJ. A prosurvival DNA damage-induced cytoplasmic interferon response is mediated by end resection factors and is limited by Trex1. Genes Dev 2017; 31(4): 353-69.
[http://dx.doi.org/10.1101/gad.289769.116] [PMID: 28279982]

[79] Bhattacharya S, Srinivasan K, Abdisalaam S, *et al.* RAD51 interconnects between DNA replication, DNA repair and immunity. Nucleic Acids Res 2017; 45(8): 4590-605.
[http://dx.doi.org/10.1093/nar/gkx126] [PMID: 28334891]

[80] Shi W, Vu T, Boucher D, *et al.* Ssb1 and Ssb2 cooperate to regulate mouse hematopoietic stem and progenitor cells by resolving replicative stress. Blood 2017; 129(18): 2479-92.
[http://dx.doi.org/10.1182/blood-2016-06-725093] [PMID: 28270450]

[81] Härtlova A, Erttmann SF, Raffi FA, *et al.* DNA damage primes the type I interferon system *via* the cytosolic DNA sensor STING to promote anti-microbial innate immunity. Immunity 2015; 42(2): 332-43.
[http://dx.doi.org/10.1016/j.immuni.2015.01.012] [PMID: 25692705]

[82] Zingoni A, Fionda C, Borrelli C, Cippitelli M, Santoni A, Soriani A. Natural Killer Cell Response to Chemotherapy-Stressed Cancer Cells: Role in Tumor Immunosurveillance. Front Immunol 2017; 8: 1194.
[http://dx.doi.org/10.3389/fimmu.2017.01194] [PMID: 28993779]

[83] Poliezhaieva T, Ermolaeva MA. DNA damage in protective and adverse inflammatory responses Friend of foe? Mech Ageing Dev 2017; 165(PtA): 47-53.

[84] Kang C, Xu Q, Martin TD, *et al.* The DNA damage response induces inflammation and senescence by inhibiting autophagy of GATA4. Science 2015; 349(6255)aaa5612
[http://dx.doi.org/10.1126/science.aaa5612] [PMID: 26404840]

[85] Ioannidou A, Goulielmaki E, Garinis GA. DNA Damage: From Chronic Inflammation to Age-Related Deterioration. Front Genet 2016; 7: 187.
[http://dx.doi.org/10.3389/fgene.2016.00187] [PMID: 27826317]

[86] Li T, Chen ZJ. The cGAS-cGAMP-STING pathway connects DNA damage to inflammation, senescence, and cancer. J Exp Med 2018; 215(5): 1287-99.
[http://dx.doi.org/10.1084/jem.20180139] [PMID: 29622565]

[87] Nakad R, Schumacher B. DNA Damage Response and Immune Defense: Links and Mechanisms. Front Genet 2016; 7: 147.

[http://dx.doi.org/10.3389/fgene.2016.00147] [PMID: 27555866]

[88] McNally JP, Millen SH, Chaturvedi V, *et al.* Manipulating DNA damage-response signaling for the treatment of immune-mediated diseases. Proc Natl Acad Sci USA 2017; 114(24): E4782-91. [http://dx.doi.org/10.1073/pnas.1703683114] [PMID: 28533414]

[89] Rodriguez E, Chevalier J, El Ghoul H, Voldum-Clausen K, Mundy J, Petersen M. DNA damage as a consequence of NLR activation. PLoS Genet 2018; 14(2)e1007235 [http://dx.doi.org/10.1371/journal.pgen.1007235] [PMID: 29462140]

[90] Chumduri C, Gurumurthy RK, Zietlow R, Meyer TF. Subversion of host genome integrity by bacterial pathogens. Nat Rev Mol Cell Biol 2016; 17(10): 659-73. [http://dx.doi.org/10.1038/nrm.2016.100] [PMID: 27534801]

[91] Adeyemi RO, Pintel DJ. The ATR signaling pathway is disabled during infection with the parvovirus minute virus of mice. J Virol 2014; 88(17): 10189-99. [http://dx.doi.org/10.1128/JVI.01412-14] [PMID: 24965470]

[92] White MK, Bellizzi A, Ibba G, Pietropaolo V, Palamara AT, Wollebo HS. The DNA damage response promotes polyomavirus JC infection by nucleus to cytoplasm NF- kappaB activation. Virol J 2017; 14(1): 31. [http://dx.doi.org/10.1186/s12985-017-0707-7] [PMID: 28202068]

[93] Edwards TG, Bloom DC, Fisher C. The ATM and Rad3-Related (ATR) Protein Kinase Pathway Is Activated by Herpes Simplex Virus 1 and Required for Efficient Viral Replication. J Virol 2018; 92(6): e01884-17. [PMID: 29263259]

[94] Rawlinson SM, Zhao T, Rozario AM, *et al.* Viral regulation of host cell biology by hijacking of the nucleolar DNA-damage response. Nat Commun 2018; 9(1): 3057. [http://dx.doi.org/10.1038/s41467-018-05354-7] [PMID: 30076298]

[95] Munoz-Fontela C, Garcia MA, Garcia-Cao I, *et al.* Resistance to viral infection of super p53 mice. Oncogene 2005; 24(18): 3059-62. [http://dx.doi.org/10.1038/sj.onc.1208477] [PMID: 17726827]

[96] Levine AJ, Oren M. The first 30 years of p53: growing ever more complex. Nat Rev Cancer 2009; 9(10): 749-58. [http://dx.doi.org/10.1038/nrc2723] [PMID: 19776744]

[97] Martinez-Zapien D, Ruiz FX, Poirson J, *et al.* Structure of the E6/E6AP/p53 complex required for HPV-mediated degradation of p53. Nature 2016; 529(7587): 541-5. [http://dx.doi.org/10.1038/nature16481] [PMID: 26789255]

[98] Frost JR, Mendez M, Soriano AM, *et al.* Adenovirus 5 E1A-Mediated Suppression of p53 *via* FUBP1. J Virol 2018; 92(14): e00439-18. [http://dx.doi.org/10.1128/JVI.00439-18] [PMID: 29743362]

[99] Soria C, Estermann FE, Espantman KC, O'Shea CC. Heterochromatin silencing of p53 target genes by a small viral protein. Nature 2010; 466(7310): 1076-81. [http://dx.doi.org/10.1038/nature09307] [PMID: 20740008]

[100] Siegl C, Rudel T. Modulation of p53 during bacterial infections. Nat Rev Microbiol 2015; 13(12): 741-8. [http://dx.doi.org/10.1038/nrmicro3537] [PMID: 26548915]

[101] González E, Rother M, Kerr MC, *et al.* Chlamydia infection depends on a functional MDM2-p53 axis. Nat Commun 2014; 5: 5201. [http://dx.doi.org/10.1038/ncomms6201] [PMID: 25392082]

[102] Drayman N, Ben-Nun-Shaul O, Butin-Israeli V, *et al.* p53 elevation in human cells halt SV40 infection by inhibiting T-ag expression. Oncotarget 2016; 7(33): 52643-60. [http://dx.doi.org/10.18632/oncotarget.10769] [PMID: 27462916]

[103] Levine AJ, Ting DT, Greenbaum BD. P53 and the defenses against genome instability caused by transposons and repetitive elements. BioEssays 2016; 38(6): 508-13. [http://dx.doi.org/10.1002/bies.201600031] [PMID: 27172878]

[104] Wylie A, Jones AE, D'Brot A, *et al.* p53 genes function to restrain mobile elements. Genes Dev 2016; 30(1): 64-77. [http://dx.doi.org/10.1101/gad.266098.115] [PMID: 26701264]

[105] Takaoka A, Hayakawa S, Yanai H, *et al.* Integration of interferon-alpha/beta signalling to p53 responses in tumour suppression and antiviral defence. Nature 2003; 424(6948): 516-23. [http://dx.doi.org/10.1038/nature01850] [PMID: 12872134]

[106] Dharel N, Kato N, Muroyama R, *et al.* Potential contribution of tumor suppressor p53 in the host defense against hepatitis C virus. Hepatology 2008; 47(4): 1136-49. [http://dx.doi.org/10.1002/hep.22176] [PMID: 18220274]

[107] Menendez D, Nguyen TA, Snipe J, Resnick MA. The Cytidine Deaminase APOBEC3 Family Is Subject to Transcriptional Regulation by p53. Mol Cancer Res 2017; 15(6): 735-43. [http://dx.doi.org/10.1158/1541-7786.MCR-17-0019] [PMID: 28232385]

[108] Maruyama W, Shirakawa K, Matsui H, *et al.* Classical NF-κB pathway is responsible for APOBEC3B expression in cancer cells. Biochem Biophys Res Commun 2016; 478(3): 1466-71. [http://dx.doi.org/10.1016/j.bbrc.2016.08.148] [PMID: 27577680]

[109] Taura M, Eguma A, Suico MA, *et al.* p53 regulates Toll-like receptor 3 expression and function in human epithelial cell lines. Mol Cell Biol 2008; 28(21): 6557-67. [http://dx.doi.org/10.1128/MCB.01202-08] [PMID: 18779317]

[110] Zhang X, Bogunovic D, Payelle-Brogard B, *et al.* Human intracellular ISG15 prevents interferon-α/β over-amplification and auto-inflammation. Nature 2015; 517(7532): 89-93. [http://dx.doi.org/10.1038/nature13801] [PMID: 25307056]

[111] Arimoto KI, Löchte S, Stoner SA, *et al.* STAT2 is an essential adaptor in USP18-mediated suppression of type I interferon signaling. Nat Struct Mol Biol 2017; 24(3): 279-89. [http://dx.doi.org/10.1038/nsmb.3378] [PMID: 28165510]

[112] Ak P, Levine AJ. p53 and NF-κB: different strategies for responding to stress lead to a functional antagonism. FASEB J 2010; 24(10): 3643-52. [http://dx.doi.org/10.1096/fj.10-160549] [PMID: 20530750]

[113] Carrà G, Crivellaro S, Taulli R, Guerrasio A, Saglio G, Morotti A. Mechanisms of p53 Functional De-Regulation: Role of the IκB-α/p53 Complex. Int J Mol Sci 2016; 17(12)E1997 [http://dx.doi.org/10.3390/ijms17121997] [PMID: 27916821]

[114] Rocha S, Campbell KJ, Perkins ND. p53- and Mdm2-independent repression of NF-kappa B transactivation by the ARF tumor suppressor. Mol Cell 2003; 12(1): 15-25. [http://dx.doi.org/10.1016/S1097-2765(03)00223-5] [PMID: 12887889]

[115] Fierabracci A, Pellegrino M. The Double Role of p53 in Cancer and Autoimmunity and Its Potential as Therapeutic Target. Int J Mol Sci 2016; 17(12)E1975 [http://dx.doi.org/10.3390/ijms17121975] [PMID: 27897991]

[116] Gudkov AV, Komarova EA. p53 and the Carcinogenicity of Chronic Inflammation. Cold Spring Harb Perspect Med 2016; 6(11)a026161 [http://dx.doi.org/10.1101/cshperspect.a026161] [PMID: 27549311]

[117] Muñoz-Fontela C, Mandinova A, Aaronson SA, Lee SW. Emerging roles of p53 and other tumour-suppressor genes in immune regulation. Nat Rev Immunol 2016; 16(12): 741-50. [http://dx.doi.org/10.1038/nri.2016.99] [PMID: 27667712]

[118] Komarova EA, Krivokrysenko V, Wang K, *et al.* p53 is a suppressor of inflammatory response in mice. FASEB J 2005; 19(8): 1030-2.

[http://dx.doi.org/10.1096/fj.04-3213fje] [PMID: 15811878]

[119] Zhu S, Pan W, Song X, *et al.* The microRNA miR-23b suppresses IL-17-associated autoimmune inflammation by targeting TAB2, TAB3 and IKK-α. Nat Med 2012; 18(7): 1077-86. [http://dx.doi.org/10.1038/nm.2815] [PMID: 22660635]

[120] Schweitzer K, Naumann M. CSN-associated USP48 confers stability to nuclear NF-κB/RelA by trimming K48-linked Ub-chains. Biochim Biophys Acta 2015; 1853(2): 453-69. [http://dx.doi.org/10.1016/j.bbamcr.2014.11.028] [PMID: 25486460]

[121] Cetkovská K, Šustová H, Uldrijan S. Ubiquitin-specific peptidase 48 regulates Mdm2 protein levels independent of its deubiquitinase activity. Sci Rep 2017; 7: 43180. [http://dx.doi.org/10.1038/srep43180] [PMID: 28233861]

[122] Yoon KW, Byun S, Kwon E, *et al.* Control of signaling-mediated clearance of apoptotic cells by the tumor suppressor p53. Science 2015; 349(6247)1261669 [http://dx.doi.org/10.1126/science.1261669] [PMID: 26228159]

[123] Lingappan K. NF-κB in Oxidative Stress. Curr Opin Toxicol 2018; 7: 81-6. [http://dx.doi.org/10.1016/j.cotox.2017.11.002] [PMID: 29862377]

[124] Herb M, Gluschko A, Wiegmann K, *et al.* Mitochondrial reactive oxygen species enable proinflammatory signaling through disulfide linkage of NEMO. Sci Signal 2019; 12(568)eaar5926 [http://dx.doi.org/10.1126/scisignal.aar5926] [PMID: 30755476]

[125] Guo Z, Kozlov S, Lavin MF, Person MD, Paull TT. ATM activation by oxidative stress. Science 2010; 330(6003): 517-21. [http://dx.doi.org/10.1126/science.1192912] [PMID: 20966255]

[126] Fang L, Choudhary S, Zhao Y, *et al.* ATM regulates NF-κB-dependent immediate-early genes *via* RelA Ser 276 phosphorylation coupled to CDK9 promoter recruitment. Nucleic Acids Res 2014; 42(13): 8416-32. [http://dx.doi.org/10.1093/nar/gku529] [PMID: 24957606]

[127] Al-Khafaji AB, Tohme S, Yazdani HO, Miller D, Huang H, Tsung A. Superoxide induces Neutrophil Extracellular Trap Formation in a TLR-4 and NOX-dependent mechanism. Mol Med 2016; 22: 621-31. [http://dx.doi.org/10.2119/molmed.2016.00054] [PMID: 27453505]

[128] Rizzi M, Carniato F, Tonello S, *et al.* Charged molecular silica trigger *in vitro* NETosis in human granulocytes *via* both oxidative and autophagic pathways. Eur Rev Med Pharmacol Sci 2018; 22(20): 7058-68. [PMID: 30402874]

[129] Baruah S, Murthy S, Keck K, *et al.* TREM-1 regulates neutrophil chemotaxis by promoting NOX-dependent superoxide production. J Leukoc Biol 2019; 105(6): 1195-207. [http://dx.doi.org/10.1002/JLB.3VMA0918-375R] [PMID: 30667543]

[130] Zhao J, Ma J, Deng Y, *et al.* A missense variant in NCF1 is associated with susceptibility to multiple autoimmune diseases. Nat Genet 2017; 49(3): 433-7. [http://dx.doi.org/10.1038/ng.3782] [PMID: 28135245]

[131] Aviello G, Knaus UG. ROS in gastrointestinal inflammation: Rescue Or Sabotage? Br J Pharmacol 2017; 174(12): 1704-18. [http://dx.doi.org/10.1111/bph.13428] [PMID: 26758851]

[132] Meyer A, Laverny G, Bernardi L, *et al.* Mitochondria: An Organelle of Bacterial Origin Controlling Inflammation. Front Immunol 2018; 9: 536. [http://dx.doi.org/10.3389/fimmu.2018.00536] [PMID: 29725325]

[133] Agod Z, Fekete T, Budai MM, *et al.* Regulation of type I interferon responses by mitochondria-derived reactive oxygen species in plasmacytoid dendritic cells. Redox Biol 2017; 13: 633-45. [http://dx.doi.org/10.1016/j.redox.2017.07.016] [PMID: 28818792]

[134] Buskiewicz IA, Montgomery T, Yasewicz EC, *et al.* Reactive oxygen species induce virus-independent MAVS oligomerization in systemic lupus erythematosus. Sci Signal 2016; 9(456): ra115. [http://dx.doi.org/10.1126/scisignal.aaf1933] [PMID: 27899525]

[135] Luo H, Mu WC, Karki R, *et al.* Mitochondrial Stress-Initiated Aberrant Activation of the NLRP3 Inflammasome Regulates the Functional Deterioration of Hematopoietic Stem Cell Aging. Cell Rep 2019; 26(4): 945-954.e4. [http://dx.doi.org/10.1016/j.celrep.2018.12.101] [PMID: 30673616]

[136] Yabal M, Calleja DJ, Simpson DS, Lawlor KE. Stressing out the mitochondria: Mechanistic insights into NLRP3 inflammasome activation. J Leukoc Biol 2019; 105(2): 377-99. [http://dx.doi.org/10.1002/JLB.MR0318-124R] [PMID: 30589456]

[137] Elliott EI, Miller AN, Banoth B, *et al.* Cutting Edge: Mitochondrial Assembly of the NLRP3 Inflammasome Complex Is Initiated at Priming. J Immunol 2018; 200(9): 3047-52. [http://dx.doi.org/10.4049/jimmunol.1701723] [PMID: 29602772]

[138] Erttmann SF, Härtlova A, Sloniecka M, *et al.* Loss of the DNA Damage Repair Kinase ATM Impairs Inflammasome-Dependent Anti-Bacterial Innate Immunity. Immunity 2016; 45(1): 106-18. [http://dx.doi.org/10.1016/j.immuni.2016.06.018] [PMID: 27421701]

[139] Lacey CA, Chambers CA, Mitchell WJ, Skyberg JA. IFN-γ-dependent nitric oxide suppresses Brucella-induced arthritis by inhibition of inflammasome activation. J Leukoc Biol 2019; 106(1): 27-34. [http://dx.doi.org/10.1002/JLB.4MIA1018-409R] [PMID: 30748031]

[140] Alberts BM, Bruce C, Basnayake K, Ghezzi P, Davies KA, Mullen LM. Secretion of IL-1β From Monocytes in Gout Is Redox Independent. Front Immunol 2019; 10: 70. [http://dx.doi.org/10.3389/fimmu.2019.00070] [PMID: 30761138]

[141] Zanoni I, Tan Y, Di Gioia M, *et al.* An endogenous caspase-11 ligand elicits interleukin-1 release from living dendritic cells. Science 2016; 352(6290): 1232-6. [http://dx.doi.org/10.1126/science.aaf3036] [PMID: 27103670]

[142] Chu LH, Indramohan M, Ratsimandresy RA, *et al.* The oxidized phospholipid oxPAPC protects from septic shock by targeting the non-canonical inflammasome in macrophages. Nat Commun 2018; 9(1): 996. [http://dx.doi.org/10.1038/s41467-018-03409-3] [PMID: 29520027]

[143] Ba X, Boldogh I. 8-Oxoguanine DNA glycosylase 1: Beyond repair of the oxidatively modified base lesions. Redox Biol 2018; 14: 669-78. [http://dx.doi.org/10.1016/j.redox.2017.11.008] [PMID: 29175754]

[144] Pan L, Zhu B, Hao W, *et al.* Oxidized Guanine Base Lesions Function in 8-Oxoguanine DNA Glycosylase-1-mediated Epigenetic Regulation of Nuclear Factor κB-driven Gene Expression. J Biol Chem 2016; 291(49): 25553-66. [http://dx.doi.org/10.1074/jbc.M116.751453] [PMID: 27756845]

[145] Visnes T, Cázares-Körner A, Hao W, *et al.* Small-molecule inhibitor of OGG1 suppresses proinflammatory gene expression and inflammation. Science 2018; 362(6416): 834-9. [http://dx.doi.org/10.1126/science.aar8048] [PMID: 30442810]

[146] Tonelli C, Chio IIC, Tuveson DA. Transcriptional Regulation by Nrf2. Antioxid Redox Signal 2018; 29(17): 1727-45. [http://dx.doi.org/10.1089/ars.2017.7342] [PMID: 28899199]

[147] Lee DF, Kuo HP, Liu M, *et al.* KEAP1 E3 ligase-mediated downregulation of NF-kappaB signaling by targeting IKKbeta. Mol Cell 2009; 36(1): 131-40. [http://dx.doi.org/10.1016/j.molcel.2009.07.025] [PMID: 19818716]

[148] Kobayashi EH, Suzuki T, Funayama R, *et al.* Nrf2 suppresses macrophage inflammatory response by blocking proinflammatory cytokine transcription. Nat Commun 2016; 7: 11624.

[http://dx.doi.org/10.1038/ncomms11624] [PMID: 27211851]

[149] O'Neill LAJ, Artyomov MN. Itaconate: the poster child of metabolic reprogramming in macrophage function. Nat Rev Immunol 2019; 19(5): 273-81.
[http://dx.doi.org/10.1038/s41577-019-0128-5] [PMID: 30705422]

[150] Mills EL, Ryan DG, Prag HA, *et al.* Itaconate is an anti-inflammatory metabolite that activates Nrf2 *via* alkylation of KEAP1. Nature 2018; 556(7699): 113-7.
[http://dx.doi.org/10.1038/nature25986] [PMID: 29590092]

[151] Olagnier D, Brandtoft AM, Gunderstofte C, *et al.* Nrf2 negatively regulates STING indicating a link between antiviral sensing and metabolic reprogramming. Nat Commun 2018; 9(1): 3506.
[http://dx.doi.org/10.1038/s41467-018-05861-7] [PMID: 30158636]

[152] Ano S, Panariti A, Allard B, O'Sullivan M, McGovern TK, Hamamoto Y, *et al.* Inflammation and airway hyperresponsiveness after chlorine exposure are prolonged by Nrf2 deficiency in mice. Free radical biology & medicine 2017; 102: 1-15.
[http://dx.doi.org/10.1016/j.freeradbiomed.2016.11.017]

[153] Suzuki T, Yamamoto M. Stress-sensing mechanisms and the physiological roles of the Keap1-Nrf2 system during cellular stress. J Biol Chem 2017; 292(41): 16817-24.
[http://dx.doi.org/10.1074/jbc.R117.800169] [PMID: 28842501]

[154] Pinto A, Bonucci A, Maggi E, Corsi M, Businaro R. Anti-Oxidant and Anti-Inflammatory Activity of Ketogenic Diet: New Perspectives for Neuroprotection in Alzheimer's Disease. Antioxidants 2018; 7(5)E63
[http://dx.doi.org/10.3390/antiox7050063] [PMID: 29710809]

[155] Tsakiri EN, Gumeni S, Iliaki KK, *et al.* Hyperactivation of Nrf2 increases stress tolerance at the cost of aging acceleration due to metabolic deregulation. Aging Cell 2019; 18(1)e12845
[PMID: 30537423]

[156] Karwaciak I, Gorzkiewicz M, Bartosz G, Pulaski L. TLR2 activation induces antioxidant defence in human monocyte-macrophage cell line models. Oncotarget 2017; 8(33): 54243-64.
[http://dx.doi.org/10.18632/oncotarget.17342] [PMID: 28903338]

[157] Flint A, Stintzi A, Saraiva LM. Oxidative and nitrosative stress defences of Helicobacter and Campylobacter species that counteract mammalian immunity. FEMS Microbiol Rev 2016; 40(6): 938-60.
[http://dx.doi.org/10.1093/femsre/fuw025] [PMID: 28201757]

[158] Rockwood N, Costa DL, Amaral EP, *et al. Mycobacterium tuberculosis* Induction of Heme Oxygenase-1 Expression Is Dependent on Oxidative Stress and Reflects Treatment Outcomes. Front Immunol 2017; 8: 542.
[http://dx.doi.org/10.3389/fimmu.2017.00542] [PMID: 28553288]

[159] Schatzman SS, Culotta VC. Chemical Warfare at the Microorganismal Level: A Closer Look at the Superoxide Dismutase Enzymes of Pathogens. ACS Infect Dis 2018; 4(6): 893-903.
[http://dx.doi.org/10.1021/acsinfecdis.8b00026] [PMID: 29517910]

[160] Komaravelli N, Tian B, Ivanciuc T, Mautemps N, Brasier AR, Garofalo RP, *et al.* Respiratory syncytial virus infection down-regulates antioxidant enzyme expression by triggering deacetylation-proteasomal degradation of Nrf2. Free radical biology & medicine 88(Pt B): 391-403.2015;
[http://dx.doi.org/10.1016/j.freeradbiomed.2015.05.043]

[161] Edwards MR, Johnson B, Mire CE, *et al.* The Marburg virus VP24 protein interacts with Keap1 to activate the cytoprotective antioxidant response pathway. Cell Rep 2014; 6(6): 1017-25.
[http://dx.doi.org/10.1016/j.celrep.2014.01.043] [PMID: 24630991]

[162] Corona Velazquez AF, Jackson WT. So Many Roads: the Multifaceted Regulation of Autophagy Induction. Mol Cell Biol 2018; 38(21): e00303-18.
[http://dx.doi.org/10.1128/MCB.00303-18] [PMID: 30126896]

[163] Kwon DH, Song HK. A Structural View of Xenophagy, a Battle between Host and Microbes. Mol Cells 2018; 41(1): 27-34.
[PMID: 29370690]

[164] Siqueira MDS, Ribeiro RM, Travassos LH. Autophagy and Its Interaction With Intracellular Bacterial Pathogens. Front Immunol 2018; 9: 935.
[http://dx.doi.org/10.3389/fimmu.2018.00935] [PMID: 29875765]

[165] Gutierrez MG, Master SS, Singh SB, Taylor GA, Colombo MI, Deretic V. Autophagy is a defense mechanism inhibiting BCG and Mycobacterium tuberculosis survival in infected macrophages. Cell 2004; 119(6): 753-66.
[http://dx.doi.org/10.1016/j.cell.2004.11.038] [PMID: 15607973]

[166] Nakagawa I, Amano A, Mizushima N, *et al.* Autophagy defends cells against invading group A Streptococcus. Science 2004; 306(5698): 1037-40.
[http://dx.doi.org/10.1126/science.1103966] [PMID: 15528445]

[167] Liu Y, Cherry S. Zika virus infection activates sting-dependent antiviral autophagy in the Drosophila brain. Autophagy 2019; 15(1): 174-5.
[http://dx.doi.org/10.1080/15548627.2018.1528813] [PMID: 30260713]

[168] Evans RJ, Sundaramurthy V, Frickel EM. The Interplay of Host Autophagy and Eukaryotic Pathogens. Front Cell Dev Biol 2018; 6: 118.
[http://dx.doi.org/10.3389/fcell.2018.00118] [PMID: 30271774]

[169] Casassa AF, Vanrell MC, Colombo MI, Gottlieb RA, Romano PS. Autophagy plays a protective role against Trypanosoma cruzi infection in mice. Virulence 2019; 10(1): 151-65.
[http://dx.doi.org/10.1080/21505594.2019.1584027] [PMID: 30829115]

[170] Thurston TL, Wandel MP, von Muhlinen N, Foeglein A, Randow F. Galectin 8 targets damaged vesicles for autophagy to defend cells against bacterial invasion. Nature 2012; 482(7385): 414-8.
[http://dx.doi.org/10.1038/nature10744] [PMID: 22246324]

[171] Harada K, Kotani T, Kirisako H, *et al.* Two distinct mechanisms target the autophagy-related E3 complex to the pre-autophagosomal structure. eLife 2019; 8e43088
[http://dx.doi.org/10.7554/eLife.43088] [PMID: 30810528]

[172] Choi J, Biering SB, Hwang S. Quo vadis? Interferon-inducible GTPases go to their target membranes *via* the LC3-conjugation system of autophagy. Small GTPases 2017; 8(4): 199-207.
[http://dx.doi.org/10.1080/21541248.2016.1213090] [PMID: 27428166]

[173] Huett A, Heath RJ, Begun J, *et al.* The LRR and RING domain protein LRSAM1 is an E3 ligase crucial for ubiquitin-dependent autophagy of intracellular Salmonella Typhimurium. Cell Host Microbe 2012; 12(6): 778-90.
[http://dx.doi.org/10.1016/j.chom.2012.10.019] [PMID: 23245322]

[174] Mishra R, Upadhyay A, Prajapati VK, *et al.* LRSAM1 E3 ubiquitin ligase: molecular neurobiological perspectives linked with brain diseases. Cell Mol Life Sci 2019; 76(11): 2093-110.
[http://dx.doi.org/10.1007/s00018-019-03055-y] [PMID: 30826859]

[175] Bian W, Guo Y, Zhang Y, Li H. The self-association and activity regulation of LRSAM1 E3 ligase. Biochem Biophys Res Commun 2017; 485(1): 95-101.
[http://dx.doi.org/10.1016/j.bbrc.2017.02.026] [PMID: 28189685]

[176] Manzanillo PS, Ayres JS, Watson RO, *et al.* The ubiquitin ligase parkin mediates resistance to intracellular pathogens. Nature 2013; 501(7468): 512-6.
[http://dx.doi.org/10.1038/nature12566] [PMID: 24005326]

[177] Polajnar M, Dietz MS, Heilemann M, Behrends C. Expanding the host cell ubiquitylation machinery targeting cytosolic *Salmonella.* EMBO Rep 2017; 18(9): 1572-85.
[http://dx.doi.org/10.15252/embr.201643851] [PMID: 28784601]

[178] Noad J, von der Malsburg A, Pathe C, Michel MA, Komander D, Randow F. LUBAC-synthesized linear ubiquitin chains restrict cytosol-invading bacteria by activating autophagy and NF-κB. Nat Microbiol 2017; 2: 17063.
[http://dx.doi.org/10.1038/nmicrobiol.2017.63] [PMID: 28481331]

[179] van Wijk SJL, Fricke F, Herhaus L, *et al.* Linear ubiquitination of cytosolic Salmonella Typhimurium activates NF-κB and restricts bacterial proliferation. Nat Microbiol 2017; 2: 17066.
[http://dx.doi.org/10.1038/nmicrobiol.2017.66] [PMID: 28481361]

[180] Ahmad L, Mostowy S, Sancho-Shimizu V. Autophagy-Virus Interplay: From Cell Biology to Human Disease. Front Cell Dev Biol 2018; 6: 155.
[http://dx.doi.org/10.3389/fcell.2018.00155] [PMID: 30510929]

[181] Mori H, Fukuhara T, Ono C, *et al.* Induction of selective autophagy in cells replicating hepatitis C virus genome. J Gen Virol 2018; 99(12): 1643-57.
[http://dx.doi.org/10.1099/jgv.0.001161] [PMID: 30311874]

[182] Chiramel AI, Best SM. Role of autophagy in Zika virus infection and pathogenesis. Virus Res 2018; 254: 34-40.
[http://dx.doi.org/10.1016/j.virusres.2017.09.006] [PMID: 28899653]

[183] Sparrer KMJ, Gack MU. TRIM proteins: New players in virus-induced autophagy. PLoS Pathog 2018; 14(2)e1006787
[http://dx.doi.org/10.1371/journal.ppat.1006787] [PMID: 29389992]

[184] Travassos LH, Carneiro LA, Ramjeet M, *et al.* Nod1 and Nod2 direct autophagy by recruiting ATG16L1 to the plasma membrane at the site of bacterial entry. Nat Immunol 2010; 11(1): 55-62.
[http://dx.doi.org/10.1038/ni.1823] [PMID: 19898471]

[185] Joubert PE, Meiffren G, Grégoire IP, *et al.* Autophagy induction by the pathogen receptor CD46. Cell Host Microbe 2009; 6(4): 354-66.
[http://dx.doi.org/10.1016/j.chom.2009.09.006] [PMID: 19837375]

[186] Kim WJ, Mai A, Weyand NJ, Rendón MA, Van Doorslaer K, So M. Neisseria gonorrhoeae evades autophagic killing by downregulating CD46-cyt1 and remodeling lysosomes. PLoS Pathog 2019; 15(2)e1007495
[http://dx.doi.org/10.1371/journal.ppat.1007495] [PMID: 30753248]

[187] Heckmann BL, Green DR. LC3-associated phagocytosis at a glance. J Cell Sci 2019; 132(5)jcs222984
[http://dx.doi.org/10.1242/jcs.222984] [PMID: 30787029]

[188] Gui X, Yang H, Li T, *et al.* Autophagy induction *via* STING trafficking is a primordial function of the cGAS pathway. Nature 2019; 567(7747): 262-6.
[http://dx.doi.org/10.1038/s41586-019-1006-9] [PMID: 30842662]

[189] Deretic V, Levine B. Autophagy balances inflammation in innate immunity. Autophagy 2018; 14(2): 243-51.
[http://dx.doi.org/10.1080/15548627.2017.1402992] [PMID: 29165043]

[190] Tschurtschenthaler M, Adolph TE, Ashcroft JW, *et al.* Defective ATG16L1-mediated removal of IRE1α drives Crohn's disease-like ileitis. J Exp Med 2017; 214(2): 401-22.
[http://dx.doi.org/10.1084/jem.20160791] [PMID: 28082357]

[191] Bel S, Pendse M, Wang Y, *et al.* Paneth cells secrete lysozyme *via* secretory autophagy during bacterial infection of the intestine. Science 2017; 357(6355): 1047-52.
[http://dx.doi.org/10.1126/science.aal4677] [PMID: 28751470]

[192] Matsuzawa-Ishimoto Y, Hwang S, Cadwell K. Autophagy and Inflammation. Annu Rev Immunol 2018; 36: 73-101.
[http://dx.doi.org/10.1146/annurev-immunol-042617-053253] [PMID: 29144836]

[193] Desai M, Fang R, Sun J. The role of autophagy in microbial infection and immunity. ImmunoTargets

Ther 2015; 4: 13-26.
[PMID: 27471708]

[194] Choy A, Dancourt J, Mugo B, *et al.* The Legionella effector RavZ inhibits host autophagy through irreversible Atg8 deconjugation. Science 2012; 338(6110): 1072-6.
[http://dx.doi.org/10.1126/science.1227026] [PMID: 23112293]

[195] Fiskin E, Bhogaraju S, Herhaus L, Kalayil S, Hahn M, Dikic I. Structural basis for the recognition and degradation of host TRIM proteins by Salmonella effector SopA. Nat Commun 2017; 8: 14004.
[http://dx.doi.org/10.1038/ncomms14004] [PMID: 28084320]

[196] Huang J, Brumell JH. Bacteria-autophagy interplay: a battle for survival. Nat Rev Microbiol 2014; 12(2): 101-14.
[http://dx.doi.org/10.1038/nrmicro3160] [PMID: 24384599]

[197] Paul P, Münz C. Autophagy and Mammalian Viruses: Roles in Immune Response, Viral Replication, and Beyond. Adv Virus Res 2016; 95: 149-95.
[http://dx.doi.org/10.1016/bs.aivir.2016.02.002] [PMID: 27112282]

[198] Thaiss CA, Levy M, Itav S, Elinav E. Integration of Innate Immune Signaling. Trends Immunol 2016; 37(2): 84-101.
[http://dx.doi.org/10.1016/j.it.2015.12.003] [PMID: 26755064]

[199] Franz KM, Kagan JC. Innate Immune Receptors as Competitive Determinants of Cell Fate. Mol Cell 2017; 66(6): 750-60.
[http://dx.doi.org/10.1016/j.molcel.2017.05.009] [PMID: 28622520]

[200] Gutschow MV, Mason JC, Lane KM, *et al.* Combinatorial processing of bacterial and host-derived innate immune stimuli at the single-cell level. Mol Biol Cell 2019; 30(2): 282-92.
[http://dx.doi.org/10.1091/mbc.E18-07-0423] [PMID: 30462580]

[201] Gaidt MM, Hornung V. The NLRP3 Inflammasome Renders Cell Death Pro-inflammatory. J Mol Biol 2018; 430(2): 133-41.
[http://dx.doi.org/10.1016/j.jmb.2017.11.013] [PMID: 29203171]

[202] Dhabhar FS. The short-term stress response - Mother nature's mechanism for enhancing protection and performance under conditions of threat, challenge, and opportunity. Front Neuroendocrinol 2018; 49: 175-92.
[http://dx.doi.org/10.1016/j.yfrne.2018.03.004] [PMID: 29596867]

[203] Martine P, Chevriaux A, Derangère V, *et al.* HSP70 is a negative regulator of NLRP3 inflammasome activation. Cell Death Dis 2019; 10(4): 256.
[http://dx.doi.org/10.1038/s41419-019-1491-7] [PMID: 30874540]

[204] McKinstry M, Chung C, Truong H, Johnston BA, Snow JW. The heat shock response and humoral immune response are mutually antagonistic in honey bees. Sci Rep 2017; 7(1): 8850.
[http://dx.doi.org/10.1038/s41598-017-09159-4] [PMID: 28821863]

[205] Nejat N, Mantri N. Plant Immune System: Crosstalk Between Responses to Biotic and Abiotic Stresses the Missing Link in Understanding Plant Defence. Curr Issues Mol Biol 2017; 23: 1-16.
[http://dx.doi.org/10.21775/cimb.023.001] [PMID: 28154243]

[206] Castrillo G, Teixeira PJ, Paredes SH, *et al.* Root microbiota drive direct integration of phosphate stress and immunity. Nature 2017; 543(7646): 513-8.
[http://dx.doi.org/10.1038/nature21417] [PMID: 28297714]

[207] Gladwyn-Ng I, Cordón-Barris L, Alfano C, *et al.* Stress-induced unfolded protein response contributes to Zika virus-associated microcephaly. Nat Neurosci 2018; 21(1): 63-71.
[http://dx.doi.org/10.1038/s41593-017-0038-4] [PMID: 29230053]

[208] Martins R, Carlos AR, Braza F, *et al.* Disease Tolerance as an Inherent Component of Immunity. Annu Rev Immunol 2019; 37: 405-37.
[http://dx.doi.org/10.1146/annurev-immunol-042718-041739] [PMID: 30673535]

[209] Shourian M, Qureshi ST. Resistance and Tolerance to Cryptococcal Infection: An Intricate Balance That Controls the Development of Disease. Front Immunol 2019; 10: 66. [http://dx.doi.org/10.3389/fimmu.2019.00066] [PMID: 30761136]

[210] Roy S, Nandi AK. Arabidopsis thaliana methionine sulfoxide reductase B8 influences stress-induced cell death and effector-triggered immunity. Plant Mol Biol 2017; 93(1-2): 109-20. [http://dx.doi.org/10.1007/s11103-016-0550-z] [PMID: 27900506]

[211] Weis S, Carlos AR, Moita MR, *et al.* Metabolic Adaptation Establishes Disease Tolerance to Sepsis. Cell 2017; 169(7): 1263-1275.e14. [http://dx.doi.org/10.1016/j.cell.2017.05.031] [PMID: 28622511]

[212] Singh RP, Jha PN. A Halotolerant Bacterium *Bacillus licheniformis* HSW-16 Augments Induced Systemic Tolerance to Salt Stress in Wheat Plant (*Triticum aestivum*). Front Plant Sci 2016; 7: 1890. [http://dx.doi.org/10.3389/fpls.2016.01890] [PMID: 28018415]

CHAPTER 11

Specialized Immune Systems of Animals

Abstract: Specialized immune systems of animals include innate and adaptive immune cells and humoral factors. Inflammation is also a specialized systemic immune response. Innate immune cells include professional phagocytes such as macrophages, dendritic cells, and neutrophils, as well as natural killer (NK) cells. Adaptive immune cells are T and B lymphocytes. Immune cells work in cooperation with humoral factors such as the complement system, secreted pattern-recognition receptors, and antibodies. Inflammation activates immune cells and generates fever. Acute inflammation is beneficial, whereas chronic inflammation may lead to pathology including atherosclerosis, type 2 diabetes, and autoimmune diseases. Recognition of pathogen-associated molecular patterns (PAMPs) by phagocytes leads to their activation, phagocytosis, respiratory burst producing reactive oxygen (ROS) and nitrogen (RNS) species, and secretion of antimicrobial peptides and pro-inflammatory cytokines. NK cells function on a different recognition principle. They express a variety of activating receptors to endogenous molecules expressed on distressed but not healthy cells. Engagement of these receptors triggers cytotoxic activity. NK cells also express inhibitory receptors that bind "self" markers such as MHCI molecules, which are expressed on all healthy cells. The balance between the engagement of activating and inhibitory receptors determines cell fate. Complement is a humoral cascade of enzymes that promotes chemotaxis, inflammation, and phagocytosis, as well as kills pathogens and infected or stressed cells. The evolutionary youngest specialized adaptive immune system consists of T and B cells, which express rearranging receptors both membrane and secreted.

Keywords: Adaptive immune receptors, Complement system, Fever, Inflammation, IL1β, IL6, Inflammatory diseases, Macrophage, NOX2, NK cells, Phagocytosis, PRR, PAMP, Respiratory burst, TNF, V(D)J recombination.

INTRODUCTION

Most animals have specialized immune systems composed of immune cells and humoral factors. In addition to fighting infections, these systems remove dead cells and cell debris from the organism, and facilitate tissue repair and wound healing. Inflammation is also a specialized system. It is induced autonomously as a reaction of an individual cell to stress or infection but can grow into a systemic response.

Specialized immune systems use two principles to detect their targets. The first

Nadia Danilova

one is the recognition of non-self or altered self. It is based on sensing pathogen-associated molecular patterns (PAMPs) or damage-associated molecular patterns (DAMPs) *via* pattern recognition receptors (PRRs) [1]. PRRs can be expressed not only as cell-associated proteins but also as secreted factors, which forms the foundation for the humoral immunity. The second principle used by immune systems is "missing self" recognition. It depends on the expression by all "self-cells" of special markers such as MHCI. These markers are absent in microbes and are downregulated in infected and transformed cells; hence natural killer (NK) cells recognize their absence as "missing self" [2].

The next major subdivision of specialized immune systems depends on the type of immune receptors they use. Mechanisms based on genomically encoded receptors are traditionally called "innate immune mechanisms". Innate immunity includes autonomous immune mechanisms, innate immune cells, humoral factors including complement system, and inflammation. Specialized immune cells recognize, phagocyte, and kill pathogens, produce regulatory factors such as IFNs and other cytokines, and induce inflammation. The complement system is involved in the promotion of phagocytosis, pathogen killing, and induction of inflammation. Inflammation mobilizes immune cells and induces fever. It also coordinates physiological responses with animal behavior for optimal recovery.

Long evolution perfected innate immune mechanisms. However, if a pathogen changes its PAMPs, the host needs to change corresponding PRRs, which can only be done by a slow process of mutation and selection. High numbers of PRRs and redundancy in PAMPs recognition decrease host vulnerability but do not eliminate it. Besides, it hinders the expansion of species to new niches that have different pathogens.

Adaptive immune systems rise above these limitations. They can develop a specific response to practically any new antigen. In adaptive lymphoid cells, immune receptors are encoded as segments that can randomly recombine during cell development [3]. This generates a practically unlimited repertoire of specificities so that every possible "non-self" antigen can be recognized by some of the receptors.

Immune cells differentiate from hematopoietic stem cells [4]. Mammalian immune cells are represented by granulocytes (neutrophils, basophils, and eosinophils), monocytes that differentiate into dendritic cells and macrophages, mast cells, and lymphoid cells. Lymphoid progenitors can differentiate into innate lymphoid cells such as natural killers (NK) cells, or into adaptive lymphoid cells, T and B. T cells are responsible for the cellular branch of the adaptive immune system, whereas B cells are responsible for the humoral branch.

Specialized immune systems incorporate autonomous immune mechanisms, metabolic, and stress signals into their responses through many layers of mutual regulation. They also respond to central physiological regulators such as cytokines, hormones, and neuropeptides [5 - 7].

11.1. Inflammation

Inflammation is both a stress response and an immunodefense mechanism [8]. It is activated in stressed, damaged, and infected cells; cytokines produced by such cells induce a generalized response. The four cardinal signs of inflammation have been defined already by the Roman doctor Cornelius Celsus. They are redness (rubor), swelling (tumor), heat (calor), and pain (dolor). Redness is associated with vasodilatation and increased blood flow that delivers plasma and immune cells to the site of damage or infection. Accumulation of fluid and immune cells leads to swelling. Fever increases body temperature by 2–4°C during acute inflammation, which restricts the proliferation of pathogens and enhances the immune response. Pain signals to the organism to withdraw from the damaging environment and restrict activity.

Inflammation takes control over homeostasis and establishes new set points necessary for the repair of damaged tissue and immune defense [9]. Normally, inflammation ceases when physiological functions are restored. Induction of pro-inflammatory state usually enhances early immune response and therefore plays an important positive role in immunity. Chronic inflammation, however, contributes to multiple autoimmune conditions like arthritis, neurodegeneration, *etc*.

11.1.1. Inflammation Triggers and Mediators

Inflammation starts with the formation of inflammasomes, which leads to the production of pro-inflammatory cytokines [10 - 12]. PRRs signaling or cytokines "prime" inflammasome by increasing the expression of its components. Next, signals from PAMPs or DAMPs during tissue injury or infection activate inflammasome assembly and production of pro-inflammatory cytokines. Necrotic cells release cytokines and DAMPs and are potent inducers of inflammation. Inflammation can also be induced in the absence of infection or tissue injury by stressors such as cold, heat, high salt, and others or by malfunction of cellular machinery due to mutations in housekeeping genes [13 - 15]. Both immune and non-immune cells can initiate inflammation [16].

The most important mediators of inflammation are members of IL1 family IL1β,

IL1α, and IL18 [17]. IL1α and IL1β signal through receptors composed of IL1R1and IL1R3 chains. Although IL1α and IL1β induce the same signaling, they differ in their structure, distribution, and activation. IL1β expression is induced mostly in response to PRR ligation in immune cells and is usually activated by pro-IL1β cleavage by caspase 1 in inflammasomes. IL1α is expressed constitutively in several cell types and is found in the cytoplasm, nucleus, and on cell surface in a membrane-bound form [18]. IL1α expression is increased during various stresses. The mature form of IL1α is produced by caspase 5 cleavage of pro-IL1α in non-canonical inflammasomes [19]. Pro-IL1α can also be cleaved by several other proteases inside and outside cells. Contrary to IL1β, cleaved and non-cleaved forms of IL1α bind IL1R1 with similar kinetics. Since IL1α is active without processing, it acts before IL1β that needs caspase cleavage for its activation; it also determines the magnitude of inflammation [20].

IL1 cytokines are released from cells undergoing inflammasome-induced pyroptotic death [21]. They can also be secreted from live macrophages with the help of the pyroptosis regulator gasdermin D [22]. At the site of the wound, coagulation system becomes activated and thrombin can cleave and activate IL1α [23]. IL1 cytokines stimulate the release of other mediators such as IL6, TNF, and prostaglandins.

IL6 cytokine plays an important role in inflammation, immuno defense, and behavior [24 - 26]. Many cell types produce IL6. It is released from macrophages during immune response but also from muscle cells during contractions. IL6 induces pleiotropic effects depending on the target cell. It signals through a receptor composed of IL6R and the signal transducer gp130, which is also used by other cytokines. Major IL6 effects are pro-inflammatory. In liver cells, IL6 stimulates the production of the acute phase response factors. They include complement factors, soluble PRRs such as mannan-binding lectin, coagulation factors such as fibrinogen, alpha 2-macroglobulin, and proteins such as ferritin that limit iron uptake by pathogens. IL6 is also involved in the induction of fever and in the maturation of B cells and neutrophils.

Many effects of inflammation are associated with tumor necrosis factor (TNF). It promotes cell death, induces ROS, stimulates NFκB, promotes IL1 secretion, induces fever, and has other effects [27]. It is produced as a transmembrane protein that needs proteolytic cleavage to generate the soluble form. TNF functions through two distinct receptors: TNFR1 promotes programmed cell death and inflammation, whereas TNFR2 mediates various homeostatic effects. TNF receptors signal through several multiprotein complexes that, depending on their composition, can activate JNK, p38, AP1, and NFκB, and induce apoptosis or necroptosis [28]. In the absence of NFκB signaling, death pathways are favored.

A knockout of NFκB in mice leads to embryonic lethality due to apoptosis of hepatocytes, which can be rescued by deletion of TNF or TNFR1 [27].

TNF is implicated in the pathogenesis of diverse disorders from autoimmune diseases to insulin resistance, hypertension, and cancer [29, 30]. Several biological agents decreasing TNF levels are used to treat inflammatory conditions [31 - 33]. TNF activation is behind the effects of anti-cancer compounds that mimic the endogenous pro-apoptotic SMAC [34]. Some cancer cells are resistant to SMAC because of poor TNF induction. Application of additional stress, *e.g.* osmotic stress, induces TNF in such cells, thus killing them [35].

TNF contributes to immunodefense by promoting the death of the infected cells and by activating NFκB and IFNs. Some bacterial effectors target components of TNF-dependent death pathways [36].

Cytokines induced during inflammation orchestrate a generalized response on the level of the organism. They stimulate the production of other molecules associated with inflammation such as chemokines, bradykinin, histamine, adhesion molecules, acute phase proteins, and eicosanoids. The main vascular effects include vasodilatation, increased vascular chemotaxis of leukocytes, and increased vascular permeability, which leads to migration of leukocytes into the tissue. Their attractants include small molecules such as H_2O_2 and ATP, and proteinous alarmins such as IL1α, HMGB1, IL33, cold-shock proteins, and Ca^{2+}-binding S100 proteins [37, 38]. Another important effect of pro-inflammatory cytokines is the induction of pain through the release of bradykinin and histamine. Metabolic, physiological, and behavioral responses associated with inflammation are known as sickness [39]. It includes fever, fatigue, anorexia, apathy, and cognitive dysfunction.

11.1.2. Fever

Fever provides surviving benefits during many infections such as flu but can be associated with worse outcome in massive inflammation such as sepsis [40, 41]. The main pyrogenic cytokine is IL6. It acts on its receptors on hypothalamic endothelial cells to induce COX2, which oxidizes arachidonic acid to produce prostaglandin E2 [42]. PGE2 then binds to its receptors on the thermoregulatory neurons, which trigger the release of noradrenaline by sympathetic neurons. This leads to increased thermogenesis in the brown adipose tissue and to vasoconstriction to prevent loss of heat. Muscle thermogenesis is also induced. Heat is generated in the mitochondria by diverting energy from ATP production. In fact, the mitochondrial temperature is physiologically maintained higher than the body temperature at about 50°C [43].

Fever-range thermal stress may help to fight infection by several mechanisms. It may induce stress in pathogens adapted to the normal host temperature, which would increase the pathogens' susceptibility to the immune attacks and decrease their propagation rates. Febrile temperature stimulates the immune system [41]. It increases neutrophil recruitment, increases NK cell recruitment and activity, stimulates migration of lymphoid cells, increases phagocytosis by macrophages and dendritic cells, and stimulates ROS production during phagocytosis. In addition to stimulation of specialized immune cells, fever-range thermal stress induces expression of Hsp70 family chaperones [44]. Extracellular Hsp70 acts as an agonist of TLR receptors and promotes inflammation. Thus, the ancient heat-shock stress response is incorporated into an evolutionary late invention —fever and contributes to immune response. Increased expression of chaperones during fever has a side effect. It can support the selection of mutations in viral proteins that escape the immune system [45]. Such mutations often decrease protein stability but since host chaperones support their function, they can form functional viruses.

Not only warm-blooded animals increase their core temperature during infection. Poikilothermic invertebrates also seek warmer environments during immune response and bees increase the hive temperature through muscle thermogenesis [46, 47]. Mitochondria of many organisms including plants and fungi contain, alongside the cytochrome-based pathway, an alternative oxidase (AOX) that catalyzes four-electron reduction of oxygen into water. AOX activity is not coupled with ATP synthesis but dissipates the energy as heat and is important for thermogenesis and stress response [48]. Some plants increase their temperature during infection, flowering, and seed dispersal [41, 49, 50].

It has been hypothesized that warm-bloodedness evolved in mammals and birds as a defense against fungal infections, which are rare in warm-blooded animals but common in poikilotherms [51]. Reduced regenerative capacity may be a tradeoff for the endothermy [52].

11.1.3. Inflammation Control and Resolution

Inflammation is self-limiting and tightly controlled. Some mechanisms of inflammation control are embedded in proinflammatory cytokines signaling through the expression of antagonists and decoy receptors. For example, IL1 signaling is regulated by its antagonist IL1RN also known as IL1RA, which binds IL1R1 receptor and prevents its association with IL1RAP co-receptor. Induction of this antagonist leads to decrease and resolution of inflammation, whereas its deficiency leads to autoimmune diseases [53, 54]. In addition, a decoy receptor can inhibit IL1 signaling [55]. The co-receptor chain IL1RAP can also be produced in a soluble form and act as a decoy receptor.

IL6 that signals through IL6R/gp130 complex is also regulated at the signaling step. IL6R receptor chain can be cleaved by ADAM17, leading to a soluble IL6R. When soluble IL6R binds IL6, this complex can bind gp130 even on those cells that do not express IL6R. This is so- called IL6 trans-signaling [56, 57]. Different types of IL6 signaling lead to a differential response in different cell types.

Several cytokines suppress inflammation. IL10 is a crucial anti-inflammatory cytokine. Proinflammatory stimuli shift cells to a metabolic program that promotes cell proliferation and secretion of various factors [58]. IL10 restricts this program by promoting a shift from glycolysis to oxidative phosphorylation, reducing glucose uptake, inhibiting mTOR, and stimulating mitophagy. In the absence of IL10 signaling, cells accumulate dysfunctional mitochondria leading to high ROS and to NLRP3 inflammasome activation. Through these mechanisms, IL10 prevents excessive inflammation.

Negative regulators of inflammasomes and NFκB also restrict inflammation [59]. Inflammasome activity is terminated by caspase-1 self-cleavage, which depends on the inflammasome size and the cell type [60]. Low molecular weight inhibitors of NLRP3 inflammasome are studied as a treatment for inflammatory diseases [61]. Many stresses lead to p53 activation and p53 suppresses NF|B preventing inflammation [62].

Polyunsaturated fatty acids, such as arachidonic acid, play an important role in inflammation. Cyclooxygenase (COX), lipoxygenase (LOX), and P450 epoxygenase metabolize arachidonic acid into eicosanoids, which include prostaglandins, leukotrienes, and thromboxanes [63]. These compounds are involved in inflammation at various stages from initiation to resolution. For example, leukotriene LTB4 forms chemotactic gradients for leukocyte migration to the site of injury [64, 65]. Prostaglandin PGE2 has multifaceted roles both in inflammation initiation and resolution [66]. Several families of pro-resolution lipid molecules named specialized pro-resolving mediators (SPM) include lipoxins, resolvins, protectins and maresins [67, 68].

Many anti-inflammatory drugs target synthesis or action of these lipid mediators. Aspirin inhibits activity of both COX and LOX enzymes, decreases the synthesis of prostaglandins and increases the production of anti-inflammatory lipoxin A4, resolvins, protectins, and maresins [69].

Stress responses cross-talks with inflammation at various levels. Accumulation of unfolded proteins in endoplasmic reticulum leads to stress, which promotes inflammation [70, 71]. On the other hand, the ER chaperone BiP is secreted during ER stress; it downregulates proinflammatory factors such as IL1 and TNF and stimulates the production of anti-inflammatory factors such as IL10 [72].

Heat shock proteins (HSP) are produced during many stresses and are induced by pathogens. They influence immune response and are implicated in both pro-inflammatory and anti-inflammatory responses [73]. HSPs inhibit inflammation through IL10 upregulation [74] and by induction of anti-inflammatory regulatory T cells [75]. HSP70 mediates degradation of the p65 subunit of NF|B [76]. Human activated monocytes respond to a recombinant HSP70 by downregulation of TNF and IL6 [77].

The ratio between pro- and anti-inflammatory cytokines varies with age, gender, and even season [78]. It can be influenced by glucocorticoids produced under psychological stress [63]. Aging is associated with a low-grade chronic inflammation that critically contributes to various age-related pathologies [79]. Long-lived people, especially centenarians, have lower levels of inflammatory markers [80]. Resolution of inflammation is hindered in many chronic conditions and has emerged as a new therapeutic target, which may be better than the current anti-inflammatory treatment [81].

11.1.4. Memory of Inflammation

The innate immune system can mount a stronger response to repeated insult. This phenomenon was termed “trained immunity”. It is mediated by epigenetic and metabolic reprogramming [82, 83]. The inflammatory response is also enhanced in response to a repeated challenge. For example, epithelial skin cells respond to tissue damage by inflammation, proliferation, and restoration of the barrier, and their response is stronger to repeated damage [84]. These cells maintain chromosomal accessibility at key stress response genes, which accelerates their response during repeated tissue damage. Western diet induces inflammation and reprogramming of immune cells, which leads to NLRP3-mediated trained inflammation [85].

11.1.5. Inflammation is a Necessary Step in Tissue Repair

Tissue injury induced by toxins or physical forces is typically repaired in four steps: clotting, inflammation, repair, and resolution [86]. Clotting of erythrocytes and platelets limits spreading of damage. Damaged, apoptotic and necrotic cells generated by damage release DAMPs and other molecules that attract immune cells. Infiltrating immune cells are mostly neutrophils and monocytes that phagocyte dead cells and cell debris. This leads to transient inflammatory response necessary for proper repair [87]. Some monocytes differentiate into macrophages. These cells seem to be especially important for tissue repair and are shown to be necessary for the survival of regenerating epithelial cells [88]. The next step is a repair. The main role in repair belongs to myofibroblasts. They deposit extracellular matrix that serves as a scaffold for regenerating cells. When repair ends, excess of the extracellular matrix is removed by macrophages and tissue structure is re-established. Invariant natural killer T cells accumulate at the end of tissue repair and produce IL4, which directs monocytes into the anti-inflammatory pathway and contributes to the resolution of inflammation [89]. Finally, immune cells migrate outside of the site of injury or are eliminated by apoptosis and the homeostasis is restored. Persistent injury or dysregulation of macrophages can lead to failure of extracellular matrix removal, fibrosis, and loss of tissue function [90].

11.1.6. Inflammation in Immunodefense

Inflammation restricts pathogens through heat stress induced by fever, increased activity of immune cells, and induction of cell death in the infected cells. In human myeloid, fibroblast, and epithelial cells exogenous IL1β induces mtDNA release in order to activate cGAS/STING/IFN pathway and restrict dengue virus infection [91].

Many pathogens suppress inflammation highlighting its importance for the immunodefense. For example, adenoviruses encode a small basic protein called protein VII that sequesters the immune danger signal HMGB1 thus decreasing inflammation [92]. Influenza virus encodes NS1 protein, which binds NLRP3 and suppresses inflammasome formation and IL1β secretion [93].

Some pathogens, on the other hand, exploit inflammation. Tuberculosis pathogen uses neutrophil inflammatory response to replicate at the site of tissue damage [94].

11.1.7. Inflammatory Diseases

Malfunction of inflammatory mechanisms may lead to auto-inflammatory conditions, autoimmunity, and immunodeficiency disorders. Auto-inflammatory conditions are associated with constitutive activation of pro-inflammatory signaling [95, 96]. Common features of auto-inflammatory disorders include recurrent episodes of fever, increased reactivity of immune cells, rash, joint pain, and other signs of inflammation in the absence of infection or autoimmunity. These episodes are usually associated with overproduction of pro-inflammatory cytokines such as IL1 and IL18. Auto-inflammatory diseases can be systemic or organ-specific, mono- or polygenic, and can be influenced by environmental factors. They include Familial Mediterranean fever caused by mutations in pyrin [97], cryopyrin-associated periodic syndromes caused by gain of function mutations in NLRP3 or defects in genes involved in NLRP3 activation and regulation [98], tumor necrosis factor receptor-associated periodic syndrome [99], and deficiency of mevalonate kinase functioning in synthesis of isoprenoids and cholesterol [100]. Autoinflammatory conditions can also be caused by other defects including deficiency of IL1RA inhibitor, IL10 receptor, deubiquitinases involved in NFκB and TNF metabolism, and deregulated autophagy [101, 102]. In many auto-inflammatory conditions, immunodeficiency is also observed [103].

Acute inflammation is mostly beneficial and serves to restore homeostasis. On the contrary, chronic inflammation may introduce homeostatic set points that lead to a pathological state. This happens in many human conditions such as obesity, type 2 diabetes, and autoimmunity that are associated with altered homeostasis and chronic inflammation [104]. Systemic lupus erythematosus (SLE) is the most severe autoimmune disease characterized by inflammation, autoantibodies, and tissue damage [105 - 107].

Obesity is a well-studied example of deregulated homeostasis [108]. Animals evolved in an environment where food was scarce, and the ability to accumulate nutrients during their abundance and store them as fat provided a survival advantage. However, in modern human life with unlimited access to food, the ability to create a surplus of nutrients is maladaptive and results in the obesity epidemic. Excess of energy and nutrients causes ER stress that activates NF|B [109]. ER stress, in turn, negatively affects mitochondrial function [104]. Stressed adipocytes upregulate p53, which induces semaphorin 3E; this promotes infiltration of macrophages expressing semaphorin receptor [110]. Saturated fatty acids may activate NLRP3 inflammasome in macrophages through K^+ efflux resulting from membrane phospholipid saturation and Na, K-ATPase disruption [111]. Myeloid cells are pro-inflammatory in obese subjects [85, 112]. Induction of chronic inflammation in adipose tissue changes glucose metabolism. TLRs and TNF signaling block insulin through JNK activation, which contributes to the development of type 2 diabetes [104]. Visceral fat, in particular, is implicated in systemic inflammation and related diseases [113].

Inflammation is involved in the pathophysiology of many other common disorders from Crohn's disease to cancer. It is well-accepted that atherosclerosis is an inflammatory disease [114]. Recent clinical trials show that antibodies to IL1β significantly decrease major cardiovascular events without effect on LDL-cholesterol, making it likely that inflammation not high cholesterol is the main cause of atherosclerosis [115]. Inflammation also contributes to aging [116] and neurodegenerative disorders [117].

Diet and exercise can significantly modulate inflammation. High fat and high carbohydrate diets promote oxidative stress and inflammation, whereas a balanced diet can provide bioactive compounds with anti-inflammatory and pro-resolving ability [85, 111, 118]. The ketone metabolite β-hydroxybutyrate induced by starvation and high-intensity exercise inhibits inflammasome by blocking K^+ efflux and preventing inflammasome activation [119]. This ketone is also produced in people on the ketogenic diet [120]. Healthy gut microbiota can suppress inflammation [121 - 124]. Exercise can also inhibit inflammation by activating vagal induction of dopamine that decreases TNF production by immune cells [125].

11.2. Phagocytosis

Phagocytosis evolved early in evolution before the emergence of multicellular organisms as a feeding mechanism. Modern members of the phylum *Amoebozoa* such as *Dictyostelium discoideum* phagocyte their bacterial prey. To kill the engulfed bacteria these organisms use mechanisms that are also used by phagocytes of higher vertebrates [126]. In multicellular animals, phagocytosis plays a dual role: the destruction of invading pathogens and removal of dead, apoptotic, defective, damaged, or transformed self-cells. These tasks are performed mostly by specialized immune cells. Among non-immune cells, phagocytosis is common in epithelial cells that have frequent contact with pathogens and have a high rate of cell turnover [127, 128].

In mammals, the major professional phagocytic cells are macrophages, dendritic cells, neutrophils, and eosinophils. Monocytes circulate in the blood, are quickly mobilized to sites of infection or injury through chemotaxis, and differentiate into macrophages. They join resident macrophages that are originated from fetal blood yolk sac progenitors [129]. Macrophages monitor tissue homeostasis and respond to the presence of PRR ligands, danger signals, and metabolic alterations. Depending on the environment, macrophages can adopt a wide range of phenotypes and roles in the maintenance of homeostasis [130, 131]. For example, in lean tissue, most macrophages have the anti-inflammatory M2 phenotype but switch to the pro-inflammatory M1 phenotype in obesity [132]. Interferon regulatory factors (IRFs) play a key role in the development of monocytes and their differentiation into macrophages. IRF1, 5, and 8 promote the proinflammatory M1 phenotype whereas IRF3 and 4 promote the anti-inflammatory M2 phenotype [133]. Other innate immune cells also demonstrate plasticity and subdivision of labor between subpopulations. Pre-dendritic cells differentiate in response to microbial or cytokine stimuli into different subpopulations, with one population specializing in IFN production and another in antigen presentation [134].

Recognition of pathogens by macrophage PRRs causes cell activation, phagocytosis, respiratory burst, and secretion of antimicrobial peptides and pro-inflammatory cytokines such as IL1β, IL6, IL12, and TNF. The precise composition of cytokines produced depends on PRR that is activated and on the tissue's environment [130]. Within seconds to minutes after PRR engagement with its ligand, macrophages secrete leukotriene B4 that forms a chemotaxis gradient for other immune cells, thus initiating abscess formation [65].

Phagocytosis of pathogens includes their binding to the cell surface *via* PRRs, their engulfment by pseudopods extension, formation and maturation of a phagosome and its fusion with a lysosome to produce a phagolysosome. Lysosomes contain hydrolases and are acidic due to the activity of the "vacuolar" ATPase, so some pathogens are killed by acidic pH [135]. The other conserved killing mechanism is respiratory burst, which produces reactive oxygen (ROS) and nitrogen (RNS) species. ROS are generated through reduction of oxygen to superoxide radicals (O_2^-) by the multisubunit enzyme NADPH oxidase (NOX); catalytic subunits in these enzymes are also called NOX. An additional source of ROS is dual oxidases (DUOXs) that produce H_2O_2. Mammals have several NOX and DUOX enzymes [136]. NOX2, known as the phagocyte oxidase, is responsible for the respiratory burst in phagocytic cells. Deficiency of NOX2 or other NADPH subunits is associated with the chronic granulomatous disease, where phagocytes can engulf bacteria but cannot generate ROS to kill them [137]. Nitric oxide (NO) is made by NO synthases (NOS) from L-arginine and oxygen. Mammals have several NOS genes, some constitutive, others such as NOS2, also known as iNOS, inducible by infections. Bacteria evolved mechanisms to counter reactive species produced by the host. For example, *Helicobacter pylori* both suppresses host NO production and upregulates its defenses against ROS [138, 139]. *Candida albicans* blocks NO production by induction of the host arginase [140]. Mycobacterium tuberculosis induces anti-inflammatory enzymes in response to ROS [141].

Additional measures to prevent bacteria from replication inside phagosomes may include pumping Fe^{2+} from bacteria-containing vacuoles while importing toxic Cu^{2+} or Zn^{2+} into the phagosomes [126, 142]. Production of lysozymes and antibacterial peptides are the killing mechanism conserved from *Dictyostelium* species to mammals [143].

The second function of phagocytosis is the removal of dead cells (efferocytosis).

Contrary to phagocytosis of pathogens, this process is mostly immunologically silent. Apoptosis is a normal part of tissue physiology and prompt removal of dead cells is essential to prevent their disintegration and activation of the immune response. Macrophages play a major role in efferocytosis. Apoptotic cells attract phagocytes through the release of "find me" factors [144]. Several molecules have been suggested as such signals including nucleotides, lysophosphatidylcholine, sphingosine-1 phosphate, and others. Dying cells also display of on their surface "eat me" signals that orchestrate engulfment of these cells. Best characterized "eat me" signals are translocation to the cell surface of the ER protein calreticulin [145] and of phosphatidylserine, which is normally located at the inner membrane leaflet [146]. Recognition of "eat me" signals leads to the formation of a phagocytic cup around the dead cell, its internalization, and processing [147].

To prevent engulfment of healthy self-cells, they express CD47 molecule on their surfaces, which serves as a marker of self — "do not eat me" — signal [148]. Its expression on cancer cells protects them from attacks by myeloid cells [149]. CD47 expression declines in senescent and damaged cells. In addition, macrophages express a receptor for the common MHC class I component beta--microglobulin. MHCI is also expressed by all healthy "self" cells. Recognition of beta-2-microglobulin inhibits maturation and function of myeloid cells [150].

Transcriptional and metabolic profile of macrophages clearing apoptotic cells differs from those that phagocyte pathogens. Efferocytic macrophages (efferocytes) upregulate glycolysis, they uptake glucose to promote further phagocytosis of dead cells and release lactate, which fosters an anti-inflammatory environment [151]. Efferocytes secrete other anti-inflammatory factors such as IL10 and peptides derived from lactoferrin, promoting resolution of inflammation [152]. Enhanced efferocytosis is correlated with better resolution of inflammation [153, 154]. Impaired clearance of apoptotic cells can lead to autoimmunity.

Macrophages also prevent induction of inflammation during phagocytosis by non-professional phagocytes such as epithelial cells. Macrophages release insulin-like growth factor 1 (IGF1), which decreases the uptake of large apoptotic bodies and increases the uptake of microvesicles by epithelial cells [155]. They also release microvesicles that are uptaken by non-professional phagocytes and inhibit their inflammatory response.

The same trigger induces different responses in macrophages and non-professional phagocytes. For example, inhalation of dust leads to lung inflammation in mice through TLRs in macrophages and ROS-dependent mechanisms in epithelial cells [156]. Macrophages responses are also not uniform but depend on their homeostatic state, cytokines in their environment, and factors released from the surrounding tissues. For example, the involvement of autophagy in phagocytosis decreases inflammatory response [157]. Mesenchymal stromal cells can secrete extracellular vesicles that induce anti-inflammatory phenotype in macrophages and promote tissue repair [158]. Various other factors such as the neuropeptide substance P would also affect the inflammatory phenotype of macrophages [159].

If phagocytes are exposed to stress, their response changes. Macrophages exposed to heat shock have increased phagocytosis, extracellular trap formation, antibacterial peptide expression, and bacterial killing but produce less pro-inflammatory cytokines and ROS [160]. Also, they produce more anti-inflammatory IL10 and resolve inflammation faster. IL10 stimulates phagocytosis [161]. Heat shock proteins downregulate TNF and IL6 in activated monocytes [77]. The main inhibitor of oxidative stress Nrf2 also enhances efferocytosis [162]. Strong stress induced by a pathogen can interfere with phagocytosis. Infection of macrophages with fungal pathogen *Cryptococcus neoformans* induces mitochondrial stress and cell cycle arrest inhibiting phagocytic functions and preventing pathogen's clearance [163].

Activation of p53 pathway in stressed macrophages exposed to pro-inflammatory stimuli can inhibit the induction of inflammation [164]. P53 has an additional function in phagocytosis. It promotes clearance of apoptotic cells by macrophages, which prevents the accumulation of dead cells and so prevents inflammation [165].

Hormones and neuromediators also regulate phagocytosis. Prolonged psychological stress leads to increased levels of norepinephrine and endocannabinoids, which decreases oxidative burst [166]. The activity of neutrophils depends on the day-night cycle and is regulated by circadian hormones [167].

11.3. Natural Killer Cells and Missing-Self Recognition

Natural killer (NK) cells use missing-self recognition to do immune surveillance [2, 168]. It is based on the expression by all "self" cells of a specific marker such as MHCI, whose absence is recognized by immune cells as "missing self". NK cells express a variety of activating receptors specific to endogenous molecules that are expressed on stressed cells but not healthy ones [169]. Engagement of these receptors triggers cytotoxic activity. NK cells also express inhibitory receptors such as Ly49 in mice or the killer-cell immunoglobulin-like receptors (KIRs) in humans, which sense the level of "self" MHCI molecules on the target cell: these receptors inhibit NK activation, thus ensuring tolerance to healthy self cells. The balance of activating and inhibitory signaling determines the fate of a target cell. Transformed, stressed, and virus-infected cells often have lower levels of MHCI expression, making them susceptible to NK killing. NK cells kill their targets by releasing the pore-forming peptide perforin and granzymes. Perforin disrupts the membranes of target cells letting granzymes enter cells and induce apoptosis. Apoptosis can also be induced through binding of death receptors (*e.g.* Fas) on target cells with their cognate ligands (*e.g.* FasL) on the surface of NK cells. NK cells can also directly kill bacteria and fungi [170].

NKG2D is the most studied activating receptor that is expressed by NK cells [171]. It recognizes several MHCI-like molecules induced by stress. The examples include: ULBP1 induced by ER stress [172], MICA and ULBP1 induced by DNA damage, MULT1 induced by heat shock, RAE induced by proliferative signals, and ULBP1 and ULBP2 induced by p53 [169].

Viral infections upregulate NKG2D ligands; such upregulation depends on the recognition of viral double-stranded RNA by RIG1/MDA5 receptors and the ensuing IFNs signaling [173]. Induction of endoplasmic reticulum stress in virus-infected cells activates NK cells and drives their expansion through IRE1α-XBP--cMyc axis [174].

In most cases, the engagement of several activating receptors is necessary to activate NK cells; this decreases the danger of improper NK activation. To avoid recognition by NK cells some pathogens express MHCI-like molecules or manipulate host expression of MHCI and ligands for NK activating receptors [175, 176].

The principle of "missing-self" recognition is incorporated into other immune mechanisms as well. For example, phagocytes express receptors for the MHC I component beta-2-microglobulin and CD47, which are present on all healthy "self" cells [148, 150]. The importance of recognition of CD47 by the phagocytic receptor SIRPA is seen in atherosclerosis, which is associated with the accumulation of dead vascular cells and cell debris due to defective phagocytosis. CD47 blocking antibodies restore phagocytosis of dead cells and prevent the disease [177]. CD47-SIRPA interaction is also involved in allorecognition. Mice without T, B, and NK cells distinguish between self and allogenic nonself by the strength of CD47-SIRPA binding. Cells from donors that differ from the recipient even in one SIRPA allele are rejected [178]. The "missing-self" principle is also used in the complement system (described below), where factor H recognizes host-specific sialic acids and prevents complement damage to host cells.

11.4. Complement System

Complement is a cascade of enzymes and other factors activated by PAMPs, tissue damage, and metabolic alterations (Fig. **11**). It promotes chemotaxis, inflammation, and phagocytosis, and can also directly kill pathogens and infected or stressed "self" cells [179, 180]. In humans, it includes more than 50 proteins. The central component of the complement system is complement component 3 (C3). C3 belongs to the conserved family of proteins that contain GCGEQ motif where the side groups of cysteine and glutamic acid form unstable thioester bond [181, 182]. In a non-activated state, the thioester bond is shielded but becomes exposed and cleaved when the protein is activated through a conformational change, elevated temperature, or proteolysis. Cleavage of the thioester bond can lead to the formation of a stable covalent bond with an adjacent substrate with an amino or hydroxyl group. C3 can be activated proteolytically through three pathways. It is cleaved spontaneously at a low rate (the alternative pathway) or by enzymes of the classical and lectin pathways. C3 cleavage leads to release of a smaller pro-inflammatory C3a fragment, whereas a large C3b fragment attaches itself to any substrate including cell surface and pathogens. The attached C3b promotes phagocytosis and activates other complement components, which may lead to the formation of a membrane attack complex (MAC) and cell lysis [183].

The sentinel of the classical pathway is a C1 complex composed of C1q subcomponent associated with serine proteases C1r and C1s [184]. C1q is a large molecule that adopts a bouquet-like shape from six collagenous stems with globular heads. C1q can bind diverse ligands including constant fragments (Fc) of antibodies in immune complexes, C-reactive protein, bacterial products such as LPS and porins, and components of stressed and dying cells including histones, annexins, and calreticulin.

C1q binding to its targets changes its conformation and induces autoactivation of C1r protease, which then activates C1s protease [185, 186]. C1s cleaves complement component C4 and the resulting product C4b complexes with C2. Cleavage of C2 within this complex leads to the formation of C4b2a, which is C3 convertase. C4b2a can associate with C3b to form C5 convertase C4b2a3b, which cleaves C5 into anaphylatoxin C5a and a substrate-binding component C5b.

The lectin pathway of C3 activation is initiated by molecules such as mannose-binding lectin (MBL) and ficolins. They are structurally similar to C1q and bind carbohydrate molecules [187, 188]. MBL-associated serine proteases (MASPs) then trigger activation of the complement cascade and formation of the same C3 convertase C4b2a as in the classical pathway.

In the alternative pathway of complement activation, a small fraction of C3 is hydrolyzed by water producing C3b [189]. C3b forms a complex with Bb serine protease than cleaves C3. In an auto-amplification loop, C3b generated by this complex binds to the proenzyme complement factor B (CFB) leading to its cleavage by complement factor D and generation of more C3bBb. C3bBb is stabilized by complement factor P (CFP), which is also called properdin and is destabilized by complement factor H (CFH). The alternative pathway of complement activation can also serve to amplify signals from the classical and lectin pathway on cell surfaces [190].

CFP/Properdin, a glycoprotein produced by leukocytes, is the main complement-activating factor [191]. Alternative pathway convertase C3bBb is unstable with only approximately 90 seconds of half-life. By binding to C3bBb properdin stabilizes it and increases the activity of the alternative pathway 5-10 fold. Properdin may also initiate the alternative pathway when it binds to the surface of microorganisms and apoptotic and necrotic cells where it serves as a docking point for C3 and C5 convertases. Properdin binds and activates innate lymphoid cells serving as a link between different branches of innate immunity [192].

CFH inhibitor is a protein containing 20 complement-control modules [193]. Its N-end binds C3b and contains regulatory domains; its C-terminal domains bind C3b and polyanions to anchor CFH on the cell surface. CFH inhibits complement through three mechanisms. It competes with Bb for binding to C3b. It promotes dissociation of Bb from C3 and C3b thus accelerating the decay of the convertases. It also serves as a cofactor for complement factor I (CFI), which cleaves C3b. The product, iC3b, cannot bind CFB preventing the production of C3bBb. iC3b can be further cleaved. CFH suppresses C3 activation in solution and on the surface of "self" cells that contain glycosaminoglycans and sialylated glycans but not on foreign surfaces. In addition to CFH, there are other CFH-related proteins and membrane complement regulators that prevent complement from targeting "self" cells [194, 195]. This system, therefore, works on the principle of "missing self" recognition. Its ability to recognize "missing self" contributes to allograft rejection in organ transplantation [196]. CFH exists in distinct redox forms that have different activity and thus connect complement with oxidative stress [197].

Spontaneous activation of C3 is very sensitive to the cellular environment. Small changes in homeostasis such as a shift in redox balance can affect it [198]. Therefore, such activation is a unique mechanism that enables complement to quickly respond to various stress-induced changes in homeostasis. This makes the complement system a hub that combines recognition of pathogens, altered "self" molecules, and homeostatic disturbances with the "missing self" strategy to generate an appropriate response.

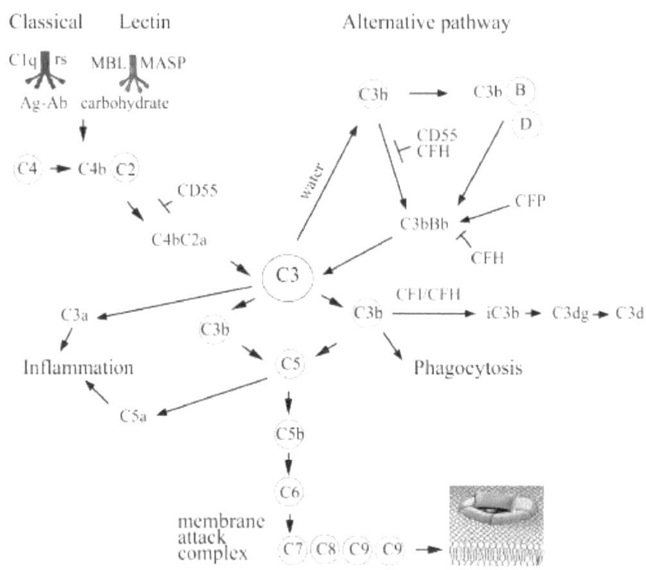

Fig. (11). A simplified schematic of the complement system. The central molecule in the complement system is C3, which can be cleaved spontaneously by the alternative pathway or by convertases C4bC2a and C3bBb. This results in the pro-inflammatory C3a fragment and C3b fragment, which attaches to any molecule with an amino or hydroxyl group. C3b attachment to cell surface promotes phagocytosis and leads to deposition of other complement components and building of a membrane attack complex. CFP is C3bBb-stabilyzing factor. CD55 and CFH are inhibitors.

The three most important outcomes of complement cascade activation are opsonization of the targets for phagocytosis, induction of inflammation, and cell lysis. Opsonization is mediated mostly by C3b and C4b, which tag cell or pathogen surface for recognition by receptors expressed on phagocytic cells. Inflammation is promoted by anaphylatoxins C3a and C5a, which signal through their receptors C3aR and C5aR1. These factors are involved in other processes from immune cells maturation and migration to neural development and metabolism [199]. C5b fragments start the assembly of the membrane attack complex (MAC). C5b is unstable but is stabilized by C6, which leads to C5bC6 complex. After binding C7, the complex anchors in the cell membrane. Incorporation of C8 starts membrane spanning and incorporation of C9 finishes the formation of a pore and leads to cell lysis [183].

The complement components are produced by the liver and by many immune cells where they act both as intracellular and secreted factors to regulate cell maturation and immune response [200]. For example, dendritic cell maturation is associated with C5a and C3ar1/C5ar1 signaling [201]. Stimulation of dendritic cells with IFNγ leads to increased production of CFH and decreased production of properdin [202]. Complement can also directly regulate T and B cells [203]. In turn, factors produced by immune cells regulate complement. For example, IL6 induces expression of complement's inhibitors CD59 and CD55 [204].

Pathogens develop many ways to avoid the complement system. They can camouflage their PAMPs to avoid recognition, use enzymes to degrade complement, produce decoy proteins that bind and deactivate complement components, produce factors that inhibit complement activation, defuse complement complexes, and hijack and manipulate its components [205, 206]. For example, *Staphylococcus aureus* produces a protein that blocks the association of C4b with C2, which prevents activation of both classical and lectin pathways [207]. Natural complement inhibitors such as CFH are frequently hijacked by pathogens [208].

Complement contributes to many human diseases [209]. Insufficient complement activity may lead to susceptibility to infection and autoimmunity. Complement enhances the uptake of apoptotic bodies by phagocytes and promotes immune tolerance. Hence, complement deficiency compromises a safe removal of apoptotic bodies and is associated with the development of systemic lupus erythematosus [210]. Overactivation of complement may lead to tissue damage and inflammation. Several hematological diseases are associated with the malfunction of the complement system including atypical hemolytic uremic syndrome, paroxysmal nocturnal hemoglobinuria, cold agglutinin disease, and others [211]. Deficiency of the complement inhibitor CD55 leads to gut diseases such as protein-losing enteropathy (CHAPLE syndrome) [212]. C3 glomerulopathies are kidney diseases caused by autoantibodies to C3 and C5 convertases that increase the half-life of these enzymes and lead to deposition of complement in tissues [213]. Overactivation of complement is associated with glaucoma, diabetic retinopathy and macular degeneration [214]. Increased eye sensitivity to complement dysregulation may be associated with chronic oxidative stress in retina cells due to exposure to constant light and oxygen. Some variants of the human complement component 4 locus are linked to risk of schizophrenia; they are associated with increased neuronal complement deposition and synapse elimination by microglia [215].

Complement status also influences the outcome of many common diseases. Complement activity is increased in atherosclerosis, heart failure, sickle cell disease, neurodegeneration, and other conditions [216 - 220]. More than 20 drugs targeting various complement pathways are in clinical trials [221].

In addition to complement, other humoral factors participate in immunodefense.

Soluble PRRs, enzymes, antibodies, and antibacterial peptides are present in blood and extracellular fluids. The coagulation system may participate in immunodefense by trapping pathogens and activation of IL1α [23, 222]. Saliva and other extracellular fluids in the body contain peroxidase, which catalyzes the reaction of hydrogen peroxide with thiocyanate to generate hypothiocyanous acid (HOSCN) [223]. HOSCN has a broad-spectrum antimicrobial activity against viruses, fungi, and bacteria.

11.5. Adaptive Immunity

Adaptive immune systems operate by creating an unlimited repertoire of lymphoid receptors by rearrangement of their gene segments. Both in jaw and jawless vertebrates, T cells are responsible for the cellular immunity, whereas B cells provide the humoral immunity by producing antibodies. Adaptive lymphoid cells cannot function independently; they need assistance from the innate immune system. Innate and adaptive cells' cross-regulation is performed by various cytokines [224, 225].

11.5.1. Adaptive Antigen Receptors

The genes encoding adaptive immune receptors are assembled from gene segments during the development of lymphoid cells [226]. Jawed and jawless vertebrates use different strategies for diversification of their receptors. In jawless lampreys and hagfish, leucine-rich repeats (LRR) based variable lymphocyte receptors (VLR) are assembled through random copying of LRRs from LRR arrays into incomplete germ-line genes by a conversion-like mechanism [227]. In jawed vertebrates, genes encoding lymphoid receptors of T and B cells (TCRs and BCRs) belong to the immunoglobulin superfamily. They are assembled by random recombination of the variable (V), diversity (D) and joining (J) gene segments that are encoded in the genome as segment arrays next to the constant (C) regions [3].

V(D)J recombination is induced when Rag1 and Rag2 proteins make double-strand breaks (DSBs) at the flanks of these gene segments [228]. DNA repair proteins process and join these segments, producing genes that encode subunits of BCRs and TCRs. The diversity is generated through the multiplicity of V, D, and J segments and through imprecision in segments' cutting and joining sites. Pairing up of different chains in BCRs and TCRs generates an additional diversity. Alpha and beta chains form TCRs of Tαβ cells, gamma and delta chains form TCRs of Tγδ cells, and immunoglobulin heavy and light chains form BCRs. B cells recognize native antigens, whereas T cells recognize antigens only in association with "self" molecules such as MHCI and MHCII on the surface of innate immune cells, which, in this capacity, are called antigen-presenting cells (APCs).

B cells develop in the bone marrow and T cells develop in the thymus [229]. Depending on cytokines and other factors in their environment, T cells differentiate into different cell types. The major types are T helpers (CD4+) specialized in regulating other immune cells and cytotoxic T cells (CD8+), which kill infected and transformed cells. Other T cell types include natural killer T cells and invariant Tγδ cells [230]. Cells with receptors recognizing self-antigens are eliminated or silenced during development, whereas those with receptors specific to pathogens are expanded [231]. Some cells survive long after infection and provide a memory response during a repeated infection. Each animal accumulates a unique repertoire of BCRs and TCRs in accordance with its life history.

Activated T cells assist B cells in antibody production, regulate other cells and kill infected and cancerous cells by releasing the pore-forming protein perforin together with serine proteases granzymes and granulysin. Dysregulation of T cells can have devastating consequences leading to autoimmune conditions like diabetes type 1 [232].

Immunoglobulin receptors of B cells can be produced in a secreted form as antibodies [233]. The specificity of BCRs is increased by somatic hypermutations [234]. Multiple rounds of mutations and antigen-mediated selection can produce B cells with very high affinity receptors/antibodies. Moreover, antibodies with the same variable regions and the same specificity can be produced with different constant regions that are specialized in different functions. In humans, antibodies can be produced as IgM, IgD, IgG(1,2,3), IgA, and IgE. Each constant region confers different properties onto the antibody such as an ability to activate complement, to be secreted into the gut lumen, *etc.*

11.5.2. DNA Damage Response in T and B Cells

To generate TCRs and BCRs by gene rearrangement, double-strand breaks (DSBs) are introduced into the DNA by Rag endonucleases. Typically, DSBs activate the DNA damage response (DDR), which includes DNA repair, cell cycle checkpoints, and apoptosis if the damage cannot be repaired. Often, the first rearrangement does not produce a functional gene and several additional rounds of rearrangement are necessary. Therefore, lymphoid cells experience prolonged genotoxic stress. Under such a condition, proapoptotic pathways may be activated threatening the survival of lymphoid cells. To prevent death during the V(D)J rearrangement, T and B cells incorporate DDR into their development [235]. First of all, they activate antiapoptotic factors. For example, during the rearrangement, the DSBs induced ATM/NFκB pathway leads to upregulation of Pim2 kinase and phosphorylation of the pro-apoptotic Bad, which inactivates Bad and inhibits apoptosis [235]. ATM kinase activates p38 MAPK, which phosphorylates and inactivates glycogen synthase kinase 3 beta (GSK3beta) also promoting cell survival [236]. Besides the canonical DDR, DSBs in T and B cells activate other cell-specific pathways that are involved in lymphoid cell development and function [237].

Somatic hypermutation and class-switch recombination of immunoglobulin are also associated with genotoxic stress. They are initiated by deamination of the cytidine to uracil by activation-induced cytidine deaminase (AID) [234]. The resulting mismatches are processed by base excision repair and MSH2-mediated mismatch repair, leading to DNA lesions and DDR activation. The associated genotoxic stress promotes B cell differentiation, proliferation, and self-renewal [238].

A proportion of natural killers (NK) cells that do not express adaptive receptors and are considered "innate" lymphocytes, express Rags during their ontogeny [239]. NK cells that do not express Rags are hyper-responsive but have a diminished survival capability after virus-driven proliferation. They also have decreased DDR. The authors propose that DDR induced by Rags DNA cleavage endow the developing cells with cellular "fitness": increased resistance to stress. These cells show better survival during periods of rapid proliferation and stress.

11.5.3. Regulation of T and B Cells

T and B cells are regulated by inputs from innate immune cells, cytokines, stress hormones, cellular stress response networks, inflammation, complement, and autonomous immune mechanisms [240 - 244]. In turn, T and B cells can regulate innate immune cells and inflammation. For example, T cells can induce opposing differentiation programs in macrophages through secretion of IL4 or IFNγ [225]. Regulatory T cells can suppress inflammation [245].

Recognition of antigens by TCRs and BCRs is not sufficient for T and B cell activation; instead, it can lead to anergy. Their activation requires a co-stimulation through the innate immune receptors. In T cell activation, a major role is played by CD28 binding to the receptors CD80 or CD86 on APCs. In turn, CD80 expression depends on TLR signaling [134]. T cells also express the inducible co-stimulatory molecule (ICOS), which binds its ligand ICOSL on APC [246]. Signaling through the innate immune receptor NLRC3 attenuates T cell activation [247]. Since the expression of the co-stimulatory molecules on APCs depends on the presence of products from pathogens or necrotic cells, these innate immune cells control T cells. The complement components produced intracellularly in T cells, control their fate by changing their metabolism and activating inflammasome [248, 249].

B cells also express several co-stimulatory receptors including CD19, CD40, the B cell-activating factor receptor, and TLRs [244]. Bacterial cyclic nucleotides directly activate B cells through STING pathway and this activation synergizes with BCR signaling [250].

Stress can affect T and B cell development and function. For example, cytotoxic T cells in a tumor are exposed to hypoxia, which is a potent inducer of ER stress and PERK kinase. PERK-mediated signaling then leads to selective upregulation of the pro-apoptotic protein CHOP, which negatively regulates T cell activity [251]. In B cells residing in the intestinal epithelium, ER stress induces T cell-independent production of a polyreactive IgA that bind various microbial factors and protect against enteric inflammation [252].

T and B cells can sense stress experienced by other cells in their environment. For example, heat shock proteins (HSPs) produced in response to various stresses, are recognized by T regulatory cells [74]. Treg cells then decrease inflammation. In addition, in dendritic cells, HSPs induce tolerogenic phenotype, which induces IL10 expression in T cells also leading to suppression of inflammation [253]. T cells express P2X7 receptor that senses the presence of extracellular ATP, which is a DAMP molecule. In the gut, this signal regulates the abundance of T helper cells in the Peyer's patches of the small intestine [254]. Microbiota produces compounds such as pentanoate that modulate metabolism in lymphocytes and reprogram their responses towards the anti-inflammatory mode [255].

Stress experienced by antigen-presenting cells also modulates T cells responses. For example, high ROS levels in dendritic cells lead to oxidation of arachidonic acid and the formation of gamma ketoaldehydes or isoketals that ligate to protein lysines. Modified peptides derived from these conjugates are presented to T cells and are recognized as nonself triggering T cells activation [256].

Glucocorticoids produced in animals under stress can affect T and B cells. They impair BCR and TLRs signaling, decrease antibody production, and promote anti-inflammatory IL10 production [243]. Glucocorticoids stimulate the expansion of Treg cells suppressing inflammation [257].

T and B cells are the newest evolutionary acquisition with the most complex regulation and dependence on the correct functioning of the innate immune system. Antigen-specific T and B cells undergo extensive proliferation during the immune response. Consequently, they are more sensitive to stress in comparison to innate immune cells and decline stronger with aging [258].

CONCLUDING REMARKS

Specialized immune systems of animals are complex and multilayered. They include cellular and humoral branches that use innate and adaptive immune receptors. Specialized immune systems work in close association with autonomous immune mechanisms and stress response mechanisms. When an organism experiences stress and infection at the same time, it allocates resources so as to increase the chance of survival. Such a decision depends on specific external and internal conditions. Stress can either activate or suppress immunity. Cellular stress often stimulates innate immune mechanisms including autonomous immune mechanisms, inflammation, and complement. In contrast, adaptive immune responses are typically suppressed by stress. Short-term stress usually induces acute inflammation and potentiates maturation, trafficking, and activity of immune cells. Conversely, chronic stress leads to persistent low-grade inflammation and inhibition of immune response. In higher animals, hormones and the nervous system also regulate stress and immune response.

CONSENT FOR PUBLICATION

Not applicable.

CONFLICT OF INTEREST

The author confirms that this chapter contents have no conflict of interest.

ACKNOWLEDGEMENTS

Declare none.

REFERENCES

[1] Janeway CA Jr, Medzhitov R. Innate immune recognition. Annu Rev Immunol 2002; 20: 197-216. [http://dx.doi.org/10.1146/annurev.immunol.20.083001.084359] [PMID: 11861602]

[2] Altmann DM. Natural killer cell transcriptional control, subsets, receptors and effector function. Immunology 2019; 156(2): 109-10. [http://dx.doi.org/10.1111/imm.13041] [PMID: 30632618]

[3] Tonegawa S. Somatic generation of antibody diversity. Nature 1983; 302(5909): 575-81. [http://dx.doi.org/10.1038/302575a0] [PMID: 6300689]

[4] Rodriguez-Fraticelli AE, Wolock SL, Weinreb CS, *et al.* Clonal analysis of lineage fate in native haematopoiesis. Nature 2018; 553(7687): 212-6. [http://dx.doi.org/10.1038/nature25168] [PMID: 29323290]

[5] Fujii T, Mashimo M, Moriwaki Y, *et al.* Expression and Function of the Cholinergic System in Immune Cells. Front Immunol 2017; 8: 1085. [http://dx.doi.org/10.3389/fimmu.2017.01085] [PMID: 28932225]

[6] Baral P, Udit S, Chiu IM. Pain and immunity: Implications for host defence. Nat Rev Immunol 2019; 19(7): 433-47.
[http://dx.doi.org/10.1038/s41577-019-0147-2] [PMID: 30874629]

[7] Seillet C, Jacquelot N. Sensing of physiological regulators by innate lymphoid cells. Cellular & molecular immunology 2019.
[http://dx.doi.org/10.1038/s41423-019-0217-1]

[8] Medzhitov R. Origin and physiological roles of inflammation. Nature 2008; 454(7203): 428-35.
[http://dx.doi.org/10.1038/nature07201] [PMID: 18650913]

[9] Medzhitov R. Inflammation 2010: new adventures of an old flame. Cell 2010; 140(6): 771-6.
[http://dx.doi.org/10.1016/j.cell.2010.03.006] [PMID: 20303867]

[10] Awad F, Assrawi E, Louvrier C, *et al.* Inflammasome biology, molecular pathology and therapeutic implications. Pharmacol Ther 2018; 187: 133-49.
[http://dx.doi.org/10.1016/j.pharmthera.2018.02.011] [PMID: 29466702]

[11] Gaidt MM, Hornung V. The NLRP3 Inflammasome Renders Cell Death Pro-inflammatory. J Mol Biol 2018; 430(2): 133-41.
[http://dx.doi.org/10.1016/j.jmb.2017.11.013] [PMID: 29203171]

[12] Alberts BM, Bruce C, Basnayake K, Ghezzi P, Davies KA, Mullen LM. Secretion of IL-1β From Monocytes in Gout Is Redox Independent. Front Immunol 2019; 10: 70.
[http://dx.doi.org/10.3389/fimmu.2019.00070] [PMID: 30761138]

[13] Prager P, Hollborn M, Steffen A, Wiedemann P, Kohen L, Bringmann A. P2Y1 Receptor Signaling Contributes to High Salt-Induced Priming of the NLRP3 Inflammasome in Retinal Pigment Epithelial Cells. PLoS One 2016; 11(10)e0165653
[http://dx.doi.org/10.1371/journal.pone.0165653] [PMID: 27788256]

[14] Liston A, Masters SL. Homeostasis-altering molecular processes as mechanisms of inflammasome activation. Nat Rev Immunol 2017; 17(3): 208-14.
[http://dx.doi.org/10.1038/nri.2016.151] [PMID: 28163301]

[15] Alimov I, Menon S, Cochran N, *et al.* Bile acid analogues are activators of pyrin inflammasome. J Biol Chem 2019; 294(10): 3359-66.
[http://dx.doi.org/10.1074/jbc.RA118.005103] [PMID: 30647128]

[16] Bautista-Hernandez LA, Gomez-Olivares JL, Buentello-Volante B, Bautista-de Lucio VM. Fibroblasts: The Unknown Sentinels Eliciting Immune Responses Against Microorganisms. European journal of microbiology & immunology 2017; 7(3): 151-7.

[17] Vasanthakumar A, Kallies A. Interleukin (IL)-33 and the IL-1 Family of Cytokines-Regulators of Inflammation and Tissue Homeostasis. Cold Spring Harb Perspect Biol 2019; 11(3)a028506
[http://dx.doi.org/10.1101/cshperspect.a028506] [PMID: 29101106]

[18] Malik A, Kanneganti TD. Function and regulation of IL-1α in inflammatory diseases and cancer. Immunol Rev 2018; 281(1): 124-37.
[http://dx.doi.org/10.1111/imr.12615] [PMID: 29247991]

[19] Wiggins KA, Parry AJ, Cassidy LD, *et al.* IL-1α cleavage by inflammatory caspases of the noncanonical inflammasome controls the senescence-associated secretory phenotype. Aging Cell 2019; 18(3)e12946
[http://dx.doi.org/10.1111/acel.12946] [PMID: 30916891]

[20] Di Paolo NC, Shayakhmetov DM. Interleukin 1α and the inflammatory process. Nat Immunol 2016; 17(8): 906-13.
[http://dx.doi.org/10.1038/ni.3503] [PMID: 27434011]

[21] Frank D, Vince JE. Pyroptosis *versus* necroptosis: similarities, differences, and crosstalk. Cell Death Differ 2019; 26(1): 99-114.

[http://dx.doi.org/10.1038/s41418-018-0212-6] [PMID: 30341423]

[22] Evavold CL, Ruan J, Tan Y, Xia S, Wu H, Kagan JC. The Pore-Forming Protein Gasdermin D Regulates Interleukin-1 Secretion from Living Macrophages. Immunity 2018; 48(1): 35-44.e6. [http://dx.doi.org/10.1016/j.immuni.2017.11.013] [PMID: 29195811]

[23] Burzynski LC, Humphry M, Pyrillou K, *et al.* The Coagulation and Immune Systems Are Directly Linked through the Activation of Interleukin-1α by Thrombin. Immunity 2019; 50(4): 1033-1042.e6. [http://dx.doi.org/10.1016/j.immuni.2019.03.003] [PMID: 30926232]

[24] Jones SA, Jenkins BJ. Recent insights into targeting the IL-6 cytokine family in inflammatory diseases and cancer. Nat Rev Immunol 2018; 18(12): 773-89. [http://dx.doi.org/10.1038/s41577-018-0066-7] [PMID: 30254251]

[25] Rose-John S. Interleukin-6 Family Cytokines. Cold Spring Harb Perspect Biol 2018; 10(2)a028415 [http://dx.doi.org/10.1101/cshperspect.a028415] [PMID: 28620096]

[26] Niraula A, Witcher KG, Sheridan JF, Godbout JP. Interleukin-6 Induced by Social Stress Promotes a Unique Transcriptional Signature in the Monocytes That Facilitate Anxiety. Biol Psychiatry 2019; 85(8): 679-89. [http://dx.doi.org/10.1016/j.biopsych.2018.09.030] [PMID: 30447911]

[27] Kalliolias GD, Ivashkiv LB. TNF biology, pathogenic mechanisms and emerging therapeutic strategies. Nat Rev Rheumatol 2016; 12(1): 49-62. [http://dx.doi.org/10.1038/nrrheum.2015.169] [PMID: 26656660]

[28] Lafont E, Draber P, Rieser E, *et al.* TBK1 and IKKε prevent TNF-induced cell death by RIPK1 phosphorylation. Nat Cell Biol 2018; 20(12): 1389-99. [http://dx.doi.org/10.1038/s41556-018-0229-6] [PMID: 30420664]

[29] Hurst LA, Dunmore BJ, Long L, *et al.* TNFα drives pulmonary arterial hypertension by suppressing the BMP type-II receptor and altering NOTCH signalling. Nat Commun 2017; 8: 14079. [http://dx.doi.org/10.1038/ncomms14079] [PMID: 28084316]

[30] Akash MSH, Rehman K, Liaqat A. Tumor Necrosis Factor-Alpha: Role in Development of Insulin Resistance and Pathogenesis of Type 2 Diabetes Mellitus. J Cell Biochem 2018; 119(1): 105-10. [http://dx.doi.org/10.1002/jcb.26174] [PMID: 28569437]

[31] Gregory AP, Dendrou CA, Attfield KE, *et al.* TNF receptor 1 genetic risk mirrors outcome of anti-TNF therapy in multiple sclerosis. Nature 2012; 488(7412): 508-11. [http://dx.doi.org/10.1038/nature11307] [PMID: 22801493]

[32] Malaviya R, Laskin JD, Laskin DL. Anti-TNFα therapy in inflammatory lung diseases. Pharmacol Ther 2017; 180: 90-8. [http://dx.doi.org/10.1016/j.pharmthera.2017.06.008] [PMID: 28642115]

[33] Eberhardson M, Hedin CRH, Carlson M, Tarnawski L, Levine YA, Olofsson PS. Towards improved control of inflammatory bowel disease. Scand J Immunol 2019; 89(3)e12745 [http://dx.doi.org/10.1111/sji.12745] [PMID: 30582196]

[34] Beug ST, Tang VA, LaCasse EC, *et al.* Smac mimetics and innate immune stimuli synergize to promote tumor death. Nat Biotechnol 2014; 32(2): 182-90. [http://dx.doi.org/10.1038/nbt.2806] [PMID: 24463573]

[35] Bittner S, Knoll G, Ehrenschwender M. Hyperosmotic stress enhances cytotoxicity of SMAC mimetics. Cell Death Dis 2017; 8(8)e2967 [http://dx.doi.org/10.1038/cddis.2017.355] [PMID: 28771230]

[36] Pearson JS, Giogha C, Ong SY, *et al.* A type III effector antagonizes death receptor signalling during bacterial gut infection. Nature 2013; 501(7466): 247-51. [http://dx.doi.org/10.1038/nature12524] [PMID: 24025841]

[37] Bertheloot D, Latz E. HMGB1, IL-1alpha, IL-33 and S100 proteins: dual-function alarmins. Cellular

& molecular immunology 2017; 14(1): 43-64.

[38] Rider P, Voronov E, Dinarello CA, Apte RN, Cohen I. Alarmins: Feel the Stress. J Immunol 2017; 198(4): 1395-402.
[http://dx.doi.org/10.4049/jimmunol.1601342] [PMID: 28167650]

[39] Harden LM, Kent S, Pittman QJ, Roth J. Fever and sickness behavior: Friend or foe? Brain Behav Immun 2015; 50: 322-33.
[http://dx.doi.org/10.1016/j.bbi.2015.07.012] [PMID: 26187566]

[40] Kluger MJ, Ringler DH, Anver MR. Fever and survival. Science 1975; 188(4184): 166-8.
[http://dx.doi.org/10.1126/science.1114347] [PMID: 1114347]

[41] Evans SS, Repasky EA, Fisher DT. Fever and the thermal regulation of immunity: the immune system feels the heat. Nat Rev Immunol 2015; 15(6): 335-49.
[http://dx.doi.org/10.1038/nri3843] [PMID: 25976513]

[42] Blomqvist A, Engblom D. Neural Mechanisms of Inflammation-Induced Fever. Neuroscientist 2018; 24(4): 381-99.
[http://dx.doi.org/10.1177/1073858418760481] [PMID: 29557255]

[43] Chrétien D, Bénit P, Ha HH, *et al.* Mitochondria are physiologically maintained at close to 50°C. PLoS Biol 2018; 16(1)e2003992
[http://dx.doi.org/10.1371/journal.pbio.2003992] [PMID: 29370167]

[44] Gupta A, Cooper ZA, Tulapurkar ME, *et al.* Toll-like receptor agonists and febrile range hyperthermia synergize to induce heat shock protein 70 expression and extracellular release. J Biol Chem 2013; 288(4): 2756-66.
[http://dx.doi.org/10.1074/jbc.M112.427336] [PMID: 23212905]

[45] Phillips AM, Ponomarenko AI, Chen K, *et al.* Destabilized adaptive influenza variants critical for innate immune system escape are potentiated by host chaperones. PLoS Biol 2018; 16(9)e3000008
[http://dx.doi.org/10.1371/journal.pbio.3000008] [PMID: 30222731]

[46] Clint E, Fessler DM. INSURMOUNTABLE HEAT: THE EVOLUTION AND PERSISTENCE OF DEFENSIVE HYPERTHERMIA. Q Rev Biol 2016; 91(1): 25-46.
[http://dx.doi.org/10.1086/685302] [PMID: 27192778]

[47] Wojda I. Temperature stress and insect immunity. Journal of thermal biology 2017; 68(Pt A): 96-103.
[http://dx.doi.org/10.1016/j.jtherbio.2016.12.002]

[48] Borovik OA, Grabelnych OI. Mitochondrial alternative cyanide-resistant oxidase is involved in an increase of heat stress tolerance in spring wheat. J Plant Physiol 2018; 231: 310-7.
[http://dx.doi.org/10.1016/j.jplph.2018.10.007] [PMID: 30368229]

[49] deBruyn RA, Paetkau M, Ross KA, Godfrey DV, Friedman CR. Thermogenesis-triggered seed dispersal in dwarf mistletoe. Nat Commun 2015; 6: 6262.
[http://dx.doi.org/10.1038/ncomms7262] [PMID: 25662062]

[50] Harrap MJ, Rands SA, Hempel de Ibarra N, Whitney HM. The diversity of floral temperature patterns, and their use by pollinators. eLife 2017; 6e31262
[http://dx.doi.org/10.7554/eLife.31262] [PMID: 29254518]

[51] Casadevall A. Fungi and the rise of mammals. PLoS Pathog 2012; 8(8)e1002808
[http://dx.doi.org/10.1371/journal.ppat.1002808] [PMID: 22916007]

[52] Hirose K, Payumo AY, Cutie S, *et al.* Evidence for hormonal control of heart regenerative capacity during endothermy acquisition. Science 2019; 364(6436): 184-8.
[http://dx.doi.org/10.1126/science.aar2038] [PMID: 30846611]

[53] Aksentijevich I, Masters SL, Ferguson PJ, *et al.* An autoinflammatory disease with deficiency of the interleukin-1-receptor antagonist. N Engl J Med 2009; 360(23): 2426-37.
[http://dx.doi.org/10.1056/NEJMoa0807865] [PMID: 19494218]

[54] Iannitti RG, Napolioni V, Oikonomou V, *et al.* IL-1 receptor antagonist ameliorates inflammasome-dependent inflammation in murine and human cystic fibrosis. Nat Commun 2016; 7: 10791. [http://dx.doi.org/10.1038/ncomms10791] [PMID: 26972847]

[55] Mora-Buch R, Dotti I, Planell N, *et al.* Epithelial IL-1R2 acts as a homeostatic regulator during remission of ulcerative colitis. Mucosal Immunol 2016; 9(4): 950-9. [http://dx.doi.org/10.1038/mi.2015.108] [PMID: 26530134]

[56] Baran P, Hansen S, Waetzig GH, *et al.* The balance of interleukin (IL)-6, IL-6·soluble IL-6 receptor (sIL-6R), and IL-6·sIL-6R·sgp130 complexes allows simultaneous classic and trans-signaling. J Biol Chem 2018; 293(18): 6762-75. [http://dx.doi.org/10.1074/jbc.RA117.001163] [PMID: 29559558]

[57] Narazaki M, Kishimoto T. The Two-Faced Cytokine IL-6 in Host Defense and Diseases. Int J Mol Sci 2018; 19(11)E3528 [http://dx.doi.org/10.3390/ijms19113528] [PMID: 30423923]

[58] Ip WKE, Hoshi N, Shouval DS, Snapper S, Medzhitov R. Anti-inflammatory effect of IL-10 mediated by metabolic reprogramming of macrophages. Science 2017; 356(6337): 513-9. [http://dx.doi.org/10.1126/science.aal3535] [PMID: 28473584]

[59] Afonina IS, Zhong Z, Karin M, Beyaert R. Limiting inflammation-the negative regulation of NF-κB and the NLRP3 inflammasome. Nat Immunol 2017; 18(8): 861-9. [http://dx.doi.org/10.1038/ni.3772] [PMID: 28722711]

[60] Boucher D, Monteleone M, Coll RC, *et al.* Caspase-1 self-cleavage is an intrinsic mechanism to terminate inflammasome activity. J Exp Med 2018; 215(3): 827-40. [http://dx.doi.org/10.1084/jem.20172222] [PMID: 29432122]

[61] Coll RC, Robertson AA, Chae JJ, *et al.* A small-molecule inhibitor of the NLRP3 inflammasome for the treatment of inflammatory diseases. Nat Med 2015; 21(3): 248-55. [http://dx.doi.org/10.1038/nm.3806] [PMID: 25686105]

[62] Ak P, Levine AJ. p53 and NF-κB: different strategies for responding to stress lead to a functional antagonism. FASEB J 2010; 24(10): 3643-52. [http://dx.doi.org/10.1096/fj.10-160549] [PMID: 20530750]

[63] Umamaheswaran S, Dasari SK, Yang P, Lutgendorf SK, Sood AK. Stress, inflammation, and eicosanoids: an emerging perspective. Cancer Metastasis Rev 2018; 37(2-3): 203-11. [http://dx.doi.org/10.1007/s10555-018-9741-1] [PMID: 29948328]

[64] Saeki K, Yokomizo T. Identification, signaling, and functions of LTB_4 receptors. Semin Immunol 2017; 33: 30-6. [http://dx.doi.org/10.1016/j.smim.2017.07.010] [PMID: 29042026]

[65] Brandt SL, Klopfenstein N, Wang S, *et al.* Macrophage-derived LTB4 promotes abscess formation and clearance of Staphylococcus aureus skin infection in mice. PLoS Pathog 2018; 14(8)e1007244 [http://dx.doi.org/10.1371/journal.ppat.1007244] [PMID: 30102746]

[66] Dejani NN, Orlando AB, Niño VE, *et al.* Intestinal host defense outcome is dictated by PGE_2 production during efferocytosis of infected cells. Proc Natl Acad Sci USA 2018; 115(36): E8469-78. [http://dx.doi.org/10.1073/pnas.1722016115] [PMID: 30127026]

[67] Serhan CN. Pro-resolving lipid mediators are leads for resolution physiology. Nature 2014; 510(7503): 92-101. [http://dx.doi.org/10.1038/nature13479] [PMID: 24899309]

[68] Wei J, Gronert K. Eicosanoid and Specialized Proresolving Mediator Regulation of Lymphoid Cells. Trends Biochem Sci 2019; 44(3): 214-25. [http://dx.doi.org/10.1016/j.tibs.2018.10.007] [PMID: 30477730]

[69] Poorani R, Bhatt AN, Dwarakanath BS, Das UN. COX-2, aspirin and metabolism of arachidonic,

eicosapentaenoic and docosahexaenoic acids and their physiological and clinical significance. Eur J Pharmacol 2016; 785: 116-32.
[http://dx.doi.org/10.1016/j.ejphar.2015.08.049] [PMID: 26335394]

[70] Frakes AE, Dillin A. The UPR^{ER}: Sensor and Coordinator of Organismal Homeostasis. Mol Cell 2017; 66(6): 761-71.
[http://dx.doi.org/10.1016/j.molcel.2017.05.031] [PMID: 28622521]

[71] Reverendo M, Mendes A, Argüello RJ, Gatti E, Pierre P. At the crossway of ER-stress and proinflammatory responses. FEBS J 2019; 286(2): 297-310.
[http://dx.doi.org/10.1111/febs.14391] [PMID: 29360216]

[72] Panayi GS, Corrigall VM. Immunoglobulin heavy-chain-binding protein (BiP): a stress protein that has the potential to be a novel therapy for rheumatoid arthritis. Biochem Soc Trans 2014; 42(6): 1752-5.
[http://dx.doi.org/10.1042/BST20140230] [PMID: 25399601]

[73] Zininga T, Ramatsui L, Shonhai A. Heat Shock Proteins as Immunomodulants. Molecules 2018; 23(11)E2846
[http://dx.doi.org/10.3390/molecules23112846] [PMID: 30388847]

[74] van Eden W, Jansen MAA, Ludwig I, van Kooten P, van der Zee R, Broere F. The Enigma of Heat Shock Proteins in Immune Tolerance. Front Immunol 2017; 8: 1599.
[http://dx.doi.org/10.3389/fimmu.2017.01599] [PMID: 29209330]

[75] Barbera Betancourt A, Lyu Q, Broere F, Sijts A, Rutten VPMG, van Eden W. T Cell-Mediated Chronic Inflammatory Diseases Are Candidates for Therapeutic Tolerance Induction with Heat Shock Proteins. Front Immunol 2017; 8: 1408.
[http://dx.doi.org/10.3389/fimmu.2017.01408] [PMID: 29123529]

[76] Tanaka T, Shibazaki A, Ono R, Kaisho T. HSP70 mediates degradation of the p65 subunit of nuclear factor κB to inhibit inflammatory signaling. Sci Signal 2014; 7(356): ra119.
[http://dx.doi.org/10.1126/scisignal.2005533] [PMID: 25515536]

[77] Ferat-Osorio E, Sánchez-Anaya A, Gutiérrez-Mendoza M, *et al.* Heat shock protein 70 down-regulates the production of toll-like receptor-induced pro-inflammatory cytokines by a heat shock factor-1/constitutive heat shock element-binding factor-dependent mechanism. J Inflamm (Lond) 2014; 11: 19.
[http://dx.doi.org/10.1186/1476-9255-11-19] [PMID: 25053922]

[78] Ter Horst R, Jaeger M, Smeekens SP, *et al.* Host and Environmental Factors Influencing Individual Human Cytokine Responses. Cell 2016; 167(4): 1111-1124.e13.
[http://dx.doi.org/10.1016/j.cell.2016.10.018] [PMID: 27814508]

[79] Monti D, Ostan R, Borelli V, Castellani G, Franceschi C. Inflammaging and human longevity in the omics era. Mech Ageing Dev 2017; 165(Pt B): 129-38.
[http://dx.doi.org/10.1016/j.mad.2016.12.008]

[80] Sotgia S, Zinellu A, Mangoni AA, *et al.* Cellular immune activation in Sardinian middle-aged, older adults and centenarians. Exp Gerontol 2017; 99: 133-7.
[http://dx.doi.org/10.1016/j.exger.2017.10.005] [PMID: 29024722]

[81] Dalli J. Does promoting resolution instead of inhibiting inflammation represent the new paradigm in treating infections? Mol Aspects Med 2017; 58: 12-20.
[http://dx.doi.org/10.1016/j.mam.2017.03.007] [PMID: 28365269]

[82] Netea MG, Joosten LA, Latz E, *et al.* Trained immunity: A program of innate immune memory in health and disease. Science 2016; 352(6284)aaf1098
[http://dx.doi.org/10.1126/science.aaf1098] [PMID: 27102489]

[83] Hamada A, Torre C, Drancourt M, Ghigo E. Trained Immunity Carried by Non-immune Cells. Front Microbiol 2019; 9: 3225.

[http://dx.doi.org/10.3389/fmicb.2018.03225] [PMID: 30692968]

[84] Naik S, Larsen SB, Gomez NC, *et al.* Inflammatory memory sensitizes skin epithelial stem cells to tissue damage. Nature 2017; 550(7677): 475-80.
[http://dx.doi.org/10.1038/nature24271] [PMID: 29045388]

[85] Christ A, Günther P, Lauterbach MAR, *et al.* Western Diet Triggers NLRP3-Dependent Innate Immune Reprogramming. Cell 2018; 172(1-2): 162-175.e14.
[http://dx.doi.org/10.1016/j.cell.2017.12.013] [PMID: 29328911]

[86] Gurtner GC, Werner S, Barrandon Y, Longaker MT. Wound repair and regeneration. Nature 2008; 453(7193): 314-21.
[http://dx.doi.org/10.1038/nature07039] [PMID: 18480812]

[87] Eming SA, Wynn TA, Martin P. Inflammation and metabolism in tissue repair and regeneration. Science 2017; 356(6342): 1026-30.
[http://dx.doi.org/10.1126/science.aam7928] [PMID: 28596335]

[88] Oishi Y, Manabe I. Macrophages in inflammation, repair and regeneration. Int Immunol 2018; 30(11): 511-28.
[PMID: 30165385]

[89] Liew PX, Lee WY, Kubes P. iNKT Cells Orchestrate a Switch from Inflammation to Resolution of Sterile Liver Injury. Immunity 2017; 47(4): 752-765.e5.
[http://dx.doi.org/10.1016/j.immuni.2017.09.016] [PMID: 29045904]

[90] Rockey DC. Fibrosis reversal after hepatitis C virus elimination. Curr Opin Gastroenterol 2019; 35(3): 137-44.
[http://dx.doi.org/10.1097/MOG.0000000000000524] [PMID: 30865043]

[91] Aarreberg LD, Esser-Nobis K, Driscoll C, Shuvarikov A, Roby JA, Gale M Jr. Interleukin-1β Induces mtDNA Release to Activate Innate Immune Signaling *via* cGAS-STING. Mol Cell 2019; 74(4): 801-815.e6.
[http://dx.doi.org/10.1016/j.molcel.2019.02.038] [PMID: 30952515]

[92] Avgousti DC, Herrmann C, Kulej K, *et al.* A core viral protein binds host nucleosomes to sequester immune danger signals. Nature 2016; 535(7610): 173-7.
[http://dx.doi.org/10.1038/nature18317] [PMID: 27362237]

[93] Moriyama M, Chen IY, Kawaguchi A, *et al.* The RNA- and TRIM25-Binding Domains of Influenza Virus NS1 Protein Are Essential for Suppression of NLRP3 Inflammasome-Mediated Interleukin-1β Secretion. J Virol 2016; 90(8): 4105-14.
[http://dx.doi.org/10.1128/JVI.00120-16] [PMID: 26865721]

[94] Mishra BB, Lovewell RR, Olive AJ, *et al.* Nitric oxide prevents a pathogen-permissive granulocytic inflammation during tuberculosis. Nat Microbiol 2017; 2: 17072.
[http://dx.doi.org/10.1038/nmicrobiol.2017.72] [PMID: 28504669]

[95] Manthiram K, Zhou Q, Aksentijevich I, Kastner DL. The monogenic autoinflammatory diseases define new pathways in human innate immunity and inflammation. Nat Immunol 2017; 18(8): 832-42.
[http://dx.doi.org/10.1038/ni.3777] [PMID: 28722725]

[96] Beck DB, Aksentijevich I. Biochemistry of Autoinflammatory Diseases: Catalyzing Monogenic Disease. Front Immunol 2019; 10: 101.
[http://dx.doi.org/10.3389/fimmu.2019.00101] [PMID: 30766537]

[97] Heilig R, Broz P. Function and mechanism of the pyrin inflammasome. Eur J Immunol 2018; 48(2): 230-8.
[http://dx.doi.org/10.1002/eji.201746947] [PMID: 29148036]

[98] Mortimer L, Moreau F, MacDonald JA, Chadee K. NLRP3 inflammasome inhibition is disrupted in a group of auto-inflammatory disease CAPS mutations. Nat Immunol 2016; 17(10): 1176-86.
[http://dx.doi.org/10.1038/ni.3538] [PMID: 27548431]

[99] Negm OH, Mannsperger HA, McDermott EM, *et al.* A pro-inflammatory signalome is constitutively activated by C33Y mutant TNF receptor 1 in TNF receptor-associated periodic syndrome (TRAPS). Eur J Immunol 2014; 44(7): 2096-110.
[http://dx.doi.org/10.1002/eji.201344328] [PMID: 24668260]

[100] Rigante D, Frediani B, Cantarini L. A Comprehensive Overview of the Hereditary Periodic Fever Syndromes. Clinical reviews in allergy & immunology 2018; 54(3): 446-53.
[http://dx.doi.org/10.1007/s12016-016-8537-8]

[101] Lee PP, Lobato-Márquez D, Pramanik N, *et al.* Wiskott-Aldrich syndrome protein regulates autophagy and inflammasome activity in innate immune cells. Nat Commun 2017; 8(1): 1576.
[http://dx.doi.org/10.1038/s41467-017-01676-0] [PMID: 29146903]

[102] Lork M, Verhelst K, Beyaert R. CYLD, A20 and OTULIN deubiquitinases in NF-κB signaling and cell death: so similar, yet so different. Cell Death Differ 2017; 24(7): 1172-83.
[http://dx.doi.org/10.1038/cdd.2017.46] [PMID: 28362430]

[103] Zinngrebe J, Rieser E, Taraborrelli L, *et al.* LUBAC deficiency perturbs TLR3 signaling to cause immunodeficiency and autoinflammation. J Exp Med 2016; 213(12): 2671-89.
[http://dx.doi.org/10.1084/jem.20160041] [PMID: 27810922]

[104] Hotamisligil GS. Inflammation, metaflammation and immunometabolic disorders. Nature 2017; 542(7640): 177-85.
[http://dx.doi.org/10.1038/nature21363] [PMID: 28179656]

[105] Bialas AR, Presumey J, Das A, *et al.* Microglia-dependent synapse loss in type I interferon-mediated lupus. Nature 2017; 546(7659): 539-43.
[http://dx.doi.org/10.1038/nature22821] [PMID: 28614301]

[106] Moulton VR, Suarez-Fueyo A, Meidan E, Li H, Mizui M, Tsokos GC. Pathogenesis of Human Systemic Lupus Erythematosus: A Cellular Perspective. Trends Mol Med 2017; 23(7): 615-35.
[http://dx.doi.org/10.1016/j.molmed.2017.05.006] [PMID: 28623084]

[107] Souyris M, Mejía JE, Chaumeil J, Guéry JC. Female predisposition to TLR7-driven autoimmunity: gene dosage and the escape from X chromosome inactivation. Semin Immunopathol 2019; 41(2): 153-64.
[http://dx.doi.org/10.1007/s00281-018-0712-y] [PMID: 30276444]

[108] Maurizi G, Della Guardia L, Maurizi A, Poloni A. Adipocytes properties and crosstalk with immune system in obesity-related inflammation. J Cell Physiol 2018; 233(1): 88-97.
[http://dx.doi.org/10.1002/jcp.25855] [PMID: 28181253]

[109] Keestra-Gounder AM, Byndloss MX, Seyffert N, *et al.* NOD1 and NOD2 signalling links ER stress with inflammation. Nature 2016; 532(7599): 394-7.
[http://dx.doi.org/10.1038/nature17631] [PMID: 27007849]

[110] Shimizu I, Yoshida Y, Moriya J, *et al.* Semaphorin3E-induced inflammation contributes to insulin resistance in dietary obesity. Cell Metab 2013; 18(4): 491-504.
[http://dx.doi.org/10.1016/j.cmet.2013.09.001] [PMID: 24093674]

[111] Gianfrancesco MA, Dehairs J, L'homme L, *et al.* Saturated fatty acids induce NLRP3 activation in human macrophages through K^+ efflux resulting from phospholipid saturation and Na, K-ATPase disruption. Biochim Biophys Acta Mol Cell Biol Lipids 2019; 1864(7): 1017-30.
[http://dx.doi.org/10.1016/j.bbalip.2019.04.001] [PMID: 30953761]

[112] Keiran N, Ceperuelo-Mallafré V, Calvo E, *et al.* SUCNR1 controls an anti-inflammatory program in macrophages to regulate the metabolic response to obesity. Nat Immunol 2019; 20(5): 581-92.
[http://dx.doi.org/10.1038/s41590-019-0372-7] [PMID: 30962591]

[113] Shin J, Syme C, Wang D, Richer L, Pike GB, Gaudet D, *et al.* Novel genetic locus of visceral fat and systemic inflammation. The Journal of clinical endocrinology and metabolism 2019; 2018: 02656.
[http://dx.doi.org/10.1210/jc.2018-02656]

[114] Ross R. Atherosclerosis--an inflammatory disease. N Engl J Med 1999; 340(2): 115-26. [http://dx.doi.org/10.1056/NEJM199901143400207] [PMID: 9887164]

[115] Koenig W. Inflammation Revisited: Atherosclerosis in The Post-CANTOS Era. Eur Cardiol 2017; 12(2): 89-91. [http://dx.doi.org/10.15420/ecr.2017:18:1] [PMID: 30416561]

[116] Camell CD, Sander J, Spadaro O, *et al.* Inflammasome-driven catecholamine catabolism in macrophages blunts lipolysis during ageing. Nature 2017; 550(7674): 119-23. [http://dx.doi.org/10.1038/nature24022] [PMID: 28953873]

[117] Dzamko N, Gysbers A, Perera G, *et al.* Toll-like receptor 2 is increased in neurons in Parkinson's disease brain and may contribute to alpha-synuclein pathology. Acta Neuropathol 2017; 133(2): 303-19. [http://dx.doi.org/10.1007/s00401-016-1648-8] [PMID: 27888296]

[118] Businaro R, Corsi M, Asprino R, *et al.* Modulation of Inflammation as a Way of Delaying Alzheimer's Disease Progression: The Diet's Role. Curr Alzheimer Res 2018; 15(4): 363-80. [http://dx.doi.org/10.2174/1567205014666170829100100] [PMID: 28847284]

[119] Youm YH, Nguyen KY, Grant RW, *et al.* The ketone metabolite β-hydroxybutyrate blocks NLRP3 inflammasome-mediated inflammatory disease. Nat Med 2015; 21(3): 263-9. [http://dx.doi.org/10.1038/nm.3804] [PMID: 25686106]

[120] Pinto A, Bonucci A, Maggi E, Corsi M, Businaro R. Anti-Oxidant and Anti-Inflammatory Activity of Ketogenic Diet: New Perspectives for Neuroprotection in Alzheimer's Disease. Antioxidants 2018; 7(5)E63 [http://dx.doi.org/10.3390/antiox7050063] [PMID: 29710809]

[121] Maynard CL, Elson CO, Hatton RD, Weaver CT. Reciprocal interactions of the intestinal microbiota and immune system. Nature 2012; 489(7415): 231-41. [http://dx.doi.org/10.1038/nature11551] [PMID: 22972296]

[122] Blaser MJ. The theory of disappearing microbiota and the epidemics of chronic diseases. Nat Rev Immunol 2017; 17(8): 461-3. [http://dx.doi.org/10.1038/nri.2017.77] [PMID: 28749457]

[123] Doonan J, Tarafdar A, Pineda MA, *et al.* The parasitic worm product ES-62 normalises the gut microbiota bone marrow axis in inflammatory arthritis. Nat Commun 2019; 10(1): 1554. [http://dx.doi.org/10.1038/s41467-019-09361-0] [PMID: 30952846]

[124] Ma Q, Xing C, Long W, Wang HY, Liu Q, Wang RF. Impact of microbiota on central nervous system and neurological diseases: the gut-brain axis. J Neuroinflammation 2019; 16(1): 53. [http://dx.doi.org/10.1186/s12974-019-1434-3] [PMID: 30823925]

[125] Shimojo G, Joseph B, Shah R, Consolim-Colombo FM, De Angelis K, Ulloa L. Exercise activates vagal induction of dopamine and attenuates systemic inflammation. Brain Behav Immun 2019; 75: 181-91. [http://dx.doi.org/10.1016/j.bbi.2018.10.005] [PMID: 30394312]

[126] Gaudet RG, Bradfield CJ, MacMicking JD. Evolution of Cell-Autonomous Effector Mechanisms in Macrophages *versus* Non-Immune Cells. Microbiology spectrum 2016; 6(6): MCHD-0050. [http://dx.doi.org/10.1128/microbiolspec.MCHD-0050-2016]

[127] Günther J, Seyfert HM. The first line of defence: insights into mechanisms and relevance of phagocytosis in epithelial cells. Semin Immunopathol 2018; 40(6): 555-65. [http://dx.doi.org/10.1007/s00281-018-0701-1] [PMID: 30182191]

[128] Penberthy KK, Lysiak JJ, Ravichandran KS. Rethinking Phagocytes: Clues from the Retina and Testes. Trends Cell Biol 2018; 28(4): 317-27. [http://dx.doi.org/10.1016/j.tcb.2018.01.004] [PMID: 29454661]

[129] Gomez Perdiguero E, Klapproth K, Schulz C, *et al.* Tissue-resident macrophages originate from yolk-sac-derived erythro-myeloid progenitors. Nature 2015; 518(7540): 547-51.
[http://dx.doi.org/10.1038/nature13989] [PMID: 25470051]

[130] Glass CK, Natoli G. Molecular control of activation and priming in macrophages. Nat Immunol 2016; 17(1): 26-33.
[http://dx.doi.org/10.1038/ni.3306] [PMID: 26681459]

[131] Silva HM, Báfica A, Rodrigues-Luiz GF, *et al.* Vasculature-associated fat macrophages readily adapt to inflammatory and metabolic challenges. J Exp Med 2019; 216(4): 786-806.
[http://dx.doi.org/10.1084/jem.20181049] [PMID: 30862706]

[132] Braga TT, Moura IC, Lepique AP, Camara NOS. Editorial: Macrophages Role in Integrating Tissue Signals and Biological Processes in Chronic Inflammation and Fibrosis. Front Immunol 2017; 8: 845.
[http://dx.doi.org/10.3389/fimmu.2017.00845] [PMID: 28785266]

[133] Chistiakov DA, Myasoedova VA, Revin VV, Orekhov AN, Bobryshev YV. The impact of interferon-regulatory factors to macrophage differentiation and polarization into M1 and M2. Immunobiology 2018; 223(1): 101-11.
[http://dx.doi.org/10.1016/j.imbio.2017.10.005] [PMID: 29032836]

[134] Alculumbre SG, Saint-André V, Di Domizio J, *et al.* Diversification of human plasmacytoid predendritic cells in response to a single stimulus. Nat Immunol 2018; 19(1): 63-75.
[http://dx.doi.org/10.1038/s41590-017-0012-z] [PMID: 29203862]

[135] Kissing S, Saftig P, Haas A. Vacuolar ATPase in phago(lyso)some biology. Int J Med Microbiol 2018; 308(1): 58-67.
[http://dx.doi.org/10.1016/j.ijmm.2017.08.007] [PMID: 28867521]

[136] Panday A, Sahoo MK, Osorio D, Batra S. NADPH oxidases: an overview from structure to innate immunity-associated pathologies. Cellular & molecular immunology 2015; 12(1): 5-23.
[http://dx.doi.org/10.1038/cmi.2014.89]

[137] Arnold DE, Heimall JR. A Review of Chronic Granulomatous Disease. Adv Ther 2017; 34(12): 2543-57.
[http://dx.doi.org/10.1007/s12325-017-0636-2] [PMID: 29168144]

[138] Gobert AP, Wilson KT. The Immune Battle against Helicobacter pylori Infection: NO Offense. Trends Microbiol 2016; 24(5): 366-76.
[http://dx.doi.org/10.1016/j.tim.2016.02.005] [PMID: 26916789]

[139] Aviello G, Knaus UG. ROS in gastrointestinal inflammation: Rescue Or Sabotage? Br J Pharmacol 2017; 174(12): 1704-18.
[http://dx.doi.org/10.1111/bph.13428] [PMID: 26758851]

[140] Wagener J, MacCallum DM, Brown GD, Gow NA. *Candida albicans* Chitin Increases Arginase-1 Activity in Human Macrophages, with an Impact on Macrophage Antimicrobial Functions. MBio 2017; 8(1)e01820
[http://dx.doi.org/10.1128/mBio.01820-16] [PMID: 28119468]

[141] Rockwood N, Costa DL, Amaral EP, *et al. Mycobacterium tuberculosis* Induction of Heme Oxygenase-1 Expression Is Dependent on Oxidative Stress and Reflects Treatment Outcomes. Front Immunol 2017; 8: 542.
[http://dx.doi.org/10.3389/fimmu.2017.00542] [PMID: 28553288]

[142] Ong CY, Berking O, Walker MJ, McEwan AG. New Insights into the Role of Zinc Acquisition and Zinc Tolerance in Group A Streptococcal Infection. Infect Immun 2018; 86(6): e00048-18.
[http://dx.doi.org/10.1128/IAI.00048-18] [PMID: 29581188]

[143] Zasloff M. Antimicrobial peptides of multicellular organisms. Nature 2002; 415(6870): 389-95.
[http://dx.doi.org/10.1038/415389a] [PMID: 11807545]

[144] Medina CB, Ravichandran KS. Do not let death do us part: 'find-me' signals in communication between dying cells and the phagocytes. Cell Death Differ 2016; 23(6): 979-89. [http://dx.doi.org/10.1038/cdd.2016.13] [PMID: 26891690]

[145] Osman R, Tacnet-Delorme P, Kleman JP, Millet A, Frachet P. Calreticulin Release at an Early Stage of Death Modulates the Clearance by Macrophages of Apoptotic Cells. Front Immunol 2017; 8: 1034. [http://dx.doi.org/10.3389/fimmu.2017.01034] [PMID: 28878781]

[146] Fadok VA, Voelker DR, Campbell PA, Cohen JJ, Bratton DL, Henson PM. Exposure of phosphatidylserine on the surface of apoptotic lymphocytes triggers specific recognition and removal by macrophages. J Immunol 1992; 148(7): 2207-16. [PMID: 1545126]

[147] Nagata S. Apoptosis and Clearance of Apoptotic Cells. Annu Rev Immunol 2018; 36: 489-517. [http://dx.doi.org/10.1146/annurev-immunol-042617-053010] [PMID: 29400998]

[148] Oldenborg PA, Zheleznyak A, Fang YF, Lagenaur CF, Gresham HD, Lindberg FP. Role of CD47 as a marker of self on red blood cells. Science 2000; 288(5473): 2051-4. [http://dx.doi.org/10.1126/science.288.5473.2051] [PMID: 10856220]

[149] Liu M, O'Connor RS, Trefely S, Graham K, Snyder NW, Beatty GL. Metabolic rewiring of macrophages by CpG potentiates clearance of cancer cells and overcomes tumor-expressed CD47-mediated 'don't-eat-me' signal. Nat Immunol 2019; 20(3): 265-75. [http://dx.doi.org/10.1038/s41590-018-0292-y] [PMID: 30664738]

[150] Barkal AA, Weiskopf K, Kao KS, *et al.* Engagement of MHC class I by the inhibitory receptor LILRB1 suppresses macrophages and is a target of cancer immunotherapy. Nat Immunol 2018; 19(1): 76-84. [http://dx.doi.org/10.1038/s41590-017-0004-z] [PMID: 29180808]

[151] Morioka S, Perry JSA, Raymond MH, *et al.* Efferocytosis induces a novel SLC program to promote glucose uptake and lactate release. Nature 2018; 563(7733): 714-8. [http://dx.doi.org/10.1038/s41586-018-0735-5] [PMID: 30464343]

[152] Lutaty A, Soboh S, Schif-Zuck S, *et al.* A 17-kDa Fragment of Lactoferrin Associates With the Termination of Inflammation and Peptides Within Promote Resolution. Front Immunol 2018; 9: 644. [http://dx.doi.org/10.3389/fimmu.2018.00644] [PMID: 29643857]

[153] Campana L, Starkey Lewis PJ, Pellicoro A, *et al.* The STAT3-IL-10-IL-6 Pathway Is a Novel Regulator of Macrophage Efferocytosis and Phenotypic Conversion in Sterile Liver Injury. J Immunol 2018; 200(3): 1169-87. [http://dx.doi.org/10.4049/jimmunol.1701247] [PMID: 29263216]

[154] Kourtzelis I, Li X, Mitroulis I, *et al.* DEL-1 promotes macrophage efferocytosis and clearance of inflammation. Nat Immunol 2019; 20(1): 40-9. [http://dx.doi.org/10.1038/s41590-018-0249-1] [PMID: 30455459]

[155] Han CZ, Juncadella IJ, Kinchen JM, *et al.* Macrophages redirect phagocytosis by non-professional phagocytes and influence inflammation. Nature 2016; 539(7630): 570-4. [http://dx.doi.org/10.1038/nature20141] [PMID: 27820945]

[156] He M, Ichinose T, Yoshida S, *et al.* PM2.5-induced lung inflammation in mice: Differences of inflammatory response in macrophages and type II alveolar cells. J Appl Toxicol 2017; 37(10): 1203-18. [http://dx.doi.org/10.1002/jat.3482] [PMID: 28555929]

[157] Heckmann BL, Green DR. LC3-associated phagocytosis at a glance. J Cell Sci 2019; 132(5)jcs222984 [http://dx.doi.org/10.1242/jcs.222984] [PMID: 30787029]

[158] Henao Agudelo JS, Braga TT, Amano MT, *et al.* Mesenchymal Stromal Cell-Derived Microvesicles Regulate an Internal Pro-Inflammatory Program in Activated Macrophages. Front Immunol 2017; 8: 881.

[http://dx.doi.org/10.3389/fimmu.2017.00881] [PMID: 28824619]

[159] Montana G, Lampiasi N. Substance P Induces HO-1 Expression in RAW 264.7 Cells Promoting Switch towards M2-Like Macrophages. PLoS One 2016; 11(12)e0167420 [http://dx.doi.org/10.1371/journal.pone.0167420] [PMID: 27907187]

[160] Vega VL, Crotty Alexander LE, Charles W, Hwang JH, Nizet V, De Maio A. Activation of the stress response in macrophages alters the M1/M2 balance by enhancing bacterial killing and IL-10 expression. J Mol Med (Berl) 2014; 92(12): 1305-17. [http://dx.doi.org/10.1007/s00109-014-1201-y] [PMID: 25163764]

[161] Mendoza-Coronel E, Ortega E. Macrophage Polarization Modulates FcγR- and CD13-Mediated Phagocytosis and Reactive Oxygen Species Production, Independently of Receptor Membrane Expression. Front Immunol 2017; 8: 303. [http://dx.doi.org/10.3389/fimmu.2017.00303] [PMID: 28396660]

[162] Helou DG, Noël B, Gaudin F, *et al.* Cutting Edge: Nrf2 Regulates Neutrophil Recruitment and Accumulation in Skin during Contact Hypersensitivity. J Immunol 2019; 202(8): 2189-94. [http://dx.doi.org/10.4049/jimmunol.1801065] [PMID: 30850475]

[163] Coelho C, Souza AC, Derengowski LdaS, *et al.* Macrophage mitochondrial and stress response to ingestion of Cryptococcus neoformans. J Immunol 2015; 194(5): 2345-57. [http://dx.doi.org/10.4049/jimmunol.1402350] [PMID: 25646306]

[164] Komarova EA, Krivokrysenko V, Wang K, *et al.* p53 is a suppressor of inflammatory response in mice. FASEB J 2005; 19(8): 1030-2. [http://dx.doi.org/10.1096/fj.04-3213fje] [PMID: 15811878]

[165] Yoon KW, Byun S, Kwon E, *et al.* Control of signaling-mediated clearance of apoptotic cells by the tumor suppressor p53. Science 2015; 349(6247)1261669 [http://dx.doi.org/10.1126/science.1261669] [PMID: 26228159]

[166] Buchheim JI, Hoskyns S, Moser D, *et al.* Oxidative burst and Dectin-1-triggered phagocytosis affected by norepinephrine and endocannabinoids: implications for fungal clearance under stress. Int Immunol 2018; 30(2): 79-89. [http://dx.doi.org/10.1093/intimm/dxy001] [PMID: 29329391]

[167] Adrover JM, Del Fresno C, Crainiciuc G, *et al.* A Neutrophil Timer Coordinates Immune Defense and Vascular Protection. Immunity 2019; 50(2): 390-402.e10. [http://dx.doi.org/10.1016/j.immuni.2019.01.002] [PMID: 30709741]

[168] Kärre K, Ljunggren HG, Piontek G, Kiessling R. Selective rejection of H-2-deficient lymphoma variants suggests alternative immune defence strategy. Nature 1986; 319(6055): 675-8. [http://dx.doi.org/10.1038/319675a0] [PMID: 3951539]

[169] Raulet DH, Marcus A, Coscoy L. Dysregulated cellular functions and cell stress pathways provide critical cues for activating and targeting natural killer cells to transformed and infected cells. Immunol Rev 2017; 280(1): 93-101. [http://dx.doi.org/10.1111/imr.12600] [PMID: 29027233]

[170] Schmidt S, Tramsen L, Rais B, Ullrich E, Lehrnbecher T. Natural killer cells as a therapeutic tool for infectious diseases - current status and future perspectives. Oncotarget 2018; 9(29): 20891-907. [http://dx.doi.org/10.18632/oncotarget.25058] [PMID: 29755697]

[171] Jelenčić V, Lenartić M, Wensveen FM, Polić B. NKG2D: A versatile player in the immune system. Immunol Lett 2017; 189: 48-53. [http://dx.doi.org/10.1016/j.imlet.2017.04.006] [PMID: 28414183]

[172] Hosomi S, Grootjans J, Tschurtschenthaler M, *et al.* Intestinal epithelial cell endoplasmic reticulum stress promotes MULT1 up-regulation and NKG2D-mediated inflammation. J Exp Med 2017; 214(10): 2985-97. [http://dx.doi.org/10.1084/jem.20162041] [PMID: 28747426]

[173] Esteso G, Guerra S, Valés-Gómez M, Reyburn HT. Innate immune recognition of double-stranded RNA triggers increased expression of NKG2D ligands after virus infection. J Biol Chem 2017; 292(50): 20472-80.
[http://dx.doi.org/10.1074/jbc.M117.818393] [PMID: 28986447]

[174] Dong H, Adams NM, Xu Y, *et al.* The IRE1 endoplasmic reticulum stress sensor activates natural killer cell immunity in part by regulating c-Myc. Nat Immunol 2019; 20(7): 865-78.
[http://dx.doi.org/10.1038/s41590-019-0388-z] [PMID: 31086333]

[175] Reyburn HT, Mandelboim O, Valés-Gómez M, Davis DM, Pazmany L, Strominger JL. The class I MHC homologue of human cytomegalovirus inhibits attack by natural killer cells. Nature 1997; 386(6624): 514-7.
[http://dx.doi.org/10.1038/386514a0] [PMID: 9087413]

[176] Lodoen MB, Abenes G, Umamoto S, Houchins JP, Liu F, Lanier LL. The cytomegalovirus m155 gene product subverts natural killer cell antiviral protection by disruption of H60-NKG2D interactions. J Exp Med 2004; 200(8): 1075-81.
[http://dx.doi.org/10.1084/jem.20040583] [PMID: 15477345]

[177] Kojima Y, Volkmer JP, McKenna K, *et al.* CD47-blocking antibodies restore phagocytosis and prevent atherosclerosis. Nature 2016; 536(7614): 86-90.
[http://dx.doi.org/10.1038/nature18935] [PMID: 27437576]

[178] Dai H, Friday AJ, Abou-Daya KI, *et al.* Donor SIRPα polymorphism modulates the innate immune response to allogeneic grafts. Sci Immunol 2017; 2(12)eaam6202
[http://dx.doi.org/10.1126/sciimmunol.aam6202] [PMID: 28783664]

[179] Ricklin D, Hajishengallis G, Yang K, Lambris JD. Complement: a key system for immune surveillance and homeostasis. Nat Immunol 2010; 11(9): 785-97.
[http://dx.doi.org/10.1038/ni.1923] [PMID: 20720586]

[180] Hajishengallis G, Reis ES, Mastellos DC, Ricklin D, Lambris JD. Novel mechanisms and functions of complement. Nat Immunol 2017; 18(12): 1288-98.
[http://dx.doi.org/10.1038/ni.3858] [PMID: 29144501]

[181] Ricklin D, Reis ES, Mastellos DC, Gros P, Lambris JD. Complement component C3 - The "Swiss Army Knife" of innate immunity and host defense. Immunol Rev 2016; 274(1): 33-58.
[http://dx.doi.org/10.1111/imr.12500] [PMID: 27782325]

[182] Shokal U, Eleftherianos I. Evolution and Function of Thioester-Containing Proteins and the Complement System in the Innate Immune Response. Front Immunol 2017; 8: 759.
[http://dx.doi.org/10.3389/fimmu.2017.00759] [PMID: 28706521]

[183] Bayly-Jones C, Bubeck D, Dunstone MA. The mystery behind membrane insertion: A review of the complement membrane attack complex. Philos Trans R Soc Lond B Biol Sci 2017; 372(1726): 20160221..
[http://dx.doi.org/10.1098/rstb.2016.0221]

[184] Reid KBM. Complement Component C1q: Historical Perspective of a Functionally Versatile, and Structurally Unusual, Serum Protein. Front Immunol 2018; 9: 764.
[http://dx.doi.org/10.3389/fimmu.2018.00764] [PMID: 29692784]

[185] Mortensen SA, Sander B, Jensen RK, *et al.* Structure and activation of C1, the complex initiating the classical pathway of the complement cascade. Proc Natl Acad Sci USA 2017; 114(5): 986-91.
[http://dx.doi.org/10.1073/pnas.1616998114] [PMID: 28104818]

[186] Almitairi JOM, Venkatraman Girija U, Furze CM, *et al.* Structure of the C1r-C1s interaction of the C1 complex of complement activation. Proc Natl Acad Sci USA 2018; 115(4): 768-73.
[http://dx.doi.org/10.1073/pnas.1718709115] [PMID: 29311313]

[187] Garred P, Genster N, Pilely K, *et al.* A journey through the lectin pathway of complement-MBL and beyond. Immunol Rev 2016; 274(1): 74-97.

[http://dx.doi.org/10.1111/imr.12468] [PMID: 27782323]

[188] Wesener DA, Dugan A, Kiessling LL. Recognition of microbial glycans by soluble human lectins. Curr Opin Struct Biol 2017; 44: 168-78.
[http://dx.doi.org/10.1016/j.sbi.2017.04.002] [PMID: 28482337]

[189] Lachmann PJ. Looking back on the alternative complement pathway. Immunobiology 2018; 223(8-9): 519-23.
[http://dx.doi.org/10.1016/j.imbio.2018.02.001] [PMID: 29525356]

[190] Harrison RA. The properdin pathway: an "alternative activation pathway" or a "critical amplification loop" for C3 and C5 activation? Semin Immunopathol 2018; 40(1): 15-35.
[http://dx.doi.org/10.1007/s00281-017-0661-x] [PMID: 29167939]

[191] Blatt AZ, Pathan S, Ferreira VP. Properdin: a tightly regulated critical inflammatory modulator. Immunol Rev 2016; 274(1): 172-90.
[http://dx.doi.org/10.1111/imr.12466] [PMID: 27782331]

[192] Narni-Mancinelli E, Gauthier L, Baratin M, *et al.* Complement factor P is a ligand for the natural killer cell-activating receptor NKp46. Sci Immunol 2017; 2(10)eaam9628
[http://dx.doi.org/10.1126/sciimmunol.aam9628] [PMID: 28480349]

[193] Blaum BS. The lectin self of complement factor H. Curr Opin Struct Biol 2017; 44: 111-8.
[http://dx.doi.org/10.1016/j.sbi.2017.01.005] [PMID: 28189794]

[194] Dopler A, Guntau L, Harder MJ, *et al.* Self *versus* Nonself Discrimination by the Soluble Complement Regulators Factor H and FHL-1. J Immunol 2019; 202(7): 2082-94.
[http://dx.doi.org/10.4049/jimmunol.1801545] [PMID: 30745459]

[195] Józsi M, Schneider AE, Kárpáti É, Sándor N. Complement factor H family proteins in their non-canonical role as modulators of cellular functions. Semin Cell Dev Biol 2019; 85: 122-31.
[http://dx.doi.org/10.1016/j.semcdb.2017.12.018] [PMID: 29305301]

[196] Casiraghi F, Azzollini N, Todeschini M, *et al.* Complement Alternative Pathway Deficiency in Recipients Protects Kidney Allograft From Ischemia/Reperfusion Injury and Alloreactive T Cell Response. Am J Transplant 2017; 17(9): 2312-25.
[http://dx.doi.org/10.1111/ajt.14262] [PMID: 28276660]

[197] Krilis M, Qi M, Qi J, Wong JWH, Guymer R, Liew G, *et al.* Dual roles of different redox forms of complement factor H in protecting against age related macular degeneration. Free radical biology & medicine 2018; 129(9): 237-46.
[http://dx.doi.org/10.1016/j.freeradbiomed.2018.09.034]

[198] Kunchithapautham K, Atkinson C, Rohrer B. Smoke exposure causes endoplasmic reticulum stress and lipid accumulation in retinal pigment epithelium through oxidative stress and complement activation. J Biol Chem 2014; 289(21): 14534-46.
[http://dx.doi.org/10.1074/jbc.M114.564674] [PMID: 24711457]

[199] Kolev M, Kemper C. Keeping It All Going-Complement Meets Metabolism. Front Immunol 2017; 8: 1.
[http://dx.doi.org/10.3389/fimmu.2017.00001] [PMID: 28149297]

[200] Lubbers R, van Essen MF, van Kooten C, Trouw LA. Production of complement components by cells of the immune system. Clin Exp Immunol 2017; 188(2): 183-94.
[http://dx.doi.org/10.1111/cei.12952] [PMID: 28249350]

[201] Sheen JH, Strainic MG, Liu J, *et al.* TLR-Induced Murine Dendritic Cell (DC) Activation Requires DC-Intrinsic Complement. J Immunol 2017; 199(1): 278-91.
[http://dx.doi.org/10.4049/jimmunol.1700339] [PMID: 28539427]

[202] Dixon KO, O'Flynn J, Klar-Mohamad N, Daha MR, van Kooten C. Properdin and factor H production by human dendritic cells modulates their T-cell stimulatory capacity and is regulated by IFN-γ. Eur J Immunol 2017; 47(3): 470-80.

[http://dx.doi.org/10.1002/eji.201646703] [PMID: 28105653]

[203] Killick J, Morisse G, Sieger D, Astier AL. Complement as a regulator of adaptive immunity. Semin Immunopathol 2018; 40(1): 37-48.
[http://dx.doi.org/10.1007/s00281-017-0644-y] [PMID: 28842749]

[204] Kapka-Skrzypczak L, Popek S, Sawicki K, Wolińska E, Czajka M, Skrzypczak M. Effect of IL-6 and IL-8 on the expression of the complement activation inhibitors MAC-inhibitory protein and decay-accelerating factor in ovarian cancer A2780 cells. Oncol Lett 2016; 12(2): 1507-12.
[http://dx.doi.org/10.3892/ol.2016.4795] [PMID: 27446461]

[205] Agrawal P, Nawadkar R, Ojha H, Kumar J, Sahu A. Complement Evasion Strategies of Viruses: An Overview. Front Microbiol 2017; 8: 1117.
[http://dx.doi.org/10.3389/fmicb.2017.01117] [PMID: 28670306]

[206] Rosbjerg A, Genster N, Pilely K, Garred P. Evasion Mechanisms Used by Pathogens to Escape the Lectin Complement Pathway. Front Microbiol 2017; 8: 868.
[http://dx.doi.org/10.3389/fmicb.2017.00868] [PMID: 28553281]

[207] Woehl JL, Ramyar KX, Katz BB, Walker JK, Geisbrecht BV. The structural basis for inhibition of the classical and lectin complement pathways by S. aureus extracellular adherence protein Protein science: a publication of the Protein Society 2017; 26(8): 1595-608.

[208] Józsi M, Factor H. Factor H Family Proteins in Complement Evasion of Microorganisms. Front Immunol 2017; 8: 571.
[http://dx.doi.org/10.3389/fimmu.2017.00571] [PMID: 28572805]

[209] Wong EKS, Kavanagh D. Diseases of complement dysregulation-an overview. Semin Immunopathol 2018; 40(1): 49-64.
[http://dx.doi.org/10.1007/s00281-017-0663-8] [PMID: 29327071]

[210] Lintner KE, Wu YL, Yang Y, *et al.* Early Components of the Complement Classical Activation Pathway in Human Systemic Autoimmune Diseases. Front Immunol 2016; 7: 36.
[http://dx.doi.org/10.3389/fimmu.2016.00036] [PMID: 26913032]

[211] Baines AC, Brodsky RA. Complementopathies. Blood Rev 2017; 31(4): 213-23.
[http://dx.doi.org/10.1016/j.blre.2017.02.003] [PMID: 28215731]

[212] Ozen A, Comrie WA, Ardy RC, *et al.* CD55 Deficiency, Early-Onset Protein-Losing Enteropathy, and Thrombosis. N Engl J Med 2017; 377(1): 52-61.
[http://dx.doi.org/10.1056/NEJMoa1615887] [PMID: 28657829]

[213] Smith RJH, Appel GB, Blom AM, *et al.* C3 glomerulopathy - understanding a rare complement-driven renal disease. Nat Rev Nephrol 2019; 15(3): 129-43.
[http://dx.doi.org/10.1038/s41581-018-0107-2] [PMID: 30692664]

[214] Clark SJ, Bishop PN. The eye as a complement dysregulation hotspot. Semin Immunopathol 2018; 40(1): 65-74.
[http://dx.doi.org/10.1007/s00281-017-0649-6] [PMID: 28948331]

[215] Sellgren CM, Gracias J, Watmuff B, *et al.* Increased synapse elimination by microglia in schizophrenia patient-derived models of synaptic pruning. Nat Neurosci 2019; 22(3): 374-85.
[http://dx.doi.org/10.1038/s41593-018-0334-7] [PMID: 30718903]

[216] Vlaicu SI, Tatomir A, Rus V, *et al.* The role of complement activation in atherogenesis: the first 40 years. Immunol Res 2016; 64(1): 1-13.
[http://dx.doi.org/10.1007/s12026-015-8669-6] [PMID: 26091721]

[217] Coan PM, Barrier M, Alfazema N, Carter RN, Marion de Proce S, Dopico XC, *et al.* Complement Factor B Is a Determinant of Both Metabolic and Cardiovascular Features of Metabolic Syndrome. Hypertension (Dallas, Tex: 1979) 2017; 117: 09242.

[218] Tatomir A, Talpos-Caia A, Anselmo F, *et al.* The complement system as a biomarker of disease

activity and response to treatment in multiple sclerosis. Immunol Res 2017; 65(6): 1103-9. [http://dx.doi.org/10.1007/s12026-017-8961-8] [PMID: 29116612]

[219] Goetzl EJ, Schwartz JB, Abner EL, Jicha GA, Kapogiannis D. High complement levels in astrocyte-derived exosomes of Alzheimer disease. Ann Neurol 2018; 83(3): 544-52. [http://dx.doi.org/10.1002/ana.25172] [PMID: 29406582]

[220] Vercellotti GM, Dalmasso AP, Schaid TR Jr, *et al.* Critical role of C5a in sickle cell disease. Am J Hematol 2019; 94(3): 327-37. [http://dx.doi.org/10.1002/ajh.25384] [PMID: 30569594]

[221] Ricklin D, Mastellos DC, Reis ES, Lambris JD. The renaissance of complement therapeutics. Nat Rev Nephrol 2018; 14(1): 26-47. [http://dx.doi.org/10.1038/nrneph.2017.156] [PMID: 29199277]

[222] Ariki S, Koori K, Osaki T, Motoyama K, Inamori K, Kawabata S. A serine protease zymogen functions as a pattern-recognition receptor for lipopolysaccharides. Proc Natl Acad Sci USA 2004; 101(4): 953-8. [http://dx.doi.org/10.1073/pnas.0306904101] [PMID: 14722355]

[223] Day BJ. The science of licking your wounds: Function of oxidants in the innate immune system. Biochem Pharmacol 2019; 163: 451-7. [http://dx.doi.org/10.1016/j.bcp.2019.03.013] [PMID: 30876918]

[224] Bal SM, Bernink JH, Nagasawa M, *et al.* IL-1β, IL-4 and IL-12 control the fate of group 2 innate lymphoid cells in human airway inflammation in the lungs. Nat Immunol 2016; 17(6): 636-45. [http://dx.doi.org/10.1038/ni.3444] [PMID: 27111145]

[225] Piccolo V, Curina A, Genua M, *et al.* Opposing macrophage polarization programs show extensive epigenomic and transcriptional cross-talk. Nat Immunol 2017; 18(5): 530-40. [http://dx.doi.org/10.1038/ni.3710] [PMID: 28288101]

[226] Boehm T, Hirano M, Holland SJ, Das S, Schorpp M, Cooper MD. Evolution of Alternative Adaptive Immune Systems in Vertebrates. Annu Rev Immunol 2018; 36: 19-42. [http://dx.doi.org/10.1146/annurev-immunol-042617-053028] [PMID: 29144837]

[227] Pancer Z, Amemiya CT, Ehrhardt GR, Ceitlin J, Gartland GL, Cooper MD. Somatic diversification of variable lymphocyte receptors in the agnathan sea lamprey. Nature 2004; 430(6996): 174-80. [http://dx.doi.org/10.1038/nature02740] [PMID: 15241406]

[228] Schatz DG, Ji Y. Recombination centres and the orchestration of V(D)J recombination. Nat Rev Immunol 2011; 11(4): 251-63. [http://dx.doi.org/10.1038/nri2941] [PMID: 21394103]

[229] Seo W, Taniuchi I. Transcriptional regulation of early T-cell development in the thymus. Eur J Immunol 2016; 46(3): 531-8. [http://dx.doi.org/10.1002/eji.201545821] [PMID: 26763078]

[230] Vavassori S, Kumar A, Wan GS, *et al.* Butyrophilin 3A1 binds phosphorylated antigens and stimulates human γδ T cells. Nat Immunol 2013; 14(9): 908-16. [http://dx.doi.org/10.1038/ni.2665] [PMID: 23872678]

[231] Mayer A, Zhang Y, Perelson AS, Wingreen NS. Regulation of T cell expansion by antigen presentation dynamics. Proc Natl Acad Sci USA 2019; 116(13): 5914-9. [http://dx.doi.org/10.1073/pnas.1812800116] [PMID: 30850527]

[232] Pugliese A. Autoreactive T cells in type 1 diabetes. J Clin Invest 2017; 127(8): 2881-91. [http://dx.doi.org/10.1172/JCI94549] [PMID: 28762987]

[233] Danilova N, Amemiya CT. Going adaptive: the saga of antibodies. Ann N Y Acad Sci 2009; 1168: 130-55. [http://dx.doi.org/10.1111/j.1749-6632.2009.04881.x] [PMID: 19566706]

[234] Pilzecker B, Jacobs H. Mutating for Good: DNA Damage Responses During Somatic Hypermutation. Front Immunol 2019; 10: 438.
[http://dx.doi.org/10.3389/fimmu.2019.00438] [PMID: 30915081]

[235] Bredemeyer AL, Helmink BA, Innes CL, *et al.* DNA double-strand breaks activate a multi-functional genetic program in developing lymphocytes. Nature 2008; 456(7223): 819-23.
[http://dx.doi.org/10.1038/nature07392] [PMID: 18849970]

[236] Thornton TM, Delgado P, Chen L, *et al.* Inactivation of nuclear GSK3β by Ser(389) phosphorylation promotes lymphocyte fitness during DNA double-strand break response. Nat Commun 2016; 7: 10553.
[http://dx.doi.org/10.1038/ncomms10553] [PMID: 26822034]

[237] Bednarski JJ, Sleckman BP. At the intersection of DNA damage and immune responses. Nat Rev Immunol 2019; 19(4): 231-42.
[http://dx.doi.org/10.1038/s41577-019-0135-6] [PMID: 30778174]

[238] Sherman MH, Kuraishy AI, Deshpande C, *et al.* AID-induced genotoxic stress promotes B cell differentiation in the germinal center *via* ATM and LKB1 signaling. Mol Cell 2010; 39(6): 873-85.
[http://dx.doi.org/10.1016/j.molcel.2010.08.019] [PMID: 20864035]

[239] Karo JM, Schatz DG, Sun JC. The RAG recombinase dictates functional heterogeneity and cellular fitness in natural killer cells. Cell 2014; 159(1): 94-107.
[http://dx.doi.org/10.1016/j.cell.2014.08.026] [PMID: 25259923]

[240] Zeng M, Hu Z, Shi X, *et al.* MAVS, cGAS, and endogenous retroviruses in T-independent B cell responses. Science 2014; 346(6216): 1486-92.
[http://dx.doi.org/10.1126/science.346.6216.1486] [PMID: 25525240]

[241] Allen EK, Randolph AG, Bhangale T, *et al.* SNP-mediated disruption of CTCF binding at the IFITM3 promoter is associated with risk of severe influenza in humans. Nat Med 2017; 23(8): 975-83.
[http://dx.doi.org/10.1038/nm.4370] [PMID: 28714988]

[242] Brown MA, Su MA. An Inconvenient Variable: Sex Hormones and Their Impact on T Cell Responses. J Immunol 2019; 202(7): 1927-33.
[http://dx.doi.org/10.4049/jimmunol.1801403] [PMID: 30885988]

[243] Franco LM, Gadkari M, Howe KN, *et al.* Immune regulation by glucocorticoids can be linked to cell type-dependent transcriptional responses. J Exp Med 2019; 216(2): 384-406.
[http://dx.doi.org/10.1084/jem.20180595] [PMID: 30674564]

[244] Rubin SJS, Bloom MS, Robinson WH. B cell checkpoints in autoimmune rheumatic diseases. Nat Rev Rheumatol 2019; 15(5): 303-15.
[http://dx.doi.org/10.1038/s41584-019-0211-0] [PMID: 30967621]

[245] Mathur AN, Zirak B, Boothby IC, *et al.* Treg-Cell Control of a CXCL5-IL-17 Inflammatory Axis Promotes Hair-Follicle-Stem-Cell Differentiation During Skin-Barrier Repair. Immunity 2019; 50(3): 655-667.e4.
[http://dx.doi.org/10.1016/j.immuni.2019.02.013] [PMID: 30893588]

[246] Wikenheiser DJ, Stumhofer JS. ICOS Co-Stimulation: Friend or Foe? Front Immunol 2016; 7: 304.
[http://dx.doi.org/10.3389/fimmu.2016.00304] [PMID: 27559335]

[247] Uchimura T, Oyama Y, Deng M, *et al.* The Innate Immune Sensor NLRC3 Acts as a Rheostat that Fine-Tunes T Cell Responses in Infection and Autoimmunity. Immunity 2018; 49(6): 1049-1061.e6.
[http://dx.doi.org/10.1016/j.immuni.2018.10.008] [PMID: 30566882]

[248] Arbore G, West EE, Spolski R, *et al.* T helper 1 immunity requires complement-driven NLRP3 inflammasome activity in CD4$^+$ T cells. Science 2016; 352(6292)aad1210
[http://dx.doi.org/10.1126/science.aad1210] [PMID: 27313051]

[249] Arbore G, West EE, Rahman J, *et al.* Complement receptor CD46 co-stimulates optimal human CD8$^+$ T cell effector function *via* fatty acid metabolism. Nat Commun 2018; 9(1): 4186.

[http://dx.doi.org/10.1038/s41467-018-06706-z] [PMID: 30305631]

[250] Walker MM, Crute BW, Cambier JC, Getahun A. B Cell-Intrinsic STING Signaling Triggers Cell Activation, Synergizes with B Cell Receptor Signals, and Promotes Antibody Responses. J Immunol 2018; 201(9): 2641-53.
[http://dx.doi.org/10.4049/jimmunol.1701405] [PMID: 30282750]

[251] Cao Y, Trillo-Tinoco J, Sierra RA, *et al.* ER stress-induced mediator C/EBP homologous protein thwarts effector T cell activity in tumors through T-bet repression. Nat Commun 2019; 10(1): 1280.
[http://dx.doi.org/10.1038/s41467-019-09263-1] [PMID: 30894532]

[252] Grootjans J, Krupka N, Hosomi S, *et al.* Epithelial endoplasmic reticulum stress orchestrates a protective IgA response. Science 2019; 363(6430): 993-8.
[http://dx.doi.org/10.1126/science.aat7186] [PMID: 30819965]

[253] Spiering R, van der Zee R, Wagenaar J, van Eden W, Broere F. Mycobacterial and mouse HSP70 have immunomodulatory effects on dendritic cells. Cell stress & chaperones 2013; 18(4): 439-6.

[254] Perruzza L, Gargari G, Proietti M, *et al.* T Follicular Helper Cells Promote a Beneficial Gut Ecosystem for Host Metabolic Homeostasis by Sensing Microbiota-Derived Extracellular ATP. Cell Rep 2017; 18(11): 2566-75.
[http://dx.doi.org/10.1016/j.celrep.2017.02.061] [PMID: 28297661]

[255] Luu M, Pautz S, Kohl V, *et al.* The short-chain fatty acid pentanoate suppresses autoimmunity by modulating the metabolic-epigenetic crosstalk in lymphocytes. Nat Commun 2019; 10(1): 760.
[http://dx.doi.org/10.1038/s41467-019-08711-2] [PMID: 30770822]

[256] McMaster WG, Kirabo A, Madhur MS, Harrison DG. Inflammation, immunity, and hypertensive end-organ damage. Circ Res 2015; 116(6): 1022-33.
[http://dx.doi.org/10.1161/CIRCRESAHA.116.303697] [PMID: 25767287]

[257] Cari L, De Rosa F, Nocentini G, Riccardi C. Context-Dependent Effect of Glucocorticoids on the Proliferation, Differentiation, and Apoptosis of Regulatory T Cells: A Review of the Empirical Evidence and Clinical Applications. Int J Mol Sci 2019; 20(5)E1142
[http://dx.doi.org/10.3390/ijms20051142] [PMID: 30845709]

[258] Nikolich-Žugich J. The twilight of immunity: emerging concepts in aging of the immune system. Nat Immunol 2018; 19(1): 10-9.
[http://dx.doi.org/10.1038/s41590-017-0006-x] [PMID: 29242543]

CHAPTER 12

Brain, Stress, and Immunity Connections

Abstract: Brain coordinates physiological, immunological, and behavioral responses to stress. A typical stress response is "freeze, fight or flight". It involves the release of catecholamines by the sympathoadrenal system and activation of the hypothalamo-pituitary-adrenal (HPA) axis to produce various hormones that prepare the organism for a particular response. Stress generally leads to a strong suppression of the adaptive immune system and a partial suppression of the innate immune system. The brain controls the immune organs through hormones produced by the HPA axis and through direct control by neurotransmitters. In turn, immune mechanisms affect behavior and brain development. Brain microglia support developing neurons, phagocyte dead and inactive cells, and participate in the formation of connections between neurons through synaptic pruning. Peripheral immune cells regulate neurons through neurotransmitters and inflammatory mediators such as IL6, TNF, PGE2, histamine, and others. These mediators induce sickness behavior during illness. They also alter the developing brain during maternal stress, maternal immune activation, and early life stress predisposing the brain to a mental illness. Immune mechanisms contribute to psychiatric and neurodegenerative disorders.

Keywords: Alpha 7 nicotinic acetylcholine receptor, Blood-brain barrier, Corticotropin-releasing hormone (CRH), Depression, Hypothalamic-pituitar--adrenal (HPA) axis, Hypothalamus, Inflammatory reflex, Microglia, Nociceptors, Neurodevelopment, Neurodegenerative diseases, Neural reflex, Psychological stress, Proopiomelanocortin, Synapse pruning, Sickness behavior, Vagus nerve.

INTRODUCTION

Brain emerges during animal evolution as a super processor that integrates all the information received by sensory organs about external and internal environment and determines which behavior is optimal for the organism's survival. In response to stress, it can adjust homeostasis almost instantaneously through the release of neurotransmitters and hormones. Immune mechanisms comprise another dynamic system that controls homeostasis. Interactions between neural and immune systems ensure normal function of the organism while miscommunication between them can cause physical and mental disorders.

Animals can experience various types of stress. Physical stress can be caused by starvation, injury, illness, infection, or by noxious environment such as heat, cold,

Nadia Danilova

irradiation, and pollution. The stress responses of primitive animals such as worms that have few neurons do not differ much from responses of unicellular organisms. As the brain becomes more complex, it increasingly controls stress responses. It also gives rise to a new type of stress, psychological stress, which is a mental response to a perceived threat or challenge. Psychological stress can be caused by injury, physical stress, or fear of a predator's attack but it also can take place in the absence of any noxious stimulus from the environment or any bodily damage. For example, the lost dog would be extremely stressed, as would its owner. Social animals are especially sensitive to their group environment. For humans, work environment and personal relationships are a major source of psychological stress.

Selye recognized three stages in animal adaptation to stress that include a short-term alarm reaction, prolong period of resistance, and the final stage of exhaustion and death [1]. He noticed that the response to physical and mental stress was associated with changes in physiology, endocrine and immune systems with the enlargement of adrenal glands, atrophy of lymphoid tissues, and peptic ulcers. His observations have been further developed into the current view of stress response as an integrated response of the whole body with neuroendocrine, physiological, metabolic, immune, and behavioral components [2, 3].

An example of stress response in higher animals is "freeze, fight or flight" response to threats from predators. It involves the release of catecholamines by the sympathoadrenal system (sympathetic-adreno-medullar (SAM) axis) and activation of hypothalamo-pituitary-adrenal (HPA) axis to produce a multitude of peptide and steroid hormones. These hormones and mediators adapt the organism to a particular response such as running or fighting. In mammals, HPA activation usually leads to a strong suppression of the adaptive immune system and partial suppression of innate immunity. Inflammation, IFN pathway, and phagocytosis can be upregulated by stress. Short-term stress lasting from minutes to hours can stimulate immune response, anti-tumor immunity, and post-surgical recovery, whereas long-term stress leads to the suppression of immune response [4].

In various other animals from corals to bees, antagonistic relationships are most often found between responses to physical stress such as heat and immune responses [5, 6]. In insects, stress hormones divert molecular resources from immune response to functions associated with fight-or-flight behaviors [7].

Multiple routes of communication between the neural system and immunity have evolved. Immune response is controlled by the brain through neuroendocrine pathways and through direct control of immune organs and cells by neurotransmitters [3, 8, 9]. In turn, immune mechanisms are involved in brain development, homeostasis, synaptic plasticity, and learning and memory. Neurons can respond to immune mediators, whereas immune cells express acetylcholine and adrenergic receptors and can produce some neurotransmitters such as acetylcholine and dopamine [3, 9]. A new scientific field has recently emerged, which studies various aspects of brain-immune interaction.

12.1. Immune System of the Brain

The brain is protected from systemic immune responses by the blood-brain barrier (BBB). It consists of endothelial cells that form walls of brain vessels, which are associated with astrocytes, neurons, and pericytes. BBB controls the movement of substances and cells from blood to the brain [10, 11]. Most proteins and immune cells cannot cross this barrier under normal conditions. But the brain has its own immune cells, microglia, which are mostly represented by resident tissue macrophages that are distributed throughout the brain and comprise approximately 10% of CNS cells [12]. These cells are highly ramified and their fine extensions continuously move to survey the local environment. Microglia respond not only to microbial products but also to neuronal activity, neurotransmitters, ATP released by injured cells, and other factors [13, 14].

Microglial cells originate from yolk sac erythromyeloid progenitors that migrate to the brain during fetal development around day 8-9.5 in mice and at 4.5 weeks in humans before the formation of the BBB [15, 16]. These progenitors also seed other tissues, forming local residential tissue macrophages such as liver Kupffer cells. Erythrocytes and other fetal blood cells also originate from these progenitors. Later during development, this so-called "primitive" wave of hematopoiesis is replaced by the production of blood cells from hematopoietic stem cells in the bone marrow. Adult-type blood cells replace most of the embryonic blood cells except some cell types. Therefore, brain microglia differ from adult macrophages; the most important difference is that microglia are maintained by self-renewal unlike macrophages, which differentiate from hematopoietic stem cells through the monocyte stage [17].

Microglia undergo three developmental stages in synchrony with neural development: early, pre-, and adult, which use different regulatory circuits [18]. In various brain regions, microglia differentiate into multiple subpopulations with different roles [19, 20]. Colony stimulating factor 1 (CSF1) signaling is necessary for microglia viability, and selective inhibitors of its receptor result in almost complete depletion of microglia in the mouse brain [21]. Microglia-depleted brains repopulate in a week with new microglia derived from microglial progenitors.

Microglia are essential for proper brain development. Microglial cells phagocyte dead cells, induce apoptosis in inactive neurons, provide trophic support to developing neurons, guide vasculature formation, and participate in the development of neural circuits through synaptic pruning [22 - 24]. Microglia even contribute to such traits as masculinization of the developing brain in male fetuses. Hormonal signaling leads to upregulation of COX2 in microglial cells and production of PGE2, which induces male-typical synaptic patterning. Ablation of microglia from the developing male brain leads to complete loss of male-typical copulatory behavior in adulthood [25].

In the adult brain, microglia continue to support developing neurons and circuit formation, guide learning-associated plasticity, maintain neuronal health, phagocyte dead cells, and perform immune surveillance [14, 26].

Although microglial cells are isolated inside the brain, they are influenced by the immune status of the whole organism through cytokines and neural signaling. Microglia even respond to changes in microbiota in a sex-specific manner [27]. Various conditions such as infection, inflammation, oxidative stress, and metabolic disturbances can lead to undesirable activation of microglia and depending on the stage of life, cause defects in brain development or cognitive decline [28]. Microglia deficiency can also have negative effects. For example, the developing brain has many more synapses than the adult brain and inefficient pruning is implicated in autism spectrum disorders and other pathologies [28, 29]. Decreased microglia function in aging brains may contribute to neurodegenerative disorders.

Microglia prune synapses by phagocytosis with the help of the classical pathway of the complement system. C1q and C3 are expressed in neuronal immature synapses while complement receptors are expressed by microglia [30]. Mice deficient in C1q, C3, or C4 have defects in synaptic connectivity pointing to the important role of complement in brain health [31]. Synapse elimination by microglia is regulated by the expression of "self" marker CD47, apoptotic markers, and "eat me" signals [32 - 34].

Mast cells are also present in the brain. They can respond to injury or infection even faster than microglia since they store preformed immune mediators and can immediately release them [35].

Non-immune cells also contribute to brain immune responses. Astrocytes can be activated into different phenotypes by neural injury, ischemia, or factors produced by microglia such as cytokines [36, 37]. In animal models of severe stress, astrocytes are the major source of pro-inflammatory cytokine IL1β in the hypothalamus [38]. Neurons also express receptors that recognize pathogens and products from injured cells as well as receptors for cytokines and can contribute to immune responses in the brain [39]. The health of brain endothelial cells is important for BBB integrity. Interference with their function or their interaction with other brain cells can disrupt BBB and induce brain inflammation [40, 41].

Disruption of the BBB during systemic inflammation may lead to an infiltration of peripheral immune cells in the brain, which can promote neuroinflammation and brain damage [42]. The meninges of the brain are populated by resident macrophages that support brain homeostasis. Peripheral monocytes can replace these macrophages during infection but cannot function properly in the new environment, which impairs brain immunity [43]. Peripheral monocytes cause TNF-mediated dendritic spine loss in mouse motor cortex, which compromises learning [44].

BBB integrity can also be compromised by stress in stress-susceptible animals. For example, in mice, chronic social stress reduces the expression of the endothelial cell tight junction protein claudin-5 (Cldn5), which leads to peripheral cytokines infiltration into the brain parenchyma [45].

12.2. Hypothalamic-pituitary-adrenal (HPA) Axis

The HPA axis governs stress response [46]. The key factor of the HPA axis is the corticotropin-releasing hormone (CRH), also known as corticotropin releasing factor (Fig. **12A**). Stress induces CRH secretion by the paraventricular nucleus of the hypothalamus. It acts both as a neurohormone and a neurotransmitter to induce endocrine, behavioral, and autonomic responses. In pituitary, CRH activates transcription of the pituitary proopiomelanocortin (POMC) gene. This gene encodes a preproprotein that can be cleaved at eight sites yielding up to ten peptides with different activities in pain alleviation, energy homeostasis, metabolism, melanocyte regulation, and immune response. One of these peptides, melanotropin alpha, has direct antimicrobial activity.

Adrenocorticotropic hormone (ACTH) is the most important product of POMC cleavage [47]. ACTH binds melanocortin 2 receptor (MC2R), which acts through cAMP as a second messenger. It regulates the production of steroid hormones by the adrenal glands, an endocrine gland located above the kidneys. ACTH can further be cleaved to produce melanocyte-stimulated hormone alpha, which increases melanin production in melanocytes and also has anorexigenic effects. Another peptide, endogenous opiate beta-endorphin, is orexigenic. Composition of POMC-derived peptides varies depending on stress type to better coordinate organismal responses to stress.

Among steroid hormones produced by the adrenal cortex are glucocorticoids (GCs) cortisol, corticosterone, and cortisone, mineralocorticoid aldosterone, and androgen precursors. GCs regulate many metabolic pathways involved in glucose, protein, and lipid metabolism, stress response, inflammation, and behavior [48 - 50]. GCs bind intracellular glucocorticoid receptor (GR), which belongs to a nuclear receptor superfamily of transcriptional factors. This releases GR from a complex with HSP90 so that it can translocate to the nucleus and activate transcription of its target genes. Among GC's target genes are many cytokines and chemokines involved in inflammation, as well as growth factors and their receptors. GR can also regulate some of its targets through direct binding. In a regulatory feedback loop, GCs inhibit CRH gene expression and secretion of the hormone in the hypothalamus.

Both GC deficiency and the excess have negative effects on the body [51]. GC deficiency is associated with weakened stress resistance, lymphoid hypertrophy, hypoglycemia, and low weight. GC excess such as in Cushing syndrome leads to hyperglycemia, obesity, and immunodeficiency. These effects can also take place during GC treatment of inflammation or immunosuppression.

A parallel mode of CRH regulation of stress response is through its effect on the arousal system. Locus coeruleus (LC) norepinephrine neurons are one of the areas responsible for arousal. They release norepinephrine (also called noradrenaline) into forebrain regions inducing behavioral response to stress. Dendrites of LC neurons have synapses with axon terminals containing CRH. Females have a higher level of CRH and norepinephrine neurons dendrites, which are associated with higher incidence of stress-associated disorders, such as PTSD and depression [52]. Cannabinoid receptor type I is expressed in the same cells as the receptor for CRH [53], its knockout increases norepinephrine release in male mice [54]. Outside the brain, norepinephrine is used as a neurotransmitter by the sympathetic nervous system, and it is also released along with epinephrine into the bloodstream by the adrenal glands. Both catecholamines bind adrenergic receptors in various tissues and mobilize the body for response to stress by increasing heart rate and blood pressure, decreasing intestinal mobility, and other adjustments.

CRH can be produced not only in the hypothalamus but also in other cells and tissues such as mast cells and the placenta where it plays specialized roles [35]. Stress-induced hormones and neurotransmitters affect every aspect of the immune response from cell activation to proliferation, migration, and effector functions. Short-term stress can enhance immune response while long-term stress is usually immunosuppressive [4].

12.3. How Brain Controls Immunity

The brain communicates with immunity not only through the HPA axis but also through other mechanisms. Several neuroimmunoregulatory circuits employ neural reflex mechanisms, which involve neurons of the autonomic nervous system [9]. Reflexes include afferent sensory neurons and efferent motor neurons. Sensory neurons detect changes in the environment and transmit this to the CNS. Somatosensory neurons innervate skin, joints, and muscles; visceral neurons innervate internal organs. Most somatosensory neurons are nociceptors, which sense noxious environments such as heat, cold, high salt, *etc*. Inflammatory mediators such as cytokines, PGE2, histamine, and others reduce the threshold for nociceptors to fire action potentials [55]. This results in hypersensitivity to pain, as it was first demonstrated for IL1β [56]. A subset of nociceptors responds only to cytokines and other products generated during inflammation. Some nociceptor neurons respond directly to factors secreted by bacteria such as N-formylated peptides and the pore-forming toxin alpha-haemolysin [57, 58].

The most important reflexes are associated with the vagus nerve, a component of the parasympathetic autonomic nervous system. It also plays a major role in regulation of immune responses [59]. Its afferent fibers transmit signals from various organs to the brainstem. A large subset of these nerves also responds to inflammatory mediators, and some express receptors for chemokines, cytokines, and IFNs [9, 60]. Vagus efferent fibers originate in the brainstem's medulla oblongata and transmit signals from the brain to organs that regulate various physiological functions such as heart rate, gut motility, and immune responses.

The efferent arc of the reflex includes the inflammatory reflex, which inhibits production of proinflammatory cytokines [9] (Fig. **12B**). In spleen, the vagus nerve stimulates a subpopulation of memory T cells that express choline acetyltransferase [61]. Acetylcholine produced by the T cells then inhibits the production of pro-inflammatory cytokines by macrophages and other cells. Acetylcholine signals through the alpha 7 nicotinic acetylcholine receptor (alpha7nAChR), which is expressed on macrophages and other immune cells. Activation of this pathway induces several effects: inhibition of the nuclear translocation of NFκB, activation of the JAK2/STAT3 pathway, stabilization of mitochondrial membranes, and inhibition of the inflammasome [9, 62, 63].

In addition to suppression of pro-inflammatory cytokines, the vagus nerve promotes resolution of inflammation [64]. Its dissection decreases peritoneal group 3 innate lymphoid cell number and reduces levels of pro-resolving mediators while the concentration of pro-inflammatory eicosanoids increases [65].

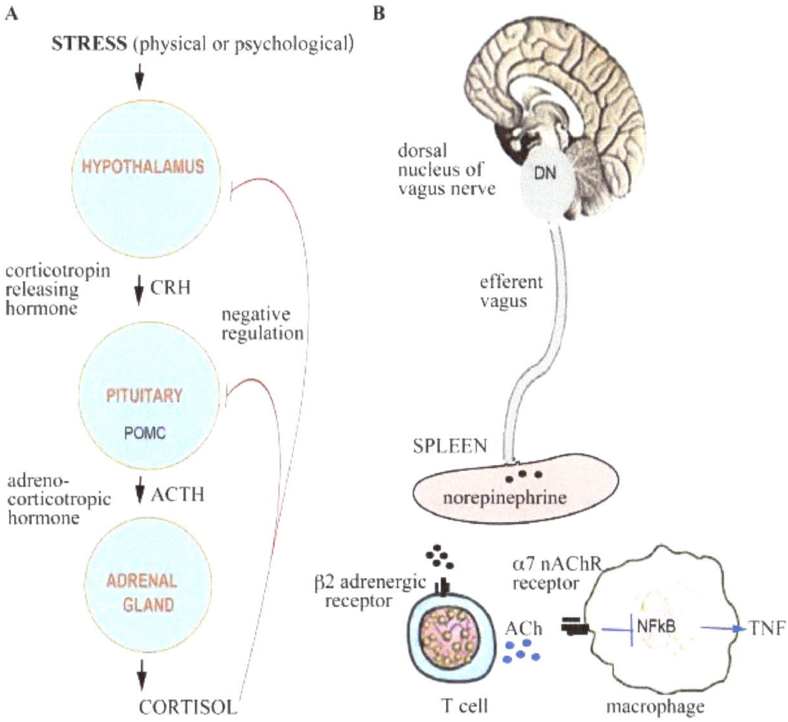

Fig. (12). A. Hypothalamic-pituitary-adrenal axis. Physical, psychological, or environmental stress activates the release of CRH. It promotes the synthesis of ACTH in the pituitary. ACTH stimulates synthesis and release of glucocorticoids such as cortisol that regulate the expression of immune genes and negatively regulates the HPA axis. **B.** The action potential travels from the brain through the efferent vagus nerve to the spleen and leads to the release of norepinephrine. Norepinephrine binds β2 adrenergic receptor on a subset of T cells that express choline acetyltransferase. These cells produce and release acetylcholine (ACh). ACh binds alpha7nAChR receptor expressed on macrophages and other immune cells and inhibits the production of inflammatory cytokines.

The sympathetic branch of the peripheral nervous system is activated through the HPA axis in response to stress and leads to the secretion of catecholamines epinephrine and norepinephrine from the adrenal glands and sympathetic nerve endings. This turns on various body stress responses such as an increase in heart rate. Sympathetic signaling regulates immunity; it usually suppresses lymphoid cells and can induce pro- or anti-inflammatory effects depending on cell type and the context [66, 67]. Loss of intestinal sympathetic innervation leads to the development of colitis due to overactivation of innate immune mechanisms [68].

Neural and immune cells form interdependent associations at barrier surfaces such as epithelial surfaces of gut and lung. Chemical signals from the environment are translated into neuronal impulses that affect local immune cells and also are propagated throughout the body including CNS [69]. Gut macrophages that reside close to lumen display proinflammatory phenotype while more distant macrophages have tissue protective phenotype, which is further enhanced during luminal infection. This effect is mediated by norepinephrine; it is released from gut sympathetic neurons and binds β2 adrenergic receptors on macrophages [70]. Acetylcholine secreted by specialized T cells stimulates immune response in intestinal epithelial cells during bacterial infection [71]. Another example includes the interaction of innate immune cells (ILCs) with neurons. Infections stimulate cholinergic neurons to produce neuropeptide neuromedin; this neuropeptide then activates ILC2 expressing a receptor for it [3]. In contrast, catecholaminergic neurons dampen this ILC2 activation.

Interestingly, activation of the dopaminergic neurons in the ventral tegmental area, a key component of the reward system, boosts both innate and adaptive immune systems with the enhanced antibacterial activity of macrophages and T cells [72]. The sympathetic nervous system is involved in this effect.

Neuro-immune interactions take place also at the level of nociceptors. Immune mediators modulate mechanical and thermal sensitivity of these receptors, whereas neural mediators modulate the activity of immune cells that reside at peripheral nerve terminals and within the spinal cord [55]. These interactions can contribute to allergic airway inflammation. Lung nociceptors are stimulated with capsaicin release neuropeptides that promote immune cell infiltration [73]. IL5 produced by activated immune cells acts on nociceptors and induces the release of the vasoactive intestinal peptide (VIP), which promotes inflammatory signaling in lymphoid cells. *Streptococcus pyogenes* hijacks nociceptor neurons by secreting streptolysin S to directly activate them and induce pain. Nociceptors, in turn, release the neuropeptide CGRP into infected tissues, which inhibits the recruitment of immune cells and prevents *S. pyogenes* killing [74].

The gateway reflex is associated with the ability of neurotransmitters to induce expression of chemokines in specific blood vessels, which serve as a gateway for immune cells to migrate into the brain or spinal cord [75]. Local brain inflammation induced by the gateway reflex can then change the regulation of organs in the periphery.

After the initial finding that direct electrical stimulation of the peripheral vagus nerve decreases TNF synthesis in the rat liver [59], several clinical trials demonstrated the efficacy of this approach for the treatment of human inflammatory diseases such as rheumatoid arthritis and inflammatory bowel disease [9, 76, 77]. In addition, pharmacological modulators such as a7nAChR agonists, acetylcholinesterase inhibitors, and β2-adrenoreceptor agonists are studied as therapeutic tools [78].

The brain can control inflammation through other mechanisms as well. For example, an increase in central serotonin decreases febrile response and the level of inflammatory cytokines [79].

12.4. Sickness Behavior

Human response to infection involves immune, physiological, and behavioral adaptations. In addition to the activation of the HPA axis, physiological changes involve suppression of the hypothalamic-pituitary-gonadal axis, decrease of gastrointestinal motility, and release from the liver of the acute phase proteins. Most acute phase proteins such as the C-reactive protein and the mannose-binding lectin recognize molecular patterns of pathogens. Acute phase proteins also include components of the complement system and hepsidin and other factors that sequester iron to prevent its use by pathogens. Behavioral adaptations to infection include fatigue, discomfort, sleepiness, hyperalgesia, and loss of interest in food, water, social activities, and sex [80, 81]. This so-called sickness behavior is observed from arthropods to birds and mammals suggesting its evolutionary conservation. Initially, it was believed that pathogens induce these symptoms. Later it was suggested that sickness behavior is an adaptive response that restricts animal activity and diverts energy to increase of body temperature and immune response [82].

An alternative hypothesis suggests that reduced transmission of infectious disease among related individuals drove the evolution of sickness behavior [83]. Indeed, decreased social contacts of a sick member decrease the spread of infection [84]. In addition to self-isolation of a sick individual, healthy members of the group avoid sick ones. Evolutionary psychologists suggest that behavior that limits contacts with infected individuals is the first line of defense against infection in humans and some other animals. This so-called 'behavioral immune system' activates aversive emotions such as disgust in response to specific cues [85, 86]. They affect person perception, sociability, prejudice against out of group people, and sexual behavior [87]. In mice, the smell of a sick mouse induces neural activity in the vomeronasal system that controls innate behaviors leading to aversive reaction [88]. Body odor is also an important clue for recognition of sick individuals in humans [89]. In addition, humans recognize facial features of sick individuals as less likable and combination of odor and photograph of a sick person has a superadditive effect. People recognize somebody as sick when the person has a swollen face, paler lips and skin, droopier corners of the mouth, hanging eyelids, redder eyes, less glossy skin, and appears more tired [90]. Everything unfamiliar is treated as infectious by default; for example, healthy out-group people are perceived psychologically as infected in-group members [91].

Reduced motility and anti-predator response in sick birds leads to their increased mortality due to predator attack [92]. In essence, the removal of sick individuals by predators may also benefit the group.

Sickness behavior is induced in the brain by proinflammatory cytokines such as IL1β, IL6, and TNF. Cytokines are carried into the brain vessels by blood or act through the primary afferent nerves such as the vagus nerve. IL1β acts on the brain mainly through fenestrated capillaries [93]. Fever and sickness behavior are induced simultaneously but independently in sick individuals [94]. Cytokines acting on brain endothelial cells induce secondary mediators such as prostaglandins. They induce both fever and anorexia although through different pathways [95]. Prostaglandins also regulate pain perception by modulating serotonin signaling [96]. PGE2 signaling decreases impulsive behavior and aggression [97]. Other features of sickness behavior such as aversion do not involve prostaglandin signaling but can be mediated by IFNs binding to their receptors expressed on brain endothelial and epithelial cells leading to production by these cells of a chemokine ligand CXCL10 [98, 99].

Increased sleep is another feature of sickness. Animal studies show that it is beneficial to survival [100]. Sleep is regulated by neurons in the lateral hypothalamic (LH) area. Sickness sleep differs from normal sleep by fragmentation and increased non-REM. In addition, normal sleep is controlled by 24 h cycle of cortisol secretion, which is disrupted during HPA activation and may also lead to deregulated sleep. Various cytokines released during stress and illness such as IL1β, TNF, and members of the epidermal growth factor (EGF) family induce sickness sleep. In *Caenorhabditis elegans*, bacterial products, heat, UV, and other stimuli lead to EGF release from stressed cells. EGF binds to a single interneuron called ALA that releases sleep-promoting neuropeptides including those with amidated arginine-phenylalanine (RFamide) C-terminal motif [101]. Two peptides with similar motif called neuropeptide Y (NPY) and neuropeptide VF (NPVF) induce sleep in zebrafish [102, 103]. NPY promotes sleep by inhibiting noradrenergic signaling. In *Drosophila*, antibacterial peptide NEMURI induces sleep and promotes survival after infection [104]. This breakthrough study establishes a direct connection between sleep and immunity. It could be that some of approximately 100 different antibacterial peptides synthesized by humans have such activity.

Sickness-associated anorexia is an evolutionary conserved response. It promotes autophagy, which can contribute to immunodefense and increase disease tolerance [105, 106]. Anorexia is mediated by the growth and differentiation factor 15

(GDF15), which is released during tissue damage and metabolic or toxin stress [107]. GDF15 receptors are expressed on neurons in the brainstem. GDF15 binding to these neurons triggers the activation of neurons regulating feeding responses in the parabrachial nucleus and amygdala, which induces anorexia.

The individual sickness responses vary depending on differences in neuroendocrine phenotypes, the environment, and other factors [81]. Women tolerate sickness better than men [108]. Susceptibility to fatigue is associated with polymorphisms in TNF, IL1β, IL4, and IL6 [109]. Other factors can influence the interaction between brain and inflammatory signaling. For example, such signaling can be attenuated by central serotonin [79] or by activation of IFNs [110]. Animals can terminate sickness behavior, for example when their energy levels drop below a minimal threshold [2].

Pathogens can manipulate sickness responses, for example, *Salmonella Typhimurium* effector SlrP suppresses anorexia *via* inhibition of IL1β-mediated signaling to the hypothalamus *via* the vagus nerve [111]. This improves host survival and increases the transmission of the pathogen.

12.5. Psychological Stress and Immunity

There is an intimate connection between emotions and immunity. Multiple studies and several meta-analyses show that acute psychological stress increases levels of IL6, IL1β, IL-10, and TNF [112]. Even events such as exams can increase the production of these cytokines [113]. Sickness symptoms caused by infection can also be increased by exposure to psychological stress [114, 115].

Social stress is a subtype of psychological stress associated with the relationship of an animal with other members of its group. Social stress can be associated with competition for territory, dominance, mate, or isolation from the group, rejection, maternal deprivation, *etc*. Many studies show co-regulation of inflammation and social behavior. Inflammation increases neural sensitivity to social experiences. When individuals are socially stressed (for example, isolated), they show increased proinflammatory activity [81]. Peer victimization is a strong social stress for human teenagers. Adolescents with more peer victimization experience show greater increases in IL6 and IL1β in response to the social stressor. The response is higher for adolescents with high levels of hopelessness [116]. The connection between psychological stress and immunity is conserved: humans, house mice, stickleback fish, and honeybees exposed to a territorial intrusion, all upregulate NFκB [117]. There is evidence complement system upregulation by stress [118].

Personality is linked to the expression of proinflammatory genes. In a cohort of 121 subjects, extraversion was associated with increased expression of pro-inflammatory genes while conscientiousness was associated with reduced expression of these genes [119]. IFNs and antibodies do not show correlation with personality type.

Psychological stress induces remodeling of the neuronal network by microglia. Chronic unpredictable stress in mice leads to increased expression of CSF1 and its receptor in the prefrontal cortex. It leads to increased phagocytosis of neuronal components by microglia, which results in depression-like behavior [120].

While proinflammatory mediators are upregulated by psychological stress, other immune mechanisms can be suppressed. Leukocytes in people who have experienced stressful life events show a reduction in telomere length [121]. TLR signaling is downregulated by academic stress [113]. Another study shows that stress induces downregulation of antimicrobial peptide expression in epidermal cells, which worsens cutaneous infections in mice [122]. Social stress increases the susceptibility to infection in the ants while their resistance against other stressors remains unchanged. This is associated with a shift from glutathione-S-transferase activities towards glutathione peroxidase activities, which increases the antioxidative capacity at the expense of immune competence [123]. Piglets that have been socially isolated are more susceptible to *Escherichia coli* infection than non-isolated piglets [115]. A number of studies report the suppression of adaptive immune response by psychological stress [124, 125].

Predator exposure induces immunosuppression despite the increased risk of wounds and infections. This was explained as trade-offs between the immune system and other systems required for anti-predator activity [126].

Psychological stress can induce other physiological and metabolic changes that can influence the immune response. For example, chronic social isolation leads to oxidative stress in the brain [127]. Epigenetic changes in response to acute psychological stress include a sharp increase in methylation of oxytocin receptor (OXTR) 10 minutes after exposure followed by a decrease in methylation 90 minutes after exposure [128].

Therefore, psychological stress can induce critical changes in body homeostasis and immunity suggesting that stressful life events can precipitate various physical, mood, and autoimmune disorders. Social skills training can decrease stress response, specifically cut the secretion of cortisol, the end-product of HPA axis by up to 51% [129]. Self-regulation also can be trained, for example using mindfulness-based techniques; patients with a generalized anxiety disorder who received mindfulness-based stress reduction training had lower levels of ACTH and proinflammatory cytokines [130]. Regular exercise stabilizes immuno-neuroendocrine connections and can likewise decrease stress response [131].

12.6. Effects of Stress and Inflammation on Brain Development and Health

The developing brain is sensitive to stress and inflammatory mediators, which makes it vulnerable to diverse environmental factors that can alter the hypothalamic-pituitary-adrenal (HPA) function [132]. They include maternal infections, maternal stress, placental insufficiency, unbalanced nutrition, medications, and adverse childhood experiences. These factors can permanently reprogram the developing brain. Maternal infections with a number of pathogens including toxoplasmosis, syphilis, varicella, mumps, parvovirus, HIV, rubella, cytomegalovirus, and herpes simplex lead to a syndrome TORCH. It is characterized by microcephaly, white matter defects, cerebral atrophy, and calcifications. Other infections such as the Zika virus also can carry a risk of microcephaly. A similar phenotype (called pseudo-TORCH) results from genetic mutations in genes associated with IFN type I signaling that lead to unbalanced upregulation of IFNs [133].

Epidemiological studies show an increased incidence of autism, schizophrenia, and several mood disorders after historic outbreaks of rubella, flu, measles, mumps, chickenpox, and polio [134]. Even in the absence of an infection, maternal autoimmune disorders, allergies, asthma, acute stress, and environmental pollution also increase the risk of mental disorders [135]. This positions maternal immune activation as a major risk factor for mental disorders. Animal models show clear associations of maternal immune activation with defects in brain development and altered behavior. In humans, maternal immune activation does not always lead to obvious defects and is thought to serve as a "neurodevelopmental disease primer" that contributes to adult mental disorders such as autism and schizophrenia combined with genetics and environmental triggers [134].

Maternal immune activation is thought to affect the fetal brain through proinflammatory cytokines such as IL6, IL17, and others. Children of mothers with higher levels of TNF in the 2nd and 3rd trimesters had lower IQ, lower visual-motor maturity, and lower Draw-a-Person test scores [136]. Higher maternal IL6 was associated with a larger right amygdala volume, increased amygdala connectivity with brain regions involved in sensory processing, integration, and learning and memory, and poorer impulse control in children at 2 years of age [137]. Activation of inflammation in the fetal brain can lead to multiple effects including excessive or insufficient synaptic pruning mediated by microglia and complement [31].

The maternal-fetal interface plays a central role in mediating the effects of maternal immune activation on the fetus. The Zika virus, which causes fetal brain defects, differs from similar viruses that do not cause such defects in that it induces inflammation in trophoblasts [138]. Knockout of IL6 signaling in trophoblasts of pregnant mice exposed to immune activation protects them from cerebellar neuropathologies and behavioral abnormalities observed in the offspring of control mice [139]. The shift of T cell repertoire to proinflammatory phenotype also may play a role. For example, maternal gut bacteria that activate IL17-producing T helper 17 (TH17) cells promote neurodevelopmental abnormalities [140].

Inadequate prenatal nutrition also affects brain development including abnormal development of the hypothalamus, which impairs the response to stress through the HPA axis in the offspring [141]. Supplementation of pregnant women with vitamins A and D decreases the risk for schizophrenia and autism in the offspring [142].

Postnatal early life stress also affects health in adults. The foundation study on the impact of early life stress on physical and mental health in adulthood was the Adverse Childhood Experiences (ACE) study. Among the most important publications from the ACE study was one reporting a correlation between adverse childhood experiences and adult health in 9508 participants. They answered questions about adverse childhood experiences in seven categories that included psychological, physical, or sexual abuse, violence against the mother, substance abuse, mental illness, or incarcerations in the family [143]. The responses then were compared with health status and risk behavior of adults. More than half of responders reported at least one and 25% reported more than two categories of childhood exposures. Persons that had four or more had a higher risk for alcoholism, drug abuse, depression, smoking, promiscuity, obesity, and poor physical health as adults. Many more subsequent studies confirmed that adverse childhood experiences are frequent and that early life experience strongly influences the physical and mental health of adults. Much has been learned about how early life stress influences brain development and function. It became clear that early life stress could impact brain development, neuronal survival, connectivity, communications between brain regions, HPA axis, and induce long-lasting inflammation and neuroinflammation. Some mechanisms are discussed below.

Maternal prenatal stress or early life stress can alter the function of HPA axis in adults. Exposure to stress can lead to epigenetic modifications in gene expression such as DNA methylation [144]. A study of genome methylation of institutionalized children in comparison to children raised by their parents shows that institutionalized children have increased methylation at 815 GpC sites and decreased methylation at 99 sites [145]. Chicks exposed to moderate heat stress 3 days post hatch have better heat stress resistance as adults *via* changes in DNA methylation [146].

The difference in maternal care such as licking and grooming of pups in rats leads to a difference in methylation of the promoter of glucocorticoid receptor NR3C1 that is involved in HPA axis [147]. Maternal prenatal stress also affects the methylation of this receptor [148]. Children who received less maternal support reported increased anxious attachment if their NR3C1gene was highly methylated [149].

Low maternal care or other early life adversity leads to increased methylation on the oxytocin receptor (OXTR) [150, 151]. Oxytocin is a small multifunctional neuropeptide that promotes social interaction and pair bonding.

Early life stress affects the function of microglia in the developing brain. Stress can alter microglial gene expression, phagocytic ability, morphology, and proliferation [152]. These changes can interfere with normal microglial functions such as trophic support of neurons, axonal growth, synaptogenesis, synaptic pruning, and myelination and lead to abnormalities in morphology and signaling in the developing brain. Microglial expression of many important factors involved in neuronal survival and neuroplasticity such as brain derived neurotrophic factor (BDNF) is strongly influenced by stress [153, 154].

Childhood maltreatment leads to the induction of pro-inflammatory mediators with long-term adult physiological and psychiatric complications [28]. A meta-analysis suggested that peripheral levels of inflammatory markers such as C-reactive protein, IL6, and TNF are elevated in individuals exposed to childhood trauma; specific types of trauma (sexual, physical or emotional abuse) impact these inflammatory markers differentially [155]. Childhood infections can also serve as a risk factor for mental disorders in adults [156]. Prophylactic treatment of the inflammatory state in individuals with childhood stress has been suggested as a means to prevent the development of more serious physical and mental diseases [157].

Because inflammatory mediators affect brain development, changes in their levels may have far-reaching consequences such as defects in masculinization [2]. Ultimately, neuronal survival and connectivity can be affected. This may impact cognitive processes involved in self-control and behavior associated with the planning and execution of specific goal-oriented actions, which is known as executive function [158]. Increased maternal IL-6 during pregnancy is associated with enlarged right amygdala and stronger left amygdala connectivity [137]. This was associated with lower impulse control in 2-year-old children. Maternal IL6-related changes in connectivity especially affect the working memory of the neonatal brain [159].

In people that survived childhood maltreatment, anxiety is often increased. This is associated with increased amygdala activation in response to stress and its abnormal connectivity with frontolimbic brain regions, including the hippocampus and the prefrontal cortex. In mouse models of unpredictable postnatal stress, the strength of the amygdala-hippocampal and amygdala-prefrontal cortex connectivity is correlated with anxiety-like behavior [160].

Females and males differ in neuro-immune function. Male brains have more microglia and higher expression of inflammatory mediators, which has been suggested to be responsible for the higher incidence of neurodevelopmental disorders in males [161]. Indeed, expression of inflammatory mediators is particularly high in males with autism [162].

Autoimmune diseases are more common in women than in men. The X chromosome has the largest number of immune-related genes, and X-chromosome deletions are associated with immunodeficiency, while patients with autoimmune conditions such as Systemic Lupus Erythematosus and Sjogren's Syndrome are more likely to have an extra X chromosome [163, 164]. Children with extra X chromosome are more vulnerable to early life stress and have problems with executive functions and social cognition [165]. Male and female embryos' responses to many stressors differ starting at the pre-implantation stage before the development of gonads and production of sex hormones [166]. An example is OXTR methylation, which only in females is associated with early life stress and anxiety in adulthood [167].

Stress-induced changes can be transmitted to the next generation both maternally and paternally [168]. Progeny of stressed male rats have increased methylation of the promoter of glucocorticoid receptor NR3C1 and lower expression of this gene, increased corticosterone levels, and more anxiety [169]. miRNAs is another means of transmitting information about stressful conditions into the next generation. Sperm carries many miRNAs and can also capture RNAs from somatic cells [170]. miRNAs in the sperm of chronically stressed male mice affect HPA axis function recapitulating paternal stress in the next generation [171]. Exposure of males to environmental toxins such as endocrine disruptors can affect miRNAs in primordial germ cells resulting in transgenerational effects on stress and immune response [172]. These examples illustrate the scope of changes that early life stress can induce ranging from brain structure alterations to changes in stress response and behavior.

12.7. Inflammation, Microglia, and Mental Illness

The recognition that inflammatory mediators produced during infection or inflammation cause sickness behavior revolutionized the field of neurobiology. Another crucial discovery was that maternal immune activation and early life stress can alter the developing brain and predispose it to mental illness [134]. Growing evidence points to the role of inflammation and microglia in neurodevelopmental, psychiatric, and neurodegenerative disorders [14, 173, 174]. Patients with Crohn's disease have changes in brain connectivity [175]. Even diet and microbiota can influence the brain, HPA axis, and neuroinflammation [176]. Some psychiatric diseases such as schizophrenia, autism spectrum disorder, bipolar disorder, and depression share common immune changes. They include microglial activation, increased levels of pro-inflammatory cytokines, autoantibodies, self-reactive T cells, and disruption of the BBB [177]. Current studies aim at understanding the role of common and unique immunological alterations in various forms of mental illness. Below we discuss the role of immunity in several mental disorders.

12.7.1. Depression

Many studies show a connection between inflammation and depression. People with inflammatory diseases such as arthritis, asthma, diabetes, and metabolic syndrome have a higher frequency of depression [178, 179]. Applications of IFN type I for treatment of hepatitis C and some cancers also lead to depression-like symptoms in about 30% of patients [180]. Sickness behavior shares a number of features with major depressive disorder (MDD) [181], which led to a hypotheses that sickness can evolve into depression in vulnerable individuals who have their balance of pro- and anti-inflammatory mediators shifted towards inflammation (more TNF or less IL10), have glucocorticoid resistance, or altered metabolism of neurotransmitters [182].

MDD is associated with altered levels of cytokines and chemokines [183]. A recent meta-analysis showed elevated peripheral levels of IL6, IL10, IL12, IL13, IL18, TNF, the soluble IL2 receptor antagonist, the IL1 receptor antagonist, and the soluble TNF receptor 2 in MDD [184]. Another meta-analysis found higher levels of neurotoxic CXCL4 and CXCL7 and lover levels of CCL4 in MDD [185]. However not all MDD patients have increased levels of proinflammatory cytokines.

Common genetic variants are involved in both immune response and depression including polymorphisms in IL1β, IL6, IL10, monocyte chemoattractant protein-1, TNF, C-reactive protein, and phospholipase A2. The severity of depression correlates with inflammatory markers. MDD patients with a weaker response to antidepressants have increased IL1β expression [186]. In individual patients, a longitudinal association exists between depression and IL6 level [187]. Upregulation of a marker of the activated microglia TSPO detected by positron emission tomography correlates with depression severity; the correlation is especially significant in patients with suicidal thoughts [188]. Expression of some TLRs is increased in the depressed-suicidal group compared with controls [189]. MDD patients with elevated inflammatory markers have decreased functional connectivity centralized in the ventral medial prefrontal cortex [190].

Antidepressant treatment decreases levels of IL6, IL10, TNF, CCL2, and C-reactive protein [191, 192]. Conversely, inhibition of inflammation by anti-inflammatory drugs often reduces depressive-like behavior in animal models, although human studies report variable effects, which points to the complexity of the relationship between inflammation and MDD [39]. However, anti-inflammatory drugs improve antidepressant treatment effects [193].

Inflammation is often associated with mitochondrial dysfunction and increased ROS, which may contribute to the induction and progression of depression. Brains of MDD patients demonstrate increased DNA oxidation and increased expression of DNA repair genes [194]. Oxidative DNA damage is repaired less efficiently in MDD patients than controls, and single nucleotide polymorphisms in some DNA damage repair genes modulate the risk for depression [195].

To explain how inflammatory mediators promote depression, many hypotheses have been proposed, including dysregulation of HPA axis, impaired neuroplasticity, and changes in serotonergic, noradrenergic, dopaminergic, and glutamatergic systems [196, 197]. Inflammatory mediators promote energy production by activating the enzyme indoleamine 2,3-dioxygenase (IDO), which directs catabolism of tryptophan along the kynurenine pathway leading to nicotinamide adenine dinucleotide (NAD) [198]. Tryptophan is the precursor of serotonin. Diversion of tryptophan metabolism from serotonin synthesis can cause a serotonin deficit. Serotonin is a major regulator of mood and its deficiency is strongly implicated in MDD. It also suppresses inflammation [79]. Importantly, pro-inflammatory cytokines increase the availability and function of serotonin transporter, which reuptakes serotonin from synapses decreasing postsynaptic serotonin signaling [199].

Many compounds from the kynurenine pathway are neuroactive [198]. 3-hydroxykynurenine and quinolinic acid act as NMDA receptors agonists. Kynurenic acid is an antagonist of NMDA, AMPA, and kainite glutamate receptors. The imbalance between kynurenine metabolites may contribute to depression through the glutamatergic system [200]. Within the brain, activated microglia produces an increased amount of quinolinic acid leading to overactivation of NMDA receptors, neurotoxicity, and oxidative stress.

Synthesis of serotonin and dopamine requires cofactor tetrahydrobiopterin (BH4), which is also a cofactor for the NO synthase. Inflammation interferes with BH4 biosynthesis and stability leading to its shortage, which affects the production of serotonin and dopamine. BH4 shortage also leads to an uncoupling of NO synthases and an increased production of reactive oxygen species (ROS), which can contribute to neural apoptosis [201].

In inflammation-mediated depression, availability and release of striatal dopamine is reduced and dopamine and its metabolites are decreased in the cerebrospinal fluid [196]. This may lead to decreased reward anticipation, decreased motivation, motor activity, and anhedonia, a loss of interest and pleasure in daily activity. Kynurenine metabolites and BH4 deficiency may also modulate dopaminergic circuitry [202].

Microglial cells can contribute both to neuropathology and neuroprotection depending on the nature of signals they receive. Activated microglia can contribute to depression and synaptic deficits observed after chronic stress by reducing dendritic spine density in neurons of the medial prefrontal cortex [120]. In contrast, microglia that migrate to the inhibitory synapses after injury and displace them from cortical neurons protects these neurons [203]. Decreases in inhibitory GABAergic synapses increase neuronal activity, stimulate the production of antiapoptotic and neurotrophic compounds, and prevent apoptosis of cortical neurons.

Neuro-immune interactions can be modulated by neuroendocrine (HPA axis) features, metabolic factors, and individual psychosocial characteristics. For example, persons with atypical depression in comparison to those with melancholic depression, have decreased HPA axis response, higher levels of inflammatory markers, smaller circadian changes in cortisol, increased triglycerides, lower levels of "good" cholesterol, and more obesity [204, 205]. Sex steroids also can modulate inflammatory responses [206]. Personality features such as conscientiousness, neuroticism, coping style, resilience, and self-esteem also can modulate the neuro-immune interaction.

In addition, cytokines such as IL1β can have non-immunological targets in the brain and can alter neuronal function directly or through other mediators [39]. Epigenetic modifications can also contribute to MDD. Several genes including those encoding the glucocorticoid receptor NR3C1 and BDNF display altered methylation patterns in chronic stress and depression [207]. Interestingly, increases in central serotonin levels leads to a decrease in the levels of corticosterone and IL6, which suggest that anti-inflammatory modulation takes place during MDD treatment with serotonin reuptake inhibitors [79].

Lasting depression can lead to neurodegeneration and cognitive dysfunction [208]. Many recently discovered mechanisms involved in MDD have been suggested as therapeutic targets. They include inflammation, oxidative stress, epigenetic changes, neural plasticity, circadian clock misalignment, and others [208]. A subset of MDD patients may benefit from non-steroidal anti-inflammatory drugs and cytokine antagonists [209]. Dietary changes, especially intake of polyunsaturated fatty acids, probiotics, and folic acid, may also have an anti-depressant effect [210]. Cervical vagus nerve stimulation is used for chronic treatment-resistant depression. Recently non-invasive transcutaneous forms of the vagus nerve stimulation have been developed. One of them stimulates the surface of the ear innervated by the vagus nerve; this treatment also improves symptoms of depression [211].

12.7.2. Anxiety

Inflammation is increased not only in MDD but also in anxiety disorders: PTSD, generalized anxiety disorders (GAD), phobias, and others [28]. Exposure to severe stress in a preclinical animal model of the PTSD is associated with an increase in hippocampal IL1β, which mediates stress-enhanced fear learning [38]. In this model, astrocytes are the major IL1β source. Blocking IL1β prevents fear learning. International students experience attachment anxiety after migration to a new country, which correlates with C-reactive protein level [212]. In a mouse model of obesity-related anxiety, anti-inflammatory treatment decreases anxiety through reducing levels of TNF in the hippocampus [213]. Other cytokines such as IL18 also may play a role in anxiety [214]. Increases in sympathetic and decreases in parasympathetic tone observed in anxiety can augment inflammation [215].

Changes in DNA methylation are a major epigenetic mechanism. Changes in methylation of a number of genes including NR3C1, BDNF, OXTR, and others are found in young adults with early life adversity as described in previous sections. Increased CpG methylation in the promoter of ASB1 hypothetically involved in cytokine regulation has been reported in people with severe anxiety and in mice subjected to social defeat stress [216].

Anxiety is associated with changes in responsiveness and connectivity of several brain regions including prefrontal cortex, insula, amygdala, and hippocampus. For example, childhood maltreatment leads to hypersensitivity of the amygdala to stress and its hyper-connectivity with the hippocampus and the prefrontal cortex [160]. The strength of the amygdala-hippocampal and amygdala-prefrontal cortex connectivity correlates with the severity of anxiety.

Anxiety disorders are more frequent in females. This may in part be due to structural differences in the male and female arousal system. The locus coeruleus, a key nucleus promoting arousal, is bigger in females and its norepinephrine neurons have more dendrites, which may mediate easier and longer lasting arousal in females [52, 217]. Other differences may be epigenetic; oxytocin receptor gene is methylated at different positions in male and females PTSD patients [218].

12.7.3. Neurodegenerative Disorders

Neurodegenerative diseases such as Alzheimer's disease (AD), Parkinson's disease (PD) and others are associated with loss of synapses and neuronal death that leads to cognitive decline and dementia, disordered movement and other functional deficits [219]. Protein aggregates is a typical feature of these diseases. They include among others β-amyloid (Aβ) plaques, aggregates of microtubule-binding protein tau, and alpha-synuclein aggregates. Alpha-synuclein, tau, and Aβ can form heteroaggregates. Protein aggregates can spread between cells and brain regions in a prion-like manner [220]. Such aggregates are also found in healthy people and not only in the brain but also in red blood cells and platelets [221]. Their amount depends on the antioxidant capability and physical activity of the subjects. Antioxidant capability is reduced in older and sedentary subjects with a proportional increase in aggregates. Accumulation of aggregates may also depend on decreased autophagy. In old age, the efficiency of autophagy decreases and cells activate an alternative Golgi-mediated degradation pathway that cannot break down toxic protein aggregates [222].

Aging is associated with an increase in the number of microglial cells, their decreased functionality and excessive or unresolved activation [16, 223]. Some neurodegenerative diseases are caused by somatic mutations in genes expressed in microglia [224]. High levels of proinflammatory cytokines including TNF, IL1β, and IL6 are found in the brain and blood of patients with neurodegenerative diseases [225]. Weakened BBB function and increased oxidative stress associated with aging also may contribute to neurodegeneration [226, 227].

Another feature of aging is the upregulation of the complement system in the brain. C1q, a sentinel of the classical complement pathway that mediates synapse pruning in the CNS, is increased more than 300 fold in aging human and mouse brains [228]. It incrcascs carlicr and more in regions affected by neurodegenerative diseases, *i.e.*, the hippocampus, the piriform cortex, and the substantia nigra. Deficiency in the frontotemporal dementia (FTD) gene progranulin (Grn) also leads to age-dependent complement activation and enhanced synaptic pruning [229]. Therefore, the complement system could be a common factor that is activated in the early stages of neurodegenerative diseases to mediate synapse loss [31]. Memory impairment in survivors of West Nile virus infection may also be caused by the loss of pre-synaptic terminals mediated by the complement [230].

12.7.3.1. Alzheimer's Disease

Alzheimer disease is very common (10–30% of people >65 years old) [231]. It is associated with extracellular β-amyloid (Aβ) plaques and intracellular aggregates of microtubule-binding protein tau. Early onset AD is caused by mutations in enzymes that process Aβ precursor protein (APP) or by APP mutations [219]. People who have the ApoE4 isoform of the apolipoprotein E have increased risk of AD; ApoE4 presence predisposes cerebral vasculature to damage including decreased blood flow and increased leakiness [232]. ApoE4 also stimulates neuronal APP production [233]. Amyloid plaques are considered a central factor in AD pathogenesis, although drugs that target Aβ plaques are only moderately effective in treating the disease. Therefore, other factors may contribute to the development of AD [234].

An alternative hypothesis of AD pathogenesis focuses on the amyloid oligomers. Low molecular weight products of APP are more neurotoxic than Aβ plaques, especially covalently cross-linked Aβ dimers [235]. Aβ oligomers can alter the conductivity of ion channels or can themselves form ion channel pores that disrupt Ca^{2+} homeostasis [236], and their localization to synapses impairs synaptic plasticity [237]. C-terminal fragment of APP is also toxic to neurons; moreover, it promotes tau aggregation [238], and recent data show that tau aggregates correlate with cognitive decline better than Aβ plaques [239].

Some data suggest that Aβ accumulation is not the earliest sign of the disease. Instead, imaging results suggest that vascular dysregulation is an early indication of the AD [240]. Increased blood levels of several complement proteins together with a complement inhibitor clusterin predict the AD onset one year later [241]. Plasma levels of C3 and C5 are increased in patients with early and progressive stages of AD [242]. Aβ and other APP metabolites have antimicrobial activity and are induced by several pathogens such as herpes simplex virus-1 and Chlamydia, which also induce neurodegeneration and cognitive deficit [243, 244]. These data suggest pathogens may be involved in AD pathogenesis.

Microglia plays a dual role in AD. Microglia have been reported to have a protective function in AD; they encapsulate amyloid deposits forming a "microglial barrier" that insulates adjacent axons and reduces neurotoxicity [245]. A membrane protein TREM2 is involved in macrophage activation. Elevated dosage of TREM2 decreases Aβ plaques, decreases dystrophy of neuritis, and improves memory [246]. R47H variant of TREM2 triples AD risk. Microglia that express it fails to associate with Aβ plaques [247].

On the other hand, activated microglia can produce ROS and cytokines that induce inflammation [248]. TNF and IL1β suppress hippocampal long-term potentiation directly at the synapse [249]. AD is associated with loss of neurons and synapses [250]. Microglia, complement, and astrocytes play a major role in synapse elimination [251]. Various products released from dying neurons have been implicated in stimulating inflammatory pathways in AD including Aβ aggregates [252]. Aβ is recognized by the receptor for the advanced glycation end product (RAGE) and the CD36 scavenger receptor [225]. Some animal AD models show microglial activation before plaque accumulation [253]. Recognition of products from damaged neurons stimulates the formation of inflammasomes that include receptors from NLR family or AIM2 and an adaptor ASC. ASC polymerize in a prion-like manner and produces a micron-sized "specks" [254]. ASC specks released by microglia bind to Aβ and increase the formation of Aβ oligomers and aggregates; intrahippocampal injection of ASC specks leads to spreading of Aβ pathology [255].

Astrocytes may also play a role in neurodegeneration. Their normal role in the brain is trophic support of neurons, but when activated by cytokines they can become neurotoxic [36]. Astrocytes from AD patients can cause neuron damage because their exosomes have increased levels of complement components C1q, C4b, C3d, factor B, factor D, Bb, C3b, and C5b-C9 [256]. Astrocytes also produce IL1β in response to stress, which contributes to brain inflammation [38].

AD is associated with increased gene copy numbers of complement component C4 and with their increased expression [257]. Complement mediates early synapse loss in Alzheimer mouse models [251, 258]. Therefore the complement system contributes to AD [31].

Many conditions associated with low-grade inflammation including stress, injury, depression, and obesity increase AD risk. Application of anti-inflammatory drugs does not improve conditions of AD patients [259]. However, decreasing inflammation may delay AD. Inflammation can be decreased not only by medications but also by lifestyle changes, which includes exercise, adequate sleep, and diet. Diets rich in antioxidants, anti-inflammatory compounds, and

NAD(+) have a neuroprotective effect [260, 261]. Keto diet also shows positive effects in AD patients [262]. Exercise increases levels of myokine irisin, which rescues synaptic plasticity and memory defects in Alzheimer models [263]. Periodic starvation can stimulate autophagy that degrades toxic APP products [264]. Stimulation of mitophagy also inhibits AD [261]. Microglia's protective functions are lost in AD, but some studies suggest they can be restored [265]. Infrared light decreases inflammation in animal models of AD [266]. Electrical stimulation of hippocampus can entrain hippocampal sharp waves and ripple oscillations involved in memory consolidation [267, 268].

12.7.3.2. Parkinson's Disease

Parkinson's disease is the second most common neurodegenerative disease after AD in the elderly (1-2%). It is associated with alpha-synuclein aggregates that form Lewy bodies, loss of dopaminergic neurons in the substantia nigra pars compacta, which results in bradykinesia, muscle rigidity and tremor, and cognitive defects [219]. Multiple genetic and environmental factors have been implicated in PD etiology. Among them, mitochondria emerge as a common theme [269].

Substantia nigra neurons that are affected in PD have very long unmyelinated axons that require high energy for the propagation of action potentials and are therefore very sensitive to perturbations in energy supply by the mitochondria. Several forms of PD are associated with mutations in genes necessary for mitochondrial functions such as Parkin, PINK1, DJ-1, FBXO7, and LRRK2 [270 - 272]. The kinase PINK1 acts as a biosensor for mitochondrial protein import efficiency, mitochondrial depolarization leads to its accumulation in the outer mitochondrial membrane followed by recruitment and phosphorylative activation of Parkin. This signaling regulates the removal of damaged mitochondrial components by mitochondria-derived vesicles, destruction of damaged mitochondria by mitophagy, and induction of mitochondrial biogenesis [273, 274]. FBXO7 is also involved in mitophagy [270]. DJ-1 repairs amino groups of proteins and nucleotides modified by glyoxals [275]. It is important for the stability of PINK1 and other proteins and mitochondrial quality in dopaminergic neurons [276].

Mitochondrial damage can lead to the release of mitochondrial DNA (mtDNA), N-formyl peptides, and cardiolipin, potent DAMPs that induce inflammation [277]. The importance of mitophagy in the prevention of inflammation has been demonstrated in Parkin and PINK1-deficient mice. These mice respond to exhaustive exercise by inflammation that depends on STING, the mediator of IFN response to cytosolic DNA [274]. Parkin-deficient mutator mice, which accumulate mutations in mtDNA during aging, lose dopaminergic neurons from the substantia nigra pars compacta and develop the motor defect, which is rescued by loss of STING, suggesting that inflammation facilitates this phenotype.

Synuclein aggregates are a characteristic feature of PD, and mutations or multiplication of alpha-synuclein gene SNCA result in familial PD [278]. SNCA mutant neurons have fragmented mitochondria with an accumulation of alpha-synuclein aggregates on the outer membrane, which leads to excessive mitophagy [279]. Incomplete autophagic degradation of phosphorylated alpha-synuclein filaments can lead to short 12.5-kDa phosphorylated alpha-synuclein fragments that also cause mitochondrial pathology; they accumulate on the mitochondrial membrane, which causes its depolarization, mitochondria fragmentation, and mitophagy [280].

In addition, alpha-synuclein oligomers interfere with neurotransmitter release at presynaptic terminals [278]. Fibrillar alpha-synuclein is recognized by TLR2; interestingly, although TLR2 is expressed both in microglia and neurons, only neuronal TLR2 expression is increased in PD and it is localized to alpha-synuclein positive Lewy bodies [281]. Both helper and cytotoxic T cells from PD patients recognize peptide epitopes derived from alpha-synuclein [282, 283]. This finding suggests T cells may contribute to immunopathology in PD. Anti alpha-synuclein antibodies are found in PD patients [284]. Some of these antibodies interfere with the formation of alpha-synuclein aggregates and may be protective.

The immune system can contribute to PD pathology in additional ways. A rare genetic form of PD is associated with activating mutations in LRRK2 (leucine-rich repeat kinase 2), which is involved in many cellular processes including inflammatory response. Mutations in LRRK2 kinase found in PD confer increased phagocytic activity to microglia [285]. LRRK2 is highly expressed in immune cells and may also promote inflammasome assembly [272].

Necroptosis may be involved in PD pathogenesis. Neural cells from PD patients with mutations in the mitochondrial OPA1 gene have severe mitochondrial fragmentation, respiration impairment, ATP deficits, and oxidative stress. They are protected from death by pharmacological inhibition of necroptosis [286]. Such treatment also protects mice PD models from neuronal cell loss.

12.7.4. Schizophrenia

Schizophrenia is a debilitating psychiatric disorder affecting cognition, perception, and motivation, with typical onset in late adolescence or early adulthood. It is associated with reduced numbers of synapses and gray matter.

Population studies identified more than 100 loci associated with schizophrenia [287]. Several immune loci including the major histocompatibility complex (MHC), B cell markers CD19 and CD20, CX3CR1 gene involved in leukocyte adhesion, and others show association with schizophrenia [288]. RNA sequencing in brain samples from individuals with schizophrenia show changes in splicing or expression of more than 25% of the transcriptome. Changes involve microglial, astrocyte, and interferon-response modules [289]. Another study found higher mRNA levels for genes involved in NFκB pathway in schizophrenic patients [290]. A meta-analysis of data on immune system involvement into schizophrenia showed the increased density of microglia in postmortem brains of schizophrenic patients and increased expression of proinflammatory genes both on transcript and protein levels [291]. Genes differentially expressed in lymphoblastoid cell lines from schizophrenia patients show enrichment for immunity-related [292].

There are other links between schizophrenia and immunity. Transgenic mice with altered neuregulin 1 expression related to schizophrenia show changes in expression of genes involved in immunity and inflammation [293]. Risk of most autoimmune diseases other than rheumatoid arthritis is increased in patients with schizophrenia [294]. And *vice versa*, autoimmune diseases have higher frequencies of neuropsychiatric diseases compared to healthy controls. Moreover, a family history of autoimmune diseases is associated with an increased risk of psychotic disorders [174].

Epidemiological and animal studies suggest that increased frequency of infections during pregnancy augment the risk for schizophrenia in the offspring [134, 295]. A subset of genetic variants associated with schizophrenia converges on genes highly expressed in placenta and involved in the placental response to stress [296]. Patients with schizophrenia have increased number of B cells and reduced number of T cells, increased CD4/CD8 T cells ratio during psychosis, and infiltration of T and B cells into the brain [294]. About 20% of patient with schizophrenia have autoantibodies against glutamate NMDA receptors that alter receptor organization and long-term potentiation at glutamatergic synapses [297]. The presence of autoantibodies to components of synapses is associated with psychosis in several neurological conditions [294, 298].

The risk for schizophrenia is associated with the MHCI locus where complement component 4 (C4) genes are located. C4 gene has two isotypes C4A and C4B, which vary in structure, copy number, and biochemical targets: C4A preferably forms amide bonds with proteins while C4B favors hydroxyl group in carbohydrates. Individual alleles contain a variable number of C4A and C4B copies. Alleles that generate a greater expression of C4A are associated with risk for schizophrenia in a dose-dependent manner [299]. C4 protein is found in neurons with higher expression at dendrites, axons, and synapses. C4 is involved in synapse elimination during postnatal development; it promotes C3 activation and attachment to the cell surface that leads to phagocytosis. Increased synapse elimination by microglia is observed in schizophrenia patient-derived models, which correlates with complement factor C4 risk variants [34]. Mice deficient in C4 have decreased C3 deposition in synapses and defective synaptic remodeling [299]. Expression C4A in blood mononuclear cells correlates with the presence and severity of delusions in schizophrenia [300]. The expression of several other schizophrenia-associated genes also localizes to synapses [287].

These data suggest that reduced numbers of synapses in individuals with schizophrenia may result from excessive complement activity. It is consistent with increased number and inflammatory state of the microglia in schizophrenia patients [291] and increased 8-iso-PGF2α, a marker of oxidative stress, which forms by non-enzymatic oxidation of the arachidonic acid [301]. Complement-microglia damage to synapses and neuronal bodies can depend on other environmental and genetic factors and may result in structural changes in the brains of schizophrenia patients as well as in psychotic symptoms [302, 303].

These findings support the notion of immune contribution to schizophrenia. In humans, extensive remodeling of synapses in evolutionary recent areas such as the prefrontal cortex takes place in adolescence and early adulthood, which coincides with the age of schizophrenia onset [304].

12.8. Physical Illnesses, the Brain, and Immunity

The brain, the body, and the immune system are tightly interconnected and change in one member of this triad causes changes in others. Physical illness, even if restricted to a particular organ, induces systemic changes and affects the brain and the immune system. It is also a major psychological stressor, especially if it prevents the normal function of the body. Hypertension is an example of physical illness that arises as a result of complex interactions of genetic factors and the environment [305]. Psychological stress also contributes to hypertension risk [306]. As was discussed earlier, psychological stress causes hypothalamic-pituitary-adrenal (HPA) axis activation that is associated with overactivation of the sympathetic branch of the autonomic nervous system and with diminished vagal tone. This increases heart rate, stimulates sodium re-absorption in the kidney, and causes vasoconstriction leading to blood pressure increase. Sympathetic signaling dominance predisposes to inflammation, which is found in hypertension [307]. Inflammation can be induced by excessive intake of dietary salt. High extracellular NaCl induces NLRP3 inflammasomes [308, 309]. Inflammation increases the formation of products of lipid peroxidation that activate immune cells [310]. Pro-inflammatory cytokines and ROS promote vascular fibrosis and thickening of arterial walls, which decreases lumen diameter, increases vascular stiffness, and impairs vascular resistance. With aging, it leads to a gradual increase in blood pressure. Inflammation inhibits the biosynthesis and stability of a cofactor for NO synthase BH_4, which leads to increased production of ROS through uncoupling of NO synthases [201]. Inflammation is also associated with increased expression of metalloproteinases, which degrade the extracellular matrix and cause uncontrolled diffusion of immune cells into organs and consequent tissue damage. Hypertension affects brain function and can cause cognitive decline [311].

Mental illness can likewise affect the physical function of the body and immune system. Depression is a considerable risk factor for the development of cardiovascular disease (CVD), stroke, obesity, diabetes, and arthritis [312]. Depressed individuals have two- to four-fold risk of developing cardiovascular disease. Depression and cardiovascular disease show bidirectional co-morbidity with the involvement of nervous, neuroendocrine, immune, vascular, and hematologic systems [313, 314]. Stress seems to be a common contributor to both cardiovascular disease and depression conditions.

The efferent vagus nerve can inhibit inflammation through acetylcholine (ACh) signaling [315]. Bioelectronic devices that modulate the inflammatory reflex have been successfully used in a clinical trial to ameliorate rheumatoid arthritis and inflammatory bowel disease [316]. This approach can be beneficial not only in immunological diseases but also in treating conditions such as hypertension, which is characterized by overactivation of the sympathetic drive to the kidney leading to hypertension and kidney inflammation [76]. Attenuation of systemic inflammation by exercise is associated with activation of the subdiaphragmatic vagus nerve leading to an increase of dopamine level and inhibition of TNF production by splenocytes [317].

Immune-neural interactions regulate other bodily functions. For example catecholamine –regulated lipolysis, which generates energy through hydrolysis of triglycerides, declines with age. This leads to visceral adiposity, decreased endurance, failure to maintain body temperature during cold stress, and decreased ability to survive starvation. A recent study showed that aging cells upregulate the monoamine oxidase, which degrades noradrenaline, in a NLRP3 inflammasome-dependent manner [318]. Deletion of NLRP3 restores lipolysis.

Dysregulation of the immune system can cause brain disorders. For example, most patients with systemic lupus erythematosus experience neuropsychiatric symptoms such as anxiety, depression, cognitive impairment, seizures, and even psychosis [319]. This is explained by IFN activation that drives microglia to trim synapses. In multiple sclerosis, IFN may cause mood abnormalities through modulation of the cannabinoid CB1 receptor [320] and complement damage [321]. Antibodies and T cells against neural antigens may cause neurological manifestations of rheumatic disorders [322]. Infections can also promote mood disorders and their effects can be modulated by sex [156]. Regular exercise can stabilize interactions between the sympathetic nervous system, hormones of HPA axis, and stress proteins in persons with deregulated inflammatory and stress feedback [131]. Other factors such as diet, sleep, and social interactions can also modulate immuno-neuroendocrine interactions [323].

12.9. Stress and Lymphoid Cells

The adaptive immune response is associated with the extensive proliferation of T and B cells, which requires a significant investment of biological resources and energy by the organism. Therefore many mechanisms guard this system against unnecessary activation. Adaptive immunity, in contrast to autonomous immunity, is often suppressed in response to physical and psychological stress. For example, heat stress dampers adaptive immunity [324]. Selye and other researches noted atrophy of lymphoid tissues in response to physical and mental stress [1].

It has long been known that stressors can increase tumor growth in humans and animals [325]. Moreover, mice that chose a passive strategy, submission and avoidance in response to aggression, have greater increases in the number of pulmonary metastases and the serum corticosterone level [326]. The same study reported that stress decreases the number of NK cells, which, together with other lymphoid cells mediate most of the antitumor immunity. Another study showed that psychological stress decreases the number of T, B, and NK cells in circulation and impairs immune defenses of B cells against lung tumor [124]. B cells fail to upregulate CD80 co-stimulatory molecule, decrease IL12 production, and do not migrate towards tumors.

Patients with major depressive disorder (MDD) often have impaired adaptive immunity with decreases in the number of T and NK cells, and poor response to vaccination and viral infection. They have an altered ratio of T and NK cell subtypes [327]. Natural regulatory T cells show an inverse association with the pro-inflammatory state of monocytes in these patients and antidepressant therapy leads to the restoration of the Treg population [328]. T cell deficits have also been found in patients with bipolar disorder [329].

Microarray data on whole blood from patients with MDD show that 90 genes are overexpressed and 75 genes are underexpressed relative to healthy controls. Most overexpressed genes are involved in inflammation and phagocytic cell function whereas underexpressed genes function in T cells, B cells, natural killers and erythroblasts [330]. Moreover, in each patient, the mean of overexpressed genes correlated with the mean of underexpressed, suggesting that these immunophenotypes are inversely coupled.

CONCLUDING REMARKS

Communication between the mind, the immune system, and the body run at multiple levels including neural signaling, hormones, and immune mediators. The brain, peripheral organs, immune cells, enteric nervous system, and even microbiota not only exchange information but also regulate each other. Interconnections between the brain, the body, and the immune system provide for fine-tuning of organismal responses. At the same time, malfunction of one component of this triad can lead to pathology in the others. One component cannot be helped without taking into account the effect on the others. This underscores the importance of a systemic approach to human health.

Inflammation is a global mechanism that is induced in response to almost any disturbance of homeostasis from psychological stress and bodily injury to infection. The paramount importance of prevention of chronic inflammation is beginning to emerge. Inflammation can be controlled in many ways besides medications, including diet, exercise, and better self-regulation in the face of stress. The mainstay one-sided approach to the treatment of human diseases with medications that target a single malfunction should give way to a more systemic and balanced approach that aims to restore homeostasis and is embedded in the lifestyle.

CONSENT FOR PUBLICATION

Not applicable.

CONFLICT OF INTEREST

The author confirms that this chapter contents have no conflict of interest.

ACKNOWLEDGEMENTS

Declare none.

REFERENCES

[1] Selye H. The Stress of Life. New York: McGraw-Hill Book 1978.

[2] Ashley NT, Demas GE. Neuroendocrine-immune circuits, phenotypes, and interactions. Horm Behav 2017; 87: 25-34.

[http://dx.doi.org/10.1016/j.yhbeh.2016.10.004] [PMID: 27765499]

[3] Veiga-Fernandes H, Artis D. Neuronal-immune system cross-talk in homeostasis. Science 2018; 359(6383): 1465-6.
[http://dx.doi.org/10.1126/science.aap9598] [PMID: 29599230]

[4] Dhabhar FS. The short-term stress response - Mother nature's mechanism for enhancing protection and performance under conditions of threat, challenge, and opportunity. Front Neuroendocrinol 2018; 49: 175-92.
[http://dx.doi.org/10.1016/j.yfrne.2018.03.004] [PMID: 29596867]

[5] McKinstry M, Chung C, Truong H, Johnston BA, Snow JW. The heat shock response and humoral immune response are mutually antagonistic in honey bees. Sci Rep 2017; 7(1): 8850.
[http://dx.doi.org/10.1038/s41598-017-09159-4] [PMID: 28821863]

[6] Zhou Z, Wu Y, Zhang C, *et al.* Suppression of NF-κB signal pathway by NLRC3-like protein in stony coral Acropora aculeus under heat stress. Fish Shellfish Immunol 2017; 67: 322-30.
[http://dx.doi.org/10.1016/j.fsi.2017.06.027] [PMID: 28606864]

[7] Adamo SA. Stress responses sculpt the insect immune system, optimizing defense in an ever-changing world. Dev Comp Immunol 2017; 66: 24-32.
[http://dx.doi.org/10.1016/j.dci.2016.06.005] [PMID: 27288849]

[8] Kipnis J. Immune system: The "seventh sense". J Exp Med 2018; 215(2): 397-8.
[http://dx.doi.org/10.1084/jem.20172295] [PMID: 29339443]

[9] Pavlov VA, Chavan SS, Tracey KJ. Molecular and Functional Neuroscience in Immunity. Annu Rev Immunol 2018; 36: 783-812.
[http://dx.doi.org/10.1146/annurev-immunol-042617-053158] [PMID: 29677475]

[10] Daneman R, Prat A. The blood-brain barrier. Cold Spring Harb Perspect Biol 2015; 7(1)a020412
[http://dx.doi.org/10.1101/cshperspect.a020412] [PMID: 25561720]

[11] Varatharaj A, Galea I. The blood-brain barrier in systemic inflammation. Brain Behav Immun 2017; 60: 1-12.
[http://dx.doi.org/10.1016/j.bbi.2016.03.010] [PMID: 26995317]

[12] Kierdorf K, Prinz M. Microglia in steady state. J Clin Invest 2017; 127(9): 3201-9.
[http://dx.doi.org/10.1172/JCI90602] [PMID: 28714861]

[13] Davalos D, Grutzendler J, Yang G, *et al.* ATP mediates rapid microglial response to local brain injury *in vivo*. Nat Neurosci 2005; 8(6): 752-8.
[http://dx.doi.org/10.1038/nn1472] [PMID: 15895084]

[14] Salter MW, Stevens B. Microglia emerge as central players in brain disease. Nat Med 2017; 23(9): 1018-27.
[http://dx.doi.org/10.1038/nm.4397] [PMID: 28886007]

[15] Gomez Perdiguero E, Klapproth K, Schulz C, *et al.* Tissue-resident macrophages originate from yolk-sac-derived erythro-myeloid progenitors. Nature 2015; 518(7540): 547-51.
[http://dx.doi.org/10.1038/nature13989] [PMID: 25470051]

[16] Li Q, Barres BA. Microglia and macrophages in brain homeostasis and disease. Nat Rev Immunol 2018; 18(4): 225-42.
[http://dx.doi.org/10.1038/nri.2017.125] [PMID: 29151590]

[17] Askew K, Li K, Olmos-Alonso A, *et al.* Coupled Proliferation and Apoptosis Maintain the Rapid Turnover of Microglia in the Adult Brain. Cell Rep 2017; 18(2): 391-405.
[http://dx.doi.org/10.1016/j.celrep.2016.12.041] [PMID: 28076784]

[18] Matcovitch-Natan O, Winter DR, Giladi A, *et al.* Microglia development follows a stepwise program to regulate brain homeostasis. Science 2016; 353(6301)aad8670
[http://dx.doi.org/10.1126/science.aad8670] [PMID: 27338705]

[19] Mrdjen D, Pavlovic A, Hartmann FJ, *et al.* High-Dimensional Single-Cell Mapping of Central Nervous System Immune Cells Reveals Distinct Myeloid Subsets in Health, Aging, and Disease. Immunity 2018; 48(2): 380-395.e6. [http://dx.doi.org/10.1016/j.immuni.2018.01.011] [PMID: 29426702]

[20] Böttcher C, Schlickeiser S, Sneeboer MAM, *et al.* Human microglia regional heterogeneity and phenotypes determined by multiplexed single-cell mass cytometry. Nat Neurosci 2019; 22(1): 78-90. [http://dx.doi.org/10.1038/s41593-018-0290-2] [PMID: 30559476]

[21] Elmore MR, Najafi AR, Koike MA, *et al.* Colony-stimulating factor 1 receptor signaling is necessary for microglia viability, unmasking a microglia progenitor cell in the adult brain. Neuron 2014; 82(2): 380-97. [http://dx.doi.org/10.1016/j.neuron.2014.02.040] [PMID: 24742461]

[22] Paolicelli RC, Bolasco G, Pagani F, *et al.* Synaptic pruning by microglia is necessary for normal brain development. Science 2011; 333(6048): 1456-8. [http://dx.doi.org/10.1126/science.1202529] [PMID: 21778362]

[23] Fourgeaud L, Través PG, Tufail Y, *et al.* TAM receptors regulate multiple features of microglial physiology. Nature 2016; 532(7598): 240-4. [http://dx.doi.org/10.1038/nature17630] [PMID: 27049947]

[24] Thion MS, Ginhoux F, Garel S. Microglia and early brain development: An intimate journey. Science 2018; 362(6411): 185-9. [http://dx.doi.org/10.1126/science.aat0474] [PMID: 30309946]

[25] McCarthy MM, Nugent BM, Lenz KM. Neuroimmunology and neuroepigenetics in the establishment of sex differences in the brain. Nat Rev Neurosci 2017; 18(8): 471-84. [http://dx.doi.org/10.1038/nrn.2017.61] [PMID: 28638119]

[26] Ayata P, Badimon A, Strasburger HJ, *et al.* Epigenetic regulation of brain region-specific microglia clearance activity. Nat Neurosci 2018; 21(8): 1049-60. [http://dx.doi.org/10.1038/s41593-018-0192-3] [PMID: 30038282]

[27] Thion MS, Low D, Silvin A, *et al.* Microbiome Influences Prenatal and Adult Microglia in a Sex-Specific Manner. Cell 2018; 172(3): 500-516.e16. [http://dx.doi.org/10.1016/j.cell.2017.11.042] [PMID: 29275859]

[28] Tay TL, Béchade C, D'Andrea I, *et al.* Microglia Gone Rogue: Impacts on Psychiatric Disorders across the Lifespan. Front Mol Neurosci 2018; 10: 421. [http://dx.doi.org/10.3389/fnmol.2017.00421] [PMID: 29354029]

[29] Mosser CA, Baptista S, Arnoux I, Audinat E. Microglia in CNS development: Shaping the brain for the future. Prog Neurobiol 2017; 149-150: 1-20. [http://dx.doi.org/10.1016/j.pneurobio.2017.01.002] [PMID: 28143732]

[30] Stevens B, Allen NJ, Vazquez LE, *et al.* The classical complement cascade mediates CNS synapse elimination. Cell 2007; 131(6): 1164-78. [http://dx.doi.org/10.1016/j.cell.2007.10.036] [PMID: 18083105]

[31] Presumey J, Bialas AR, Carroll MC. Complement System in Neural Synapse Elimination in Development and Disease. Adv Immunol 2017; 135: 53-79. [http://dx.doi.org/10.1016/bs.ai.2017.06.004] [PMID: 28826529]

[32] Györffy BA, Kun J, Török G, *et al.* Local apoptotic-like mechanisms underlie complement-mediated synaptic pruning. Proc Natl Acad Sci USA 2018; 115(24): 6303-8. [http://dx.doi.org/10.1073/pnas.1722613115] [PMID: 29844190]

[33] Vilalta A, Brown GC. Neurophagy, the phagocytosis of live neurons and synapses by glia, contributes to brain development and disease. FEBS J 2018; 285(19): 3566-75. [http://dx.doi.org/10.1111/febs.14323] [PMID: 29125686]

[34] Sellgren CM, Gracias J, Watmuff B, *et al.* Increased synapse elimination by microglia in schizophrenia patient-derived models of synaptic pruning. Nat Neurosci 2019; 22(3): 374-85. [http://dx.doi.org/10.1038/s41593-018-0334-7] [PMID: 30718903]

[35] Kempuraj D, Selvakumar GP, Thangavel R, *et al.* Mast Cell Activation in Brain Injury, Stress, and Post-traumatic Stress Disorder and Alzheimer's Disease Pathogenesis. Front Neurosci 2017; 11: 703. [http://dx.doi.org/10.3389/fnins.2017.00703] [PMID: 29302258]

[36] Liddelow SA, Guttenplan KA, Clarke LE, *et al.* Neurotoxic reactive astrocytes are induced by activated microglia. Nature 2017; 541(7638): 481-7. [http://dx.doi.org/10.1038/nature21029] [PMID: 28099414]

[37] Rothhammer V, Borucki DM, Tjon EC, *et al.* Microglial control of astrocytes in response to microbial metabolites. Nature 2018; 557(7707): 724-8. [http://dx.doi.org/10.1038/s41586-018-0119-x] [PMID: 29769726]

[38] Jones ME, Lebonville CL, Paniccia JE, Balentine ME, Reissner KJ, Lysle DT. Hippocampal interleukin-1 mediates stress-enhanced fear learning: A potential role for astrocyte-derived interleukin-1β. Brain Behav Immun 2018; 67: 355-63. [http://dx.doi.org/10.1016/j.bbi.2017.09.016] [PMID: 28963000]

[39] Liu X, Quan N. Microglia and CNS Interleukin-1: Beyond Immunological Concepts. Front Neurol 2018; 9: 8. [http://dx.doi.org/10.3389/fneur.2018.00008] [PMID: 29410649]

[40] Niu J, Tsai HH, Hoi KK, *et al.* Aberrant oligodendroglial-vascular interactions disrupt the blood-brain barrier, triggering CNS inflammation. Nat Neurosci 2019; 22(5): 709-18. [http://dx.doi.org/10.1038/s41593-019-0369-4] [PMID: 30988524]

[41] Thurgur H, Pinteaux E. Microglia in the Neurovascular Unit: Blood-Brain Barrier-microglia Interactions After Central Nervous System Disorders. Neuroscience 2019; 405: 55-67. [http://dx.doi.org/10.1016/j.neuroscience.2018.06.046] [PMID: 31007172]

[42] Prinz M, Priller J. The role of peripheral immune cells in the CNS in steady state and disease. Nat Neurosci 2017; 20(2): 136-44. [http://dx.doi.org/10.1038/nn.4475] [PMID: 28092660]

[43] Rua R, Lee JY, Silva AB, *et al.* Infection drives meningeal engraftment by inflammatory monocytes that impairs CNS immunity. Nat Immunol 2019; 20(4): 407-19. [http://dx.doi.org/10.1038/s41590-019-0344-y] [PMID: 30886419]

[44] Garré JM, Silva HM, Lafaille JJ, Yang G. CX3CR1^{+} monocytes modulate learning and learning-dependent dendritic spine remodeling *via* TNF-α. Nat Med 2017; 23(6): 714-22. [http://dx.doi.org/10.1038/nm.4340] [PMID: 28504723]

[45] Menard C, Pfau ML, Hodes GE, *et al.* Social stress induces neurovascular pathology promoting depression. Nat Neurosci 2017; 20(12): 1752-60. [http://dx.doi.org/10.1038/s41593-017-0010-3] [PMID: 29184215]

[46] Spencer RL, Deak T. A users guide to HPA axis research. Physiol Behav 2017; 178: 43-65. [http://dx.doi.org/10.1016/j.physbeh.2016.11.014] [PMID: 27871862]

[47] Gallo-Payet N. 60 YEARS OF POMC: Adrenal and extra-adrenal functions of ACTH. J Mol Endocrinol 2016; 56(4): T135-56. [http://dx.doi.org/10.1530/JME-15-0257] [PMID: 26793988]

[48] Barnes PJ. Glucocorticosteroids. Handb Exp Pharmacol 2017; 237: 93-115. [http://dx.doi.org/10.1007/164_2016_62] [PMID: 27796513]

[49] Vandewalle J, Luypaert A, De Bosscher K, Libert C. Therapeutic Mechanisms of Glucocorticoids. Trends in endocrinology and metabolism. Trends Endocrinol Metab 2018; 29(1): 42-54. [http://dx.doi.org/10.1016/j.tem.2017.10.010] [PMID: 29162310]

[50] Walker DJ, Spencer KA. Glucocorticoid programming of neuroimmune function. Gen Comp Endocrinol 2018; 256: 80-8.
[http://dx.doi.org/10.1016/j.ygcen.2017.07.016] [PMID: 28728884]

[51] Vegiopoulos A, Herzig S. Glucocorticoids, metabolism and metabolic diseases. Mol Cell Endocrinol 2007; 275(1-2): 43-61.
[http://dx.doi.org/10.1016/j.mce.2007.05.015] [PMID: 17624658]

[52] Bangasser DA, Wiersielis KR, Khantsis S. Sex differences in the locus coeruleus-norepinephrine system and its regulation by stress. Brain Res 2016; 1641(Pt B): 177-88.
[http://dx.doi.org/10.1016/j.brainres.2015.11.021]

[53] Wyrofsky R, Reyes BAS, Van Bockstaele EJ. 2017; Co-localization of the cannabinoid type 1 receptor with corticotropin-releasing factor-containing afferents in the noradrenergic nucleus locus coeruleus: implications for the cognitive limb of the stress response. Brain structure & function 2017; 222(7): 3007-23.
[http://dx.doi.org/10.1007/s00429-017-1381-7]

[54] Wyrofsky RR, Reyes BAS, Yu D, Kirby LG, Van Bockstaele EJ. Sex differences in the effect of cannabinoid type 1 receptor deletion on locus coeruleus-norepinephrine neurons and corticotropin releasing factor-mediated responses. Eur J Neurosci 2018; 48(5): 2118-38.
[http://dx.doi.org/10.1111/ejn.14103] [PMID: 30103253]

[55] Pinho-Ribeiro FA, Verri WA Jr, Chiu IM. Nociceptor Sensory Neuron-Immune Interactions in Pain and Inflammation. Trends Immunol 2017; 38(1): 5-19.
[http://dx.doi.org/10.1016/j.it.2016.10.001] [PMID: 27793571]

[56] Ferreira SH, Lorenzetti BB, Bristow AF, Poole S. Interleukin-1 beta as a potent hyperalgesic agent antagonized by a tripeptide analogue. Nature 1988; 334(6184): 698-700.
[http://dx.doi.org/10.1038/334698a0] [PMID: 3137474]

[57] Chiu IM, Heesters BA, Ghasemlou N, *et al.* Bacteria activate sensory neurons that modulate pain and inflammation. Nature 2013; 501(7465): 52-7.
[http://dx.doi.org/10.1038/nature12479] [PMID: 23965627]

[58] Blake KJ, Baral P, Voisin T, *et al.* Staphylococcus aureus produces pain through pore-forming toxins and neuronal TRPV1 that is silenced by QX-314. Nat Commun 2018; 9(1): 37.
[http://dx.doi.org/10.1038/s41467-017-02448-6] [PMID: 29295977]

[59] Borovikova LV, Ivanova S, Zhang M, *et al.* Vagus nerve stimulation attenuates the systemic inflammatory response to endotoxin. Nature 2000; 405(6785): 458-62.
[http://dx.doi.org/10.1038/35013070] [PMID: 10839541]

[60] Wang J, Kollarik M, Ru F, *et al.* Distinct and common expression of receptors for inflammatory mediators in vagal nodose *versus* jugular capsaicin-sensitive/TRPV1-positive neurons detected by low input RNA sequencing. PLoS One 2017; 12(10)e0185985
[http://dx.doi.org/10.1371/journal.pone.0185985] [PMID: 28982197]

[61] Rosas-Ballina M, Olofsson PS, Ochani M, *et al.* Acetylcholine-synthesizing T cells relay neural signals in a vagus nerve circuit. Science 2011; 334(6052): 98-101.
[http://dx.doi.org/10.1126/science.1209985] [PMID: 21921156]

[62] Guarini S, Altavilla D, Cainazzo MM, *et al.* Efferent vagal fibre stimulation blunts nuclear factor-kappaB activation and protects against hypovolemic hemorrhagic shock. Circulation 2003; 107(8): 1189-94.
[http://dx.doi.org/10.1161/01.CIR.0000050627.90734.ED] [PMID: 12615800]

[63] de Jonge WJ, van der Zanden EP, The FO, *et al.* Stimulation of the vagus nerve attenuates macrophage activation by activating the Jak2-STAT3 signaling pathway. Nat Immunol 2005; 6(8): 844-51.
[http://dx.doi.org/10.1038/ni1229] [PMID: 16025117]

[64] Mirakaj V, Dalli J, Granja T, Rosenberger P, Serhan CN. Vagus nerve controls resolution and pro-

resolving mediators of inflammation. J Exp Med 2014; 211(6): 1037-48. [http://dx.doi.org/10.1084/jem.20132103] [PMID: 24863066]

[65] Dalli J, Colas RA, Arnardottir H, Serhan CN. Vagal Regulation of Group 3 Innate Lymphoid Cells and the Immunoresolvent PCTR1 Controls Infection Resolution. Immunity 2017; 46(1): 92-105. [http://dx.doi.org/10.1016/j.immuni.2016.12.009] [PMID: 28065837]

[66] Bellinger DL, Lorton D. Sympathetic Nerve Hyperactivity in the Spleen: Causal for Nonpathogenic-Driven Chronic Immune-Mediated Inflammatory Diseases (IMIDs)? Int J Mol Sci 2018; 19(4)E1188 [http://dx.doi.org/10.3390/ijms19041188] [PMID: 29652832]

[67] Körner A, Schlegel M, Kaussen T, *et al.* Sympathetic nervous system controls resolution of inflammation *via* regulation of repulsive guidance molecule A. Nat Commun 2019; 10(1): 633. [http://dx.doi.org/10.1038/s41467-019-08328-5] [PMID: 30733433]

[68] Willemze RA, Welting O, van Hamersveld P, *et al.* Loss of intestinal sympathetic innervation elicits an innate immune driven colitis. Mol Med 2019; 25(1): 1. [http://dx.doi.org/10.1186/s10020-018-0068-8] [PMID: 30616543]

[69] Yoo BB, Mazmanian SK. The Enteric Network: Interactions between the Immune and Nervous Systems of the Gut. Immunity 2017; 46(6): 910-26. [http://dx.doi.org/10.1016/j.immuni.2017.05.011] [PMID: 28636959]

[70] Gabanyi I, Muller PA, Feighery L, Oliveira TY, Costa-Pinto FA, Mucida D. Neuro-immune Interactions Drive Tissue Programming in Intestinal Macrophages. Cell 2016; 164(3): 378-91. [http://dx.doi.org/10.1016/j.cell.2015.12.023] [PMID: 26777404]

[71] Ramirez VT, Godinez DR, Brust-Mascher I, *et al.* T-cell derived acetylcholine aids host defenses during enteric bacterial infection with Citrobacter rodentium. PLoS Pathog 2019; 15(4)e1007719 [http://dx.doi.org/10.1371/journal.ppat.1007719] [PMID: 30973939]

[72] Ben-Shaanan TL, Azulay-Debby H, Dubovik T, *et al.* Activation of the reward system boosts innate and adaptive immunity. Nat Med 2016; 22(8): 940-4. [http://dx.doi.org/10.1038/nm.4133] [PMID: 27376577]

[73] Talbot S, Abdulnour RE, Burkett PR, *et al.* Silencing Nociceptor Neurons Reduces Allergic Airway Inflammation. Neuron 2015; 87(2): 341-54. [http://dx.doi.org/10.1016/j.neuron.2015.06.007] [PMID: 26119026]

[74] Pinho-Ribeiro FA, Baddal B, Haarsma R, *et al.* Blocking Neuronal Signaling to Immune Cells Treats Streptococcal Invasive Infection. Cell 2018; 173(5): 1083-1097.e22. [http://dx.doi.org/10.1016/j.cell.2018.04.006] [PMID: 29754819]

[75] Kamimura D, Murakami M. Neural stimulations regulate the infiltration of immune cells into the CNS. J Intern Med 2019; 286(3): 259-67. [http://dx.doi.org/10.1111/joim.12912] [PMID: 30955210]

[76] Okusa MD, Rosin DL, Tracey KJ. Targeting neural reflex circuits in immunity to treat kidney disease. Nat Rev Nephrol 2017; 13(11): 669-80. [http://dx.doi.org/10.1038/nrneph.2017.132] [PMID: 28970585]

[77] Inoue T, Tanaka S, Rosin DL, Okusa MD. Bioelectronic Approaches to Control Neuroimmune Interactions in Acute Kidney Injury. Cold Spring Harb Perspect Med 2019; 9(6)a034231 [http://dx.doi.org/10.1101/cshperspect.a034231] [PMID: 30126836]

[78] Quadri M, Bagdas D, Toma W, *et al.* The Antinociceptive and Anti-Inflammatory Properties of the alpha7 nAChR Weak Partial Agonist p-CF3 N,N-diethyl-N'-phenylpiperazine. J Pharmacol Exp Ther 2018; 367(2): 203-14. [http://dx.doi.org/10.1124/jpet.118.249904] [PMID: 30111636]

[79] Mota CMD, Rodrigues-Santos C, Fernández RAR, *et al.* Central serotonin attenuates LPS-induced systemic inflammation. Brain Behav Immun 2017; 66: 372-81. [http://dx.doi.org/10.1016/j.bbi.2017.07.010] [PMID: 28723348]

[80] Harden LM, Kent S, Pittman QJ, Roth J. Fever and sickness behavior: Friend or foe? Brain Behav Immun 2015; 50: 322-33.
[http://dx.doi.org/10.1016/j.bbi.2015.07.012] [PMID: 26187566]

[81] Eisenberger NI, Moieni M, Inagaki TK, Muscatell KA, Irwin MR. Sickness and in Health: The Co-Regulation of Inflammation and Social Behavior. Neuropsychopharmacology : official publication of the American College of Neuropsychopharmacology. 2017; 42: pp. (1)242-53.

[82] Hart BL. Biological basis of the behavior of sick animals. Neurosci Biobehav Rev 1988; 12(2): 123-37.
[http://dx.doi.org/10.1016/S0149-7634(88)80004-6] [PMID: 3050629]

[83] Shakhar K, Shakhar G. Why Do We Feel Sick When Infected--Can Altruism Play a Role? PLoS Biol 2015; 13(10)e1002276
[http://dx.doi.org/10.1371/journal.pbio.1002276] [PMID: 26474156]

[84] Lopes PC, Block P, König B. Infection-induced behavioural changes reduce connectivity and the potential for disease spread in wild mice contact networks. Sci Rep 2016; 6: 31790.
[http://dx.doi.org/10.1038/srep31790] [PMID: 27548906]

[85] Schaller M, Murray DR, Bangerter A. Implications of the behavioural immune system for social behaviour and human health in the modern world. Philos Trans R Soc Lond B Biol Sci 2015; 370(1669): 20140105.
[http://dx.doi.org/10.1098/rstb.2014.0105]

[86] Murray DR, Prokosch ML, Airington Z. PsychoBehavioroimmunology: Connecting the Behavioral Immune System to Its Physiological Foundations. Front Psychol 2019; 10: 200.
[http://dx.doi.org/10.3389/fpsyg.2019.00200] [PMID: 30804853]

[87] Sawada N, Auger E, Lydon JE. Activation of the Behavioral Immune System: Putting the Brakes on Affiliation. Personality & social psychology bulletin 2018; 44(2): 224-37.

[88] Boillat M, Challet L, Rossier D, Kan C, Carleton A, Rodriguez I. The vomeronasal system mediates sick conspecific avoidance. Curr Biol 2015; 25(2): 251-5.
[http://dx.doi.org/10.1016/j.cub.2014.11.061] [PMID: 25578906]

[89] Regenbogen C, Axelsson J, Lasselin J, *et al.* Behavioral and neural correlates to multisensory detection of sick humans. Proc Natl Acad Sci USA 2017; 114(24): 6400-5.
[http://dx.doi.org/10.1073/pnas.1617357114] [PMID: 28533402]

[90] Axelsson J, Sundelin T, Olsson MJ, Sorjonen K, Axelsson C, Lasselin J, *et al.* Identification of acutely sick people and facial cues of sickness. Proceedings Biological sciences 2018; 285(1870): 20172430.
[http://dx.doi.org/10.1098/rspb.2017.2430]

[91] Petersen MB. Healthy Out-Group Members Are Represented Psychologically as Infected In-Group Members. Psychol Sci 2017; 28(12): 1857-63.
[http://dx.doi.org/10.1177/0956797617728270] [PMID: 29048976]

[92] Adelman JS, Mayer C, Hawley DM. Infection reduces anti-predator behaviors in house finches. J Avian Biol 2017; 48(4): 519-28.
[http://dx.doi.org/10.1111/jav.01058] [PMID: 29242677]

[93] Knoll JG, Krasnow SM, Marks DL. Interleukin-1β signaling in fenestrated capillaries is sufficient to trigger sickness responses in mice. J Neuroinflammation 2017; 14(1): 219.
[http://dx.doi.org/10.1186/s12974-017-0990-7] [PMID: 29121947]

[94] Corrard F, Copin C, Wollner A, *et al.* Sickness behavior in feverish children is independent of the severity of fever. An observational, multicenter study. PLoS One 2017; 12(3)e0171670
[http://dx.doi.org/10.1371/journal.pone.0171670] [PMID: 28278190]

[95] Nilsson A, Wilhelms DB, Mirrasekhian E, Jaarola M, Blomqvist A, Engblom D. Inflammation-induced anorexia and fever are elicited by distinct prostaglandin dependent mechanisms, whereas

conditioned taste aversion is prostaglandin independent. Brain Behav Immun 2017; 61: 236-43. [http://dx.doi.org/10.1016/j.bbi.2016.12.007] [PMID: 27940259]

[96] Singh AK, Zajdel J, Mirrasekhian E, *et al.* Prostaglandin-mediated inhibition of serotonin signaling controls the affective component of inflammatory pain. J Clin Invest 2017; 127(4): 1370-4. [http://dx.doi.org/10.1172/JCI90678] [PMID: 28287401]

[97] Umamaheswaran S, Dasari SK, Yang P, Lutgendorf SK, Sood AK. Stress, inflammation, and eicosanoids: an emerging perspective. Cancer Metastasis Rev 2018; 37(2-3): 203-11. [http://dx.doi.org/10.1007/s10555-018-9741-1] [PMID: 29948328]

[98] Blank T, Detje CN, Spieß A, *et al.* Brain Endothelial- and Epithelial-Specific Interferon Receptor Chain 1 Drives Virus-Induced Sickness Behavior and Cognitive Impairment. Immunity 2016; 44(4): 901-12. [http://dx.doi.org/10.1016/j.immuni.2016.04.005] [PMID: 27096319]

[99] Fritz M, Klawonn AM, Jaarola M, Engblom D. Interferon-ɣ mediated signaling in the brain endothelium is critical for inflammation-induced aversion. Brain Behav Immun 2018; 67: 54-8. [http://dx.doi.org/10.1016/j.bbi.2017.08.020] [PMID: 28864260]

[100] Davis KC, Raizen DM. A mechanism for sickness sleep: lessons from invertebrates. J Physiol 2017; 595(16): 5415-24. [http://dx.doi.org/10.1113/JP273009] [PMID: 28028818]

[101] Iannacone MJ, Beets I, Lopes LE, *et al.* The RFamide receptor DMSR-1 regulates stress-induced sleep in *C. elegans.* eLife 2017; 6e19837 [http://dx.doi.org/10.7554/eLife.19837] [PMID: 28094002]

[102] Lee DA, Andreev A, Truong TV, *et al.* Genetic and neuronal regulation of sleep by neuropeptide VF. eLife 2017; 6e25727 [http://dx.doi.org/10.7554/eLife.25727] [PMID: 29106375]

[103] Singh C, Rihel J, Prober DA, Neuropeptide Y. Neuropeptide Y Regulates Sleep by Modulating Noradrenergic Signaling. Curr Biol 2017; 27(24): 3796-3811.e5. [http://dx.doi.org/10.1016/j.cub.2017.11.018] [PMID: 29225025]

[104] Toda H, Williams JA, Gulledge M, Sehgal A. A sleep-inducing gene, *nemuri*, links sleep and immune function in *Drosophila.* Science 2019; 363(6426): 509-15. [http://dx.doi.org/10.1126/science.aat1650] [PMID: 30705188]

[105] van Niekerk G, Loos B, Nell T, Engelbrecht AM. Autophagy--A free meal in sickness-associated anorexia. Autophagy 2016; 12(4): 727-34. [http://dx.doi.org/10.1080/15548627.2016.1147672] [PMID: 27050464]

[106] McCarville JL, Ayres JS. Disease tolerance: concept and mechanisms. Curr Opin Immunol 2018; 50: 88-93. [http://dx.doi.org/10.1016/j.coi.2017.12.003] [PMID: 29253642]

[107] Hsu JY, Crawley S, Chen M, *et al.* Non-homeostatic body weight regulation through a brainstem-restricted receptor for GDF15. Nature 2017; 550(7675): 255-9. [http://dx.doi.org/10.1038/nature24042] [PMID: 28953886]

[108] Lasselin J, Lekander M, Paues-Göranson S, Olsson MJ, Axelsson J. Communication of health in experimentally sick men and women: A pilot study. Psychoneuroendocrinology 2018; 87: 188-95. [http://dx.doi.org/10.1016/j.psyneuen.2017.10.024] [PMID: 29102898]

[109] Wang T, Yin J, Miller AH, Xiao C. A systematic review of the association between fatigue and genetic polymorphisms. Brain Behav Immun 2017; 62: 230-44. [http://dx.doi.org/10.1016/j.bbi.2017.01.007] [PMID: 28089639]

[110] Blank T, Prinz M. Type I interferon pathway in CNS homeostasis and neurological disorders. Glia 2017; 65(9): 1397-406. [http://dx.doi.org/10.1002/glia.23154] [PMID: 28519900]

[111] Rao S, Schieber AMP, O'Connor CP, Leblanc M, Michel D, Ayres JS. Pathogen-Mediated Inhibition of Anorexia Promotes Host Survival and Transmission. Cell 2017; 168(3): 503-516.e12. [http://dx.doi.org/10.1016/j.cell.2017.01.006] [PMID: 28129542]

[112] Marsland AL, Walsh C, Lockwood K, John-Henderson NA. The effects of acute psychological stress on circulating and stimulated inflammatory markers: A systematic review and meta-analysis. Brain Behav Immun 2017; 64: 208-19. [http://dx.doi.org/10.1016/j.bbi.2017.01.011] [PMID: 28089638]

[113] Maydych V, Claus M, Dychus N, *et al.* Impact of chronic and acute academic stress on lymphocyte subsets and monocyte function. PLoS One 2017; 12(11)e0188108 [http://dx.doi.org/10.1371/journal.pone.0188108] [PMID: 29145439]

[114] Brydon L, Walker C, Wawrzyniak A, *et al.* Synergistic effects of psychological and immune stressors on inflammatory cytokine and sickness responses in humans. Brain Behav Immun 2009; 23(2): 217-24. [http://dx.doi.org/10.1016/j.bbi.2008.09.007] [PMID: 18835437]

[115] Tuchscherer M, Puppe B, Tuchscherer A, Kanitz E. Psychosocial stress sensitizes neuroendocrine and inflammatory responses to *Escherichia coli* challenge in domestic piglets. Brain Behav Immun 2018; 68: 274-87. [http://dx.doi.org/10.1016/j.bbi.2017.10.026] [PMID: 29102802]

[116] Giletta M, Slavich GM, Rudolph KD, Hastings PD, Nock MK, Prinstein MJ. Peer victimization predicts heightened inflammatory reactivity to social stress in cognitively vulnerable adolescents. J Child Psychol Psychiatry 2018; 59(2): 129-39. [http://dx.doi.org/10.1111/jcpp.12804] [PMID: 28892126]

[117] Rittschof CC, Bukhari SA, Sloofman LG, *et al.* Neuromolecular responses to social challenge: common mechanisms across mouse, stickleback fish, and honey bee. Proc Natl Acad Sci USA 2014; 111(50): 17929-34. [http://dx.doi.org/10.1073/pnas.1420369111] [PMID: 25453090]

[118] Föcking M, Sabherwal S, Cates HM, *et al.* Complement pathway changes at age 12 are associated with psychotic experiences at age 18 in a longitudinal population-based study: evidence for a role of stress. Mol Psychiatry 2019. [http://dx.doi.org/10.1038/s41380-018-0306-z] [PMID: 30635638]

[119] Vedhara K, Gill S, Eldesouky L, *et al.* Personality and gene expression: Do individual differences exist in the leukocyte transcriptome? Psychoneuroendocrinology 2015; 52: 72-82. [http://dx.doi.org/10.1016/j.psyneuen.2014.10.028] [PMID: 25459894]

[120] Wohleb ES, Terwilliger R, Duman CH, Duman RS. Stress-Induced Neuronal Colony Stimulating Factor 1 Provokes Microglia-Mediated Neuronal Remodeling and Depressive-like Behavior. Biol Psychiatry 2018; 83(1): 38-49. [http://dx.doi.org/10.1016/j.biopsych.2017.05.026] [PMID: 28697890]

[121] Lopizzo N, Tosato S, Begni V, *et al.* Transcriptomic analyses and leukocyte telomere length measurement in subjects exposed to severe recent stressful life events. Transl Psychiatry 2017; 7(2)e1042 [http://dx.doi.org/10.1038/tp.2017.5] [PMID: 28221367]

[122] Aberg KM, Radek KA, Choi EH, *et al.* Psychological stress downregulates epidermal antimicrobial peptide expression and increases severity of cutaneous infections in mice. J Clin Invest 2007; 117(11): 3339-49. [http://dx.doi.org/10.1172/JCI31726] [PMID: 17975669]

[123] Schneider SA, Scharffetter C, Wagner AE, *et al.* Social stress increases the susceptibility to infection in the ant Harpegnathos saltator. Sci Rep 2016; 6: 25800. [http://dx.doi.org/10.1038/srep25800] [PMID: 27161621]

[124] Jones HP, Aldridge B, Boss-Williams K, Weiss JM. A role for B cells in facilitating defense against an NK cell-sensitive lung metastatic tumor is revealed by stress. J Neuroimmunol 2017; 313: 99-108. [http://dx.doi.org/10.1016/j.jneuroim.2017.10.017] [PMID: 29153616]

[125] Gimsa U, Tuchscherer M, Kanitz E. Psychosocial Stress and Immunity-What Can We Learn From Pig Studies? Front Behav Neurosci 2018; 12: 64. [http://dx.doi.org/10.3389/fnbeh.2018.00064] [PMID: 29666573]

[126] Adamo SA, Easy RH, Kovalko I, *et al.* Predator exposure-induced immunosuppression: trade-off, immune redistribution or immune reconfiguration? J Exp Biol 2017; 220(Pt 5): 868-75. [PMID: 28011823]

[127] Filipovic D, Todorovic N, Bernardi RE, Gass P. Oxidative and nitrosative stress pathways in the brain of socially isolated adult male rats demonstrating depressive- and anxiety-like symptoms. Brain structure & function 2017; 222(1): 1-20.

[128] Unternaehrer E, Luers P, Mill J, *et al.* Dynamic changes in DNA methylation of stress-associated genes (OXTR, BDNF) after acute psychosocial stress. Transl Psychiatry 2012; 2e150 [http://dx.doi.org/10.1038/tp.2012.77] [PMID: 22892716]

[129] Engert V, Kok BE, Papassotiriou I, Chrousos GP, Singer T. Specific reduction in cortisol stress reactivity after social but not attention-based mental training. Sci Adv 2017; 3(10)e1700495 [http://dx.doi.org/10.1126/sciadv.1700495] [PMID: 28983508]

[130] Hoge EA, Bui E, Palitz SA, *et al.* The effect of mindfulness meditation training on biological acute stress responses in generalized anxiety disorder. Psychiatry Res 2018; 262: 328-32. [http://dx.doi.org/10.1016/j.psychres.2017.01.006] [PMID: 28131433]

[131] Ortega E. The "bioregulatory effect of exercise" on the innate/inflammatory responses. J Physiol Biochem 2016; 72(2): 361-9. [http://dx.doi.org/10.1007/s13105-016-0478-4] [PMID: 26979741]

[132] Matthews S, McGowan P. Developmental programming of the HPA axis and related behaviours: epigenetic mechanisms. J Endocrinol 2019. JOE-19-0057.R1.

[133] Meuwissen ME, Schot R, Buta S, *et al.* Human USP18 deficiency underlies type 1 interferonopathy leading to severe pseudo-TORCH syndrome. J Exp Med 2016; 213(7): 1163-74. [http://dx.doi.org/10.1084/jem.20151529] [PMID: 27325888]

[134] Estes ML, McAllister AK. Maternal immune activation: Implications for neuropsychiatric disorders. Science 2016; 353(6301): 772-7. [http://dx.doi.org/10.1126/science.aag3194] [PMID: 27540164]

[135] Depino AM. Perinatal inflammation and adult psychopathology: From preclinical models to humans. Semin Cell Dev Biol 2018; 77: 104-14. [http://dx.doi.org/10.1016/j.semcdb.2017.09.010] [PMID: 28890420]

[136] Ghassabian A, Albert PS, Hornig M, *et al.* Gestational cytokine concentrations and neurocognitive development at 7 years. Transl Psychiatry 2018; 8(1): 64. [http://dx.doi.org/10.1038/s41398-018-0112-z] [PMID: 29531226]

[137] Graham AM, Rasmussen JM, Rudolph MD, *et al.* Maternal Systemic Interleukin-6 During Pregnancy Is Associated With Newborn Amygdala Phenotypes and Subsequent Behavior at 2 Years of Age. Biol Psychiatry 2018; 83(2): 109-19. [http://dx.doi.org/10.1016/j.biopsych.2017.05.027] [PMID: 28754515]

[138] Luo H, Winkelmann ER, Fernandez-Salas I, *et al.* Zika, dengue and yellow fever viruses induce differential anti-viral immune responses in human monocytic and first trimester trophoblast cells. Antiviral Res 2018; 151: 55-62. [http://dx.doi.org/10.1016/j.antiviral.2018.01.003] [PMID: 29331320]

[139] Wu WL, Hsiao EY, Yan Z, Mazmanian SK, Patterson PH. The placental interleukin-6 signaling

controls fetal brain development and behavior. Brain Behav Immun 2017; 62: 11-23. [http://dx.doi.org/10.1016/j.bbi.2016.11.007] [PMID: 27838335]

[140] Kim S, Kim H, Yim YS, *et al.* Maternal gut bacteria promote neurodevelopmental abnormalities in mouse offspring. Nature 2017; 549(7673): 528-32. [http://dx.doi.org/10.1038/nature23910] [PMID: 28902840]

[141] Grissom NM, George R, Reyes TM. The hypothalamic transcriptional response to stress is severely impaired in offspring exposed to adverse nutrition during gestation. Neuroscience 2017; 342: 200-11. [http://dx.doi.org/10.1016/j.neuroscience.2015.07.022] [PMID: 26215917]

[142] Freedman R, Hunter SK, Hoffman MC. Prenatal Primary Prevention of Mental Illness by Micronutrient Supplements in Pregnancy. Am J Psychiatry 2018; 175(7): 607-19. [http://dx.doi.org/10.1176/appi.ajp.2018.17070836] [PMID: 29558816]

[143] Felitti VJ, Anda RF, Nordenberg D, *et al.* Relationship of childhood abuse and household dysfunction to many of the leading causes of death in adults. The Adverse Childhood Experiences (ACE) Study. Am J Prev Med 1998; 14(4): 245-58. [http://dx.doi.org/10.1016/S0749-3797(98)00017-8] [PMID: 9635069]

[144] Denhardt DT. Effect of stress on human biology: Epigenetics, adaptation, inheritance, and social significance. J Cell Physiol 2018; 233(3): 1975-84. [http://dx.doi.org/10.1002/jcp.25837] [PMID: 28158904]

[145] Naumova OY, Lee M, Koposov R, Szyf M, Dozier M, Grigorenko EL. Differential patterns of whole-genome DNA methylation in institutionalized children and children raised by their biological parents. Dev Psychopathol 2012; 24(1): 143-55. [http://dx.doi.org/10.1017/S0954579411000605] [PMID: 22123582]

[146] Cramer T, Rosenberg T, Kisliouk T, Meiri N. PARP Inhibitor Affects Long-term Heat-stress Response *via* Changes in DNA Methylation. Neuroscience 2019; 399: 65-76. [http://dx.doi.org/10.1016/j.neuroscience.2018.12.018] [PMID: 30579833]

[147] Weaver IC, Cervoni N, Champagne FA, *et al.* Epigenetic programming by maternal behavior. Nat Neurosci 2004; 7(8): 847-54. [http://dx.doi.org/10.1038/nn1276] [PMID: 15220929]

[148] Sosnowski DW, Booth C, York TP, Amstadter AB, Kliewer W. Maternal prenatal stress and infant DNA methylation: A systematic review. Dev Psychobiol 2018; 60(2): 127-39. [http://dx.doi.org/10.1002/dev.21604] [PMID: 29344930]

[149] Bosmans G, Young JF, Hankin BL. NR3C1 methylation as a moderator of the effects of maternal support and stress on insecure attachment development. Dev Psychol 2018; 54(1): 29-38. [http://dx.doi.org/10.1037/dev0000422] [PMID: 29058930]

[150] Unternaehrer E, Meyer AH, Burkhardt SC, *et al.* Childhood maternal care is associated with DNA methylation of the genes for brain-derived neurotrophic factor (BDNF) and oxytocin receptor (OXTR) in peripheral blood cells in adult men and women. Stress 2015; 18(4): 451-61. [http://dx.doi.org/10.3109/10253890.2015.1038992] [PMID: 26061800]

[151] Fujisawa TX, Nishitani S, Takiguchi S, Shimada K, Smith AK, Tomoda A. Oxytocin receptor DNA methylation and alterations of brain volumes in maltreated children. Neuropsychopharmacology : official publication of the American College of Neuropsychopharmacology 2019. [http://dx.doi.org/10.1038/s41386-019-0414-8]

[152] Johnson FK, Kaffman A. Early life stress perturbs the function of microglia in the developing rodent brain: New insights and future challenges. Brain Behav Immun 2018; 69: 18-27. [http://dx.doi.org/10.1016/j.bbi.2017.06.008] [PMID: 28625767]

[153] McGregor NW, Dimatelis JJ, Van Zyl PJ, *et al.* A translational approach to the genetics of anxiety disorders. Behav Brain Res 2018; 341: 91-7. [http://dx.doi.org/10.1016/j.bbr.2017.12.030] [PMID: 29288745]

[154] Hing B, Sathyaputri L, Potash JB. A comprehensive review of genetic and epigenetic mechanisms that regulate BDNF expression and function with relevance to major depressive disorder. American journal of medical genetics Part B, Neuropsychiatric genetics : the official publication of the International Society of Psychiatric Genetics 2018; 177(2): 143-67.

[155] Baumeister D, Akhtar R, Ciufolini S, Pariante CM, Mondelli V. Childhood trauma and adulthood inflammation: a meta-analysis of peripheral C-reactive protein, interleukin-6 and tumour necrosis factor-α. Mol Psychiatry 2016; 21(5): 642-9.
[http://dx.doi.org/10.1038/mp.2015.67] [PMID: 26033244]

[156] Ajdacic-Gross V, Aleksandrowicz A, Rodgers S, *et al.* Infectious, atopic and inflammatory diseases, childhood adversities and familial aggregation are independently associated with the risk for mental disorders: Results from a large Swiss epidemiological study. World J Psychiatry 2016; 6(4): 419-30.
[http://dx.doi.org/10.5498/wjp.v6.i4.419] [PMID: 28078206]

[157] Danese A. Psychoneuroimmunology of Early-Life Stress: The Hidden Wounds of Childhood Trauma? Neuropsychopharmacology : official publication of the American College of Neuropsychopharmacology 2017; 42(1): 99-114.

[158] Peters AT, Ren X, Bessette KL, *et al.* Interplay between pro-inflammatory cytokines, childhood trauma, and executive function in depressed adolescents. J Psychiatr Res 2019; 114: 1-10.
[http://dx.doi.org/10.1016/j.jpsychires.2019.03.030] [PMID: 30978658]

[159] Rudolph MD, Graham AM, Feczko E, *et al.* Maternal IL-6 during pregnancy can be estimated from newborn brain connectivity and predicts future working memory in offspring. Nat Neurosci 2018; 21(5): 765-72.
[http://dx.doi.org/10.1038/s41593-018-0128-y] [PMID: 29632361]

[160] Johnson FK, Delpech JC, Thompson GJ, *et al.* Amygdala hyper-connectivity in a mouse model of unpredictable early life stress. Transl Psychiatry 2018; 8(1): 49.
[http://dx.doi.org/10.1038/s41398-018-0092-z] [PMID: 29463821]

[161] McCarthy MM. Sex differences in neuroimmunity as an inherent risk factor. Neuropsychopharmacology : official publication of the American College of Neuropsychopharmacology 2019; 44(1): 38-44.
[http://dx.doi.org/10.1038/s41386-018-0138-1]

[162] Carter CJ. Autism genes and the leukocyte transcriptome in autistic toddlers relate to pathogen interactomes, infection and the immune system. A role for excess neurotrophic sAPPα and reduced antimicrobial Aβ. Neurochem Int 2019; 126: 36-58.
[http://dx.doi.org/10.1016/j.neuint.2019.03.007] [PMID: 30862493]

[163] Liu K, Kurien BT, Zimmerman SL, Kaufman KM, Taft DH, Kottyan LC, *et al.* X Chromosome Dose and Sex Bias in Autoimmune Diseases: Increased Prevalence of 47,XXX in Systemic Lupus Erythematosus and Sjogren's Syndrome. Arthritis & rheumatology (Hoboken, NJ) 2016; 68(5): 1290-300.

[164] Ho B, Greenlaw K, Al Tuwaijri A, *et al.* X chromosome dosage and presence of SRY shape sex-specific differences in DNA methylation at an autosomal region in human cells. Biol Sex Differ 2018; 9(1): 10.
[http://dx.doi.org/10.1186/s13293-018-0169-7] [PMID: 29463315]

[165] van Rijn S, Barneveld P, Descheemaeker MJ, Giltay J, Swaab H. The effect of early life stress on the cognitive phenotype of children with an extra X chromosome (47,XXY/47,XXX). Child neuropsychology : a journal on normal and abnormal development in childhood and adolescence 2018; 24(2): 277-86.

[166] Pérez-Cerezales S, Ramos-Ibeas P, Rizos D, Lonergan P, Bermejo-Alvarez P, Gutiérrez-Adán A. Early sex-dependent differences in response to environmental stress. Reproduction 2018; 155(1): R39-51.
[http://dx.doi.org/10.1530/REP-17-0466] [PMID: 29030490]

[167] Gouin JP, Zhou QQ, Booij L, *et al.* Associations among oxytocin receptor gene (OXTR) DNA methylation in adulthood, exposure to early life adversity, and childhood trajectories of anxiousness. Sci Rep 2017; 7(1): 7446.
[http://dx.doi.org/10.1038/s41598-017-07950-x] [PMID: 28785027]

[168] Batchelor V, Pang TY. HPA axis regulation and stress response is subject to intergenerational modification by paternal trauma and stress. General and comparative endocrinology 2019; S0016-6480(18): 30520-.
[http://dx.doi.org/10.1016/j.ygcen.2019.04.010]

[169] Niknazar S, Nahavandi A, Peyvandi AA, Peyvandi H, Roozbahany NA, Abbaszadeh HA. Hippocampal NR3C1 DNA methylation can mediate part of preconception paternal stress effects in rat offspring. Behav Brain Res 2017; 324: 71-6.
[http://dx.doi.org/10.1016/j.bbr.2017.02.014] [PMID: 28212940]

[170] Spadafora C. Sperm-Mediated Transgenerational Inheritance. Front Microbiol 2017; 8: 2401.
[http://dx.doi.org/10.3389/fmicb.2017.02401] [PMID: 29255455]

[171] Rodgers AB, Morgan CP, Leu NA, Bale TL. Transgenerational epigenetic programming *via* sperm microRNA recapitulates effects of paternal stress. Proc Natl Acad Sci USA 2015; 112(44): 13699-704.
[http://dx.doi.org/10.1073/pnas.1508347112] [PMID: 26483456]

[172] Brieño-Enríquez MA, García-López J, Cárdenas DB, *et al.* Exposure to endocrine disruptor induces transgenerational epigenetic deregulation of microRNAs in primordial germ cells. PLoS One 2015; 10(4)e0124296
[http://dx.doi.org/10.1371/journal.pone.0124296] [PMID: 25897752]

[173] Skaper SD, Facci L, Zusso M, Giusti P. An Inflammation-Centric View of Neurological Disease: Beyond the Neuron. Front Cell Neurosci 2018; 12: 72.
[http://dx.doi.org/10.3389/fncel.2018.00072] [PMID: 29618972]

[174] Jeppesen R, Benros ME. Autoimmune Diseases and Psychotic Disorders. Front Psychiatry 2019; 10: 131.
[http://dx.doi.org/10.3389/fpsyt.2019.00131] [PMID: 30949074]

[175] Thomann AK, Reindl W, Wüstenberg T, *et al.* Aberrant brain structural large-scale connectome in Crohn's disease. Neurogastroenterol Motil 2019; 31(6)e13593
[http://dx.doi.org/10.1111/nmo.13593] [PMID: 30983094]

[176] Farzi A, Fröhlich EE, Holzer P. Gut Microbiota and the Neuroendocrine System. Neurotherapeutics 2018; 15(1): 5-22.
[http://dx.doi.org/10.1007/s13311-017-0600-5] [PMID: 29380303]

[177] Pape K, Tamouza R, Leboyer M, Zipp F. Immunoneuropsychiatry - novel perspectives on brain disorders. Nat Rev Neurol 2019; 15(6): 317-28.
[http://dx.doi.org/10.1038/s41582-019-0174-4] [PMID: 30988501]

[178] Pryce CR, Fontana A. Depression in Autoimmune Diseases. Curr Top Behav Neurosci 2017; 31: 139-54.
[http://dx.doi.org/10.1007/7854_2016_7] [PMID: 27221625]

[179] Chan KL, Cathomas F, Russo SJ. Central and Peripheral Inflammation Link Metabolic Syndrome and Major Depressive Disorder. Physiology (Bethesda) 2019; 34(2): 123-33.
[http://dx.doi.org/10.1152/physiol.00047.2018] [PMID: 30724127]

[180] Hoyo-Becerra C, Schlaak JF, Hermann DM. Insights from interferon-α-related depression for the pathogenesis of depression associated with inflammation. Brain Behav Immun 2014; 42: 222-31.
[http://dx.doi.org/10.1016/j.bbi.2014.06.200] [PMID: 25066466]

[181] Harrison NA. Brain Structures Implicated in Inflammation-Associated Depression. Curr Top Behav Neurosci 2017; 31: 221-48.
[http://dx.doi.org/10.1007/7854_2016_30] [PMID: 27272069]

[182] Dantzer R, O'Connor JC, Freund GG, Johnson RW, Kelley KW. From inflammation to sickness and depression: when the immune system subjugates the brain. Nat Rev Neurosci 2008; 9(1): 46-56. [http://dx.doi.org/10.1038/nrn2297] [PMID: 18073775]

[183] Himmerich H, Patsalos O, Lichtblau N, Ibrahim MAA, Dalton B. Cytokine Research in Depression: Principles, Challenges, and Open Questions. Front Psychiatry 2019; 10: 30. [http://dx.doi.org/10.3389/fpsyt.2019.00030] [PMID: 30792669]

[184] Köhler CA, Freitas TH, Maes M, *et al.* Peripheral cytokine and chemokine alterations in depression: a meta-analysis of 82 studies. Acta Psychiatr Scand 2017; 135(5): 373-87. [http://dx.doi.org/10.1111/acps.12698] [PMID: 28122130]

[185] Leighton SP, Nerurkar L, Krishnadas R, Johnman C, Graham GJ, Cavanagh J. Chemokines in depression in health and in inflammatory illness: a systematic review and meta-analysis. Mol Psychiatry 2018; 23(1): 48-58. [http://dx.doi.org/10.1038/mp.2017.205] [PMID: 29133955]

[186] Barnes J, Mondelli V, Pariante CM. Genetic Contributions of Inflammation to Depression. Neuropsychopharmacology : official publication of the American College of Neuropsychopharmacology 2017; 42(1): 81-98. [http://dx.doi.org/10.1038/npp.2016.169]

[187] Lamers F, Milaneschi Y, Smit JH, Schoevers RA, Wittenberg G, Penninx BWJH. Longitudinal Association Between Depression and Inflammatory Markers: Results From the Netherlands Study of Depression and Anxiety. Biol Psychiatry 2019; 85(10): 829-37. [http://dx.doi.org/10.1016/j.biopsych.2018.12.020] [PMID: 30819515]

[188] Holmes SE, Hinz R, Conen S, *et al.* Elevated Translocator Protein in Anterior Cingulate in Major Depression and a Role for Inflammation in Suicidal Thinking: A Positron Emission Tomography Study. Biol Psychiatry 2018; 83(1): 61-9. [http://dx.doi.org/10.1016/j.biopsych.2017.08.005] [PMID: 28939116]

[189] Pandey GN, Rizavi HS, Bhaumik R, Ren X. Innate immunity in the postmortem brain of depressed and suicide subjects: Role of Toll-like receptors. Brain Behav Immun 2019; 75: 101-11. [http://dx.doi.org/10.1016/j.bbi.2018.09.024] [PMID: 30266463]

[190] Yin L, Xu X, Chen G, Mehta ND, Haroon E, Miller AH, *et al.* Inflammation and decreased functional connectivity in a widely-distributed network in depression: centralized effects in the ventral medial prefrontal cortex. Brain Behav Immun 2019; S0889-1591(19): 30352-6. [http://dx.doi.org/10.1016/j.bbi.2019.05.011]

[191] Köhler CA, Freitas TH, Stubbs B, *et al.* Peripheral Alterations in Cytokine and Chemokine Levels After Antidepressant Drug Treatment for Major Depressive Disorder: Systematic Review and Meta-Analysis. Mol Neurobiol 2018; 55(5): 4195-206. [PMID: 28612257]

[192] Jha MK, Minhajuddin A, Chin-Fatt C, Greer TL, Carmody TJ, Trivedi MH. Sex differences in the association of baseline c-reactive protein (CRP) and acute-phase treatment outcomes in major depressive disorder: Findings from the EMBARC study. J Psychiatr Res 2019; 113: 165-71. [http://dx.doi.org/10.1016/j.jpsychires.2019.03.013] [PMID: 30959227]

[193] Köhler-Forsberg O, N Lydholm C, Hjorthøj C, Nordentoft M, Mors O, Benros ME. Efficacy of anti-inflammatory treatment on major depressive disorder or depressive symptoms: meta-analysis of clinical trials. Acta Psychiatr Scand 2019; 139(5): 404-19. [http://dx.doi.org/10.1111/acps.13016] [PMID: 30834514]

[194] Szebeni A, Szebeni K, DiPeri TP, *et al.* Elevated DNA Oxidation and DNA Repair Enzyme Expression in Brain White Matter in Major Depressive Disorder. Int J Neuropsychopharmacol 2017; 20(5): 363-73. [PMID: 28034960]

[195] Czarny P, Wigner P, Galecki P, Sliwinski T. The interplay between inflammation, oxidative stress, DNA damage, DNA repair and mitochondrial dysfunction in depression. Progress in neuro-psychopharmacology & biological psychiatry 2018; 80(Pt C): 309-21.

[196] Felger JC, Treadway MT. Inflammation Effects on Motivation and Motor Activity: Role of Dopamine. Neuropsychopharmacology : official publication of the American College of Neuropsychopharmacology 2017; 42(1): 216-41.

[197] Liu CS, Adibfar A, Herrmann N, Gallagher D, Lanctôt KL. Evidence for Inflammation-Associated Depression. Curr Top Behav Neurosci 2017; 31: 3-30.
[http://dx.doi.org/10.1007/7854_2016_2] [PMID: 27221622]

[198] Savitz J. The kynurenine pathway: a finger in every pie. Mol Psychiatry 2019.
[http://dx.doi.org/10.1038/s41380-019-0414-4] [PMID: 30980044]

[199] Krishnadas R, Nicol A, Sassarini J, *et al.* Circulating tumour necrosis factor is highly correlated with brainstem serotonin transporter availability in humans. Brain Behav Immun 2016; 51: 29-38.
[http://dx.doi.org/10.1016/j.bbi.2015.08.005] [PMID: 26255693]

[200] Barone P. The 'Yin' and the 'Yang' of the kynurenine pathway: excitotoxicity and neuroprotection imbalance in stress-induced disorders. Behavioural pharmacology 2019; 30(2 and 3 - Special Issue): 163-86.
[http://dx.doi.org/10.1097/FBP.0000000000000477]

[201] Gamal M, Moawad J, Rashed L, Morcos MA, Sharawy N. Possible involvement of tetrahydrobiopterin in the disturbance of redox homeostasis in sepsis - Induced brain dysfunction. Brain Res 2018; 1685: 19-28.
[http://dx.doi.org/10.1016/j.brainres.2018.02.008] [PMID: 29428597]

[202] Pan Z, Rosenblat JD, Swardfager W, McIntyre RS. Role of Proinflammatory Cytokines in Dopaminergic System Disturbances, Implications for Anhedonic Features of MDD. Curr Pharm Des 2017; 23(14): 2065-72.
[http://dx.doi.org/10.2174/1381612823666170111144340] [PMID: 28078987]

[203] Chen Z, Jalabi W, Hu W, *et al.* Microglial displacement of inhibitory synapses provides neuroprotection in the adult brain. Nat Commun 2014; 5: 4486.
[http://dx.doi.org/10.1038/ncomms5486] [PMID: 25047355]

[204] Lamers F, Vogelzangs N, Merikangas KR, de Jonge P, Beekman AT, Penninx BW. Evidence for a differential role of HPA-axis function, inflammation and metabolic syndrome in melancholic *versus* atypical depression. Mol Psychiatry 2013; 18(6): 692-9.
[http://dx.doi.org/10.1038/mp.2012.144] [PMID: 23089630]

[205] Juruena MF, Bocharova M, Agustini B, Young AH. Atypical depression and non-atypical depression: Is HPA axis function a biomarker? A systematic review. J Affect Disord 2018; 233: 45-67.
[http://dx.doi.org/10.1016/j.jad.2017.09.052] [PMID: 29150144]

[206] Slowik A, Lammerding L, Hoffmann S, Beyer C. Brain inflammasomes in stroke and depressive disorders: Regulation by oestrogen. J Neuroendocrinol 2018; 30(2)
[http://dx.doi.org/10.1111/jne.12482] [PMID: 28477436]

[207] Bakusic J, Schaufeli W, Claes S, Godderis L. Stress, burnout and depression: A systematic review on DNA methylation mechanisms. J Psychosom Res 2017; 92: 34-44.
[http://dx.doi.org/10.1016/j.jpsychores.2016.11.005] [PMID: 27998510]

[208] Brown GM, McIntyre RS, Rosenblat J, Hardeland R. Depressive disorders: Processes leading to neurogeneration and potential novel treatments. Progress in neuro-psychopharmacology & biological psychiatry 2018; 80(Pt C): 189-204.

[209] Raison CL. The Promise and Limitations of Anti-Inflammatory Agents for the Treatment of Major Depressive Disorder. Curr Top Behav Neurosci 2017; 31: 287-302.
[http://dx.doi.org/10.1007/7854_2016_26] [PMID: 27278642]

[210] Hastings CN, Sheridan H, Pariante CM, Mondelli V. Does Diet Matter? The Use of Polyunsaturated Fatty Acids (PUFAs) and Other Dietary Supplements in Inflammation-Associated Depression. Curr Top Behav Neurosci 2017; 31: 321-38.
[http://dx.doi.org/10.1007/7854_2016_31] [PMID: 27431396]

[211] Kong J, Fang J, Park J, Li S, Rong P. Treating Depression with Transcutaneous Auricular Vagus Nerve Stimulation: State of the Art and Future Perspectives. Front Psychiatry 2018; 9: 20.
[http://dx.doi.org/10.3389/fpsyt.2018.00020] [PMID: 29459836]

[212] Gouin JP, MacNeil S. Attachment style and changes in systemic inflammation following migration to a new country among international students. Attachment & human development 2019; 21(1): 38-56.
[http://dx.doi.org/10.1080/14616734.2018.1541515]

[213] Fourrier C, Bosch-Bouju C, Boursereau R, *et al.* Brain tumor necrosis factor-α mediates anxiety-like behavior in a mouse model of severe obesity. Brain Behav Immun 2019; 77: 25-36.
[http://dx.doi.org/10.1016/j.bbi.2018.11.316] [PMID: 30508579]

[214] Lisboa SF, Issy AC, Biojone C, *et al.* Mice lacking interleukin-18 gene display behavioral changes in animal models of psychiatric disorders: Possible involvement of immunological mechanisms. J Neuroimmunol 2018; 314: 58-66.
[http://dx.doi.org/10.1016/j.jneuroim.2017.11.008] [PMID: 29195684]

[215] Michopoulos V, Powers A, Gillespie CF, Ressler KJ, Jovanovic T. Inflammation in Fear- and Anxiety-Based Disorders: PTSD, GAD, and Beyond. Neuropsychopharmacology : official publication of the American College of Neuropsychopharmacology 2017; 42(1): 254-70.

[216] Emeny RT, Baumert J, Zannas AS, *et al.* Anxiety Associated Increased CpG Methylation in the Promoter of Asb1: A Translational Approach Evidenced by Epidemiological and Clinical Studies and a Murine Model. Neuropsychopharmacology 2018; 43(2): 342-53.
[http://dx.doi.org/10.1038/npp.2017.102] [PMID: 28540928]

[217] Valentino RJ, Bangasser DA. Sex-biased cellular signaling: molecular basis for sex differences in neuropsychiatric diseases. Dialogues Clin Neurosci 2016; 18(4): 385-93.
[PMID: 28179810]

[218] Nawijn L, Krzyzewska IM, van Zuiden M, Henneman P, Koch SBJ, Mul AN, *et al.* Oxytocin receptor gene methylation in male and female PTSD patients and trauma-exposed controls. European neuropsychopharmacology : the journal of the European College of Neuropsychopharmacology 2019; 29(1): 147-55.
[http://dx.doi.org/10.1016/j.euroneuro.2018.10.006]

[219] Dugger BN, Dickson DW. Pathology of Neurodegenerative Diseases. Cold Spring Harb Perspect Biol 2017; 9(7)a028035
[http://dx.doi.org/10.1101/cshperspect.a028035] [PMID: 28062563]

[220] Ayers JI, Giasson BI, Borchelt DR. Prion-like Spreading in Tauopathies. Biol Psychiatry 2018; 83(4): 337-46.
[http://dx.doi.org/10.1016/j.biopsych.2017.04.003] [PMID: 28506438]

[221] Daniele S, Pietrobono D, Fusi J, *et al.* α-Synuclein Aggregates with β-Amyloid or Tau in Human Red Blood Cells: Correlation with Antioxidant Capability and Physical Exercise in Human Healthy Subjects. Mol Neurobiol 2018; 55(3): 2653-75.
[http://dx.doi.org/10.1007/s12035-017-0523-5] [PMID: 28421539]

[222] Barthet VJA, Ryan KM. Autophagy in Neurodegeneration: Can't Digest It, Spit It Out! Trends Cell Biol 2018; 28(3): 171-3.
[http://dx.doi.org/10.1016/j.tcb.2018.01.001] [PMID: 29395716]

[223] Bachiller S, Jiménez-Ferrer I, Paulus A, *et al.* Microglia in Neurological Diseases: A Road Map to Brain-Disease Dependent-Inflammatory Response. Front Cell Neurosci 2018; 12: 488.
[http://dx.doi.org/10.3389/fncel.2018.00488] [PMID: 30618635]

[224] Mass E, Jacome-Galarza CE, Blank T, *et al.* A somatic mutation in erythro-myeloid progenitors causes neurodegenerative disease. Nature 2017; 549(7672): 389-93. [http://dx.doi.org/10.1038/nature23672] [PMID: 28854169]

[225] Labzin LI, Heneka MT, Latz E. Innate Immunity and Neurodegeneration. Annu Rev Med 2018; 69: 437-49. [http://dx.doi.org/10.1146/annurev-med-050715-104343] [PMID: 29106805]

[226] Cobley JN, Fiorello ML, Bailey DM. 13 reasons why the brain is susceptible to oxidative stress. Redox Biol 2018; 15: 490-503. [http://dx.doi.org/10.1016/j.redox.2018.01.008] [PMID: 29413961]

[227] Erickson MA, Banks WA. Age-Associated Changes in the Immune System and Blood‑Brain Barrier Functions. Int J Mol Sci 2019; 20(7)E1632 [http://dx.doi.org/10.3390/ijms20071632] [PMID: 30986918]

[228] Stephan AH, Madison DV, Mateos JM, *et al.* A dramatic increase of C1q protein in the CNS during normal aging. J Neurosci 2013; 33(33): 13460-74. [http://dx.doi.org/10.1523/JNEUROSCI.1333-13.2013] [PMID: 23946404]

[229] Lui H, Zhang J, Makinson SR, *et al.* Progranulin Deficiency Promotes Circuit-Specific Synaptic Pruning by Microglia *via* Complement Activation. Cell 2016; 165(4): 921-35. [http://dx.doi.org/10.1016/j.cell.2016.04.001] [PMID: 27114033]

[230] Vasek MJ, Garber C, Dorsey D, *et al.* A complement-microglial axis drives synapse loss during virus-induced memory impairment. Nature 2016; 534(7608): 538-43. [http://dx.doi.org/10.1038/nature18283] [PMID: 27337340]

[231] Scheltens P, Blennow K, Breteler MM, *et al.* Alzheimer's disease. Lancet 2016; 388(10043): 505-17. [http://dx.doi.org/10.1016/S0140-6736(15)01124-1] [PMID: 26921134]

[232] Tai LM, Thomas R, Marottoli FM, *et al.* The role of APOE in cerebrovascular dysfunction. Acta Neuropathol 2016; 131(5): 709-23. [http://dx.doi.org/10.1007/s00401-016-1547-z] [PMID: 26884068]

[233] Huang YA, Zhou B, Wernig M, Südhof TC. ApoE2, ApoE3, and ApoE4 Differentially Stimulate APP Transcription and Aβ Secretion. Cell 2017; 168(3): 427-441.e21. [http://dx.doi.org/10.1016/j.cell.2016.12.044] [PMID: 28111074]

[234] Morris GP, Clark IA, Vissel B. Questions concerning the role of amyloid-β in the definition, aetiology and diagnosis of Alzheimer's disease. Acta Neuropathol 2018; 136(5): 663-89. [http://dx.doi.org/10.1007/s00401-018-1918-8] [PMID: 30349969]

[235] Brinkmalm G, Hong W, Wang Z, *et al.* Identification of neurotoxic cross-linked amyloid-β dimers in the Alzheimer's brain. Brain 2019; 142(5): 1441-57. [http://dx.doi.org/10.1093/brain/awz066] [PMID: 31032851]

[236] Bode DC, Baker MD, Viles JH. Ion Channel Formation by Amyloid-β42 Oligomers but Not Amyloid-β40 in Cellular Membranes. J Biol Chem 2017; 292(4): 1404-13. [http://dx.doi.org/10.1074/jbc.M116.762526] [PMID: 27927987]

[237] He Y, Wei M, Wu Y, *et al.* Amyloid β oligomers suppress excitatory transmitter release *via* presynaptic depletion of phosphatidylinositol-4,5-bisphosphate. Nat Commun 2019; 10(1): 1193. [http://dx.doi.org/10.1038/s41467-019-09114-z] [PMID: 30867420]

[238] Kametani F, Hasegawa M. Reconsideration of Amyloid Hypothesis and Tau Hypothesis in Alzheimer's Disease. Front Neurosci 2018; 12: 25. [http://dx.doi.org/10.3389/fnins.2018.00025] [PMID: 29440986]

[239] Huber CM, Yee C, May T, Dhanala A, Mitchell CS. Cognitive Decline in Preclinical Alzheimer's Disease: Amyloid-Beta *versus* Tauopathy. J Alzheimers Dis 2018; 61(1): 265-81. [http://dx.doi.org/10.3233/JAD-170490] [PMID: 29154274]

[240] de la Torre J. The Vascular Hypothesis of Alzheimer's Disease: A Key to Preclinical Prediction of Dementia Using Neuroimaging. J Alzheimers Dis 2018; 63(1): 35-52. [http://dx.doi.org/10.3233/JAD-180004] [PMID: 29614675]

[241] Shi L, Baird AL, Westwood S, *et al.* A Decade of Blood Biomarkers for Alzheimer's Disease Research: An Evolving Field, Improving Study Designs, and the Challenge of Replication. J Alzheimers Dis 2018; 62(3): 1181-98. [http://dx.doi.org/10.3233/JAD-170531] [PMID: 29562526]

[242] Lin CN, Huang CC, Huang KL, Lin KJ, Yen TC, Kuo HC. A metabolomic approach to identifying biomarkers in blood of Alzheimer's disease. Ann Clin Transl Neurol 2019; 6(3): 537-45. [http://dx.doi.org/10.1002/acn3.726] [PMID: 30911577]

[243] Al-Atrache Z, Lopez DB, Hingley ST, Appelt DM. Astrocytes infected with Chlamydia pneumoniae demonstrate altered expression and activity of secretases involved in the generation of β-amyloid found in Alzheimer disease. BMC Neurosci 2019; 20(1): 6. [http://dx.doi.org/10.1186/s12868-019-0489-5] [PMID: 30786875]

[244] De Chiara G, Piacentini R, Fabiani M, *et al.* Recurrent herpes simplex virus-1 infection induces hallmarks of neurodegeneration and cognitive deficits in mice. PLoS Pathog 2019; 15(3)e1007617 [http://dx.doi.org/10.1371/journal.ppat.1007617] [PMID: 30870531]

[245] Condello C, Yuan P, Grutzendler J. Microglia-Mediated Neuroprotection, TREM2, and Alzheimer's Disease: Evidence From Optical Imaging. Biol Psychiatry 2018; 83(4): 377-87. [http://dx.doi.org/10.1016/j.biopsych.2017.10.007] [PMID: 29169609]

[246] Lee CYD, Daggett A, Gu X, *et al.* Elevated TREM2 Gene Dosage Reprograms Microglia Responsivity and Ameliorates Pathological Phenotypes in Alzheimer's Disease Models. Neuron 2018; 97(5): 1032-1048.e5. [http://dx.doi.org/10.1016/j.neuron.2018.02.002] [PMID: 29518357]

[247] Song WM, Joshita S, Zhou Y, Ulland TK, Gilfillan S, Colonna M. Humanized TREM2 mice reveal microglia-intrinsic and -extrinsic effects of R47H polymorphism. J Exp Med 2018; 215(3): 745-60. [http://dx.doi.org/10.1084/jem.20171529] [PMID: 29321225]

[248] Hansen DV, Hanson JE, Sheng M. Microglia in Alzheimer's disease. J Cell Biol 2018; 217(2): 459-72. [http://dx.doi.org/10.1083/jcb.201709069] [PMID: 29196460]

[249] Prieto GA, Tong L, Smith ED, Cotman CW. TNFα and IL-1β but not IL-18 Suppresses Hippocampal Long-Term Potentiation Directly at the Synapse. Neurochem Res 2019; 44(1): 49-60. [http://dx.doi.org/10.1007/s11064-018-2517-8] [PMID: 29619614]

[250] Selkoe DJ. Alzheimer's disease is a synaptic failure. Science 2002; 298(5594): 789-91. [http://dx.doi.org/10.1126/science.1074069] [PMID: 12399581]

[251] Henstridge CM, Tzioras M, Paolicelli RC. Glial Contribution to Excitatory and Inhibitory Synapse Loss in Neurodegeneration. Front Cell Neurosci 2019; 13: 63. [http://dx.doi.org/10.3389/fncel.2019.00063] [PMID: 30863284]

[252] Venegas C, Heneka MT. Danger-associated molecular patterns in Alzheimer's disease. J Leukoc Biol 2017; 101(1): 87-98. [http://dx.doi.org/10.1189/jlb.3MR0416-204R] [PMID: 28049142]

[253] Boza-Serrano A, Yang Y, Paulus A, Deierborg T. Innate immune alterations are elicited in microglial cells before plaque deposition in the Alzheimer's disease mouse model 5xFAD. Sci Rep 2018; 8(1): 1550. [http://dx.doi.org/10.1038/s41598-018-19699-y] [PMID: 29367720]

[254] Gambin Y, Giles N, O'Carroll A, Polinkovsky M, Hunter D, Sierecki E. Single-Molecule Fluorescence Reveals the Oligomerization and Folding Steps Driving the Prion-like Behavior of ASC. J Mol Biol 2018; 430(4): 491-508. [http://dx.doi.org/10.1016/j.jmb.2017.12.013] [PMID: 29288634]

[255] Venegas C, Kumar S, Franklin BS, *et al.* Microglia-derived ASC specks cross-seed amyloid-β in Alzheimer's disease. Nature 2017; 552(7685): 355-61.
[http://dx.doi.org/10.1038/nature25158] [PMID: 29293211]

[256] Goetzl EJ, Schwartz JB, Abner EL, Jicha GA, Kapogiannis D. High complement levels in astrocyte-derived exosomes of Alzheimer disease. Ann Neurol 2018; 83(3): 544-52.
[http://dx.doi.org/10.1002/ana.25172] [PMID: 29406582]

[257] Zorzetto M, Datturi F, Divizia L, *et al.* Complement C4A and C4B Gene Copy Number Study in Alzheimer's Disease Patients. Curr Alzheimer Res 2017; 14(3): 303-8.
[PMID: 27758680]

[258] Hong S, Beja-Glasser VF, Nfonoyim BM, *et al.* Complement and microglia mediate early synapse loss in Alzheimer mouse models. Science 2016; 352(6286): 712-6.
[http://dx.doi.org/10.1126/science.aad8373] [PMID: 27033548]

[259] Ozben T, Ozben S. Neuro-inflammation and anti-inflammatory treatment options for Alzheimer's disease. Clinical biochemistry 2019; S0009-9120(19): 30153-5.
[http://dx.doi.org/10.1016/j.clinbiochem.2019.04.001]

[260] Businaro R, Corsi M, Asprino R, *et al.* Modulation of Inflammation as a Way of Delaying Alzheimer's Disease Progression: The Diet's Role. Curr Alzheimer Res 2018; 15(4): 363-80.
[http://dx.doi.org/10.2174/1567205014666170829100100] [PMID: 28847284]

[261] Fang EF. Mitophagy and NAD^+ inhibit Alzheimer disease. Autophagy 2019; 15(6): 1112-4.
[http://dx.doi.org/10.1080/15548627.2019.1596497] [PMID: 30922179]

[262] Pinto A, Bonucci A, Maggi E, Corsi M, Businaro R. Anti-Oxidant and Anti-Inflammatory Activity of Ketogenic Diet: New Perspectives for Neuroprotection in Alzheimer's Disease. Antioxidants 2018; 7(5)E63
[http://dx.doi.org/10.3390/antiox7050063] [PMID: 29710809]

[263] Lourenco MV, Frozza RL, de Freitas GB, *et al.* Exercise-linked FNDC5/irisin rescues synaptic plasticity and memory defects in Alzheimer's models. Nat Med 2019; 25(1): 165-75.
[http://dx.doi.org/10.1038/s41591-018-0275-4] [PMID: 30617325]

[264] González AE, Muñoz VC, Cavieres VA, *et al.* Autophagosomes cooperate in the degradation of intracellular C-terminal fragments of the amyloid precursor protein *via* the MVB/lysosomal pathway. FASEB J 2017; 31(6): 2446-59.
[http://dx.doi.org/10.1096/fj.201600713R] [PMID: 28254759]

[265] Daria A, Colombo A, Llovera G, *et al.* Young microglia restore amyloid plaque clearance of aged microglia. EMBO J 2017; 36(5): 583-603.
[http://dx.doi.org/10.15252/embj.201694591] [PMID: 28007893]

[266] Johnstone DM, Moro C, Stone J, Benabid AL, Mitrofanis J. Turning On Lights to Stop Neurodegeneration: The Potential of Near Infrared Light Therapy in Alzheimer's and Parkinson's Disease. Front Neurosci 2016; 9: 500.
[http://dx.doi.org/10.3389/fnins.2015.00500] [PMID: 26793049]

[267] Iaccarino HF, Singer AC, Martorell AJ, *et al.* Gamma frequency entrainment attenuates amyloid load and modifies microglia. Nature 2016; 540(7632): 230-5.
[http://dx.doi.org/10.1038/nature20587] [PMID: 27929004]

[268] Khan IS, D'Agostino EN, Calnan DR, Lee JE, Aronson JP. Deep Brain Stimulation for Memory Modulation: A New Frontier. World neurosurgery 2019; S1878-8750(19): 30056-7.
[http://dx.doi.org/10.1016/j.wneu.2018.12.184]

[269] Helley MP, Pinnell J, Sportelli C, Tieu K. Mitochondria: A Common Target for Genetic Mutations and Environmental Toxicants in Parkinson's Disease. Front Genet 2017; 8: 177.
[http://dx.doi.org/10.3389/fgene.2017.00177] [PMID: 29204154]

[270] Delgado-Camprubi M, Esteras N, Soutar MP, Plun-Favreau H, Abramov AY. Deficiency of Parkinson's disease-related gene Fbxo7 is associated with impaired mitochondrial metabolism by PARP activation. Cell Death Differ 2017; 24(1): 120-31.
[http://dx.doi.org/10.1038/cdd.2016.104] [PMID: 27689878]

[271] Hauser DN, Mamais A, Conti MM, *et al.* Hexokinases link DJ-1 to the PINK1/parkin pathway. Mol Neurodegener 2017; 12(1): 70.
[http://dx.doi.org/10.1186/s13024-017-0212-x] [PMID: 28962651]

[272] Alessi DR, Sammler E. LRRK2 kinase in Parkinson's disease. Science 2018; 360(6384): 36-7.
[http://dx.doi.org/10.1126/science.aar5683] [PMID: 29622645]

[273] Mouton-Liger F, Jacoupy M, Corvol JC, Corti O. PINK1/Parkin-Dependent Mitochondrial Surveillance: From Pleiotropy to Parkinson's Disease. Front Mol Neurosci 2017; 10: 120.
[http://dx.doi.org/10.3389/fnmol.2017.00120] [PMID: 28507507]

[274] Sliter DA, Martinez J, Hao L, *et al.* Parkin and PINK1 mitigate STING-induced inflammation. Nature 2018; 561(7722): 258-62.
[http://dx.doi.org/10.1038/s41586-018-0448-9] [PMID: 30135585]

[275] Richarme G, Abdallah J, Mathas N, Gautier V, Dairou J. Further characterization of the Maillard deglycase DJ-1 and its prokaryotic homologs, deglycase 1/Hsp31, deglycase 2/YhbO, and deglycase 3/YajL. Biochem Biophys Res Commun 2018; 503(2): 703-9.
[http://dx.doi.org/10.1016/j.bbrc.2018.06.064] [PMID: 29932913]

[276] Junn E, Jang WH, Zhao X, Jeong BS, Mouradian MM. Mitochondrial localization of DJ-1 leads to enhanced neuroprotection. J Neurosci Res 2009; 87(1): 123-9.
[http://dx.doi.org/10.1002/jnr.21831] [PMID: 18711745]

[277] Boyapati RK, Tamborska A, Dorward DA, Ho GT. Advances in the understanding of mitochondrial DNA as a pathogenic factor in inflammatory diseases. F1000 Res 2017; 6: 169.
[http://dx.doi.org/10.12688/f1000research.10397.1] [PMID: 28299196]

[278] Bridi JC, Hirth F. Mechanisms of α-Synuclein Induced Synaptopathy in Parkinson's Disease. Front Neurosci 2018; 12: 80.
[http://dx.doi.org/10.3389/fnins.2018.00080] [PMID: 29515354]

[279] Ryan T, Bamm VV, Stykel MG, *et al.* Cardiolipin exposure on the outer mitochondrial membrane modulates α-synuclein. Nat Commun 2018; 9(1): 817.
[http://dx.doi.org/10.1038/s41467-018-03241-9] [PMID: 29483518]

[280] Grassi D, Howard S, Zhou M, *et al.* Identification of a highly neurotoxic α-synuclein species inducing mitochondrial damage and mitophagy in Parkinson's disease. Proc Natl Acad Sci USA 2018; 115(11): E2634-43.
[http://dx.doi.org/10.1073/pnas.1713849115] [PMID: 29487216]

[281] Dzamko N, Gysbers A, Perera G, *et al.* Toll-like receptor 2 is increased in neurons in Parkinson's disease brain and may contribute to alpha-synuclein pathology. Acta Neuropathol 2017; 133(2): 303-19.
[http://dx.doi.org/10.1007/s00401-016-1648-8] [PMID: 27888296]

[282] Sulzer D, Alcalay RN, Garretti F, *et al.* T cells from patients with Parkinson's disease recognize α-synuclein peptides. Nature 2017; 546(7660): 656-61.
[http://dx.doi.org/10.1038/nature22815] [PMID: 28636593]

[283] Garretti F, Agalliu D, Lindestam Arlehamn CS, Sette A, Sulzer D. Autoimmunity in Parkinson's Disease: The Role of α-Synuclein-Specific T Cells. Front Immunol 2019; 10: 303.
[http://dx.doi.org/10.3389/fimmu.2019.00303] [PMID: 30858851]

[284] Li X, Koudstaal W, Fletcher L, *et al.* Naturally occurring antibodies isolated from PD patients inhibit synuclein seeding *in vitro* and recognize Lewy pathology. Acta Neuropathol 2019; 137(5): 825-36.
[http://dx.doi.org/10.1007/s00401-019-01974-5] [PMID: 30805666]

[285] Kim KS, Marcogliese PC, Yang J, *et al.* Regulation of myeloid cell phagocytosis by LRRK2 *via* WAVE2 complex stabilization is altered in Parkinson's disease. Proc Natl Acad Sci USA 2018; 115(22): E5164-73.
[http://dx.doi.org/10.1073/pnas.1718946115] [PMID: 29760073]

[286] Iannielli A, Bido S, Folladori L, *et al.* Pharmacological Inhibition of Necroptosis Protects from Dopaminergic Neuronal Cell Death in Parkinson's Disease Models. Cell Rep 2018; 22(8): 2066-79.
[http://dx.doi.org/10.1016/j.celrep.2018.01.089] [PMID: 29466734]

[287] Pardiñas AF, Holmans P, Pocklington AJ, *et al.* Common schizophrenia alleles are enriched in mutation-intolerant genes and in regions under strong background selection. Nat Genet 2018; 50(3): 381-9.
[http://dx.doi.org/10.1038/s41588-018-0059-2] [PMID: 29483656]

[288] Pouget JG. The Emerging Immunogenetic Architecture of Schizophrenia. Schizophr Bull 2018; 44(5): 993-1004.
[http://dx.doi.org/10.1093/schbul/sby038] [PMID: 29701842]

[289] Gandal MJ, Zhang P, Hadjimichael E, *et al.* Transcriptome-wide isoform-level dysregulation in ASD, schizophrenia, and bipolar disorder. Science 2018; 362(6420)eaat8127
[http://dx.doi.org/10.1126/science.aat8127] [PMID: 30545856]

[290] Volk DW, Moroco AE, Roman KM, Edelson JR, Lewis DA. The Role of the Nuclear Factor-κB Transcriptional Complex in Cortical Immune Activation in Schizophrenia. Biol Psychiatry 2019; 85(1): 25-34.
[http://dx.doi.org/10.1016/j.biopsych.2018.06.015] [PMID: 30082065]

[291] van Kesteren CF, Gremmels H, de Witte LD, *et al.* Immune involvement in the pathogenesis of schizophrenia: a meta-analysis on postmortem brain studies. Transl Psychiatry 2017; 7(3)e1075
[http://dx.doi.org/10.1038/tp.2017.4] [PMID: 28350400]

[292] Sanders AR, Drigalenko EI, Duan J, *et al.* Transcriptome sequencing study implicates immune-related genes differentially expressed in schizophrenia: new data and a meta-analysis. Transl Psychiatry 2017; 7(4)e1093
[http://dx.doi.org/10.1038/tp.2017.47] [PMID: 28418402]

[293] Deakin IH, Godlewska BR, Walker MA, *et al.* Altered hippocampal gene expression and structure in transgenic mice overexpressing neuregulin 1 (Nrg1) type I. Transl Psychiatry 2018; 8(1): 229.
[http://dx.doi.org/10.1038/s41398-018-0288-2] [PMID: 30348978]

[294] Al-Diwani AAJ, Pollak TA, Irani SR, Lennox BR. Psychosis: an autoimmune disease? Immunology 2017; 152(3): 388-401.
[http://dx.doi.org/10.1111/imm.12795] [PMID: 28704576]

[295] Lydholm CN, Köhler-Forsberg O, Nordentoft M, *et al.* Parental Infections Before, During, and After Pregnancy as Risk Factors for Mental Disorders in Childhood and Adolescence: A Nationwide Danish Study. Biol Psychiatry 2019; 85(4): 317-25.
[http://dx.doi.org/10.1016/j.biopsych.2018.09.013] [PMID: 30446204]

[296] Ursini G, Punzi G, Chen Q, *et al.* Convergence of placenta biology and genetic risk for schizophrenia. Nat Med 2018; 24(6): 792-801.
[http://dx.doi.org/10.1038/s41591-018-0021-y] [PMID: 29808008]

[297] Jézéquel J, Johansson EM, Dupuis JP, *et al.* Dynamic disorganization of synaptic NMDA receptors triggered by autoantibodies from psychotic patients. Nat Commun 2017; 8(1): 1791.
[http://dx.doi.org/10.1038/s41467-017-01700-3] [PMID: 29176681]

[298] Najjar S, Steiner J, Najjar A, Bechter K. A clinical approach to new-onset psychosis associated with immune dysregulation: the concept of autoimmune psychosis. J Neuroinflammation 2018; 15(1): 40.
[http://dx.doi.org/10.1186/s12974-018-1067-y] [PMID: 29433523]

[299] Sekar A, Bialas AR, de Rivera H, *et al.* Schizophrenia risk from complex variation of complement

component 4. Nature 2016; 530(7589): 177-83.
[http://dx.doi.org/10.1038/nature16549] [PMID: 26814963]

[300] Melbourne JK, Rosen C, Feiner B, Sharma RP. 2018; C4A mRNA expression in PBMCs predicts the presence and severity of delusions in schizophrenia and bipolar disorder with psychosis. Schizophrenia research 2018; S0920-9964(18): 30038-0.
[http://dx.doi.org/10.1016/j.schres.2018.01.018]

[301] Jordan W, Dobrowolny H, Bahn S, *et al.* Oxidative stress in drug-naïve first episode patients with schizophrenia and major depression: effects of disease acuity and potential confounders. Eur Arch Psychiatry Clin Neurosci 2018; 268(2): 129-43.
[http://dx.doi.org/10.1007/s00406-016-0749-7] [PMID: 27913877]

[302] Howes OD, McCutcheon R. Inflammation and the neural diathesis-stress hypothesis of schizophrenia: a reconceptualization. Transl Psychiatry 2017; 7(2)e1024
[http://dx.doi.org/10.1038/tp.2016.278] [PMID: 28170004]

[303] Nimgaonkar VL, Prasad KM, Chowdari KV, Severance EG, Yolken RH. The complement system: a gateway to gene-environment interactions in schizophrenia pathogenesis. Mol Psychiatry 2017; 22(11): 1554-61.
[http://dx.doi.org/10.1038/mp.2017.151] [PMID: 28761078]

[304] Petanjek Z, Judaš M, Šimic G, *et al.* Extraordinary neoteny of synaptic spines in the human prefrontal cortex. Proc Natl Acad Sci USA 2011; 108(32): 13281-6.
[http://dx.doi.org/10.1073/pnas.1105108108] [PMID: 21788513]

[305] Oparil S, Acelajado MC, Bakris GL, *et al.* Hypertension. Nat Rev Dis Primers 2018; 4: 18014.
[http://dx.doi.org/10.1038/nrdp.2018.14] [PMID: 29565029]

[306] Pantell MS, Prather AA, Downing JM, Gordon NP, Adler NE. Association of Social and Behavioral Risk Factors With Earlier Onset of Adult Hypertension and Diabetes. JAMA Netw Open 2019; 2(5)e193933
[http://dx.doi.org/10.1001/jamanetworkopen.2019.3933] [PMID: 31099868]

[307] Norlander AE, Madhur MS, Harrison DG. The immunology of hypertension. J Exp Med 2018; 215(1): 21-33.
[http://dx.doi.org/10.1084/jem.20171773] [PMID: 29247045]

[308] Ip WK, Medzhitov R. Macrophages monitor tissue osmolarity and induce inflammatory response through NLRP3 and NLRC4 inflammasome activation. Nat Commun 2015; 6: 6931.
[http://dx.doi.org/10.1038/ncomms7931] [PMID: 25959047]

[309] Rucker AJ, Rudemiller NP, Crowley SD. Salt, Hypertension, and Immunity. Annu Rev Physiol 2018; 80: 283-307.
[http://dx.doi.org/10.1146/annurev-physiol-021317-121134] [PMID: 29144825]

[310] Barbaro NR, Foss JD, Kryshtal DO, *et al.* Dendritic Cell Amiloride-Sensitive Channels Mediate Sodium-Induced Inflammation and Hypertension. Cell Rep 2017; 21(4): 1009-20.
[http://dx.doi.org/10.1016/j.celrep.2017.10.002] [PMID: 29069584]

[311] Iadecola C, Gottesman RF. Neurovascular and Cognitive Dysfunction in Hypertension. Circ Res 2019; 124(7): 1025-44.
[http://dx.doi.org/10.1161/CIRCRESAHA.118.313260] [PMID: 30920929]

[312] Bica T, Castello R, Toussaint LL, Monteso-Curto P. Depression as a Risk Factor of Organic Diseases:An International Integrative Review. Journal of nursing scholarship : an official publication of Sigma Theta Tau International Honor Society of Nursing 2017; 49(4): 389-99.

[313] Halaris A. Inflammation-Associated Co-morbidity Between Depression and Cardiovascular Disease. Curr Top Behav Neurosci 2017; 31: 45-70.
[http://dx.doi.org/10.1007/7854_2016_28] [PMID: 27830572]

[314] Larsen JB, Iversen VC, Reitan SK. Association of psychosis, affective disorders and diseases affecting

the immune system. Nord J Psychiatry 2018; 72(2): 145-9.
[http://dx.doi.org/10.1080/08039488.2017.1402952] [PMID: 29141491]

[315] Hoover DB. Cholinergic modulation of the immune system presents new approaches for treating inflammation. Pharmacol Ther 2017; 179: 1-16.
[http://dx.doi.org/10.1016/j.pharmthera.2017.05.002] [PMID: 28529069]

[316] Chavan SS, Pavlov VA, Tracey KJ. Mechanisms and Therapeutic Relevance of Neuro-immune Communication. Immunity 2017; 46(6): 927-42.
[http://dx.doi.org/10.1016/j.immuni.2017.06.008] [PMID: 28636960]

[317] Shimojo G, Joseph B, Shah R, Consolim-Colombo FM, De Angelis K, Ulloa L. Exercise activates vagal induction of dopamine and attenuates systemic inflammation. Brain Behav Immun 2019; 75: 181-91.
[http://dx.doi.org/10.1016/j.bbi.2018.10.005] [PMID: 30394312]

[318] Camell CD, Sander J, Spadaro O, *et al.* Inflammasome-driven catecholamine catabolism in macrophages blunts lipolysis during ageing. Nature 2017; 550(7674): 119-23.
[http://dx.doi.org/10.1038/nature24022] [PMID: 28953873]

[319] Bialas AR, Presumey J, Das A, *et al.* Microglia-dependent synapse loss in type I interferon-mediated lupus. Nature 2017; 546(7659): 539-43.
[http://dx.doi.org/10.1038/nature22821] [PMID: 28614301]

[320] Mandolesi G, Bullitta S, Fresegna D, *et al.* Interferon-γ causes mood abnormalities by altering cannabinoid CB1 receptor function in the mouse striatum. Neurobiol Dis 2017; 108: 45-53.
[http://dx.doi.org/10.1016/j.nbd.2017.07.019] [PMID: 28757328]

[321] Tatomir A, Talpos-Caia A, Anselmo F, *et al.* The complement system as a biomarker of disease activity and response to treatment in multiple sclerosis. Immunol Res 2017; 65(6): 1103-9.
[http://dx.doi.org/10.1007/s12026-017-8961-8] [PMID: 29116612]

[322] Lvovich S, Goldsmith DP. Neurological Complications of Rheumatic Disease. Semin Pediatr Neurol 2017; 24(1): 54-9.
[http://dx.doi.org/10.1016/j.spen.2016.12.007] [PMID: 28779866]

[323] Lasselin J, Alvarez-Salas E, Grigoleit JS. Well-being and immune response: a multi-system perspective. Curr Opin Pharmacol 2016; 29: 34-41.
[http://dx.doi.org/10.1016/j.coph.2016.05.003] [PMID: 27318753]

[324] Moriyama M, Ichinohe T. High ambient temperature dampens adaptive immune responses to influenza A virus infection. Proc Natl Acad Sci USA 2019; 116(8): 3118-25.
[http://dx.doi.org/10.1073/pnas.1815029116] [PMID: 30718396]

[325] Holden C. Cancer and the mind: how are they connected? Science 1978; 200(4348): 1363-9.
[http://dx.doi.org/10.1126/science.200.4348.1363] [PMID: 17736309]

[326] Azpiroz A, De Miguel Z, Fano E, Vegas O. Relations between different coping strategies for social stress, tumor development and neuroendocrine and immune activity in male mice. Brain Behav Immun 2008; 22(5): 690-8.
[http://dx.doi.org/10.1016/j.bbi.2007.10.007] [PMID: 18061400]

[327] Suzuki H, Savitz J, Kent Teague T, *et al.* Altered populations of natural killer cells, cytotoxic T lymphocytes, and regulatory T cells in major depressive disorder: Association with sleep disturbance. Brain Behav Immun 2017; 66: 193-200.
[http://dx.doi.org/10.1016/j.bbi.2017.06.011] [PMID: 28645775]

[328] Grosse L, Hoogenboezem T, Ambrée O, *et al.* Deficiencies of the T and natural killer cell system in major depressive disorder: T regulatory cell defects are associated with inflammatory monocyte activation. Brain Behav Immun 2016; 54: 38-44.
[http://dx.doi.org/10.1016/j.bbi.2015.12.003] [PMID: 26674997]

[329] Vogels RJ, Koenders MA, van Rossum EF, Spijker AT, Drexhage HA. T Cell Deficits and

Overexpression of Hepatocyte Growth Factor in Anti-inflammatory Circulating Monocytes of Middle-Aged Patients with Bipolar Disorder Characterized by a High Prevalence of the Metabolic Syndrome. Front Psychiatry 2017; 8: 34.
[http://dx.doi.org/10.3389/fpsyt.2017.00034] [PMID: 28373847]

[330] Leday GGR, Vértes PE, Richardson S, *et al.* Replicable and Coupled Changes in Innate and Adaptive Immune Gene Expression in Two Case-Control Studies of Blood Microarrays in Major Depressive Disorder. Biol Psychiatry 2018; 83(1): 70-80.
[http://dx.doi.org/10.1016/j.biopsych.2017.01.021] [PMID: 28688579]

SUBJECT INDEX

A

Ablation of microglia 383
Acetyltransferases 229
Acids 14, 15, 21, 48, 56, 71, 75, 81, 110, 112, 113, 127, 128, 158, 166, 239, 262, 266, 267, 275, 353, 358, 399, 400
acetic 112
bile 266, 267
folic 400
galacturonic 110
glutamic 110
hypochlorous 166
hypothiocyanous 358
lactic 56
lipoic 15
lipoteichoic 48
polysialic 128
quinolinic 399
retinoic 275
sapienic 21
sialic 353
teichoic 127
teichuronic 110
tricarboxylic 113, 239
uric 15, 262
Activated AMPK phosphorylates 193
Adaptive immune systems 48, 133, 141, 241, 257, 309, 340, 358, 380, 381, 388
Adenine 131, 132, 169
DNA glycosylase 131
methylates 131
methylation 169
Adenosine 3, 81, 126, 165, 184, 270, 271, 282
deaminase 270
diphosphate 165
monophosphate 271
synthase 126, 282
triphosphate 3
Adrenocorticotropic hormone 384
Adverse childhood experiences (ACE) 393, 394
Aicardi-Goutières syndrome (AGS) 270
Aminoacylation 12
Amino-acyl-tRNA 165
AMP-activated protein kinase 12, 185
Anorexia 343, 390, 391
Antibiotic resistance genes 44, 137, 158, 170
disseminate 158
Antibiotics 21, 22, 47, 52, 77, 78, 79, 86, 87, 88, 89, 126, 127, 155, 156, 169
catabolize 21
multiple diverse 21
resist 47
treatment failure 88
trimethoprim 21
Antidepressants 398, 409
therapy 409
treatment 398
Antifungal peptide drosomycin 260
Antigen-presenting cells (APCs) 359, 360, 361
Antioxidant 191, 319, 401
capability 401
enzymes 319
factors 191
Antioxidant genes 195, 317
canonical 195
Antioxidative mechanisms 317
Anxiety disorders, generalized 392, 400
Apolipoprotein 402
Apoptosis 87, 185, 195, 196, 198, 199, 200, 201, 224, 227, 233, 236, 280, 313, 315, 353, 399
damage-induced 236
-like death (ALD) 87
neural 399
-promoting regulatory factors 185
suppressing 280
Apoptosome 200, 281
oligomeric 200
Apoptotic cells 201, 316, 351, 352
clearing 351
display 201

APP 402
metabolites 402
mutations 402
Arabidopsis thaliana response 323
Asthma 56, 393, 397
Autoimmune 265, 268, 359, 392, 396
conditions 265, 268, 359, 396
disorders 392
Autophagic 238, 321
degradation 238
machinery 321
removal 321
response 321
Autophagosome 189, 320
biogenesis 189, 320
formation 189

B

Bacillus cereus 132
Bacillus cereus toxin forms pores 267
Base excision repair (BER) 14, 23, 360
mechanism 14
Biofilm 53, 72, 75, 77, 85, 86, 88, 89, 127, 157, 158, 159, 167
formation 72, 75, 77, 85, 86, 88, 89, 127, 157, 158, 167
growth 53
matrix 159
Biogenesis 161, 225, 404
mitochondrial 404
ribosomal 225
Blood-brain barrier (BBB) 380, 382, 384, 397
Blood cells 20, 55, 382
embryonic 382
fetal 382
infected red 55
Blood glucose level 8, 9
Body homeostasis 392
Bovine 43, 140
spongiform encephalopathy 43
viral diarrhea virus 140
Bradykinesia 404
Bradykinin 343
Brain 225, 384, 395, 396, 400
damage 384
derived neurotrophic factor (BDNF) 395, 400
parenchyma 384
regions, frontolimbic 396
stem cells 225
Brainstem's medulla oblongata 386
BREX system 159
Brown adipose tissue (BAT) 204, 343

C

Calcitonin gene-related peptide (CGRP) 51
Calcium carbonate 111
Caldopentamine 104
cAMP receptor protein (CRP) 136, 163
Cancer 195, 234, 237, 238, 239, 276, 343, 351
cells 195, 234, 237, 238, 239, 343, 351
genomes 276
growth 239
Candida albicans 37
Capsaicin 17, 203
Carbohydrate metabolism 57, 73
Carbon biopolymers 116
Carbonylation 14
Cardiac glycosides 22
Cardiolipin 21, 281, 317, 404
mitochondrial lipid 317
Catabolic 12, 190, 193
enzymes 190
processes 12, 193
Catalases 15, 79, 106, 194, 195
Catecholamines 9, 380, 381, 385, 387, 408
epinephrine 387
Cell-autonomous responses 202
Cell cycle 74, 186, 199, 232, 237, 359
checkpoints 359
inhibitors 74, 232
progression 186, 232
regulation 199, 237
Cell cycle arrest 79, 84, 85, 202, 224, 225, 226, 227, 231, 232, 234, 235, 236, 237, 315
and elimination of stressed cells 315
induction of 225, 226, 236
irreversible 232
permanent 202, 225
Cerebellar neuropathologies 394
Chaperones 1, 73, 75, 76, 82, 86, 89, 182, 186, 188, 190, 191, 196
mitochondrial 191
molecular 10, 103, 203
periplasmic 76
CHAPLE syndrome 357
Chemiosmotic hypothesis 4

Chemotaxis 49, 51, 72, 201, 339, 343, 349, 354
increased vascular 343
Choline acetyltransferase 386, 387
Chromoblastomycosis 54
Chromosomal 273, 346
accessibility 346
DNA mutations 273
Chronic 239, 275, 392
inflammation predisposes 239
rhinosinusitis, developing 275
unpredictable stress in mice 392
Coccidioidomycosis 54
Cold-inducible proteins (CIPs) 18
Cold shock response (CSR) 18, 71, 73, 74
Complement factor 355, 356
B (CFB) 355
P (CFP) 355, 356
proenzyme 355
Copper-binding chaperone 113
Corticotropin-releasing hormone (CRH) 380, 384, 385, 387
Creutzfeldt-Jakob disease in humans 43
Cro gene 41
Crohn's disease 316, 348, 397
Cryopyrin-associated periodic syndromes 268, 348
Cyanobacteria 2, 3, 48, 87, 102, 107
photosynthetic 48
Cycle 21, 81, 139
catalytic 81
pathological 21
reproduction 139
Cyclic 271, 312
-di-adenosine monophosphate 312
dinucleotides 271
guanosine monophosphate 271
Cyclin-dependent kinases (CDKs) 232
Cyclooxygenase 345
Cysteine 14, 167
biosynthesis 167
proteome 14
Cytidine 275, 276, 360
deaminate 275
triphosphate 276
Cytochrome 5, 13, 82, 200, 238, 281
non-proton pumping 82
P450 enzymes 13
reduced 5
releasing 281
Cytokines 257, 264, 340, 341, 342, 343, 245, 246, 350, 351, 358, 359, 360, 383, 386, 387, 389, 390, 403
anti-inflammatory 345, 346
inflammatory 387, 389
Cytomegalovirus 278, 393
Cytoplasm 18, 20, 81, 82, 107, 110, 113, 191, 192, 270, 271, 272, 281, 282, 312, 313
acidic 113
neutral 113
Cytoplasmic 18, 82, 261
biopolymers 18
chaperone Hsp31 82
chaperones DnaK 82
immunoreceptors 261
Cytosolic 185, 190, 200, 268, 272, 274
dsDNA 272
Smac 200
Cytotoxic HIV-1 protein TAT damages 191

D

Damage 13, 15, 82, 84, 88, 109, 182, 186, 191, 225, 233, 238, 257, 262, 263, 271, 273, 278, 281, 308, 312, 319, 323, 340, 341, 346
acidic 82
-associated molecular pattern (DAMPs) 191, 257, 262, 263, 273, 281, 308, 340, 341
cellular 233
collateral 323
escapes DNA repair machinery 225
genomic 238
mitigate 182, 186
oxidative 109
Deacetylases 229, 231
sirtuin 229
Deadenylation 258
Death-inducing signaling complex (DISC) 201
Dehydrogenases 7
Deoxyadenosine 126, 282
Deoxycytidine 138
Deoxynucleosides 276
Deoxyribonuclease 50
Depolarization, mitochondrial 404
Depression 380, 385, 394, 397, 398, 399, 400, 403, 408
and cardiovascular disease 408
chronic treatment-resistant 400

inflammation-mediated 399
melancholic 399
Depriving cells of cystine 202
Deregulated production 265
Derivatives, toxic 22
Detoxification 22, 77, 82, 167, 187, 319
enzymes 319
radical 167
Diabetic retinopathy 357
Diarrhea, massive life-threatening 51
Dictyostelium discoideum phagocyte 349
Diseases 38, 45, 55, 56, 58, 185, 225, 272, 316, 339, 343, 344, 348, 357, 380, 388, 389, 396, 401, 402, 406, 408
autoimmune 58, 316, 339, 343, 344, 348, 396, 406
chronic granulomatous 166, 350
cold agglutinin 357
genetic 45
hematological 357
immunological 408
infectious 38, 272, 389
inflammatory bowel 56, 388, 408
intestinal 55
kidney 357
metabolic 56
neurodegenerative 185, 225, 380, 401, 402
sickle cell 357
Disorders 195, 196, 199, 200, 202, 262, 275, 347, 349, 380, 383, 385, 396, 397, 401, 408, 409
aging-associated 202
autism spectrum 383, 397
auto-inflammatory 347
cause brain 408
developmental 199
human 195, 275
immune 200
neurodegenerative 196, 349, 380, 383, 397, 401
neurodevelopmental 396
rheumatic 409
stress-associated 385
Disorganization 103
Diverse 194, 279
electrophiles oxidize 194
immune mechanisms 279
Diversion of tryptophan metabolism 398
Diversity 43, 49, 88, 103, 128, 129, 168, 358
-generating retroelements (DGRs) 128, 129
genetic 49
of thermophiles 103
parasite 168
DNA 16, 17, 21, 39, 42, 44, 46, 48, 84, 104, 130, 131, 132, 134, 135, 137, 138, 156, 163, 199, 198, 235, 236, 271, 272, 312, 404
adenovirus 312
and formyl peptides 262chromosomal 42, 46, 48
complementary 44
cytosolic 272, 404
-dependent DNA synthesis 44
glucosyl-hydroxymethylated 131
kinase senses 198
methylase 132
methyltransferase 130
methylated 132
microbial 271
modifying cellular 163
parasitic 163
polymerase 198
topoisomerases 21
DNA-binding 75, 131, 226, 231, 236
domain (DBD) 226, 231, 236
motif 226
protein 75
specificity domain 131
DNA damage 22, 23, 49, 84, 85, 155, 156, 163, 164, 165, 198, 199, 200, 224, 225, 227, 228, 230, 232, 307, 312, 313, 314, 316, 322, 323, 359, 360
and associated DNA repair 155
chromosomal 312
increased 313
repair of 49, 163, 199
response (DDR) 198, 199, 227, 230, 232, 307, 312, 313, 314, 316, 322, 323, 359, 360
unrepairable 23
DNA lesions 23, 199, 313, 360
and DDR activation 360
unrepaired 313
DNA methylation 16, 45, 132, 167, 187, 194, 275, 314, 394, 395, 400
and loss of epigenetic control 45
change 275
landscape 167
radiation perturbs 16
DNA repair 87, 169, 232, 313, 358, 398

genes 87, 398
genes promoters 232
malfunction 169
proteins 313
proteins process 358
DNA replication 21, 23, 47, 78, 85, 86, 108, 161, 235, 237, 275, 276
genes 47
DNA sequence 130, 131, 194
consensus 194
foreign 130
DNAses 46, 282
producing 282
DNA synthesis 232, 235, 239
mitochondrial 239
translesion 235
Double-strand breaks (DSBs) 16, 131, 134, 198, 236, 241, 312, 313, 358, 359, 360
Downregulate RNA targets 258
Downstream 134, 200, 261, 264
effector caspases Casp3 200
kinase RIP2 261
DREAM 232
complex 232
formation 232
DUOX enzymes 350

E

Earth's magnetic field by magnetotactic bacteria 49
Effective multicellularity 224
Effector proteins 165, 273
bacterial 273
Efferocytic macrophages 351
Efferocytosis 280, 281, 316, 350, 351, 352
Eicosanoids 343, 345
Electrical stimulation of hippocampus 404
Electron transport chain 4, 6, 22, 75, 81, 190, 191, 238
mitochondrial 238
rotenone toxin targets mitochondria 22
proteins 75
Encode 39, 44, 47, 71, 86, 87, 105, 137, 140, 156, 165, 188, 347
adenoviruses 347
amoebae 39
prokaryotic genomes 86
Endocannabinoids 352
Endonucleases 45, 131, 164, 313
Endothelial cells 239, 343, 382, 384, 390
hypothalamic 343
Endothermic homeotherms 17
Energy 1, 2, 3, 6, 7, 83, 107, 109, 116, 193, 238, 309, 384
homeostasis 384
metabolism 83, 193, 238
sensor 309
sources 1, 2, 3, 6, 7, 107, 109, 116
Environmental stress response (ESR) 185, 186
Enzymes 11, 15, 21, 44, 50, 52, 54, 78, 79, 80, 88, 103, 106, 112, 113, 130, 131, 132, 166, 187, 192, 193, 229, 238, 276, 319, 350, 354
anti-inflammatory 350
antioxidative 106
deubiquitinating 229
digestive 54, 88, 166
glutathione-related 319
glycolytic 238
hydrolytic 52
inositol-requiring 187
lysosomal 166
metabolic 193
prokaryotic 192
thermophilic 103
Eosinophils 56, 340, 349
Epidermal growth factor (EGF) 390
Epigenetic modulators 233
ER-associated degradation 187
Erythroblasts 410
Erythrocytes 346, 382
Escherichia coli infection 392
Esoinophils 311
Ethanol and stress response element (ESRE) 308
Eukaryotic 36, 43, 46, 319
pathogens 36, 319
RNA viruses 43
spliceosome 46
Exopolysaccharide capsules 127
coating 127
enveloping cells 127
Extraction 111, 134
metal 111

F

Fas receptor 201
Fast 16, 313

division of activated immune cells 313
proliferating cells 16
Fatigue 343, 389, 391
Fatty acid(s) 5, 9, 57, 82, 102, 104, 105, 110, 112, 114, 161, 186, 193, 201, 204, 345, 348, 400
long-chain 104
oxidation 204
metabolism 186
polyunsaturated 105, 201, 345, 400
saturated 104, 348
short-chain 57
short-chain saturated 110
unsaturated 82, 102, 104, 114
Fear learning 400
stress-enhanced 400
Features, distinctive metabolic 202
Flavin adenine dinucleotide (FAD) 4
Flavobacterium columnare 137
Fluids 105, 197, 113, 204, 341, 358, 399
cerebrospinal 399
extracellular 197, 358
hydrothermal 113
Foreign DNA 130, 132, 133, 138, 157, 159, 163, 164, 168, 267, 271, 272, 312
acquired 130
destruction 133
methyltransferases 132
unmodified 163
Fragmentation 198, 201, 390, 405
apoptosis-initiated chromatin 198
mitochondrial 405
Frontotemporal dementia 402

G

Gametocytes 55
immature 55
Gene expression 7, 77, 87, 106, 128, 156, 157, 170, 187, 235, 272, 318, 394, 395
activating lytic 170
disrupt 128
early viral 272
microglial 395
phage's lytic 156
regulators adjust 7
Generalized anxiety disorders (GAD) 400
General stress response sigma factor 166
Genomic evolvability 159
Genotoxic 228, 229, 230, 241, 281, 312, 313, 315, 360
factors 313
stress 228, 229, 230, 241, 281, 312, 315, 360
Germ cells 203, 224, 236, 260, 274, 396
primordial 396
protection 236
specification 260
Germline micronucleus 260
Glucocorticoids 346, 361, 385, 387, 395, 399
receptor (GR) 385, 395, 396, 399
stimulate 361
Glucokinase 192
Gluconeogenesis 8, 192
Glucose 4, 9, 12, 56, 111, 136, 190, 192, 193, 195, 197, 227, 237, 238
homeostasis 56
intake 192
metabolism 192, 195
sensor 192
starvation 193
transporters 12, 238
uptake 9, 238
Glucosylated hydroxymethylcytosine 132
Glutamate 7, 81
decarboxylase 81
deprotonation 7
protonated 7
protonates 7
Glutathione 15, 79, 195, 239, 319
antioxidant 239
oxidized 15
peroxidase 195
reductase 15, 195
Glycans 274, 319, 355
sialylated 355
viral O-linked 274
Glycation 14
Glycine 106, 114
Glycogen utilization 186
Glycolysis 4, 162, 231, 237, 238, 265, 315, 345, 351
aerobic 237
upregulate 351
Glycoprotein 278, 355
influenza virus 278
Glycosaminoglycans 355
Glycosylases 14, 21, 131
Glycosylated proteins 127 187

misfolded 187
Gut microbiota 56, 57, 88, 266
human commensal 88

H

Haloenzymes 108
Halophiles 101, 107, 108, 109
extreme 107
moderate 107
Halophilic archaea halolamina 108
Heat shock 10, 17, 22, 71, 73, 74, 114, 155, 196, 203, 205, 262, 346, 352, 361
factors 17
protein genes 205
proteins (HSPs) 10, 17, 114, 155, 196, 203, 262, 346, 352, 361
response (HSR) 17, 22, 71, 73, 74, 196
Heat stress 1, 37, 82, 158, 182, 196, 197, 203, 347, 395, 409
dampers 409
moderate 395
resistance 395
response 196, 197
Helicases 137, 261 309, 313
cytosolic RNA 261
Helicobacter pylori 38, 314, 350
Hemolytic uremic syndrome 165, 357
atypical 357
Hepatocytes 55, 343
Herpes simplex virus-1 402
Heterochromatin 199, 314
Heterotrimeric GTPase 183
Histoplasmosis 54
Homeostasis 8, 9, 10, 11, 20, 186, 188, 189, 190, 192, 198, 199, 263, 265, 307, 308, 310, 311, 347, 348, 349, 355
adaptive 10
calcium 190
cellular 198, 199, 263, 265
deregulated 348
macrophages monitor tissue 349
nutrient 192
Homeostatic 8, 9, 10, 308
perturbations 308
regulation 8
signals 9, 10
Homing endonucleases (HEs) 46, 47
Horizontal gene transfer (HGT) 43, 46, 47, 48, 49, 50, 54, 126, 128, 130, 136, 137, 155, 156
Hormones 192, 193, 204, 380, 381, 384, 385
corticotropin-releasing 380, 384
insulin 192
steroid 381, 384, 385
thyroid 193, 204
Hosphorylcholine 318
Hyaluronidase 50
Hydrogenases 82
Hydrolases 57, 79, 350
glycoside 57
small alarmone 79
Hydrolysis 17, 46, 130, 199, 408
spontaneous 17
Hydroperoxides 14, 201
lipid 14, 201
Hyperosmolarity 19, 323
Hypo-osmotic shock 18
Hypothalamo-pituitary-adrenal (HPA) 380, 381, 384, 393, 407

I

IFN regulatory factors (IRFs) 264, 349
Immune 257, 275, 307, 308, 310, 340, 341, 360, 362, 382, 410
mechanisms, autonomous 257, 275, 307, 308, 340, 341, 360, 362
mediators 310, 382, 410
Immune receptors 51, 261, 262, 317, 339, 340, 358, 360, 362
adaptive 339, 358, 362
innate 360
Immune response 43, 125, 155, 262, 263, 271, 272, 273, 307, 308, 309, 321, 322, 323, 339, 341, 344, 362, 381, 383, 384, 385, 386, 392, 409
adaptive 321, 362, 392, 409
autonomous 307, 323
early 341
enhanced 263
in prokaryotes 155
innate 43
specialized systemic 339
systemic 382
Immunity-related GTPase (IRG) 278
Immunocompromised hosts 37, 159
Immunodeficiency disorders 347

Immunoglobulin superfamily 358
Immunostimulatory effect of transient DDR 313
Indoleamine 2,3-dioxygenase 276
Inflammasome activation 267, 268, 271, 308, 317, 322, 349
  efflux channel triggering NLRP3 267
  histamine modulate NLRP6 268
  preventing 349
Inflammatory reflexes 380, 386, 408
Inflow, dietary glucose 9
Inhibition 18, 21, 86, 87, 182, 185, 188, 228, 231, 233, 235, 238, 241, 265, 323 386
  oxidative 323
  pathogen growth 265
Inhibitors 42, 88, 200, 225, 230, 231, 241, 268, 278, 313, 315, 318, 356, 388, 400
  acetylcholinesterase 388
  nucleotide synthesis 275
  quorum signaling 88
  serotonin reuptake 400
  synthetic 318
Integrated 41, 182, 183, 184, 185, 188, 202, 307, 309, 310, 311, 322
  prophages 41
  stress response (ISR) 182, 183, 184, 185, 188, 202, 307, 309, 310, 311, 322
Integration of stress and immune signaling 322
Interferon(s) 184, 263, 264, 265, 269, 272, 273, 275, 277, 278, 309, 310, 313, 318, 320, 321, 349
  genes 273
  -induced plasminogen activator inhibitor 278
  -induced tetratricopeptide 277
  -inducible GTPases 278, 320
  -inducible protein 272
  infected animal cells 264
  -stimulated genes (ISGs) 265, 269, 275, 310
  upregulate itaconate production 318
Internal ribosome entry sites (IRES) 184, 228
Intestine dysfunction 50
Intrachromosomal gene shuffling 168
Intrahippocampal injection 403
Intrinsic apoptosis 233, 280
Iron 6, 12, 20, 83, 111, 116, 157, 166, 167, 201, 202
  chelators 202
  storage 83

K

Kinases 7, 72, 77, 81, 182, 183, 184, 185, 187, 189, 195, 196, 197, 198, 229, 231, 232, 234, 261, 268, 269, 348, 360
  cell cycle controller CDK1 197
  cyclin-dependent 232
  extracellular signal-regulated 195
  glycogen synthase 195, 360
  heme-regulated inhibitor 183
  histidine 72
  inner membrane histidine 77
  mammalian 269
  mevalonate 348
  phosphatidylinositol 3-kinase-like 198
  sensor 81
  sensor histidine 7, 72
  tyrosine 195, 261
Klebsiella pneumoniae 128
Kuru and Creutzfeldt-Jakob disease 43
Kynurenic acid 398

L

Lactococcus lactis 139, 161
Lactoferrin 351
Lactoperoxidase 280
Latent HIV-1 infection 42
Leucine-rich repeat (LRR) 261, 266, 273, 320, 358, 405
Ligases 11, 320, 322
  ubiquitin-protein 11
Lipases 50, 127
Lipid peroxidation 87, 201, 233, 407
  iron-dependent 87, 201
Lipoxygenases 201, 345
Listeria monocytogenes 51, 83, 85, 167, 278, 310
  FrvA 83
Listeriosis 85
Liver gluconeogenesis 323
Locus coeruleus (LC) 385, 401
Long terminal repeats (LTRs) 44, 45
Lymphocytosis 241
Lymphoid cells 340, 344, 355, 358, 359, 387, 388, 409
  innate 340, 355
Lysosomal degradation 267

Lysozymes 78, 116, 140, 166, 308, 321, 350
phage-associated 140

M

Macroautophagy 189, 190
Major 50, 353, 397, 398, 399, 400, 406, 409
depressive disorder (MDD) 397, 398, 399, 400, 409
histocompatibility complex (MHC) 50, 353, 406
Maladaptive DNA repair 21
Malfunction 191, 241, 275, 347
of inflammatory mechanisms 347
of mitochondria 191, 275
of NFkB system 241
Mannose-binding lectin (MBL) 354, 389
Mannosidases 187
MAVS 270, 279, 316, 317
expression 317
oligomerization 317
mitochondrial signaling molecule 316
polymerization 270
MBL-associated serine proteases (MASPs) 354
Melanocortin 189, 384
Melanocytes 384
Membrane(s) 4, 5, 6, 7, 18, 73, 89, 104, 105, 108, 109, 110, 112, 140, 160, 161, 188, 320
alkaliphile 110
autophagosomal 320
stress proteins 161
transport systems 109
Memory 186, 230, 265, 382, 386, 393, 396, 402, 403, 404
consolidation 404
epigenetic transcriptional 265
impairment 402
Mental disorders 380, 393, 395, 397
Mesenchymal stromal cells 352
Metabolic 54, 170, 397
gene clusters (MGCs) 54
imbalance 170
syndrome 397
Metalloproteinases 408
Methylguanosine 269
Microglia 357, 380, 382, 383, 392, 393, 395, 396, 397, 399, 401, 403, 404, 405, 406, 407
synapse elimination by 357, 383
Mobile genetic elements (MGEs) 36, 43, 44, 46, 47, 48, 155, 156, 157, 158, 159, 165, 168, 169, 170
Multiple sclerosis (MS) 58, 409
Multisystem disorder 195
Mutagenesis 169, 170
inducing 170
treatment-induced 169
Mycobacterium smegnatis 87
Mycobacterium tuberculosis 48, 51, 320, 350
Mycotoxins 22, 54
Myeloid cells 348, 351
Myeloperoxidase 282
Myofibroblasts 347

N

Natural killer (NK) 313, 339, 340, 347, 352, 353, 359, 360, 410
Necroptosis 200, 201, 202, 280, 281, 342, 405
Necrotic cells release cytokines and DAMPs 341
Neisseria gonorrhoeae 129, 321
Neisseria meningitides 129, 159
Neural degeneration 199
Neurons 202, 203, 204, 343, 380, 381, 382, 383, 385, 386, 388, 390, 391, 399, 401, 402, 403, 404, 405
cholinergic 388
cold-sensitive 204
dopaminergic 388, 404, 405
efferent motor 386
norepinephrine 385, 401
thermoregulatory 343
Neuropeptides 341, 388, 390
capsaicin release 388
sleep-promoting 390
Neuroticism 399
Neurotransmitters 380, 381, 382, 384, 385, 388, 397
Neutrophil(s) 50, 51, 78, 126, 129, 166, 274, 281, 282, 316, 339, 340, 342, 346, 347, 349, 352
extracellular traps (NETs) 78, 166, 281, 282
lysis 50
myeloperoxidase 166
migration 316

Nicotinamide adenine dinucleotide (NAD) 4, 141, 398, 404
Nitric oxide synthase 166
Nitrogen 13, 80, 162
species, reactive 13, 80
starvation 162
Non-homologous end joining (NHEJ) 23
Norepinephrine neurons dendrites 385
Nucleases 125, 126, 127, 130, 134, 135, 164, 235, 258
crRNA-guided 135
extracellular 125, 126
periplasmic 130
translesion 235
Nucleic acid(s) 21, 49, 81, 104, 126, 127, 133, 135, 257, 274, 275
endogenous 275
modifying 21
-stabilizing proteins 81
Nucleotide 23, 84, 108, 193, 261, 266, 273
-binding 193
-binding domain (NBD) 261, 266, 273
dimerization 23
excision repair (NER) 23, 84, 108

O

Okadaic acid 22
Oligoadenylate 269
phosphodiesterases 269
synthetases 269
Oligodeoxynucleotides 272
Oligomerization domain (OD) 226, 231
Oligomers 261, 269, 270, 272, 277, 402
amyloid 402
forms filamentous 270, 272
oligoadenylate 269
Onychomycosis 54
Operon, leucine biosynthesis 136
Osmolyte trimethylamine *N*-oxide 115
Oxidases 5, 6, 7, 20, 238, 280, 344, 350, 408
alternative 6, 344
binding cytochrome 20
dual 280, 350
monoamine 408
Oxidation 2, 4, 6, 14, 15,83, 111, 194, 201, 280, 282, 316, 361
anaerobic 6
iron-mediated histidine 83
Oxidative 15 318, 398
damages, main 318
DNA damage 15, 398
Oxidoreductases 167
Oxygen 12, 71
catastrophe 12
-derived radicals 71
Oxytocin receptor gene 401

P

Parkinson's disease (PD) 43, 401, 404, 405
Paroxysmal nocturnal hemoglobinuria 357
Partner-switching systems (PSS) 72, 73
Pathogen 257, 260, 261, 262, 263, 278, 307, 308, 320, 339, 340, 341, 354, 357
-associated molecular patterns (PAMPs) 257, 260, 261, 262, 263, 307, 308, 339, 340, 341, 354, 357
-containing vacuoles (PCVs) 278, 320
Pattern-recognition receptors (PRRs) 257, 260, 261, 262, 265, 266, 274, 275, 282, 307, 319, 320, 339, 340, 350
Pentaphosphates 205
Peptic ulcers 381
Peptidases 21, 127, 316
Peptidoglycan 48, 77, 87, 127, 140, 159, 273
bacterial cell-wall 140
biosynthesis 77
cell wall synthesis 87
Peptidyl arginine deiminase 281
Peripheral 52, 384
biofilm cells 52
monocytes 384
Periplasmic 52, 77, 113
copper-binding proteins rusticyanin 113
foldases 77
predators 52
Peroxidases 15, 280, 319, 358
bacterial 280
Peroxides 13, 15
organic 13
Peroxiredoxin 15, 195
Peroxynitrite 13, 166
Phage 167, 140
shock protein response 167
superinfection exclusion systems 140
Phagocytes 126, 127, 201, 280, 282, 339, 316, 340, 346, 349, 350, 351, 352, 353, 380, 383
microglial cells 383

NADPH oxidase 316
oxidase 350
Phagocytosis 51, 83, 339, 340, 344, 349, 350, 351, 352, 353, 354, 356, 381, 383
blocking antibodies restore 353
of pathogens 83, 350, 351
Phagolysosomes 81, 166, 350
Phagosomes 166, 278, 320, 350
deplete 278
Phenol-soluble modulins (PSMs) 50
Phenotypes 11, 88, 201, 239, 311, 349, 352, 383, 393, 405
anti-inflammatory 352
inflammatory 311, 352
Phosphatases 72, 75, 231, 264
Phosphatidylinositol phosphates 281
Phosphoanhydride bonds 3
Phospholipid(s) 21, 48, 75, 108, 268, 318
bilayer 48, 75
charged acidic 21
oxidized 268, 318
Phosphorothioation 131
Phosphorylated alpha-synuclein filaments 4, 405
reactive substrates 4
Phosphorylates 75, 188, 240, 261, 281, 360
activated PERK 188
Phosphorylation 3, 4, 6, 7, 13, 20, 115, 183, 184, 224, 229, 230, 231, 234, 240, 241, 267, 268, 270, 271, 315, 317, 245
activating 241
blocking 317
differential 234
oxidative 4, 6, 13, 20, 115, 315, 345
substrate level 4
Phosphotyrosine 280
Photosynthesis 3, 6, 12, 101
oxygenic 12
plants use 3
Piezophiles 113, 114
archaeal 114
eukaryotic 114
psychrophilic 114
Plant 282, 323
pathogens escape 282
susceptibility 323
Poikilothermic invertebrates 344
Polyamines 104, 110
branched RNA-binding 104
long linear DNA-binding 104
Polyextremophiles 101, 115
eukaryotic 115
Polymerase chain reaction (PCR) 102
Polymorphic toxin systems (PTSs) 126
Polynucleoid filament 52
POMC-derived peptides 384
Porins 76, 112, 354
Posttranscriptional 205, 229, 240, 258
mechanisms 205
-transcriptional processing 258
modifications 229, 240
enzymatic 229
Post-translational processing 105
Preimplantation blastocysts 45
Prions 43
Progenitors 349, 382
fetal blood yolk sac 349
microglial 382
yolk sac erythromyeloid 382
Programmed cell death (PCD) 71, 86, 138, 139, 168, 184, 200, 202, 224, 236, 342
Pro-inflammatory cytokines 257, 281, 309, 339, 341, 343, 350, 352, 386, 397, 398
Proteases 17, 22, 50, 76, 77, 132, 182, 186, 191, 200, 274, 277, 354
complement-degrading 274
lysosomal 277
Proteasomal 11, 194, 224, 225, 229, 232, 240, 274, 314, 319, 322
degradation 11, 194, 224, 225, 229, 232, 240, 314, 319, 322
rapid 274
Proteasome-mediated degradation 193
Protective polysaccharide coating 112
Proteinaceous toxins 21
Proteins 1, 10, 11, 13, 14, 16, 17, 18, 51, 73, 74, 75, 76, 82, 83, 103, 108, 113, 114, 115, 128, 132, 133, 141, 162, 186, 188, 189, 190, 196, 226, 240, 262, 278, 310, 311, 314, 319, 343, 354, 361, 389
acid-denatured 82
acute phase 240, 311, 343, 389
antibody-binding 51
binding immunoglobulin 188
cold-inducible RNA-binding 18, 74
conserved autophagy-related 189
cytoplasmic 319
cytosolic 190
denatured 186
eukaryotic chaperone heat-shock 162

expression 128
folding 10, 17, 73, 103, 196, 310, 311
chaperones guide 10
glycated 262
guanylate-binding 278
lysines 361
thermophilic 103
Proteobacteria 2
Proteolysis 14, 72, 74, 86, 282, 311, 354
Proton-motive force (PMF) 6, 7, 104, 110, 111, 161
Protospacer-adjacent motif (PAM) 134, 135, 137
Protospacers, crRNA-matching 137
Pseudomonas aeruginosa 37, 167, 191
Psychiatric disorder 405
Pulmonary metastases 409
Pyroptosis-induced intracellular traps 281
Pyroptotic death, inflammasome-induced 342

Q

Quorum sensing (QS) 52, 86, 136, 157, 167, 274

R

Reactive 6, 13, 14, 16, 79, 80, 156, 166, 201, 280, 308, 316, 317, 339, 350, 399, 408
lipid alkoxy radicals 201
nitrogen species (RNS) 13, 14, 80, 166, 280, 316, 317, 339, 350
oxygen species (ROS) 6, 13, 14, 16, 79, 80, 156, 166, 280, 308, 316, 317, 350, 399, 408
Reductase 5, 239, 323
methionine sulfoxide 323
ribonucleotide 239
Regulated 87, 189 199, 200, 201, 202, 224, 233, 257, 280, 281, 282, 307, 311
cell death (RCD) 87, 199, 200, 201, 202, 224, 233, 257, 280, 281, 282
IRE1-dependent decay (RIDD) 189, 307, 311
Regulation 72, 136, 137, 187, 190, 192, 194, 204, 205, 226, 227, 228, 229, 314, 315, 384, 400
cytokine 400
melanocyte 384
Regulators 75, 77, 79, 80, 130, 136, 158, 186, 191, 229, 230, 238, 307, 308, 341
central homeostatic 238
central physiological 341
transcriptional 158
Regulatory systems 72, 78, 129
Repair 15, 16, 17, 84, 199
DNA 16, 84
efficiency 199
macromolecules 15
protein misfolding 17
Resistance 46, 47, 49, 77, 84, 158, 165
antibiotic 46, 49, 77, 84, 158
antibiotic multidrug 47
antimicrobial 165
Respiration 7, 158, 162, 197, 237
aerobic 162, 197, 237
anaerobic 7, 158
Respiratory syncytial virus (RSV) 319
Restriction 112, 132, 164, 167, 192, 320
dietary 192
host-mediated 167
Retrotransposons 44, 45, 46, 235, 260
silence 260
Retrotransposon transcription 236
Reverse 45, 275, 276, 279
transcriptase 45
transcription 45, 275, 276, 279
Ribosomal protein genes 187
Ribosomes 17, 18, 39, 40, 73, 78, 80, 114, 165, 185, 190, 194, 197, 230, 275, 277, 314
assembly 18, 277
biogenesis 185, 197, 230, 275, 314
hybrid 185
modification 194
RNA 259, 276
-dependent RNA polymerases 259, 276
-directed RNA polymerase 259
RNA-induced silencing 258
complex (RISC) 258
RNA interference (RNAi) 138, 257, 258, 259, 275, 282
machinery 138

S

Saccharomyces cerevisiae 185
Senescence-associated secretory phenotype (SASP) 202, 232

Sense 188, 308
protein-to-lipid ratio 188
RNA viruses 308
Sensors 9, 188, 264, 265, 269, 271, 272, 273, 311,
activated 188
cytoplasmic RNA 311
cytosolic DNA 271
energy stress 75
immune 264, 265
intracellular 9
nuclear dsDNA 272
nucleic acid 273
of autonomous immune system 269
Serotonin 398
synthesis 398
transporter 398
Signaling 9, 13, 185, 188, 189, 198, 201, 203, 204, 233, 234, 261, 263, 264, 265, 266, 273, 276, 280, 311, 315, 316, 353, 356, 360, 387, 391, 393, 395, 398, 404
apoptotic 189
cascades 13, 198, 261, 263
cytokine 273, 316
death-inducing 201
decreasing postsynaptic serotonin 398
increased mitogenic 276
inhibitory 353
proapoptotic 188
proteins 185
serotonergic 203
subvert pathogen 280
sympathetic 9, 387
sympathetic nervous system 204
transcellular chaperone 203
Signal(s) 7, 8, 9, 11, 76, 109, 138, 193, 224, 225, 240, 263, 264, 265, 271, 274, 315, 322, 341, 342, 349, 351, 386
acetylcholine 386
additional 193
danger 271, 274, 349
t ransducers 322
transduction 11, 109
transduction cascades 8
trunsduction 138
Silence RNA targets 138
Single strand DNA viruses 125
Sjogren's Syndrome 396
Small molecule imiquimod 267
Social 391, 392
skills training 392
stress 391, 392
Sodium permeability 108
SOS 84, 85, 156
genes 84
-independent mechanisms of response 85
-sensitive repressor 156
signaling 84
systems 84
SOS response 84, 85, 87, 155, 156, 157, 164, 165, 169
antibiotic-induced 169
bacterial 157
Spermine oxidase 280
Sphingolipids 188
Sphingomyelinase 50
Spindle assembly checkpoint (SAC) 234
Splenocytes 408
Spliceosomes 187, 198
Spontaneous activation 355
Sporadic DNA damage 157
Stimulation 18, 84, 132, 273, 282, 308, 309, 310, 313, 314, 322, 388, 404
direct electrical 388
of mitophagy 404
Stored curvature elastic (SCE) 77, 160
Stress 12, 42, 75, 85, 115, 157, 167, 182, 185, 186, 187, 228, 308, 310, 392, 395, 396, 409
cause 12
cell 187, 308
childhood 395
granules (SGs) 185, 310
proteins 228, 409
resistance 42, 75, 156, 168, 205
response genes, viral infection downregulates 155
Stressors 57, 72, 85, 115, 157, 167, 169, 182, 185, 186, 266, 315, 391, 392, 396, 407
environmental 169, 266
external 315
multiple 186
psychological 407
social 391
Stringent response and acid stress 165
Synapses 383, 385, 398, 399, 401, 402, 403, 406, 407
glutamatergic 406
inhibitory 399
inhibitory GABAergic 399

neuronal immature 383
Synaptic 382, 383, 402
connectivity pointing 383
patterning, male-typical 383
plasticity 382, 402
Synaptogenesis 395
Synucleinopathies 43
Systemic 319, 348, 357, 396, 408
activation of Nrf2 319
lupus erythematosus (SLE) 348, 357, 396, 408

T

Tau 401, 402
microtubule-binding protein 401, 402
aggregates 402
Thermogenesis 204, 343, 344
increased 343
Thermostable enzyme endo-β-glucanase 113
Thiocyanate 358
Thioredoxin 15, 20, 79, 195, 319
dissociation 195
reductase 15, 20, 79, 195, 319
TNF 343, 388
activation 343
synthesis 388
Tolerance 22, 56, 85, 88, 112, 196, 205, 323, 352, 357
antibiotic 88
extreme acid 112
immune 357
impaired glucose 56
Toll-like receptors (TLRs) 257, 260, 261, 264, 270, 348, 351, 361, 398
Toxic domains encode 127
Toxic shock syndrome toxin (TSST) 51
Toxins 21, 22, 51, 308
diphtheria 22
diverse 21
producing pore-forming 308
toxic shock syndrome 51
Toxoplasmosis 393
Transcriptases 45
encode reverse 45
Transcriptional 194, 225, 351
and metabolic profile of macrophages 351
Transcriptional factors 74, 89, 193, 205, 225, 240, 241, 308, 385
archaeal 89
heterodimeric 240
Transient receptor potential (TRP) 9, 17, 203
Translation73, 183, 184, 231, 277
inhibition of 183, 231
mediated 184
mRNA 73, 277
Translation and post-translational processing 105
Translational 39, 80, 194
elongation 80
machinery 39
Transposases 44, 48
Tumor necrosis factor (TNF) 339, 342, 343, 346, 348, 350, 390, 391, 393, 395, 397, 398, 400, 401
Tumor suppression programs 237
Tyrosine 14, 47
nitration 14
recombinase 47

U

Ubiquitin 10, 187, 195, 232
ligases polyubiquitinate 232
-proteasomal system (UPS) 10, 195
-proteasome machinery 187
Unfolded protein response (UPR) 182, 187, 188, 189, 191, 307, 308, 310, 311, 322, 323
Upregulates 74, 81, 83, 157, 197, 232, 234, 238, 258, 350, 408
aging cells 408
DNA repair genes 234
transcription 74
transporters 238
Upregulating expression 186, 315
UV-induced DNA damage repair 205

V

Vacuoles 20, 189, 278, 320, 350
bacteria-containing 350
pathogen-containing 278, 320
phagocytic 20
Vagus nerve 380, 386, 387, 388, 390, 391, 400, 408
efferent 387, 408
peripheral 388
subdiaphragmatic 408
Variable lymphocyte receptors (VLR) 358

Vasoactive intestinal peptide (VIP) 388

W

Water-oxidizing organisms 6

X

Xenopus oogenesis apoptosis 230

Y

Yeasts 8, 53, 54, 115, 187, 189, 192, 193, 194, 196, 197, 198, 200
black 115
budding 197
introns 187
microscopic 53
response 194, 197
Yersinia pestis 37, 278

Z

Zika virus 37, 276, 320, 393, 394

www.ingramcontent.com/pod-product-compliance
Lightning Source LLC
LaVergne TN
LVHW070944210926
848482LV00002B/87